THE CALIFORNIA NITROGEN ASSESSMENT

THE CALIFORNIA NITROGEN ASSESSMENT

CHALLENGES AND SOLUTIONS FOR PEOPLE, AGRICULTURE, AND THE ENVIRONMENT

Edited by

THOMAS P. TOMICH

SONJA B. BRODT

RANDY A. DAHLGREN

KATE M. SCOW

Collaborating Institutions

AGRICULTURAL SUSTAINABILITY INSTITUTE AT UC DAVIS

UC ANR SUSTAINABLE AGRICULTURE RESEARCH AND
EDUCATION PROGRAM

UC ANR KEARNEY FOUNDATION OF SOIL SCIENCE

UC ANR AGRICULTURAL ISSUES CENTER

UC ANR CALIFORNIA INSTITUTE FOR WATER RESOURCES

WATER SCIENCE AND POLICY CENTER AT UC RIVERSIDE

UNIVERSITY OF CALIFORNIA PRESS

University of California Press, one of the most distinguished university presses in the United States, enriches lives around the world by advancing scholarship in the humanities, social sciences, and natural sciences. Its activities are supported by the UC Press Foundation and by philanthropic contributions from individuals and institutions. For more information, visit www .ucpress.edu.

University of California Press
Oakland, California

Library of Congress Cataloging-in-Publication Data

The California nitrogen assessment : challenges and solutions for people, agriculture, and the environment / edited by Thomas P. Tomich, Sonja B. Brodt, Randy A. Dahlgren, Kate M. Scow.
  pages cm
  Includes bibliographical references and index.
  "Collaborating Institutions: Agricultural Sustainability Institute at UC Davis, UC ANR Sustainable Agriculture Research and Education Program, UC ANR Kearney Foundation of Soil Science, UC ANR Agricultural Issues Center, UC ANR California Institute for Water Resources, Water Science and Policy Center at UC Riverside."
  ISBN 978-0-520-28712-9 (pbk. : alk. paper)—ISBN 0-520-28712-6 (pbk. : alk. paper)—ISBN 978-0-520-96223-1 (ebook)—ISBN 0-520-96223-0 (ebook)
  1. Nitrogen—Environmental aspects—California.
  2. Nitrogen cycle—Environmental aspects—California.
  3. Nitrogen fertilizers—Government policy—California.
  4. Biodiversity conservation—California. I. Tomich, Thomas P. (Thomas Patrick), 1956– editor.
  TD196.N55C35 2016
  631.8′409794—dc23

2015035774

Manufactured in China
25 24 23 22 21 20 19 18 17 16
10 9 8 7 6 5 4 3 2 1

The paper used in this publication meets the minimum requirements of ANSI/NISO Z39.48-1992 (R 2002) (*Permanence of Paper*).

CONTENTS

LIST OF ONLINE APPENDICES

Online appendices may be accessed at http://asi.ucdavis.edu/nitrogen

LIST OF BOXES

LIST OF FIGURES

LIST OF TABLES

LIST OF MAPS

PREFACE

Nitrogen is essential to all life on Earth. In its reactive forms, nitrogen plays vital roles in the health of humanity, our planet's ecosystems, and our increasingly global economy. As a key component of synthetic and organic fertilizers, nitrogen feeds plants and is critical to the success of agriculture. The invention of the Haber–Bosch process over a century ago, which enabled cheap nitrogen synthesis on an industrial scale, has been a cornerstone of the so-called "Green Revolution" (more aptly, the "Seed-Fertilizer Revolution"). Billions of people are alive today because of the abundant harvests made possible by synthetic nitrogen fertilizer and fertilizer-responsive crop varieties. And yet, there are significant costs associated with human intervention in the nitrogen cycle: water and air pollution, climate change, and detrimental effects for human health, biodiversity, and natural habitat. Too little nitrogen limits ecosystem processes that are key to our food supply, while too much transforms ecosystems profoundly, with adverse consequences for human well-being.

Striking the right balance between the benefits and costs resulting from human influences on the nitrogen cycle necessitates a critical look at the trade-offs involved; this is all the more urgent with projections for human population exceeding 9 billion by the middle of the twenty-first century. Tracking trends in nitrogen flows, understanding key inputs and outputs, evaluating management options, and devising new policies and institutions that address the multimedia interactions inherent in the nitrogen cycle will be essential to ensuring this balance is attained. While much research has addressed elements of these trade-offs, far less effort has gone into integration across the disparate forms of scientific information encompassing all nitrogen flows (not just agriculture) needed to strike an appropriate balance.

In 2008, the Agricultural Sustainability Institute (ASI) at the University of California (UC) Davis conducted a stakeholder survey requesting views on the top agricultural sus-

tainability issues for California. Many of these—climate change, water scarcity, petroleum dependence, and air and water pollution—are directly related to nitrogen, particularly synthetic nitrogen, and illustrated the need to approach these issues from a systems perspective to better understand linkages, complementarities, and trade-offs. This survey also underscored the differences in opinions between farmers and researchers.

At the same time, the multi-stakeholder Climate Action Team was developing plans for implementation of California's Global Warming Solutions Act (AB32). While AB32 focused on carbon dioxide (CO_2), it became clear that nitrous oxide (N_2O), with roughly 300 times the effect of CO_2, was a potentially important omission. In discussions with the Action Team, researchers at UC Davis realized that both leading scientists and policy-makers did not have the complete N story for California. Given growing international awareness of the effects of humans on the global nitrogen cycle (e.g., MA, 2005b), ASI proposed to adapt integrated ecosystem assessment methodologies from the global scale to policy and practice in California concerning nitrogen. ASI's External Advisory Board, representing a wide range of relevant stakeholder groups, endorsed the strategic relevance of the proposed nitrogen assessment, both for ASI's programmatic goals regarding the future vitality of California's agriculture and as an area of growing interest for California as a globally recognized innovator in agriculture, climate change mitigation strategy, and environmental policy.

The California Nitrogen Assessment (CNA) is the first comprehensive accounting of nitrogen drivers, flows, impacts, practices, and policies for California. Broadly, the goals of the assessment were the following:

- Gain a comprehensive view of nitrogen flows in California, with a necessary emphasis on agriculture, which dominates the state's N cycle.

- Provide useful insights for stakeholders into the balance between the benefits of agricultural nitrogen and the effects of surplus nitrogen in the environment.
- Compare options, including practices and policies, for improving the management of nitrogen and mitigating the negative impacts of surplus nitrogen in the environment.
- Move beyond "academic business-as-usual" to more effectively link science with action and to produce information that informs both policy and field-level practice.

Within the time frame of the CNA, another important assessment focused on nitrate contamination of groundwater in the Tulare Basin and Salinas Valley of California (Harter et al., 2012). Although the time frame, geographic and thematic scope, funding, assessment process, and purpose of what is now known as the "Harter Report" differs in important ways from the CNA, a number of colleagues contributed to both of these efforts. Because of the collegiality among team members and similarity in methods, the Harter et al. effort produced important data, methods, and insights (e.g., regarding denitrification in aquifers that is a key area of uncertainty addressed in Chapter 4) that have shaped this volume.

The CNA followed established protocols for integrated ecosystem assessments, a research approach that differs from other scientific methods (Ash et al., 2010). Considerable information already existed on nitrogen in California, but had never been considered as a whole. The CNA synthesized this large body of data, used it to analyze patterns and trends, and assessed the quality of information and knowledge about key issues. In line with international assessment practice, the CNA was developed through a complex participatory design, working with stakeholders to guide the research agenda to ensure its process and outputs are considered legitimate by a broad range of stakeholders, as well as practically useful, policy relevant, and scientifically credible. This involved engagement with over 350 individuals across 50 organizations. Early consultations with stakeholders resulted in the identification of over 100 questions, which then were synthesized into major topics to frame the assessment's research priorities. As detailed in the Acknowledgments and in Chapter 1, this volume was subjected to a rigorous two-part review process: scientific peer review by dozens of leading scientists, followed by public review in which comments were invited from more than 400 stakeholders. All review comments and responses can be accessed online (www.asi.ucdavis.edu/nitrogen).

Inspired by decades of international assessment practice (Chapter 1; Ash et al., 2010), the chapters of this book follow a logical sequential structure of drivers (Chapters 2–3), current states (Chapters 4–5), plausible future scenarios (Chapter 6), and both technological and policy responses (Chapters 7–8). Similarly, the structure within chapters also follows international practice. To orient readers, each chapter opens with a succinct statement "What Is This Chapter About?" and list of the stakeholder questions that will be addressed. This is followed by "Main Messages," in italic font, which are intended to be a self-contained summary of the conclusions of that chapter. The main body of the chapter is the assessment itself, which assesses the scientific evidence available to address stakeholder questions and which is the basis for the "Main messages." Extensive supplemental materials, summaries of key findings targeted to specific audiences, and information on other nitrogen-related research are available on the website of the ASI (www.asi.ucdavis.edu/nitrogen). While this assessment serves as a benchmark of our knowledge about nitrogen in California at the turn of the twenty-first century, the website will continue to evolve to reflect changing information and new issues.

Publication of this volume comes at a pivotal time for California's policies regarding nitrogen. Our intention is for this assessment to be a resource in debate on public policy options as well as to inspire a new generation of researchers to fill gaps and reduce uncertainty in key aspects of these important issues. Building on findings of this assessment, ASI's Sustainable Agriculture Research and Education Program, which is a statewide program of the UC Division of Agriculture and Natural Resources, has launched a "Solution Center for Nutrient Management" (ucanr.edu/sites/Nutrient_Management_Solutions/). This program is intended to link knowledge generated by the CNA in a collaborative effort with farmers, extension specialists, advisors, and other practitioners to develop solutions involving the stakeholder partnerships that have been launched during the seven-year assessment process.

Thomas P. Tomich
Sonja B. Brodt
Randy A. Dahlgren
Kate M. Scow
Davis, California, USA
July 31, 2015

ACKNOWLEDGMENTS

The California Nitrogen Assessment (CNA) has been a 7-year project (2008–2015) led by the Agricultural Sustainability Institute (ASI) at the University of California (UC) Davis and the UC Division of Agriculture and Natural Resources (UC ANR) Sustainable Agriculture Research and Education Program (UC SAREP), in collaboration with the UC ANR Kearney Foundation of Soil Science, the UC ANR Agricultural Issues Center, the UC ANR California Institute for Water Resources, and the Water Science and Policy Center at UC Riverside. Major funding for the California Nitrogen Assessment was provided by a grant to ASI from the David and Lucile Packard Foundation. Further financial support has been provided by the UC Division of Agriculture and Natural Resources (UC ANR), including UC SAREP and the UC ANR Kearney Foundation of Soil Science, United States Department of Agriculture Hatch project CA-D-XXX-7766-H funds for the project entitled "California Agroecosystem Assessment," and the WK Kellogg Endowed Chair in Sustainable Food Systems at UC Davis. At the Packard Foundation, Kai Lee, Walt Reid, and Jamie Dean (now with the 11th Hour Project) each provided important input as "boundary spanning" colleagues above and beyond the financial support. (Kai, we are happy we finished this before you retired!) At its inception, the Packard Foundation arranged for a review of the project by William C. Clark, of Harvard University, and Steven R. Carpenter, of the University of Wisconsin, Madison. Their expertise and helpful comments contributed to our thinking at that formative stage. Bill Frost, UC ANR Associate Vice President, has been steadfast throughout in his support for these efforts and his encouragement of this wildly collaborative approach that has engaged so many UC ANR faculty, extension specialists, farm advisors, and other colleagues. All of the collaborating institutions and the host institutions of coauthors, review editors, reviewers, and other contributors to this assessment acknowledged below contributed significantly in-kind through their time, attention, talent, and experience.

Please see immediately following these acknowledgments the lists of contributors' names and affiliations. The assessment was a collaborative effort involving 43 lead and contributing authors and 64 other distinguished colleagues who provided guidance and reviewed various sections. The CNA followed a rigorous two-part review process—scientific peer review comprising dozens of leading scientists, followed by a public review period in which comments were invited from more than 400 stakeholders. Jointly, these processes evaluated the comprehensiveness of the assessment, balance, and clarity in presentation of evidence, and validity and appropriateness of the interpretations. Chapter review editors acted as "referees" to ensure all comments received appropriate attention. The assessment team gratefully acknowledges Alan Townsend, formerly of the University of Colorado and now dean of the Nicholas School at Duke University, who provided early encouragement of this daunting enterprise and served as the overall review editor for the assessment; eight other internationally renowned chapter review editors for their thorough and judicious oversight of both the scientific peer review and the stakeholder review processes; 51 subject experts who undertook scientific reviews of specific sections in their areas of specialization; three external reviewers and one UC Press board member for their review and useful suggestions on the full manuscript; and the nine members of our Technical Advisory Committee who provided direction from the inception of the assessment.

We particularly want to acknowledge the support the assessment has received from hundreds of our stakeholders across California who attended consultations to aid the development of the assessment's priorities and framing of assessment questions, and who provided valuable data and other information, gave feedback at numerous conferences on preliminary presentations of the assessment, and submitted comments on draft chapters through the formal stakeholder review process. We particularly wish to recognize the indispensable input from the 33 members of the Stakeholder Advisory Committee. This stakeholder engagement was a critical part of the assessment and these diverse voices of our constituents shaped and significantly improved our results in countless ways. The breadth of our stakeholders reflects the complexity of the issues and the wide-ranging implications nitrogen has for the environment and human well-being within the distinctive political economy of California.

The core team of the assessment included a number of researchers, postdoctoral fellows, and communications specialists whose contributions are reflected in their authorship of various chapters, but who also contributed to the design and implementation of the assessment process overall: V. Ryan Haden (now with the Ohio State University), Todd Rosenstock (now with World Agroforestry Center, Nairobi, Kenya), Dan Liptzin (now with the University of Colorado, Boulder), and Antoine Champetier de Ribes (now with the Swiss Federal Technical Institute, Zurich). Our communication specialists for the assessment, Colin Bishop, Stephanie Ogburn, and Aubrey White, helped us create innovative strategies for engagement, communication, and outreach that led to development of the Solution Center for Nutrient Management, which is hosted by UC SAREP, a unit of ASI.

Our assessment team also included a remarkable group of other institute staff and students. Among these, two stand out for their overall leadership of the process. Karen Thomas (now with the University of New Orleans) was instrumental in coordinating the successful proposal to fund the assessment and in the launch of the assessment process. Later, we were fortunate to work with Mariah Coley, who brought her superb organizational, editing, communication, and graphic skills to the final stages of this process. Without the dedication and commitment of Dr. Thomas and Ms. Coley, it is difficult to imagine how the assessment could have been completed successfully. Others who made specific contributions include the following: Prescott "Scottie" Alexander, who provided custom software tools for organizing the bibliography; Cesca Wright, who conducted a formative evaluation of the assessment process; Ryan Murphy, Prashant Hedao, and Allan Hollander who prepared several of the maps for this volume, and, in alphabetic order, a number of others who have contributed their talents to the successful completion of this project, including Karen Curley, Melissa Haworth, Ji Hoon Im, Caitlin Kiley, Ruthie Musker, Bev Ransom, Courtney Riggle, Shea Robinson, Dianne Stassi, and Michele Tobias. In closing, we also wish to thank our two UC Press editors, Blake Edgar and Merrik Bush-Pirkle, for their encouragement and guidance.

LIST OF CONTRIBUTORS

KENNETH BAERENKLAU School of Public Policy, UC Riverside

JEFF BARNUM Magenta Studios

COLIN BISHOP Agricultural Sustainability Institute at UC Davis

SONJA B. BRODT Agricultural Sustainability Institute at UC Davis

ANTOINE CHAMPETIER DE RIBES Swiss Federal Institute of Technology in Zurich

MARIAH COLEY Agricultural Sustainability Institute at UC Davis

RANDY A. DAHLGREN Department of Land, Air and Water Resources, UC Davis

SAMIRA DAROUB University of Florida

ANN DREVNO UC Santa Cruz

GAIL FEENSTRA Agricultural Sustainability Institute at UC Davis

VAN RYAN HADEN The Ohio State University

GERALD HARRIS Quantum Planning Group

THOMAS HARTER Department of Land, Air and Water Resources, UC Davis

BENJAMIN HOULTON Department of Land, Air and Water Resources, UC Davis

DAVID KANTER New York University

CATHY KLING Iowa State University

MARCIA KREITH UC Agricultural Issues Center

TIMOTHY LANG University of Florida

HAROLD LEVERENZ Department of Civil and Environmental Engineering, UC Davis

C-Y CYNTHIA LIN Department of Agricultural and Resource Economics, UC Davis

DANIEL LIPTZIN INSTAAR, University of Colorado

DEANNE MEYER UC Cooperative Extension

CONNY MITTERHOFER State of California, State Water Resources Control Board

DANIEL MUNK UC Agriculture and Natural Resources

ALLISON OLIVER University of Alberta

DOUG PARKER UC California Institute for Water Resources

PAUL LARRY PHELAN The Ohio State University

DANIEL PRESS UC Santa Cruz

JENNIFER RIDDELL Mendocino College Sustainable Technology Program

JOHN THOMAS ROSEN-MOLINA US Department of State

TODD S. ROSENSTOCK World Agroforestry Centre (ICRAF)

KURT SCHWABE School of Public Policy, UC Riverside

KATE M. SCOW Department of Land, Air and Water Resources, UC Davis

JOHAN SIX Swiss Federal Institute of Technology in Zurich

DANIEL SUMNER UC Agricultural Issues Center

KAREN THOMAS University of New Orleans

THOMAS P. TOMICH Agricultural Sustainability Institute at UC Davis

ZDRAVKA TZANKOVA UC Santa Cruz

JAMES VANDERSLICE University of Utah

JINGJING WANG University of New Mexico

AUBREY WHITE Agricultural Sustainability Institute at UC Davis

BOON-LING YEO Department of Environmental Science and Policy, UC Davis

CHRISTINA ZAPATA Institute of Transportation Studies, UC Davis

MEMBERS OF THE TECHNICAL
ADVISORY COMMITTEE

Randy A. Dahlgren, Russell L. Rustici Endowed Chair in Rangeland Watershed Sciences; Director *emeritus*, University of California Kearney Foundation of Soil Science; Professor, Land, Air and Water Resources Department, University of California, Davis.

Thomas Harter, ANR CE Specialist, Department of Land, Air and Water Resources, University of California, Davis.

Ermias Kebreab, Professor, Department of Animal Science, and Deputy Director, Agricultural Sustainability Institute, University of California, Davis.

Frank Mitloehner, Professor and ANR CE Specialist, Department of Animal Sciences, University of California, Davis.

Dan Putnam, ANR CE Specialist, Department of Plant Sciences, University of California, Davis; University of California Agricultural and Natural Resources.

Kate Scow, Director, Russell Ranch Sustainable Agriculture Facility of the Agricultural Sustainability Institute at UC Davis; Director *emeritus*, University of California Kearney Foundation of Soil Science; Professor, Land, Air and Water Resources Department University of California, Davis.

Johan Six, Professor, ETH Zurich.

Daniel Sumner, Professor, Department of Agricultural and Resource Economics, and Director, Agricultural Issues Center, University of California.

Thomas P. Tomich, W.K. Kellogg Endowed Chair in Sustainable Food Systems; Director, Agricultural Sustainability Institute at the University of California, Davis; Director, University of California Sustainable Agriculture Research and Education Program; Professor, Departments of Human Ecology and Environmental Science and Policy, University of California, Davis.

MEMBERS OF THE STAKEHOLDER
ADVISORY COMMITTEE

Pelayo Alvarez, Program Director, California Rangeland Conservation Coalition, Defenders of Wildlife.

Ted Batkin, President (former), Citrus Research Board.

Steve Beckley, Executive Director, Organic Fertilizer Association of California.

Don Bransford, Chairman (former), CA Rice Producer's Group, California Rice Commission, and President, Bransford Farms; member of California State Board of Food and Agriculture.

Renata Brillinger, Executive Director, California Climate and Agriculture Network.

Cynthia Cory, Director, Environmental Affairs, California Farm Bureau Federation.

Bob Curtis, Associate Director of Agricultural Affairs, Almond Board of California.

Michael Dimock, President, Roots of Change.

Laurel Firestone, Co-Executive Director, Community Water Center.

Hank Giclas, Sr. Vice President, Strategic Planning, Science and Technology, Western Growers Association.

Joseph Grant, Farm Advisor, University of California Cooperative Extension, San Joaquin County.

Tim Johnson, President-CEO, California Rice Commission.

Matthew Keeling, California Regional Water Quality Control Board, Central Coast Region.

David Lighthall, Health Science Advisor, San Joaquin Valley Air Pollution Control District.

Woody Loftis, EPA Liaison, United States Department of Agriculture – Natural Resources Conservation Service.

Karl Longley, Coordinator of Water Resources Programs, California Water Institute.

Jim Lugg, Consultant, Fresh Express/Chiquita.

Paul Martin, Western United Dairymen (former).

Albert Medvitz, McCormack Sheep and Grain.

Rob Mikkelsen, Director, Western North America, International Plant Nutrition Institute.

Belinda Morris, Climate and Land Use Program Officer, David and Lucille Packard Foundation.

Alberto Ortiz, General Manager, Ag Services (Salinas).

Renee Pinel, President/CEO, Western Plant Health Association.

Brise Tencer, Executive Director, Organic Farming Research Foundation (OFRF).

Bruce Rominger, Owner, Rominger Brothers Farms.

David Runsten, Policy Director, Community Alliance with Family Farmers.

Ann Thrupp, Executive Director, Berkeley Food Institute (BFI) at UC Berkeley.

Kathy Viatella, California Water Foundation Program Manager, Resources Legacy Fund.

Allen Dusault, Program Director, Sustainable Conservation.

Ian Greene, Research Programs Manager, California Strawberry Commission.

Larry Glashoff, Horticultural Tech Manager, Hines Nursery.

Edward Hard, CDFA Fertilizer Research and Education Program.

Don Hodge, Environmental Protection Specialist, US Environmental Protection Agency.

Claudia Reid, California Certified Organic Farmers.

LIST OF REVIEW EDITORS

These review editors served as independent "referees" to oversee the process that incorporated comments from scientific and stakeholder reviewers, ensuring that every review comment was considered by the lead authors and received appropriate attention and a response from the assessment team. Chapter Review Editors provided feedback for a specific chapter(s), while the Review Editor provided oversight for the whole assessment.

Review Editor

Alan Townsend, Duke University

Chapter Review Editors

Chapter One: Introducing the California Nitrogen Assessment—Neville Ash, United Nations Environment Program.

Chapter Two: Underlying Drivers of Nitrogen Flows in California—Eric Lambin, Stanford University.

Chapter Three: Direct Drivers of California's Nitrogen Cycle—Eric Lambin, Stanford University.

Chapter Four: A California Nitrogen Mass Balance for 2005—Jim Galloway, University of Virginia.

Chapter Five: Ecosystem Services and Human Well-Being—Peter Vitousek, Stanford University; Paul English, California Department of Public Health.

Chapter Six: Scenarios for the Future of Nitrogen Management in California Agriculture—Monika Zurek, Environmental Change Institute, University of Oxford.

Chapter Seven: Responses: Technologies and Practices—Cliff Snyder, International Plant Nutrition Institute.

Chapter Eight: Responses: Policies and Institutions—David Zilberman, University of California, Berkeley.

LIST OF SCIENTIFIC REVIEWERS

Stanford Bronwen, University of California, Santa Cruz.

Patrick Brown, University of California, Davis.

Karen Burow, US Geological Survey.

Klaus Butterbach-Bahl, Karlsruhe Institute of Technology.

Marsha Campbell-Matthews, University of California Cooperative Extension.

Sarah Carvill, University of California, Santa Cruz.

Alejandro R. Castillo, University of California Cooperative Extension, Merced.

Jana Compton, US Environmental Protection Agency, Western Ecology Division.

Jeannie Darby, University of California, Davis.

Mark David, University of Illinois at Urbana-Champaign.

Ann Drevno, University of California, Santa Cruz.

Walter Falcon, Stanford University.

Mark Fenn, US Forest Service, Pacific Southwest Research Station.

Peter Groffman, Cary Institute of Ecosystem Studies.

Baojing Gu, Zhejiang University.

Jane Hall, California State University, Fullerton.

Thomas Harter, University of California, Davis.

Timothy Hartz, University of California, Davis.

Greg Hitzhusen, Ohio State University.

Don Horneck, Oregon State University.

Casey Hoy, Ohio State University.

Lars Stoumann Jensen, University of Copenhagen.

Vivian Jensen, University of California, Davis.

Thomas Jordan, Smithsonian Environmental Research Center.

Kathleen Lask, University of California, Berkeley.

Matt Liebman, Iowa State University.

Bruce Linquist, University of California, Davis.

Nicolas Juan Lucas, The Nature Conservancy, Argentina.

Marsha Mathews, University of California Cooperative Extension.

Jean Moran, California State University, East Bay.

Geoff Morrison, University of California, Davis.

Jay E. Noel, California Polytechnic State University.

Doug Parker, California Institute for Water Resources, University of California Agriculture and Natural Resources.

Changhui Peng, University of Quebec at Montreal.

G. Stuart Pettygrove, University of California, Davis.

Robert Pinder, US Environmental Protection Agency, Office of Research and Development.

Daniel Press, University of California, Santa Cruz.

Sasha Reed, US Geological Survey.

Bradley Rickard, Cornell University.

Pam Rittelmeyer, University of California, Santa Cruz.

Thomas Ryerson, National Oceanic and Atmospheric Administration, Chemical Sciences Division.

Hugh Safford, USDA Forest Service.

Bob Scholes, Wits University.

Mindy Selman, World Resources Institute.

David Simpson, National Center for Environmental Economics, US Environmental Protection Agency.

Benjamin Sleeter, US Geological Survey.

Heidi Stallman, University of Missouri.

Zdravka Tzankova, University of California, Santa Cruz.

Chris van Kessel, University of California, Davis.

Mary Ward, National Institutes of Health.

Arnim Wiek, Arizona State University.

COMMON ACRONYMS AND ABBREVIATIONS

AIC — Agricultural Issues Center (University of California)

ANR — University of California Division of Agriculture and Natural Resources

ASAE — American Society of Agricultural Engineers

ASI — Agricultural Sustainability Institute at UC Davis

BMP — best management practices

BNF — biological nitrogen fixation

BOE — Board of Equalization

C — carbon

CA DOF — California Department of Finance

CA DWR — California Department of Water Resources

CalTrans — California Department of Transportation

CARB — California Air Resources Board

CASA — California Association of Sanitary Agencies

CA SWRCB — California State Water Resources Control Board

CDC LRP — Division of Land Resource Protection

CDFA — California Department of Food and Agriculture

CDFA FREP — Fertilizer Research and Education Program

CEC — California Energy Commission

CIWMB — California Integrated Waste Management Board

CNA — California Nitrogen Assessment

COPD — chronic obstructive pulmonary disease

CWTRC — California Wastewater Training & Research Center

DOF — US Department of Finance

DOL BLS — US Department of Labor, Bureau of Labor Statistics

DOL WHD — US Department of Labor, Wage, and Hour Division

EF — emission factor

EPA — US Environmental Protection Agency

FRAP — Fire Resource and Assessment Program

GHGs — greenhouse gases

GTP — global temperature potential

GWP — global warming potential

IPCC — Intergovernmental Panel on Climate Change

MA — Millennium Ecosystem Assessment

MCL — maximum contaminant level

NAAQS — National Ambient Air Quality Standard

NANI — net anthropogenic nitrogen inputs

Nr — reactive nitrogen

NRC — National Research Council

NUE — nitrogen use efficiency

OECD — Organization for Economic Co-Operation and Development

OWTS — onsite wastewater treatment systems

PM — particulate matter

PNB — partial nutrient balance

POTW — publically owned treatment works

RF — radiative forcing

SWRCB — State Water Resources Control Board

TMDL — total maximum daily load

UC — University of California

UN FAO	Food and Agriculture Organization of the United Nations
US BEA	Bureau of Economic Analysis
US EIA	Energy Information Administration
USDA	US Department of Agriculture
USDA ARS	Agricultural Research Service
USDA ERS	Economic Research Service
USDA FAS	Foreign Agricultural Service
USDA FSA	Farm Service Agency
USDA NASS	National Agricultural Statistics Service
USDI	US Department of the Interior
USGS	US Geological Survey
VOC	volatile organic compounds
WWTP	wastewater treatment plant

CHEMICAL FORMULAS

CH_4	methane		N_2O	nitrous oxide
CO	carbon monoxide		NO, NO_x	nitric oxides
CO_2	carbon dioxide		NO_2	nitrogen dioxide
N_2	dinitrogen		NO_2^-	nitrite
NH_3	ammonia		NO_3^-	nitrate
NH_4^+	ammonium		$NO_3^-\text{-}N$	nitrogen in the form of nitrate
$NH_4^+\text{-}N$	nitrogen in the form of ammonium		O_3	ozone

CHAPTER ONE

Introducing the California Nitrogen Assessment

Lead Authors:
K. THOMAS, D. LIPTZIN, AND T. P. TOMICH

Contributing Authors:
M. COLEY, R. DAHLGREN, B. HOULTON, K. SCOW, AND A. WHITE

What Is This Chapter About?

This chapter provides background information to understand the issues related to nitrogen use and management in California agriculture, presents an overview of the assessment approach and implementation, and outlines the goals of the California Nitrogen Assessment (CNA). Assessments emphasize legitimacy, credibility, and relevancy; the research and stakeholder engagement process is as important as the results and outputs produced. An assessment looks at existing knowledge and reduces complexity by synthesizing what is known and widely accepted and differentiating it from that which is unknown or not agreed upon. This chapter describes how the CNA engaged with stakeholders to establish research priorities and outputs that meet the needs of a wide range of perspectives, including farmers, government, and environmental and health organizations.

Main Messages

Nitrogen is necessary to sustain all life and is often the primary nutrient limiting productivity. Nitrogen is also a critical nutrient required to sustain agriculture in California and the global food supply. As a component of synthetic and organic fertilizer, nitrogen is critical to plant growth and the expansion of crop production. With estimates for worldwide population growth and rising per capita incomes and agricultural production, the demand for N is likely to intensify in coming years to meet growing global demands for food. This is also true for California; the state produces more than half of the fruits, nuts, and vegetables grown in the United States and 21% of the dairy commodities.

Since 1960, the amount of nitrogen used in agriculture has doubled on the planet, as has food production, with excess nitrogen being released to the environment. The majority of nitrogen imports to California are in the forms of fertilizer, imported animal feed, fossil fuel combustion, and biological nitrogen fixation. While much of that nitrogen contributes to productive agriculture, excess nitrogen from those sources contributes to surface water and groundwater contamination and air pollutants such as ammonia and nitrous oxide, a potent greenhouse gas. Striking the right balance between the benefits nitrogen provides to our food supply and the costs it can have on our health and environment demands a critical look at the trade-offs involved.

This is the first comprehensive accounting of nitrogen flows, practices, and policies for California agriculture at the statewide level. The assessment identifies key drivers of nitrogen use decisions and examines how these drivers influence the statewide mass balance of nitrogen—how much enters the state through new sources and, ultimately, the multiple ways these compounds enter and affect the environment. The assessment tracks nitrogen's impacts on environmental health and human well-being, and examines technological and policy options to minimize nitrogen leakage while sustaining the vitality of agriculture.

Assessments reduce complexity through the synthesis and integration of a large body of existing information, providing a valuable method for focusing efforts and systematically calling out uncertainty. Assessments are designed to be *legitimate* in the eyes of key stakeholders, *relevant* to decision-makers' needs, and scientifically *credible*.

In an assessment, understanding what is not known is just as important as assessing what is known. We evaluated the quality of the data and note when results are based on very reliable information or data that are less reliable (e.g., due to gaps in information or disagreement in the literature). We employed "reserved wording" to describe uncertainty. Questions lacking good data were highlighted where more research and record keeping are needed. The online

Appendices[1] describe the sources and approaches used in the assessment, evaluate the level of uncertainty of this information, and highlight where information gaps exist.

The assessment used multiple avenues to engage with more than 350 stakeholders across 50 organizations. Through a participatory research design, stakeholders identified more than 100 nitrogen-related questions that were used to direct research and synthesize priorities, provided data and examples of on-the-ground practices and management options, and developed four "scenarios" for the future of nitrogen in California agriculture.

The assessment's findings underwent a multistage peer review process. This included consecutive reviews by (1) over 50 scientific experts, (2) the Stakeholder Advisory Committee (SAC), and (3) an open public comment period. A group of 10 review editors (9 chapter review editors and 1 overall review editor) ensured all comments received appropriate attention and responses from authors.

1.0. The California Nitrogen Assessment

Since 1960, the amount of nitrogen (N) used in agriculture has doubled on the planet, as has food production (Galloway et al., 2004; MA, 2005a; Vitousek et al., 1997). With expectations for continued growth in the human population and per capita consumption, and intensifying utilization of natural resources such as freshwater, striking a balance between the benefits and costs of human nitrogen use will require significant trade-offs and a critical investigation of technological and policy options to minimize nitrogen losses to the environment while sustaining the vitality of agriculture.

Tracking trends in nitrogen flows, understanding key inputs and outputs, evaluating management options, and devising new regulatory structures that are sensitive to cross-system interactions will be essential to ensuring this balance is attained. Despite increasing awareness of the importance of these trade-offs, a lack of cohesive knowledge that gives a big-picture view of California's nitrogen system still hampers effective decision-making from policy options to field-level practices.

The CNA is a comprehensive evaluation of existing knowledge about nitrogen science, practice, and policy in the state. Broadly, the goals of the assessment were to:

1. Gain a comprehensive view of N flows in California, with emphasis on agriculture's roles.
2. Provide useful insights for stakeholders into the balance between the benefits of agricultural nitrogen and the effects of excess nitrogen in the environment.
3. Compare options, including practices and policies, for improving the management of nitrogen and

mitigating the negative impacts of excess nitrogen in the environment.
4. Move beyond "academic business-as-usual" to more effectively link science with action and to produce information that informs both policy and field-level practices.

The assessment is targeted towards a broad audience with diverse and often conflicting perspectives. Throughout the assessment, the level of detail varies, with emphasis placed on those issues and topics identified as being of greatest interest to our stakeholders. We do not offer recommendations, but rather endeavor to synthesize the current scientific understanding, point out gaps in knowledge, and present a balanced understanding of the issues and options for moving forward on key concerns.

1.1. Understanding Nitrogen and Its Global Implications

Dinitrogen gas (N_2) comprises 78% of Earth's atmosphere. However, this form of N is largely unreactive. While all biological species require nitrogen for growth and development, the transformation of N_2 into reactive nitrogen forms (Nr) is required to make it biologically available (Box 1.1). The biological and physical transformations that comprise the global nitrogen cycle can be categorized into three groups, all of which can result from biotic and abiotic processes: (1) the fixation of atmospheric N_2, (2) the transformations among forms of solid, dissolved, and gaseous forms of Nr in the air, land, and water, and (3) the production of N_2, largely as a result of denitrification (Box 1.2).

Nitrogen gas is converted to Nr naturally in ecosystems through the process of nitrogen fixation. A small amount of nitrogen is fixed abiotically by lightning in the atmosphere and during vegetation fires, but the majority of natural nitrogen fixation is a result of biological activity (Chapter 4). Once Nr has been fixed in an ecosystem, the N can be assimilated by organisms and converted into living biomass. When the organisms die, the various organic forms of N are transformed back into ammonia (ammonification) and nitrate (nitrification) by specialized groups of microorganisms. Other microorganisms denitrify the nitrate to N_2 with nitrogen oxide gases (NO and N_2O) as intermediate products of the series of enzymatic reactions. In addition to denitrification, ecosystems can also lose N during leaching or surface water runoff containing dissolved forms of organic N and nitrate (NO_3^-). While most ecosystems tend to retain most of their N, some N is transported between ecosystems in groundwater and surface water and in the atmosphere. Superimposed on the natural flows of N, humans created several new flows of Nr and altered the magnitude of others. Fossil fuel combustion, synthetic fertilizer use, increased concentration of livestock, sewage dis-

---

1. Appendices are available online at http://asi.ucdavis.edu/nitrogen.

BOX 1.1. COMMON FORMS OF NITROGEN

With the Exception of Dinitrogen, All Other Forms Are Considered Reactive N (Nr)

Nitrogen Compound	Form
Dinitrogen: N_2	Gas
Nitrous oxide: N_2O	Gas
Nitrogen oxides: $NO, NO_2 (NO_x)$	Gas
Ammonia: NH_3	Gas
Ammonium: NH_4^+	Water soluble ion
Nitrite: NO_2^-	Water soluble ion
Nitrate: NO_3^-	Water soluble ion
Organic N: various	Solid, dissolved, or gas

BOX 1.2. MAJOR STEPS IN THE NITROGEN CYCLE

Process	Function
Fixation	Process by which atmospheric nitrogen (N_2) is converted into ammonia (NH_3). N_2 is largely inert and does not easily react with other chemicals to form new compounds. Nitrogen fixation can result from biological or industrial activity. In the case of biological activity, the N is fixed by microorganisms that may or may not be symbiotically affiliated with plants. Fixation also can refer to the various physical conversions of N_2 to Nr, such as the production of nitrogen oxides (NO_x) by lightning or fossil fuel combustion.
Ammonification (or mineralization)	Process by which organic forms of N are converted to ammonium during the decomposition of organic matter. When a plant or animal dies, or an animal expels waste, the initial form of N is organic. Bacteria, or in some cases fungi, convert the organic N within the remains back into ammonium (NH_4^+).
Nitrification	Process by which ammonium is oxidized to nitrite and then nitrate by microorganisms under aerobic conditions.
Denitrification	The reduction of nitrate into N_2 via a series of enzymatic reactions by microorganisms in anaerobic environments. Denitrification may also occur under aerobic conditions with co-respiration of oxygen and nitrate. Both denitrification pathways involve formation of intermediate gases (NO and N_2O) that may be lost to the atmosphere.
Immobilization	The uptake of Nr by microorganisms or plants from the surrounding environment to meet their metabolic needs.

charge, cultivation of legumes, irrigation, frequency and magnitude of wildfires, and groundwater pumping all alter amount, distribution, and flows of Nr in the environment (figure 1.1). These flows have both positive and negative consequences for ecosystem health and human well-being. The N cascade conceptual model illustrates that a single molecule of N can have multiple positive and negative effects from the time that it is fixed to the time it is returned to the atmosphere as N_2. Depending on the particular fate of the N that escapes its intended purpose, the costs to ecosystems and human health will vary (see Chapter 5; Birch et al. 2011; Compton et al., 2011).

Human activity has dramatically reshaped the global N cycle in terms of the magnitude of the production of Nr, availability of N in ecosystems, and rates of N cycling. Currently, humans create more Nr than do all of the planet's natural processes combined (Rockström et al., 2009), and it has been suggested that the anthropogenic changes to the N cycle have already crossed a "planetary boundary," or threshold for stability of Earth system processes (Rockström et al., 2009). However, the timescale of these changes is small relative to their magnitude. Historically, farmers in the United States relied on planting legumes, local recycling of human and animal wastes, or import of inorganic nitrates to replace N loss in soils. Limited by the amount of naturally occurring Nr available for food production, many realized that the natural rate of Nr replenishment in soils would not match the rate of global population growth. This interest in sourcing additional Nr for food production led to the invention of the Haber–Bosch process in 1913. This technology gave humans the ability to produce Nr on an industrial scale and removed the need to rely solely on naturally occurring Nr for food production (Galloway et al., 2004).

The Haber–Bosch process is one of three pathways by which humans produce Nr. Additionally, fossil fuel combustion creates gaseous forms of Nr because there is N in the fuel (in the case of coal) and because the combustion process can lead to reactions between N_2 and oxygen in the atmosphere. Finally, humans plant vast acreage of leguminous crops for several reasons—many take advantage of legumes' N-fixing properties to enrich soil. Leguminous crops also are harvested widely for both human and animal consumption.

Roughly half the human population on earth is supported by Haber–Bosch-produced nitrogen fertilizer (Davidson et al., 2012), making this one of the greatest innovations ever. There has been an exponential increase in the use of the Haber–Bosch process since the 1940s, revolutionizing agriculture and allowing for abundant fertilizer supplies to support growing human populations. Davidson et al. (2012) found that for 2008, 56% of the major sources of natural and anthropogenic nitrogen inputs to the United States came from agriculture (synthetic N fertilizer [Galloway et al., 2008] and crop biological N fixation). Globally, anthropogenic sources of newly fixed nitrogen now exceed

natural terrestrial sources by at least 50% (Galloway et al., 2008). With continued growth in human population forecasted at 9.6 billion people by 2050 (Gerland et al., 2014), the implications for increased food production and nitrogen use, as well as resulting effects on the environment and human well-being, are substantial.

Nitrogen's effect on the broader environment also includes unintended consequences for human and ecosystem health. Only 55% of the intentionally fixed N in the United States makes its way into an intended product (i.e., food, fiber, energy, industry) (Houlton et al., 2013). The N efficiency of the most common grain crops is typically less than 50% (Cassman et al., 2002), which allows the remaining N to escape from the soil as nitrate or in various gaseous forms. This escaped N can alter ecosystem services and damage human health: eutrophication and anoxic "dead zones" in surface waters and coastal areas; harmful algae blooms; high fluxes of nitrous oxide, a potent greenhouse gas; loss of plant biodiversity; enhancing competition from invasive species; and nitrate contamination of drinking water (see Chapter 5). While nitrogen interacts with human health and well-being in a variety of ways, the trade-offs involved in agricultural nitrogen use highlight many of the key sustainability issues related to the challenges of the twenty-first century: global climate change, depletion of fossil fuels, and mounting pressure on land, air, and water resources from growing human population and rising incomes. For example, work in Europe estimates that the environmental and human health costs of excess N now exceed the annual benefits of N use for crop production (Sutton et al., 2011b). Houlton et al. (2013) estimate that agricultural N spillover results in air-quality damages in the United States that exceed $16 billion each year. Recent studies suggest that an increase in public and private funding on the order of $17–34 million per year over many decades will be needed to implement required nitrate mitigation projects for water systems in the Tulare Lake Basin and Salinas Valley (Honeycutt et al., 2012).

1.2. Why California?

California provides an excellent location to study nitrogen because of its diversity. Its ecosystems range from deserts to alpine tundra. Its population is concentrated in large metropolitan areas, but the majority of the state is rural. California agriculture has both a large livestock and crop component. Further, California is the source of the majority of production for many fruit, nut, and vegetable crops for the United States, and thus carries a lot of the nitrogen burden for many non-Californians. In addition, California is actively dealing with many of the challenges confronting agriculture throughout the United States and internationally: population growth and urbanization (Landis and Reilly, 2004; Williams et al., 2005); changing demographics in rural communities (Bradshaw and Muller, 1998); flood control and water demand for irrigation

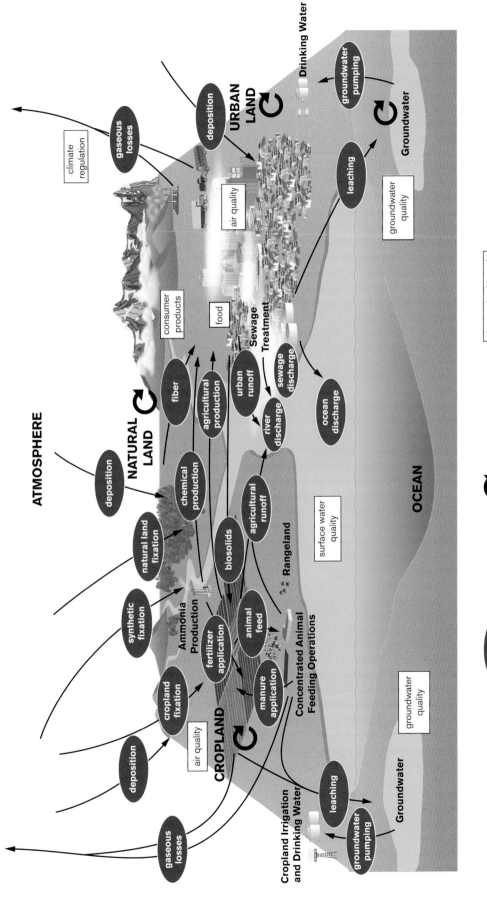

ATMOSPHERE

climate regulation

gaseous losses

deposition

NATURAL LAND

deposition

URBAN LAND

Drinking Water

groundwater pumping

air quality

Groundwater

groundwater quality

leaching

consumer products

food

fiber

agricultural production

Sewage Treatment

urban runoff

sewage discharge

river discharge

ocean discharge

chemical production

natural land fixation

synthetic fixation

Ammonia Production

CROPLAND

cropland fixation

deposition

gaseous losses

air quality

fertilizer application

biosolids

animal feed

manure application

Concentrated Animal Feeding Operations

Rangeland

agricultural runoff

surface water quality

OCEAN

groundwater quality

Cropland Irrigation and Drinking Water

leaching

groundwater pumping

Groundwater

human health and environmental effects

flow

N storage

(Tanaka and Sato, 2005); maintaining air, soil, and water quality; coping with climate change (Cavagnaro et al., 2006; Hayhoe et al., 2004); responding to domestic and international markets (AIC, 2006); and facing increasing regulation. Thus, California exemplifies the biophysical and social context in which this assessment will be both locally relevant and nationally and internationally significant—it encompasses extreme diversity in its agricultural production systems and climatic regions and landscapes, is subject to enormous population pressures and complex social problems, and has pioneered innovative environmental policies.

With the largest and most diverse agriculture in the United States, California epitomizes the successes and dilemmas of higher productivity in agriculture. In 2013, California generated $46.4 billion from agriculture, representing 14.7% of national agricultural production with less than 4% of the nation's farmland (USDA NASS, 2014a, 2015a). The state is vital for domestic consumption with two-thirds of production consumed outside the state. The state produces 21% of the nation's dairy commodities and more than half of the fruits, nuts, and vegetables grown in the United States (CDFA, 2010). Agricultural land encompasses about 9.9 million acres, of which 74% is nonwoody crops (annual crops such as grains, vegetables, cotton, etc.) and 26% woody crops (orchards, vineyards, etc.) (USDA NASS, 2006). Land area in pasture and rangeland for livestock grazing contributes another 6 million acres. Due to the wet winter/dry summer Mediterranean climate and reliance on irrigation during the main growing season, the way nitrogen is managed in the state is closely tied to water management.

N management is increasingly viewed as one of the great global challenges of the twenty-first century. It is also garnering increasing attention in California. For example, the California Global Warming Solutions Act of 2006 (Assembly Bill 32) requires the reduction of three major greenhouse gases—carbon dioxide (CO_2), nitrous oxide, and methane (CH_4)—to 1990 levels by 2020 and a further 80% by 2050. This includes mandatory reporting for the largest sectors (e.g., oil, gas, power, cement, landfills), but little attention has been paid to how agriculture fits into this framework. Further, since excess nitrogen can be stored in groundwater (used commonly as drinking water in the state) and is associated with human health problems, clean water policies are becoming an increasing concern, producing a growing trend of regional water boards enforcing stricter regulations on agriculture. With groundwater contamination by nitrates (principally from fertilizers) becoming a major water-quality issue, in February 2013, the State Water Resources Control Board released 15 recommendations that underscore the need for cooperative and comprehensive solutions, including new potential legislation. While there are other agricultural regions around the world that must actively work to replace the N and other nutrients extracted by agriculture (see, for instance, Cruzate and

Casas, 2012), in California an excess (rather than a depletion) of N is the significant issue.

1.3. Approach: An Integrated Assessment of Nitrogen across Science, Policy, and Practice

An assessment is a critical evaluation of information for purposes of guiding decisions on complex public issues with topics defined by stakeholders (MA, 2005a). Conducted in a transparent manner, assessments improve clarity by synthesizing what is known and widely accepted from that which is unknown or not agreed upon. They generate results that can only be produced when a large body of existing information is examined. However, an assessment is not just about the results—"getting the process right, from the early stages through to the communication of findings, is essential in order to have an impact" (Ash et al., 2010) (Box 1.3).

The CNA followed established protocols for integrated ecosystem assessments, an approach that differs from other scientific methods (Box 1.4). The CNA adapted these methods to California and found the established methodologies to be even more applicable to a state-level assessment. For example, a key distinction of assessments is responsiveness to stakeholders' needs and it was much more feasible to get a wide range of perspectives "at the table" in a single state.

Rather than generating new primary data, the assessment looked at existing knowledge to identify what is well known about agricultural nitrogen and that which is more speculative. A large amount of information already existed on agriculture-related nitrogen in California, which had never been examined as a whole. The CNA adds value by collect-

BOX 1.3. KEY CHARACTERISTICS OF ASSESSMENTS

- *Guides decisions* on complex issues.
- *Critical evaluation* of information.
- Conducted in a *credible, useful, legitimate* process (open, transparent).
- *Engages stakeholders and responds to users' needs.*
- *Adds value by sorting, summarizing, synthesizing, translating, and communicating* what is known and widely accepted from what is not known (or well published); includes evaluation of uncertainty.
- *Policy relevant, not policy prescriptive* (assesses options; "if, then" approach).
- *Peer reviewed* (by researchers and stakeholders).

SOURCE: MA (2005a).

Intergovernmental Panel on Climate Change (IPCC, 1990)	*Millennium Ecosystem Assessments (MA, 2005)*	*European Nitrogen Assessment (ENA, 2011)*
• Synthesis of knowledge on climate science and options for mitigation adaptation. • Four assessment reports since 1990. • Methodology reports and guidelines to estimate emissions. • Creates scenarios. • Addresses uncertainty. • Suggests potential actions. • 1,250 authors; 195 countries. • Over 2,500 reviewers.	• Synthesis of knowledge of trends in earth's ecosystems, including agriculture. • Creates future scenarios. • Addresses uncertainty. • Subsequent 18 MA-approved sub-global assessments. • Builds capacity for conducting scientific assessments. • 1,360 authors; 95 countries. • 850 reviewers.	• Synthesis of knowledge of nitrogen sources, impacts, and interactions across Europe. • Creates future scenarios. • Analysis of environmental costs versus economic benefits of nitrogen in agriculture. • Provides recommendations and policy options. • 200 authors; 21 countries. • International peer review.

ing and synthesizing this large body of data, using it to analyze patterns and trends, assessing the quality of information and knowledge about key issues, and translating and communicating key information.

The assessment was comprised of four distinct phases: (1) design; (2) implementation; (3) review; and (4) dissemination of results (Table 1.1). At all stages, the assessment engaged with stakeholders and collaborators in a variety of ways to foster constructive dialogue and promote transparency in the assessment process.

1.3.1. Ensuring Credibility, Relevance, and Legitimacy

Assessment methodologies differ from other scientific research approaches; legitimacy, credibility, and relevance receive equal emphasis, making stakeholder engagement a critical part of the process from start to finish (Box 1.3). Keys to achieving these core values were (1) a strategic stakeholder outreach process, with early engagement designed to help shape the assessment's approach and ensure outputs meet the needs of varied users; (2) fostering an inclusive process comprising a broad range of views and interests; (3) rigorous and transparent review processes; and (4) systematic treatment of uncertainty in available data and current scientific knowledge.

1.3.1.1. A STAKEHOLDER-DRIVEN APPROACH

The assessment was designed to lay the foundation for a network of partnerships, across the scientific community, policymakers, farmers and ranchers, and nongovernmental organizations (NGOs). We employed "strategic listening" (Pidgeon and Fischhoff, 2011) throughout the assessment to

ensure we produced credible information that is relevant to target audiences and supports decision-making, and to create information and products that address partners' needs.

The CNA engaged stakeholders from a broad spectrum of backgrounds and interests, presenting a balance between nitrogen's integral role in food production and related benefits to society while also acknowledging potential environmental and human health effects. However, we found that many stakeholders do not think about nitrogen in these terms. Thus, one of the major outcomes (Box 1.5) of the CNA's stakeholder engagement activities was awareness-building and fostering understanding of both the benefits and the problems associated with nitrogen. Throughout this process, we used a wide range of engagement mechanisms (Table 1.1) in an effort to provide a variety of avenues for stakeholder participation that engaged the diversity of perspectives across the state. As detailed below, this resulted in bidirectional flows of information and feedback from stakeholders that directed much of the assessment.

The SAC, comprising 29 representatives (a total of 33 have served throughout the assessment process) from government agencies, environmental and health nonprofit organizations, and producers and agricultural commodity groups, represents the constituent groups the assessment targeted (Appendix 1.1). The SAC was intentionally weighted on the side of producers and users due to this group's large role in nitrogen use and management. The SAC provided input on key topics and helped to prioritize the assessment's focal issues, commented on early drafts of key assessment documents, and acted as a liaison between the CNA and members' constituencies. The committee also participated in a facilitated group scenario-building exercise envisioning the future of nitrogen in California agriculture (see Chapter 6).

TABLE 1.1

Key stakeholder engagement activities

Assessment stage	Engagement mechanism	Targeted stakeholders	Outcomes
Design: setting the agenda	Stakeholder forums and grower meetings	Growers, agricultural industry, government, environmental and health organizations	Suggestions for research questions; feedback on conceptual framework; network and database of stakeholders invested in N management in California.
	Email and phone consultations	Growers, agricultural industry, government, environmental and health organizations	Suggestions for research questions; network and database of stakeholders invested in N management in California.
	Survey	UCCE Farm Advisors	Data on current cropping practices, opinions, and level of awareness.
Implementation: research and dialogue	Field visits	Growers and agricultural industry	Information on agricultural practices and management options; audio and video clips of N management issues and solutions from growers' perspective (available at www.asi.ucdavis.edu/nitrogen).
	Email and phone consultations	Growers, agricultural industry, government, researchers (UC Davis and elsewhere)	Information and data on flows of N, agricultural practices, and management options; suggestions on methodological approaches; network of researchers and stakeholders invested in N management.
	Participatory scenarios workshop	Members of Stakeholder Advisory Committee	Raise awareness of key drivers of N use and brainstorm plausible futures of N in California; identify differences in stakeholder perspectives and build shared understanding; create network of stakeholders invested in N management in California.
	Online comment forum	Members of Stakeholder Advisory Committee	Feedback on selected materials to guide assessment research.
		General public/stakeholders	Comments on assessment (e.g., input on data needs and key issues).
	Seminar class	Students and researchers at UC Davis	Build capacity for assessment methodologies and application of these methods to issues on agriculture, nitrogen, and climate change in California.
	Academic conferences	Researchers, government, environmental and health organizations	Raise awareness of CNA and obtain feedback on research goals, methodologies, and preliminary findings.
	Presentations to commodity boards and government	Government, agricultural industry, growers	Raise awareness of CNA and obtain feedback on research goals, methodologies, and preliminary findings.

	Activity	Participants	Objective
	Speaker series	Students and researchers at UC Davis	Knowledge transfer with seven scientific experts from across the United States and internationally; raise awareness of CNA and N management issues in California and elsewhere.
	N symposium	UC researchers and UC Cooperative extension specialists	Knowledge transfer on nitrogen-related issues in California; raise awareness of the CNA and build network of researchers across a range of disciplines.
Review process	*Stage 1: scientific review* – online review template	9 members from Technical Advisory Committee 40 scientific experts from academia, government, and nongovernment organizations; 10 review editors	Ensure final assessment report has scientific credibility, reflects the state of knowledge fairly and adequately, and is attentive to varied stakeholder perspectives.
	Stage 2A: SAC review – preliminary online review *Stage 2B: public review* – online review forum	29 members of the Stakeholder Advisory Committee; 10 review editors Participants in past CNA meetings/events; general public; 10 review editors	Archive of all comments, authors' responses, and editors' feedback posted to www.asi.ucdavis.edu/nitrogen.
Dissemination of results	Nitrogen website – http://asi.ucdavis.edu/nitrogen	Key stakeholder groups involved in assessment; general public; researchers	Raise awareness of CNA process and findings; provide point of access for stakeholders interested in N management in California.
	Targeted summaries, policy briefs, white papers	Policymakers, farmers, educators/students, agricultural industry	Raise awareness of CNA and main messages: key findings presented in easily digestible formats targeted to specific interests/perspectives.
	Stakeholder forum	Growers, agricultural industry, government, environmental and health organizations	Raise awareness of CNA and main messages; develop shared understanding of N issues in California and potential solutions; build network of varied interests to move forward.

In addition to interacting with the SAC, the CNA met with groups of growers, farm advisors, government agencies, and environmental and health organizations around California (Map 1.1). Key engagement mechanisms included multi-stakeholder workshops, small-group consultations with growers, farm visits, industry field trips, and individual contact via phone and email. Altogether, the assessment interacted directly with more than 350 individuals (figure 1.2). Further, the assessment contacted almost 110 organizations (Appendix 1.2) to participate in events and to solicit information. Throughout the assessment, updates and information on nitrogen-related events were available to stakeholders and the public through the website of the Agricultural Sustainability Institute (see www.asi.ucdavis.edu/nitrogen).

Collectively, the assessment's engagement with stakeholders generated over 100 questions about nitrogen use and its impacts in California that were then synthesized into five overarching research areas to guide the assessment: biogeochemistry, management practices, economics and policy, public health, and communications (figure 1.3). The synthesized list of questions set the research priorities and acted as a point of reference throughout the assessment. While these questions framed the assessment, it was just as important to flag questions or topics that could not be answered due to lack of evidence. Identifying these gaps in information and problems in available data is an important contribution to understanding the state of knowledge about nitrogen in California and what is needed to move forward.

To ensure scientific credibility, the assessment engaged a broad range of scientists at the University of California (UC) Davis and other institutions through one-on-one contact, focus group meetings, and coauthoring relationships (Appendix 1.2). A nine-member, multidisciplinary Technical Advisory Committee also provided guidance through-

out the project. Relationships with contributing authors helped to advance our understanding of issues on a diversity of topics. As noted above, the SAC played a vital role in linking research and end users.

By including key stakeholders in the governance structure and at the outset to help frame priorities, the assessment established communication channels that provide a framework for long-term collaborations on nitrogen management in California (figure 1.4; see also Appendix 1.1). The breadth of stakeholders involved in the assessment reflects the complexity of the issues and the wide-ranging implications agricultural nitrogen management has on the environment, human health, and the political and economic climate.

1.3.1.2. REVIEW PROCESS

The Agricultural Sustainability Institute at UC Davis is the convening organization of the CNA, with a mandate from its External Advisory Board to launch the assessment. The CNA followed the basic peer review methodology of the Intergovernmental Panel on Climate Change (IPCC, 1999), which is a transparent and inclusive process of writing and review. Each chapter of the assessment underwent two stages of external review: (1) scientific review and (2) stakeholder review. First, because the assessment covers a wide range of topics, the CNA relied on 40 subject experts to evaluate the comprehensiveness of the research, balance in presentation of evidence, and validity of the interpretations. After a period of author response, the revised document underwent a second stage of stakeholder review. This included feedback from the members of the SAC as well as an open public comment period in which stakeholders involved in prior engagement activities played a key role. Throughout this process, chapter review editors with expertise in relevant areas served as independent "referees" for each chapter to ensure that all comments received appropriate attention and response from the assessment authors, and an overall review editor also provided feedback and oversight across chapters. The complete iterative process, including review comments, authors' responses, editors' feedback, and Appendices are available at http://asi.ucdavis.edu/nitrogen.

1.3.1.3. DEALING WITH UNCERTAINTY

Inherent to the assessment process is considerable uncertainty in many of the data sets and research sources. Throughout the CNA, we have evaluated the quality of the data and we note when results are based upon very reliable data or data that are less reliable (i.e., due to gaps in information or disagreement in the literature). Following the models of the IPCC and Millennium Ecosystem Assessment (MA), the assessment employed reserve wording to quantify areas of uncertainty (Box 1.6). Those were highlighted as areas where more research and data collection are needed.

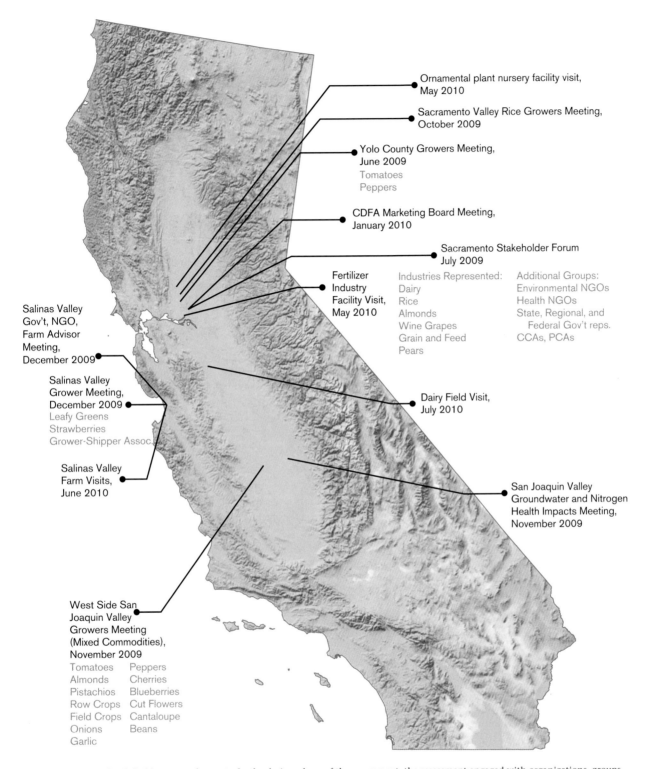

Ornamental plant nursery facility visit,
May 2010

Sacramento Valley Rice Growers Meeting,
October 2009

Yolo County Growers Meeting,
June 2009
Tomatoes
Peppers

CDFA Marketing Board Meeting,
January 2010

Sacramento Stakeholder Forum
July 2009

Fertilizer
Industry
Facility Visit,
May 2010

Industries Represented: Additional Groups:
Dairy   Environmental NGOs
Rice   Health NGOs
Almonds   State, Regional, and
Wine Grapes   Federal Gov't reps.
Grain and Feed   CCAs, PCAs
Pears

Salinas Valley
Gov't, NGO,
Farm Advisor
Meeting,
December 2009

Salinas Valley
Grower Meeting,
December 2009
Leafy Greens
Strawberries
Grower-Shipper Assoc.

Salinas Valley
Farm Visits,
June 2010

Dairy Field Visit,
July 2010

San Joaquin Valley
Groundwater and Nitrogen
Health Impacts Meeting,
November 2009

West Side San
Joaquin Valley
Growers Meeting
(Mixed Commodities),
November 2009
Tomatoes   Peppers
Almonds   Cherries
Pistachios   Blueberries
Row Crops   Cut Flowers
Field Crops Cantaloupe
Onions   Beans
Garlic

MAP 1.1. Location of stakeholder outreach events. In the design phase of the assessment, the assessment engaged with organizations, groups, and individuals from a broad range of perspectives through stakeholder forums, grower consultations, and field visits. This map does not include numerous additional meetings held at UC Davis, national and international nitrogen meetings, and stakeholder engagement via telephone interviews and email surveys (map by R. Murphy).

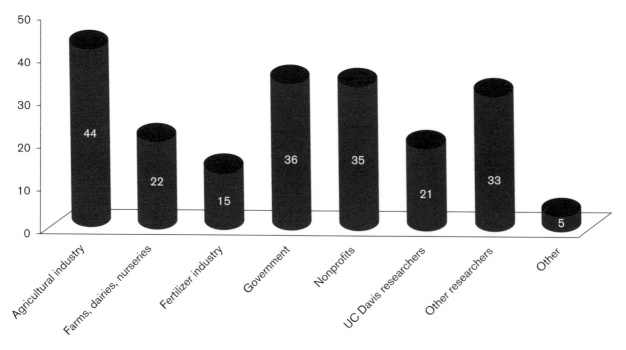

FIGURE 1.2. Stakeholder engagement. Includes participants at 12 stakeholder forums, grower consultations, and field visits between June 2009 and July 2010, as well as individuals contacted by phone or email. In addition to these activities, the assessment engaged with 151 farm advisors and extension specialists with University of California Cooperative Extension (UCCE) and the members of the Stakeholder Advisory Committee (SAC).

To support the statements and conclusions in the assessment, we developed online Supplemental Data Tables (Appendix 4.1). These contain (1) summaries of the sources of data and approaches or methodologies used in the assessment and (2) a systematic evaluation of the quality of available information. The data tables are organized by broad categories and each lists the specific indicators used in the assessment related to that topic. To help readers track information, relevant sections of the assessment are noted for each indicator in the table.

1.4. What Is the Purpose of the California Nitrogen Assessment?

Throughout the assessment a conceptual framework was used to structure the research scope and the CNA underwent adaptations based on changing scientific knowledge and emerging needs of our stakeholders. The breadth of our stakeholder engagement activities, in terms of both the diversity and inclusiveness of perspectives involved and the range of outreach efforts, resulted in outcomes well situated to address the complexity of nitrogen issues in California agriculture and meet the varying needs of these audiences.

1.4.1. The Conceptual Framework: Roadmap to the Assessment

Modified from the MA, the conceptual framework (figure 1.5) is a snapshot of the big-picture approach of the CNA,

focusing on the reasons that nitrogen flows the way it does, including the human decisions and responses that are involved, as well as the science behind how nitrogen moves through the N cycle. Strategic actions can take place at almost any point in this framework to respond to negative changes or to augment positive changes (Alcamo and Bennett, 2003). This framework was developed collaboratively, moving through several models as the research progressed and through consultations with stakeholders.

The conceptual framework places the nitrogen challenges in California within the context of ecosystem services and human well-being. We first identified important *underlying drivers* (Chapter 2) of nitrogen use decisions—the economic, political, and technological processes that influence human decision-making in such a way as to affect nitrogen's passage through California ecosystems. These drivers indirectly affect ecosystems and can cross both temporal and spatial scales. Examples include growth in global demand for California commodities and prices of fuel and fertilizers. These underlying drivers, in turn, influence *direct drivers* (Chapter 3)—the human and natural processes that directly alter the nitrogen cycle. The assessment focused on the six activities in the state that influence, and will continue to influence, N use and emissions, including historical trends in these activities. This includes growth in acreage of high nitrogen-demanding crops, concentration of animals in feedlot dairies, and fossil fuel combustion in vehicles.

These drivers, in turn, affect the statewide *mass balance* of nitrogen (Chapter 4)—the quantification of how much N enters the state through new sources, the quantity of

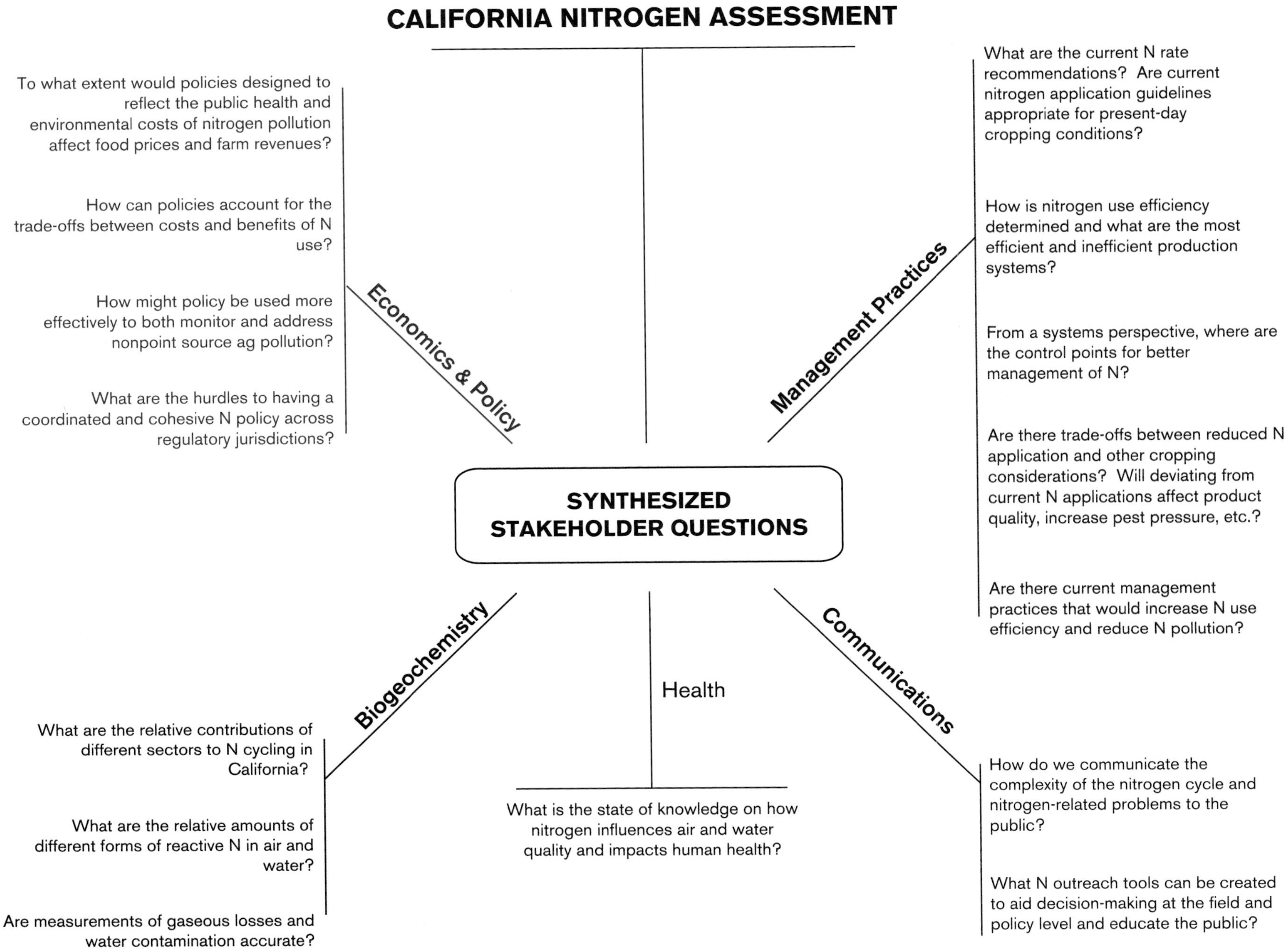

CALIFORNIA NITROGEN ASSESSMENT

SYNTHESIZED STAKEHOLDER QUESTIONS

Management Practices

What are the current N rate recommendations? Are current nitrogen application guidelines appropriate for present-day cropping conditions?

How is nitrogen use efficiency determined and what are the most efficient and inefficient production systems?

From a systems perspective, where are the control points for better management of N?

Are there trade-offs between reduced N application and other cropping considerations? Will deviating from current N applications affect product quality, increase pest pressure, etc.?

Are there current management practices that would increase N use efficiency and reduce N pollution?

Communications

How do we communicate the complexity of the nitrogen cycle and nitrogen-related problems to the public?

What N outreach tools can be created to aid decision-making at the field and policy level and educate the public?

Health

What is the state of knowledge on how nitrogen influences air and water quality and impacts human health?

Economics & Policy

To what extent would policies designed to reflect the public health and environmental costs of nitrogen pollution affect food prices and farm revenues?

How can policies account for the trade-offs between costs and benefits of N use?

How might policy be used more effectively to both monitor and address nonpoint source ag pollution?

What are the hurdles to having a coordinated and cohesive N policy across regulatory jurisdictions?

Biogeochemistry

What are the relative contributions of different sectors to N cycling in California?

What are the relative amounts of different forms of reactive N in air and water?

Are measurements of gaseous losses and water contamination accurate?

FIGURE 1.3. Major categories of stakeholder-generated questions about nitrogen. These have been consolidated from over 100 questions generated by stakeholders (see Appendix 1.4 for a list of stakeholder questions). These five overarching thematic areas represent the major issues identified as most important for nitrogen in California agriculture; these synthesized questions were used to guide the assessment's research priorities.

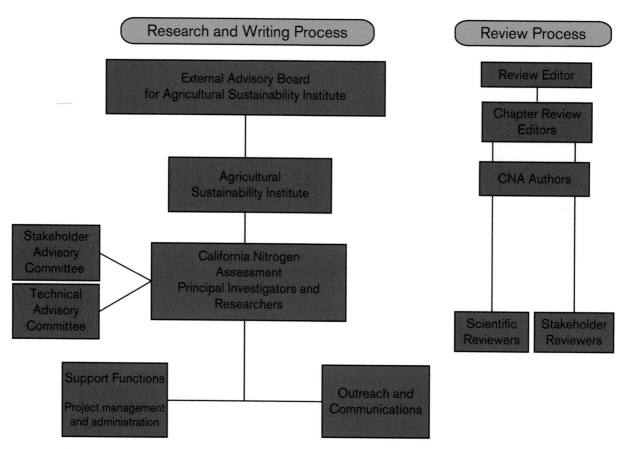

FIGURE 1.4. Governance structure of the California Nitrogen Assessment.

nitrogen-containing compounds transformed from one form to another, and, ultimately, how much of these varied compounds enter the environment. Conducted for the year 2005, the mass balance calculates the magnitude of nitrogen flows at the statewide level as well as for eight subsystems: cropland, livestock, urban land, people and pets, natural land, atmosphere, surface water, and groundwater. A mass balance approach is useful not only to compare the size of nitrogen flows, but also to identify gaps in understanding about the size and directions of these flows. Knowledge of the relative magnitude of flows should inform management and policy decisions targeting reductions in nitrogen.

The resulting changes in ecosystems affect human well-being. For example, some of the nitrogen used in agriculture "leaks" from farms, dairies, and fertilized landscapes, and fuel combustion also results in significant releases of nitrogen into the atmosphere, resulting in costs to the environment and human health. We investigated the impacts on *ecosystem health and human well-being* (Chapter 5) of these varied forms of nitrogen flows on five major ecosystem services: healthy food and other agricultural products, clean drinking water, clean air, climate regulation, and cultural and spiritual values (e.g., fishing, swimming, and biodiversity). Where data exist, we reviewed trends in nitrogen inputs and effects on these ecosystem services, linkages to

human health and social equity, and the corresponding economic impacts of these changes to ecosystem services.

Scenarios can help stakeholders deal with controversy and complexity, and they are particularly useful in cases where there is a large amount of uncertainty, as is the case in this assessment. The CNA worked with stakeholders to develop scenarios to facilitate dialogue and consensus on the issues and help focus potential technical and policy responses (Chapter 6). This exercise led to the development of four distinct scenarios of how nitrogen-relevant technologies and policies might unfold in the next 20 years and how these would affect nitrogen use and impacts. Although the starting perspectives were quite diverse, two areas of uncertainty were emphasized in this process: future profitability of California agriculture and the future course of policy and mechanisms for implementation.

Finally, we examined the state of the science around the most promising *technical and policy solutions* to minimize nitrogen leakage while sustaining vitality of agriculture. The assessment does not include prescriptive recommendations. Rather, it utilizes an "if/then" approach, considering practices and policy options and the potential effects each would have on agriculture, the environment, and human health. Based on the California nitrogen mass balance, nine critical areas for intervention in the nitrogen cascade were identified. Chapter 7 reviews these critical control

BOX 1.6. COMMUNICATING UNCERTAINTY

Quantitative Analyses The following reserved wording was used for statements that lent themselves to formal statistical treatment, or for judgments where broad probability ranges could be assigned:

Virtually certain	Greater than 99% chance of being true or occurring
Very likely	90–99% chance of being true or occurring
Likely	66–90% chance of being true or occurring
Medium likelihood	33–66% chance of being true or occurring
Very unlikely	1–33% chance of being true or occurring
Exceptionally unlikely	Less than 1% chance of being true or occurring

Qualitative Analyses The following reserved wording was used for more qualitative statements:

<table>
<tr><th colspan="5">Amount of Evidence</th></tr>
<tr><th colspan="2"></th><th>Limited</th><th>Medium</th><th>High</th></tr>
<tr><td rowspan="3">Level of Agreement</td><td>*High*</td><td>Agreed but unproven</td><td>Agreed but incompletely documented</td><td>Well established</td></tr>
<tr><td>*Medium*</td><td>Tentatively agreed by most</td><td>Provisionally agreed by most</td><td>Generally accepted</td></tr>
<tr><td>*Low*</td><td>Suggested but unproven</td><td>Speculative</td><td>Alternate explanations</td></tr>
</table>

From Ash et al. (2010). Copyright © 2010 by the authors. Reproduced by permission of Island Press, Washington, DC.

points and evaluates related mitigative strategies and technological options to reduce emissions of nitrogen. Chapter 8 examines a variety of policy instruments and approaches and their potential to balance the costs and benefits of nitrogen use, and catalogs the barriers to creating a coordinated and cohesive policy to manage nitrogen.

1.4.2. Target Audience

The assessment contributes to the international understanding of nitrogen cycles, balances (i.e., key inputs and outputs), sources, and effects on the environment and human health. It also clarifies options for technological changes and policymaking. An assessment is a model of research currently underutilized and the CNA provides a practical example that can be used from the local to global scale. For policymakers, this approach may be of particular interest given the high level of stakeholder engagement and focus on transparency and legitimacy in the research process and findings.

As detailed above, the CNA conducted a great deal of outreach to key stakeholder groups in California as part of the project's design and implementation—from the development of research questions and priorities to data collection and feedback on methodologies and findings. Through this process we have developed a network of contacts invested in nitrogen science and management across the state, representing the diversity of the target audience. Primary target audiences include the following:

- Policymakers and governmental agencies (California, United States, and globally)
- Agribusiness (processors, distributors, fertilizer producers/distributors)
- Cooperative extension farm advisors and specialists
- NGOs—environment, health, and social justice
- Commodity organizations and farmer associations
- Farmers/growers and ranchers
- California citizens
- Scientists and researchers (California, United States, and globally)

Mechanisms to ensure products and outcomes are relevant to the varying, and often conflicting, needs of these stakeholders have been built into the assessment (Table 1.1). Communication and dissemination of the assessment findings will be the final stage in the assessment process. The nitrogen website will be used as an ongoing outreach tool, with interactive information on key findings designed to facilitate broad understanding of the issues and linkages across the chapters. This will also serve as a dynamic source

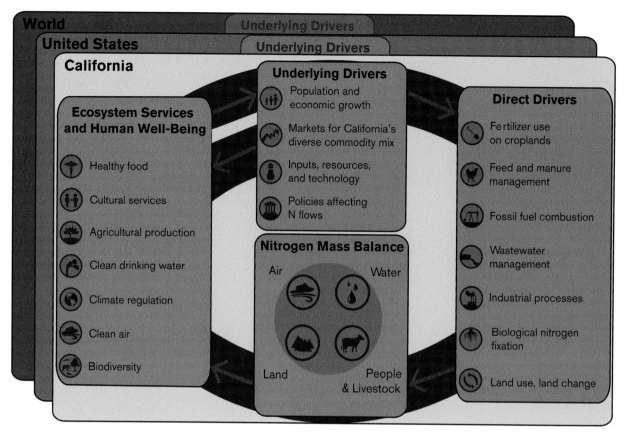

FIGURE 1.5. The conceptual framework.

of emerging information as agricultural nitrogen issues develop in California.

1.5. Global Implications: Is Nitrogen the Next Carbon?

From the perspective of ecosystem services and human well-being, nitrogen shares many attributes that have made carbon—specifically the disruption of the global carbon cycle and resulting potential destabilization of our global climate and related systems, including food production (IPCC, 2014; Vermeulen et al., 2012)—the focus of an intense and highly polarized international policy debate. For example, while various forms of each element indisputably play essential roles in natural and social systems, evidence is mounting that for both carbon and nitrogen:

- It is *well established* that human activity now dominates the global biogeochemical cycles for both C and N (Gruber and Galloway, 2008; IPCC, 2007a; MA, 2005a).
- Thresholds for unintended consequences for the global climate system and environment are controversial and uncertain; it has been *suggested (but remains unproven)* that emissions of both C and N

have crossed the boundaries of "safe operating space" for Earth's life support systems (Rockström et al., 2009).

- Shifts in flows within these cycles matter, because their multiple forms and media have very different environmental impacts, with some benign and others damaging. This is *well established* for nitrogen, with large amounts of evidence ranging across scales from the field level (e.g., Gardner and Drinkwater, 2009) to global biogeochemical cycles (Canfield et al., 2010; Galloway et al., 2008).
- It is *generally accepted* that consequences of these shifts—both intended and unintended—are closely linked to and indeed driven by human population growth and prosperity (MA, 2005a; Steffen et al., 2011).
- It is *well established* that significant environmentally damaging consequences of carbon emissions and nitrogen leakages are intrinsic features of currently dominant technologies (Canfield et al., 2010; IPCC, 2007a; Pelletier et al., 2011; Vermeulen et al., 2012).
- It also is *generally accepted* that better balance between intended economic benefits and unintended environmental damage is unlikely under "business as usual" development and requires fundamental changes in major technologies (for

TABLE I.2

Comparison with some key elements of the European Nitrogen Assessment (ENA), organized by chapter of the California Nitrogen Assessment (CNA)

CNA chapter	California Nitrogen Assessment	European Nitrogen Assessment
1. Assessment process, engagement, and communication	Statewide assessment modeled on the Millennium Ecosystem Assessment (Ash et al., 2010), driven by stakeholder questions; subjected to two rounds of review (scientific review and stakeholder review).	Continental-scale research and assessment project, developed through a series of five main workshops and subject to international scientific review (see Sutton et al., 2011a, Chapters 1, 2, 5, and 26).
2. Underlying drivers	Prosperity and population growth in California and across the world; export demand for high-quality agricultural products.	Prosperity and population growth within Europe (see Sutton et al., 2011a, pp.552–557).
3. Direct drivers	Synthetic N fertilizer use, dairy production, fossil fuel combustion.	Livestock production, synthetic N fertilizer use, fossil fuel combustion (see Sutton et al., 2011a, pp.552–557).
4. Mass balance	Largest inflow: synthetic nitrogen fertilizer Largest outflow: nitrate leaching to groundwater	Largest inflow: synthetic nitrogen fertilizer. Largest outflow: gaseous N to the atmosphere[1] (see Sutton et al., 2011a, Part III, especially Chapter 16).
5. Ecosystem services and human well-being	Key benefits: nationally (and in some cases globally) important producer of vegetables, fruit, and tree nuts. Main costs: air and water pollution.	Key benefits: food production for domestic population and export. Main costs: air pollution dominant, also water pollution (see Sutton et al., 2011a, Chapters 3, pp.17–21, and Chapter 22 on cost/benefit analysis).
6. Scenarios	None of the scenarios by themselves led to sufficient improvement in groundwater quality to fully address human health concerns.	Short/intermediate term scope for more integrated approach to N policy plus longer term prospects for eventual population decline and dietary shifts (see Sutton et al., 2011a, Chapter 24).
7. Responses: practices and technologies	Technologies and practices that can reduce nitrogen pollution from agriculture certainly do exist, but they typically are costly (in money and management); thus, voluntary adoption tends to be low.	ENA control points are virtually identical to CNA: N use efficiency in crop and livestock agriculture, manure management, transport and energy, waste water treatment, dietary choices (Sutton et al., 2011a, pp.xxxii–xxxiii).
8. Responses: policies and institutions	For communities where drinking water supplies are unsafe because of high nitrate concentrations, point-of-use treatment or some other approach will be needed to assure safe drinking water for all California communities.	Parallel recognition of need for integrated policies; cost/benefit analysis indicates air pollution is greatest social cost of N leakage and that these costs are significant (€150–170 per capita per year); internalizing these costs would lower fertilizer rates significantly (see Sutton et al., 2011a, Chapters 5, 22, 23 and 25; especially Chapter 22, p.513).

1. ENA includes coastal zones within its N budget, while CNA does not.

example, vehicles, power generation, nitrogen synthesis, agricultural practices) (Canfield et al., 2010; IPCC, 2007a; MA, 2005a; Steffen et al., 2011). The Haber–Bosch process enabled a shift from large-scale natural sources of nitrogen, such as biological fixation and guano (Davis, 2013), to ammonia synthesis based on fossil hydrocarbons (natural gas) as a feedstock, which has become a force in both the C and N cycles globally. New approaches to displace the dominant Haber–Bosch process in ammonia synthesis, while minimizing greenhouse gas emissions, still have "a long way to go" (Service, 2014).

· Moreover, it is *well established* that significant portions of these environmentally damaging emissions and leakages result from land use and land cover change, meaning they are spatially dispersed and often difficult to monitor—creating particular challenges in the design of mitigation policies that are both effective and efficient (IPCC, 2007a; MA, 2005a; Sutton et al., 2011b; Vermeulen et al., 2012). These unintended consequences of human-caused changes in carbon and nitrogen cycles are central to some of the greatest challenges of the twenty-first century, including global climate change, depletion of natural biodiversity, and other mounting pressures on land, air, and water resources. Nitrogen's roles in these processes are most apparent at regional and continental scales, while destabilization of the Earth's climate system is in many ways the iconic global issue of our time.

Readers of this volume also should be aware of the European Nitrogen Assessment (ENA) (Sutton et al., 2011a), which is a comparable integrated assessment of N flows at policy-relevant scales. Table 1.2 compares and contrasts these sister assessments, organized by chapter of the CNA and with corresponding chapter references in the ENA. Despite significant differences in the range of climatic and edaphic conditions and in their economies, policies, and many other dimensions, it is striking that these two nitrogen assessments show many more commonalities than contrasts. Agriculture looms large in shaping nitrogen flows in each case, followed by combustion of fossil fuels. In each, there are abundant technological options for N management and the control points of technological and policy leverage discussed in Chapter 7 of this volume are essentially identical to those identified by the ENA. In Europe as in California, policy has been much more effective in controlling pollution from fuel combustion, while effective policy regarding N leakage from agriculture has been much

more elusive. Because of a substantial scientific literature on cost–benefit analyses relevant to nitrogen flows in Europe, the ENA is able to go much further in assessing policy options (Sutton et al., 2011a, Chapter 22) than is possible in Chapter 8 of this volume. While there can be no substitute for conducting these analyses for California as a basis for state policy, it is noteworthy that the ENA concludes that air pollution produces the greatest social costs of N leakage and that internalizing these human health costs would lower fertilizer application rates significantly (Sutton et al., 2011a, p.513). As shown in Chapter 4 of this volume, perhaps the key contrast between California and Europe taken as a whole (but perhaps not with the European nations bordering the Mediterranean) is that, in California, groundwater is the main sink for N leakages. This leads to the particular policy priority for California of some sort of point-of-use intervention to ensure safe drinking water for all of its citizens (discussed in Chapters 7 and 8). Nevertheless, both assessments call for an integrated approach to N policy because of its multimedia nature.

1.6. Conclusion: Complex Issues Require Sophisticated Understanding

Substances like nitrogen pose unique challenges to traditional regulatory frameworks because their effects cross traditional media-specific regulatory boundaries (e.g., from air to water). The CNA shows that both the challenges and opportunities for improved nitrogen management depend crucially on context. Due to the mobility of nitrogen and its occurrence in multiple forms, the most effective strategy to address nitrogen leakage in agriculture will likely be one that integrates across multiple biogeochemical processes, spatiotemporal scales, and nitrogen sources. Such integration will require careful consideration of the technical potential of different farming practices to limit total nitrogen inputs and increase nitrogen use efficiency, while also accounting for the potential trade-offs that can occur with practices that may limit one form of nitrogen pollution while increasing another. Technology implementation will need to be coupled with smart policies that address such trade-offs, while also recognizing the complex, multiobjective context in which farm managers make decisions to maintain the viability of their operations. By acknowledging and addressing the needs of multiple stakeholders, the CNA identifies areas where we can make progress on these complex issues and provides scientific grounding for decision-makers (policy and in the field) to move forward, utilizing collaborative and integrative approaches necessary to sustain agricultural productivity, economic prosperity, environmental health, and human well-being.

CHAPTER TWO

Underlying Drivers of Nitrogen Flows in California

Lead Authors:
A. CHAMPETIER, D. SUMNER, AND T. P. TOMICH

Contributing Authors:
S. BRODT, M. COLEY, V. R. HADEN, M. KREITH,
J. T. ROSEN-MOLINA, AND K. THOMAS

What Is This Chapter About?

To understand the stocks and flows of nitrogen in California, we first identify important underlying drivers—the economic, political, and technological processes that influence human decision-making in such a way as to affect nitrogen's presence in and passage through California ecosystems. These drivers encompass a range of temporal and spatial scales and, in turn, influence direct drivers of nitrogen use and, ultimately, the statewide mass balance of nitrogen. This chapter examines four key underlying drivers affecting nitrogen use decisions in California: (1) human population and economic growth; (2) market opportunities for California commodities; (3) agricultural production costs and technological change; and (4) policies targeting nitrogen in California.

Main Messages

Forces affecting levels of agricultural production and fossil fuel combustion have been the dominant drivers of the nitrogen (N) cycle in California.

California's agriculture ships a large share of its products to other states and regions of the world. For 2009, almost 50% of production went to Europe and Canada, and another 27% to Mexico, China, and Japan. Long-term reduction of transportation costs and reduction of international trade barriers have increased access to international markets for California producers. Thus California carries a lot of the nitrogen burden for many non-Californians.

Over the last 50 years, world population doubled and global income quadrupled. The resulting increase in global demand for food has been a fundamental driver of expansion of agricultural production in California.

Demand for many of California's main agricultural exports (pistachios, almonds, rice, walnuts, and oranges) is driven by rising per capita incomes and perceptions of quality. Accordingly, population growth of high-income countries and increases in household incomes in regions such as East Asia have been the dominant underlying drivers of demand for food and other agricultural commodities produced in California.

Long-term decline in nitrogen fertilizer prices resulted in a large increase in fertilizer use in California from the 1950s through the 1970s. Thereafter, fertilizer prices were relatively stable compared to the prices of crops until 2000. Fertilizer price increases between 2001 and 2011 exceeded increases in crop prices.

California's population doubled over the last 50 years, while income more than doubled over the same period. The growth of California's economy has resulted in a growth in nonagricultural activities that generate nitrogen emissions, including fossil fuel combustion and wastewater creation. In addition, population and economic growth in California have increased nonagricultural use of resources such as land and water.

Value of development for housing and other urban land uses drove land use change in California for most of the twentieth century. Historically, financial returns to agriculture have been much less than these land development alternatives; hence levels of farm revenues have had little or no influence on conversion of land to nonfarm uses. These relationships have attenuated since the mid-2000s. The contraction in home construction brought by the Great Recession lowered demand for conversion of agricultural land to housing and other forms of development. Over the same period, increases in tree nut and other export commodity prices have driven significant increases in California agricultural land prices; it remains to be seen what effect the drought (still ongoing in 2015) will have on farmland values.

In comparison to the effects of economic growth on fossil fuel combustion or the increase in fertilizer use, policies targeting nitrogen pollution have had small effects on nitrogen flows in California to date.

The bottom line: short of catastrophe, demand side fundamentals driven by growth in population and income in the rest of the world suggest that nitrogen flows in California agriculture are unlikely to decrease and indeed are likely to continue to grow. In short, California agriculture is unlikely to disappear; in fact, on balance, it seems more likely to continue growing. Moreover, while there is considerable uncertainty about future climate, water supply, energy prices, and labor costs, the history of innovation in California agriculture gives some tentative (but unproven) reasons to believe that technological change and other forms of adaptation will enable California agriculture to continue to grow in value and employment. Since these underlying drivers on balance portend continued growth in agriculture and attendant nitrogen flows for the foreseeable future, we proceed to assess the direct drivers, relative magnitudes of N flows, and their consequences for the state's ecosystems and the well-being of California's inhabitants in Chapters 3, 4, and 5, respectively. As long as the direct benefits of the system are so big, it is not likely that the attendant external costs (environmentally or socially) will be mitigated on their own. The main sources of uncertainty regarding the future balance of costs and benefits of nitrogen flows in California agriculture concern policy choices regarding trade and exchange rates determined in national and international policy arenas and regarding environmental and public health policies largely shaped within California. The implications of these uncertainties and their interactions regarding opportunities for profitable agricultural exports, the balance of costs and benefits—for the state as a whole and for the profitability of the agriculture sector in particular—of different policy strategies, and the prospects for technological and institutional innovation necessary for adaptation, are explored in the scenarios in Chapter 6.

2.0. Introduction

The remarkable increase in human population over the last 100 years and the even more dramatic growth in average wealth per capita have been the two dominant underlying drivers of changes in ecosystem services around the world (MA, 2005b). Ecosystem services related to nitrogen (N) in California are no exception. Many of the underlying drivers of changes in nitrogen flows within California have originated outside the state because of California's economic connections with the rest of the world. The underlying drivers considered in this chapter are emphasized because of their importance in shaping the direct drivers covered in Chapter 3. Agriculture and fossil fuel combustion are the human activities that have brought the largest increases in flows of nitrogen in California over the past 50–60 years. Apart from these and biological nitrogen fixation, other activities that have significantly shaped Califor-

nia's nitrogen flows are sewage treatment and, to a lesser extent, land use change.

Specifically, in this chapter we review four key underlying drivers of changes in nitrogen flows arising from agricultural production and fossil fuel combustion in California: human population and economic growth, market opportunities for California agriculture, agricultural production costs and technological change, and public policies, though few have targeted nitrogen pollution directly.

- *Global increases in human population and income* have driven up global demand for food, creating market opportunities for agricultural products (Section 2.1).
- *Increasing demand for food in the United States and elsewhere* has been particularly strong for agricultural products in which California excels (Section 2.2).
- Meanwhile, economic growth *within the state* has affected the costs of California's land and water resources. Competition for these limited resources between agriculture and other uses has played a central role in shaping the economic incentives facing California farmers. Fortunately, *agricultural research and development* (R&D) have greatly enhanced agricultural productivity in California, helping to preserve the state's comparative advantage in a wide range of commodities (Section 2.3).
- Particularly since the 1970s, *federal, state, and local environmental policies and regulations* have curbed some of the unintended flows of nitrogen—most significantly regarding surface water and air pollution. Most of the regulations that affect nitrogen in California (either directly or indirectly) arise from regional or federal policies (Section 2.4).

We can draw on well-established data series on human population, global economic growth, and patterns of food demand. Although there is some uncertainty going forward, it is likely that all of these will continue to drive nitrogen flows higher in California. On the other hand, future prospects for agricultural R&D and for environmental policy, particularly federal and/or state-level regulations aimed specifically at nitrogen pollution, hold the greatest uncertainty. In combination, the mix of agricultural innovations and public policies will play powerful roles in determining levels and management of nitrogen in California in the decades ahead; these interacting areas of uncertainty are the focus of the scenarios presented in Chapter 6.

2.1. Human Population and Economic Growth

Worldwide increases in population and economic activity have increased global demand for food and, with that, a corresponding demand for nutrients such as N. Large increases in per capita income in parts of the world have resulted in shifts in diet composition towards more protein and, in particular, more animal protein, which also affects N flows both through greater derived demand for feed grains and through increasing animal manure production. The extent to which agricultural producers in California are affected by the global rise in food demand depends on the response of producers in other parts of the world, the United States included, as well as factors affecting trade, including both transportation costs and trade policies (see Sections 2.2 and 2.3). Levels and relative magnitudes of N flows are calculated in Chapter 4.

In the 50 years between 1960 and 2010, world population more than doubled, increasing from 3.03 to 6.92 billion people (United Nations, 2011). By 2050, the medium variant projection for global population exceeds 9.5 billion (the range—low and high variants—of these UN projections is 8.3–10.9 billion). Much of the population growth on the planet has been in East and South Asia, which, with a combined 3.9 billion people, constituted the most populous region on Earth in 2010. This pattern reflects population growth rates that have been and are forecasted to remain higher in Asia than in other regions of the world except sub-Saharan Africa (United Nations, 2013).

In addition to population size, gross domestic product (GDP) is a fundamental indicator of the size of economies and, as discussed below, also is a key determinant of food demand. Between 1970 and 2013, the world's GDP increased sixfold, from $11.9 to $65.1 trillion (constant 2005 US$) (World Bank, 2015). Most of this economic growth has been located in Europe, the United States, and Asia. The rapid growth in the GDP of Asia seen during the last decade is forecast to continue, which could bring Asia's share of world economic activity to about one-third by 2030 (World Bank, 2015).

2.1.1. Income Growth and Patterns of Demand for Food

Income affects food consumption. In general, and especially in developing countries, increases in per capita income increase demand for food measured both in expenditure and in calories (see Appendix 2.1.1). Income increases also tend to change diet, including increases in protein consumption, increases in the share of animal protein in total protein consumption, and other changes related to perceived diet quality (Alderman, 1986; Grigg, 1995).

The well-established negative relationship between income and share of food in household expenditures is known as Engel's law (i.e. the *share* of food in total expenditures decreases as income increases). One consequence of Engel's law is that although increases in per capita income can lead to large increases in demand for food at very low incomes, this effect attenuates as income grows. In the United States, where shares of disposable income spent on food fell from about 18% in 1960 to about 10% in 2009 (USDA ERS, n.d., "Food Expenditures"), most of the overall decrease in share of food expenditure reflected a reduction in the share of food eaten at home, whereas the expenditure on food eaten away from home increased between 1960 and 2012 (figure 2.1).

In addition to the trend toward eating out, other expected changes in diet (increases in consumption of fruit, vegetables and meat, as well as luxuries such as wine) still are unfolding in the United States, despite already high income levels. Per capita consumption of fresh fruit and vegetables in the United States increased moderately since 1970 (figure 2.2). Consumption of wine and tree nuts, two commodity groups in which California leads the nation, almost doubled over the same period. The composition of animal products consumed in the United States also has changed significantly (figure 2.3). Chicken consumption per capita more than doubled between 1970 and 2012, whereas the consumption of other meats diminished slightly over that period. Dairy consumption per capita has remained relatively constant. Overall these figures show fairly typical patterns of demand for a high-income country.

The current mix of California commodities corresponds predominantly to the diet of regions with high-income per capita. Accordingly, the dominant underlying drivers of food demand facing California are to be found in the population growth of high-income countries and the increase in the proportion of relatively higher-income households in regions such as East Asia.

2.1.2. Population and Economic Growth in California

The tremendous growth in California's human population and economy over the last century has resulted in large increases in both intended and unintended flows of nitrogen. The conversion of land to urban uses and the treatment of sewage and other urban waste, as well as the fixation of nitrogen during fossil fuel combustion, are the main drivers of nitrogen flows that have resulted from a larger and wealthier California economy.

Furthermore, increased use of land for urban purposes has increased not only the cost of land as an input for agriculture but also the occurrence of externalities between land uses. For example, the conflict between the Chino dairy industry and urban residents, in which residents protested against degraded air quality, is one of the most prominent cases involving nitrogen pollution and the demand for environmental quality (see Hughes et al., 2002).

Between 1960 and 2010, the population of California increased from 15.7 to 38.7 million (United States Census Bureau, 2013). This growth has been the combined result of a birth rate exceeding a death rate, migration from other states, and immigration from other countries, both legal and illegal. The Department of Homeland Security estimated that in 2011, 2.8 million unauthorized immigrants resided in California (Hoefer et al., 2012).

Although the increase in California's population has been concentrated in the areas around Los Angeles, the San Francisco Bay, and parts of the Central Valley, the

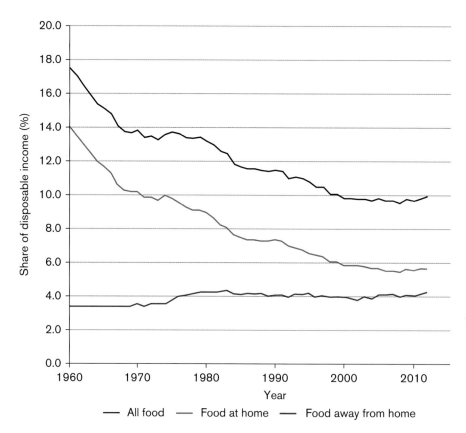

FIGURE 2.1. Food expenditures as a percentage of US disposable personal income, 1960–2012.
DATA FROM: USDA ERS (n.d., *Food Expenditures*).

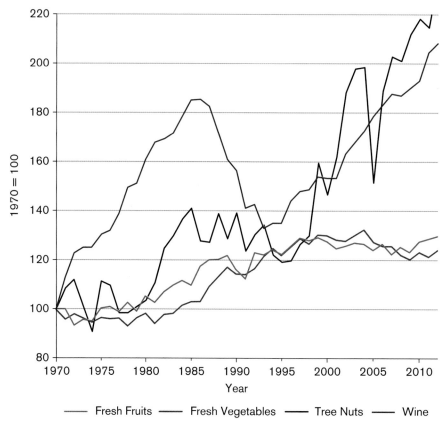

FIGURE 2.2. Index of fruit, vegetable, nut, and wine consumption per capita in the United States, 1970–2012 (1970 = 100).
DATA FROM: USDA ERS (2013).

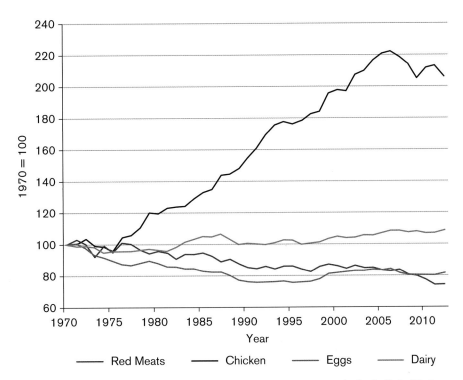

FIGURE 2.3. Index of red meat, chicken, egg, and dairy consumption per capita in the United States, 1970–2012 (1970 = 100).

DATA FROM: USDA ERS (2013).

population of every one of the 58 counties increased during the last 50 years (United States Census Bureau, 2013).

According to 2008 estimates of the California Department of Finance, California's population will increase to about 46 million by 2025, with 30% of the state's population born in foreign countries (PPIC, 2008). Available data since 1985 demonstrate that income per capita in California has been slightly above that of the US average (figure 2.4). High incomes per capita and the demand for labor have contributed to sustained legal and illegal immigration into California from other states and other countries, with Mexico providing a large share of immigrants.

2.1.3. Global Population and Incomes Are Increasing: So What?

Rising population and especially rising incomes in the rest of the world will continue to drive up demand for food, particularly for commodities in which California agriculture excels. This is very likely to be reinforced by growing population and incomes within California. These drivers will tend toward expanding agricultural production in California, and hence toward continued increases in nitrogen flows.

2.2. Markets for California's Diverse Commodity Mix

Given available natural and human resources, market incentives (as conditioned by regulations) drive agricultural

production in California and, hence, shape important N flows. The diversity of California's agriculture reflects the diversity of marketing opportunities for its products as well as the diversity of its soils and climates. Over the last 50 years, large changes in market prices for the commodities that can be produced in California have resulted in correspondingly large changes in the composition of California's production. In addition, some reductions in transportation costs and in government-set barriers to trade have increased marketing opportunities for California commodities.

This section presents indicators for the parallel changes in commodity prices and production mix that have occurred in California over the last 50 years. The patterns of trade that underlie marketing opportunities for California's commodities, as well as the factors that have resulted in a reduction in trade barriers, are then described.

2.2.1. Market Prices and California's Commodity Mix

California's agriculture is diverse and responsive to changes in market incentives—the ranking of the top 15 commodities by cash receipts has changed significantly and rapidly over the last 50 years (Table 2.1). Although the ranking of some commodities, such as grapes and dairy products, has remained relatively stable over time, the ranking of many other commodities has changed. Cash receipts for some commodities such as almonds, greenhouse and nursery products, and strawberries have risen rapidly, whereas those for others, such as cotton, oranges, potatoes, or barley, have decreased.

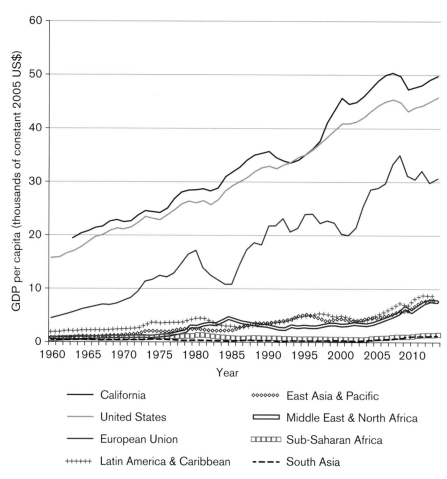

FIGURE 2.4. Inflation-adjusted gross domestic product per capita in California, the United States, and world by region, 1960–2013.

DATA FROM: World Bank (2015); CA DOF (2005, 2011, 2013, 2014).

TABLE 2.1
Ranking of California's commodities by cash receipts in 1960, 1980, 2000, and 2010.

Rank	1960	1980	2000	2010
1	Cattle and calves	Milk, wholesale	Milk, wholesale	Milk and cream
2	Cotton	Cattle and calves	Greenhouse/nursery	Grapes
3	Milk, wholesale	Cotton	Grapes	Almonds
4	Chicken eggs	Grapes	Lettuce	Nursery
5	Grapes	Greenhouse/nursery	Cattle and calves	Cattle and calves
6	Oranges	Hay	Tomatoes	Strawberries
7	Hay	Tomatoes	Misc. vegetables	Lettuce
8	Tomatoes	Misc. vegetables	Strawberries, Spring	Tomatoes
9	Greenhouse/nursery	Almonds	Almonds	Pistachios
10	Potatoes	Rice	Cotton	Hay
11	Lettuce	Lettuce	Broccoli	Walnuts
12	Turkeys	Chicken eggs	Oranges	Flowers and foliage
13	Plums and prunes	Sugar beets	Hay	Rice
14	Barley	Wheat	Avocados	Chickens
15	Milk, retail	Broilers	Celery	Oranges

DATA FROM: USDA ERS. n.d. Farm Income and Wealth Statistics.

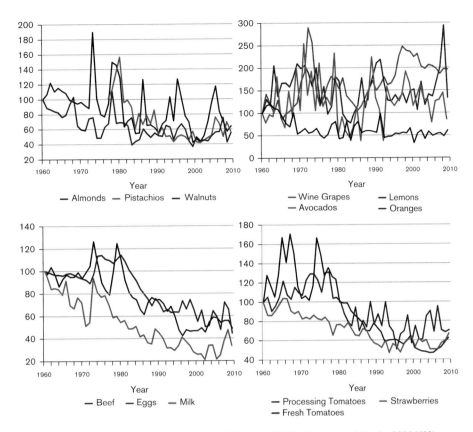

FIGURE 2.5. Index (1960 = 100) of prices received for select California commodities (in 2000 US$), 1960–2009.

DATA FROM: USDA NASS (2010); US BEA (2015).

These changes in California's commodity mix reflect changes in farm profits more than the patterns in commodity prices. Prices for agricultural commodities have generally fallen relative to prices for other products and services over the last several decades (Anderson, 1987). The prices of commodities that have risen in importance, such as almonds and strawberries, have seen smaller declines than the prices of commodities such as oranges (figure 2.5). These cross-commodity shifts affect N flows because of different fertilizer use rates and management practices. The indexed prices for beef and milk have fallen more than crop indexed prices. However, dairy products, and to a lesser extent cattle and calves, have maintained their ranking through large increases in production (see Chapter 3).

2.2.2. International Trade in California's Commodities

A large share of California's agricultural products is consumed outside of California, both in other states and in other countries. There are however no data on California's consumption of food production; available data are only nationwide and sometimes broken down by demographic group. Brunke et al. (2004) estimated that using these demographic data to correct for differences in food consumption related to the demographic characteristics of California did not generate a significantly different estimate than simply assuming that California's consumption patterns resemble national patterns. Accordingly, we calculated that about 13% of consumption occurs in the state for each commodity. Of course, even when California produces more than 13% of the national total of one commodity, California ships food products both in and out, reflecting differences in seasonality and specific food characteristics. For instance, table grapes are imported from Mexico when not in season in California. Long grain rice is shipped from the South of the United States, whereas California exports short and medium grain rice around the world.

A significant share of the value of California's agricultural production is exported outside the United States, but the shares of exports range from a low of about 2% for hay to about two-thirds or more for almonds (Table 2.2 AIC, 2012a). In 2012, almost half of California's international exports went to Canada and the European Union and another 35% went to Mexico, Japan, and China (figure 2.6). Export patterns vary from crop to crop, reflecting differences in transportation costs, among other factors. For instance, in 2012 Europe represented 31% of almond exports, whereas almost half of the hay exports were destined to Japan. Agricultural export earnings totaled about $18.2 billion in 2012. More than half of the state's production of almonds, walnuts, pistachios, beans, plums, and cotton was exported in 2012, and California produces a significant share of the world's tree nuts (AIC, 2012a).

TABLE 2.2

Cash receipts, share of California receipts, California share of US value, ratio of exports to production, and share of the
United States in world production for major California commodities, 2008–2009 averages

Rank	Commodity	Value of receipts[1] ($1,000)	Share of California receipts[1] (%)	California share of US value[1] (%)	Ratio of exports to production[2] (%)	Share of US in world production[3] (%)
1	Dairy products	5,730,646	15.6	19.4	18.7	15.3
2	Greenhouse/nursery	3,794,823	10.4	23.4	NA	NA
3	Grapes, all	3,095,432	8.5	88.0	29.8	5.8
4	Almonds	2,318,350	6.3	100.0	65.7	83.1
5	Cattle and calves	1,780,517	6.3	5.0	6.7	12.4
6	Lettuce	1,653,315	4.4	77.3	8.2	NA
7	Strawberries	1,651,704	4.4	79.7	10.9	NA
8	Poultry/eggs	1,384,002	3.7	3.9	NA	NA
9	Hay	1,205,391	3.8	20.9	2.3	NA
10	Tomatoes, process	1,037,772	2.2	72.3	19.1	NA
11	Rice	877,158	2.1	24.4	54.9	1.4
12	Broccoli	680,848	2.1	106.9	14.2	NA
13	Walnuts	648,305	1.9	107.7	48.7	32.1
14	Oranges	607,397	1.7	30.6	42.8	15.6
15	Pistachios	581,375	1.6	102.0	96.6	39.7
	All commodities	36,624,028	100.0	12.2	22.0	

DATA FROM: 1. USDA ERS (USDA ERS. n.d. Farm Income and Wealth Statistics. Accessed June 22, 2011. http://www.ers.usda.gov/data-products
/farm-income-and-wealth-statistics/cash-receipts-by-commodity.aspx.); 2. Matthews et al. (2011); 3. USDA FAS (n.d.).

For the commodities for which California is a large producer nationally or internationally, the prices received by producers are driven by changes in national or global demand, conditioned by trade barriers. In contrast, when California farmers face competition from producers from other states or countries, market prices result from both demand changes and changes in the response of these competing producers.

The market competitors of California agriculture are dispersed all over the world (Table 2.3). European countries are large producers of several commodities such as wine and dairy, countries with Mediterranean climates are competitors for almonds, and China is a large producer of many crops grown in California, with large shares of the world's production of lettuce and processed tomatoes. The geographic diversity of the competition facing California agriculture reflects the diversity of its commodity mix. This diversity in competition has made the demand facing California growers dependent on the economic growth of many disparate regions of the world.

2.2.2.1. THE IMPORTANCE OF EXCHANGE RATES

Bilateral exchange rates measure fluctuations between the US dollar and foreign currencies and have a powerful effect on the competitiveness of US agriculture (including California). When the US dollar appreciates, prices for US exports, including agricultural products from California, become less competitive in world markets. The top four destinations for California agricultural exports are Canada, the European Union, China, and Japan (AIC, 2012a). Since 1999, there have been large fluctuations in the exchange rates for all these destinations, some of which have been joint movements and some seemingly independent (figures 2.7 and 2.8). The Canadian dollar and the Euro both appreciated significantly against the US dollar. Canada receives a variety of California's agricultural exports, especially fresh fruits and vegetables, and Euro zone countries are major importers of tree nuts and wine, among other products. After fluctuating over the first 7 years of the period, the Japanese Yen has appreciated against the US dollar since January 2012. Japan is a major destination for tree nuts, citrus, and rice, which are shipped under an import arrangement and not sensitive to price. The Hong Kong dollar has been pegged to the US dollar during this period, as was the Chinese Renminbi (RMB) until the middle of 2005. Since that time, the RMB has appreciated about 21% against the US dollar. Exchange rates affect not only bilateral trade between regions, but also trade patterns with third countries. For example, a falling US dollar relative to the Korean

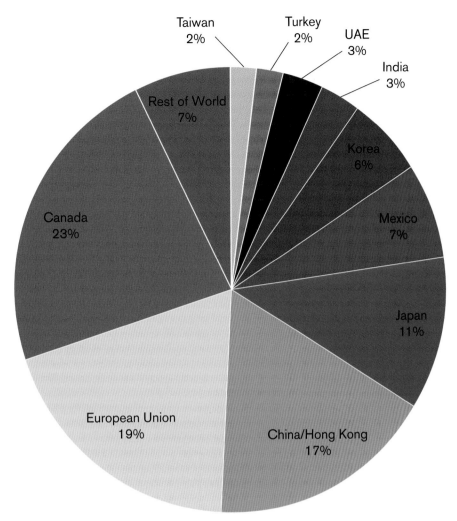

FIGURE 2.6. California agricultural exports to the top 10 destinations, by value, 2012.
DATA FROM: AIC (2012a).

Won helped the competitive position of US beef exports to Korea relative to the Australian exporters, because the Australian dollar has been strong relative to the Won.

2.2.2.2. TRANSPORTATION COSTS FOR AGRICULTURAL COMMODITIES

The reduction of transportation costs resulting from technological improvements has often been cited as a large driver of the increase in international trade since the 1950s (Hummels, 1999). However, data on transportation costs have not provided unconditional support for that hypothesis (Hummels, 2007). For example, an examination of maritime transportation prices from 1950 to 2004 reveals that the index based on the US GDP deflator indicates a large decrease in transportation costs, while the index that is deflated on commodity prices reveals no visible downward or upward trend (Hummels, 2007). Although a ton of wheat became cheaper to ship, a dollar worth of wheat did not (Hummels, 2007). That is, the real price of wheat fell even faster than the real price of shipping over the last half-century.

For more recent trends, the US Bureau of Labor Statistics (BLS) publishes national producer price index data for truck, rail, air, and deep-sea transportation (figure 2.9). For all four modes of transportation, shipping costs increased over the period covered by the data. Although most exports to Asia and Europe are shipped by sea, a few high-value crops, such as cut flowers and strawberries, are also air-shipped (California Rural Policy Task Force, 2003).

BLS transportation price data do not capture variations that affect California or food products specifically. For instance, shipping costs from California to Asia tend to be lower than average shipping costs for comparable distances because of the backhaul of ships importing Chinese products into the United States. The USDA publishes transportation data per commodity, but the span of these data is insufficient to evaluate trends. There is no study or dataset available that reports the time series pattern of transportation costs that affect California agricultural commodities specifically.

International shipping costs, at least by sea, often represent a relatively low share of commodity prices and

TABLE 2.3
Production shares for top six producing countries of major California commodities, 2000–2009 averages

	Dairy (%)		Lettuce and chicory (%)
United States (California's share of United States)	18.5 (19)	China	49.2
Germany	8.8	United States (California's share of US)	20.9 (73)
France	8.5	Spain	4.7
India	7.8	Italy	4.4
New Zealand	4.0	India	3.7
Netherlands	3.8	Japan	2.6
	Wine (%)		Strawberries (%)
France	18.6	United States (California's share of United States)	28.1 (61)
Italy	17.8	Spain	8.3
Spain	13.5	Republic of Korea	5.6
United States (California's share of United States)	8.8 (89)	Japan	5.5
Argentina	5.2	Russian Federation	5.3
China	4.8	Turkey	5.3
	Almonds (%)		Tomatoes (%)
United States (California's share of United States)	48.2 (100)	China	24.1
Spain	12.1	United States (California's share of United States)	10.4 (52)
Syrian Arab Republic	6.0	Turkey	7.9
Italy	5.9	India	7.1
Iran (Islamic Republic of)	5.2	Egypt	6.4
Morocco	4.1	Italy	5.4

SOURCES: Author's calculations of data from FAOSTAT; data from UN FAO (n.d.).

therefore play only a secondary role in California agricultural trade patterns. For instance, shipping costs for almonds represented 2.4% of cargo value when going to Hong Kong and 5.4% when shipped to the United Kingdom. For bottled wine, however, the share of shipping costs in cargo value was 10% for Hong Kong and 22% for the United Kingdom.[1]

2.2.3. Agricultural and Trade Policies Affecting California Commodities

California agriculture has been affected by federal trade policies including those of the Farm Bill and of federal legislation implementing trade agreements. Commodity subsidies have focused on grains, cotton, and oilseeds, and have had

a small effect in California relative to other states because these crops accounted for less than 5% of the value of production in California in 2008 (USDA ERS, 2009). Analysis of the implications for the 2014 Farm Bill by Lee and Sumner (2014), which they refer to as "business as usual," reconfirmed this conclusion. Other programs such as crop insurance, specialty crop block grants, soil conservation programs, and school nutrition have likely had some effects on California agriculture. These effects are not well established, but crop insurance is considered briefly in Section 2.2.3.2.

2.2.3.1. COMMODITY POLICIES OF THE UNITED STATES AND MAJOR TRADING PARTNERS

Agricultural policies in the United States have supported farm prices for US farmers and ranchers since the 1930s. Yet California's most important crops in terms of value are specialty crops for which there are few subsidies. In California agriculture, rice, cotton, and dairy operations are the most influenced by commodity programs. In addition, livestock

---

1. Author's calculations from shipping cost information obtained at 'https://www.freight-calculator.com/ex_apxocean_cal.asp' and price information obtained from the United States International Trade Commission (USITC) for 2009 and 2010. All shipping costs were calculated from California.

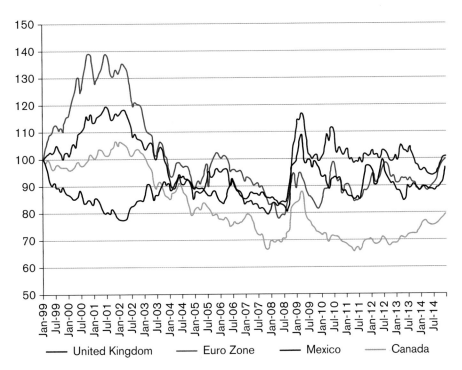

FIGURE 2.7. Indexed exchange rates for Canadian dollars, Euros, British pounds, and Mexican pesos against US dollar, monthly: January 1999–December 2014.

DATA FROM: USDA ERS (n.d., *Agricultural Exchange Rate*).

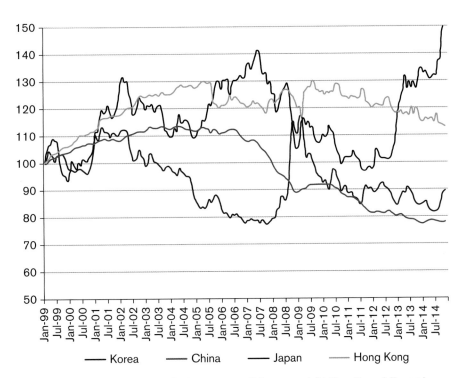

FIGURE 2.8. Indexed exchange rates for Japanese yen, Chinese renminbi, Hong Kong dollar, and Korean won against US dollar, monthly January 1999–December 2014 (January 1999 = 100).

DATA FROM: USDA ERS (n.d., *Agricultural Exchange Rate*).

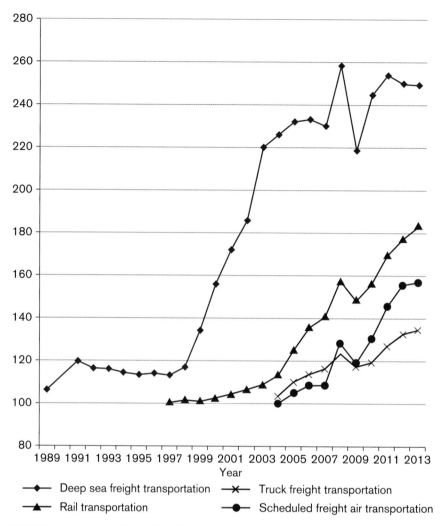

FIGURE 2.9. Deep sea, trucking, rail, and air transportation price indexes, 1989–2013. Reference years for indexes: deep sea (1988 = 100), trucking (2004 = 100), rail (1997 = 100), and air (2004 = 100).

DATA FROM: DOL BLS (2015a).

in California is indirectly affected by the programs and mandates for biofuels that influence the prices of grains, oilseeds, and grain commodities.

In the United States, government payments to agriculture have continued to increase over the last few decades but at a slower pace than total agricultural revenue, resulting in a decrease in the ratio of subsidies per dollar of revenue (figure 2.10). In addition, the nature of these payments and their effect on farmers' production incentives has changed with development and expansion of payments that are not based on current production or prices. A second important trend in payment composition is the growth of funding for subsidies with environmental linkages such as the Environmental Quality Incentives Program (EQIP). Such programs provide fewer direct incentives for production, but they also may stimulate agricultural production, for example, by helping cover the costs of complying with regulations that farmers face whether these subsidies are in place or not.

No estimate of the effects of federal farm support on the size and composition of California agriculture has been published, but these effects are likely relatively small, with the exception of a few commodities such as rice and cotton.

Agricultural subsidy rates and composition in other developed regions such as Europe, Canada, Japan, Korea, and Australia have followed trends similar to the ones in the United States (Tangermann et al., 2010). The effects of subsidy reductions in other countries on California agriculture as a whole are likely to be positive and small, given that these subsidy decreases and decoupling, although not complete, have reduced the production incentives of some of the competitors of California farmers.

2.2.3.2. US CROP INSURANCE POLICY

Subsidies for crop insurance encourage the planting of crops with more variable yields and returns. However, no evidence

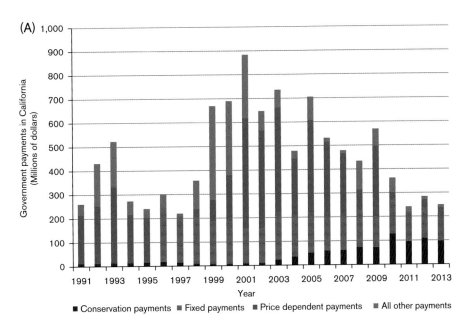

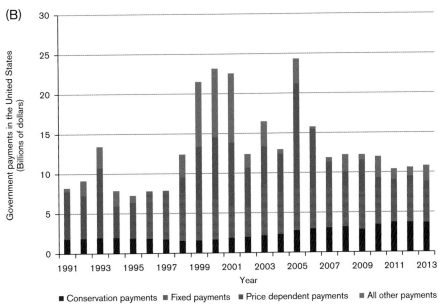

FIGURE 2.10. Government payments to farmers in (A) California and (B) the United States by payment type, 1991–2013 fiscal years. All other payments include emergency payments. Conservation payments include the Conservation Reserve Programs and NRCS programs such as EQIP.

DATA FROM: USDA ERS (n.d., *Farm Income and Wealth*).

is available on specific impacts on cropping patterns within California. Despite high subsidy rates, participation in the program varies widely across crops in California with 13% participation rate in rice and less than 40% for most other crops (Table 2.4). In contrast, participation is almost universal in regions growing rain fed crops such as the Midwest.

2.2.3.3. INTERNATIONAL TRADE BARRIERS

Reductions in trade barriers, such as those facilitated by multilateral trade agreements, generally have positive impacts on the prices of California commodities and on the revenues of California producers. Lower trade barriers open new market opportunities for agricultural exports. However, such agreements can also result in increased competition on domestic markets from foreign producers. Because of the diversity of California's commodity mix and export destinations, single agreements have differential effects on different crops. For instance, the North American Free Trade Agreement of 1994 had a positive effect on California strawberry and lettuce producers and a negative effect on California avocado producers (Brunke

TABLE 2.4

Federal crop insurance participation rates in California in 1999 (percentage of acres in crop)

Annual crops	%
Tomatoes (fresh and canning)	35
Sugar beets	26
Wheat (durum only)	14
Rice	13
Cotton (upland)	12
Total annual crops	11
Perennial crops	
Raisins (industry estimates)	80
Prunes	45
Almonds	34
Figs	27
Navels and Valencia oranges	26
Peaches (cling)	14
Lemons	11
Plums	11
Grapefruit	10
Total perennial crops	16
Total annual and perennial crops	12

SOURCE: Adapted from Lee (1999).

and Sumner, 2002). The effect of the general trend towards trade liberalization on California agriculture has not been evaluated. A full model of the details of California would need to be embedded in models of global agriculture, such as the one developed by Rae and Strutt (2004), in order to accurately assess the magnitude of the production and price effects and the corresponding nitrogen balances of trade agreements on California agriculture.

2.2.4. California's Agricultural Sector is Expanding: So What?

trends in trade policies—both domestically and internationally—have generally accommodated expansion of California's agricultural sector. This supportive export environment for California agriculture could reverse quickly if global trading regimes unraveled, but it is difficult to determine whose overall interests would be served by this and it is impossible to predict. Hence, trade policy is an important source of uncertainty regarding future prospects for California's agricultural exports. Similarly, currency exchange rates have a powerful effect on profitability of California exports, but these are driven by monetary and political factors outside the

agricultural sector (and outside California) that are difficult (or impossible) to predict. Thus, opportunities for trade and the profitability of trade, as conditioned by exchange rates, are a major source of uncertainty regarding the future of California agriculture and, in turn, the drivers of attendant nitrogen flows. Because of this high level of uncertainty, these issues are taken up in the scenarios in Chapter 6.

2.3. Inputs, Resources, and Technology in California Agriculture

In addition to commodity market prices, returns to agricultural production in California depend on the cost of the inputs and resources that are used in growing crops and raising livestock, as well as on the technologies, such as breeds and varieties, that are available to farmers.

Inputs of California's resources, such as land and water, and their cost to farmers are driven by economic and regulatory forces at work within California. In contrast, inputs such as fertilizer and fuel are, for the most part, imported into California and their cost is mainly driven by their global demand and supply. For instance, fertilizer prices have depended on the relationship between global fertilizer demand and supply. For traded inputs, market prices provide good indicators of financial costs although other components of opportunity cost, such as the farmer's management effort, may be important.

Agricultural research and development by both individual farmers and organized institutions have shaped the technologies, varieties, and breeds available for agricultural production in California. Research and development efforts within California and externally have had important impacts on technological improvements, although the predominance of "specialty" crops in California's agriculture has made the transfer of technologies from other regions less immediate and widespread than in agricultural regions that grow "commodity" crops such as corn and soybeans.

Changes in the costs of specific inputs that represent a large share of production costs have correspondingly great consequences on agricultural production and practices. Hired labor (expenditure share of almost 30%) and purchased feed (expenditure share of about 12%) represented the two largest expenditures between 1994 and 2007 (Table 2.5). Other inputs such as fertilizer, fuel, pesticides, and land each represented between 3% and 6% of farm expenditures. These average shares across all agricultural commodities mask very large variations that exist among commodities. Furthermore, it is especially difficult to measure the average cost of irrigation water and its cost share for California as a whole even though some of the energy and capital expenditure reported in Table 2.5 account for water pumping costs.

Changes in the cost of inputs trigger substitutions between inputs for a given commodity. Moreover, changes in input costs trigger shifts in the commodity mix towards commodities that make the most productive use of more expensive inputs. Although these effects have been well studied and

TABLE 2.5
Shares of farm expenditures in California, 1994–2007, in year-2000 inflation-adjusted dollars

	1994 (%)	1999 (%)	2004 (%)	2007 (%)	Average (%)
Inputs and utilities					
Feed purchased	12.1	11.5	12.6	14.7	12.7
Livestock and poultry purchased	3.5	2.7	3.2	2.7	3.0
Seed purchased	2.6	3.4	3.9	3.6	3.4
Fertilizers and lime	4.1	3.6	3.8	4.3	4.0
Pesticides	4.9	4.8	4.3	4.5	4.6
Petroleum fuel and oils	2.3	2.3	2.9	4.6	3.1
Electricity	3.2	2.9	2.5	2.5	2.8
Total labor					
Contract labor	5.6	5.6	6.8	6.8	6.2
Employee compensation (total hired labor)	19.0	23.1	23.1	19.6	21.2
Marketing, custom work, other					
Repair and maintenance of capital items	4.2	4.2	4.7	4.4	4.4
Machine hire and custom work	4.7	4.7	3.1	2.4	3.1
Marketing, storage, and transportation	9.7	7.8	6.5	8.3	8.1
Miscellaneous expenses	11.1	11.8	11.6	10.5	11.2
Rent, taxes, interest, and fees					
Net rent received by non-operator landlords	2.9	2.1	2.7	1.5	2.3
Real estate and non-real estate interest	7.1	6.4	5.4	5.9	6.2
Property taxes, motor vehicle registration, and licensing	3.1	3.1	3.0	3.6	3.2
Total farm expenditures	100	100	100	100	100

SOURCE: Author's calculations of data. Data from AIC (2009).

models are available to estimate them, there are few published studies that assess the impact of input costs on nitrogen flows related to agricultural production. Available estimates suggest that nitrogen fertilizer prices would have to increase a great deal in order to have a significant effect on N pollution. In the Tulare Basin, for example, a recent modeling effort suggests that a tax on nitrogen of nearly 150% would be necessary to induce a 25% reduction in N leakage to the environment (Medellín-Azuara et al., 2013).

2.3.1. Cost of Agricultural Land

Availability of land for agricultural use in California is constrained by the spread of urban and residential areas and the degradation of land through increases in soil salinity in some regions (e.g., near the Salton Sea and some zones of the Cen-

tral Valley). Land conversion from agricultural uses to urban uses is driven by population and economic growth as conditioned by zoning policies. Salinity-related degradation is the result of agricultural production and water management.

The average real value of an acre of farm real estate in California has been higher and increased more rapidly than the national average (figure 2.11). This pattern reflects the suitability of California's soils and climate for the production of high-value crops as well as some effect of capitalized development value. Analysis by Fischer (2006) indicated, unsurprisingly, that "climate-related variables such as degree days and available irrigation water" have the potential to affect California farmland values; however, no published evidence has emerged of negative effects of the current drought (going into its fourth year as this assessment is completed) on farmland prices.

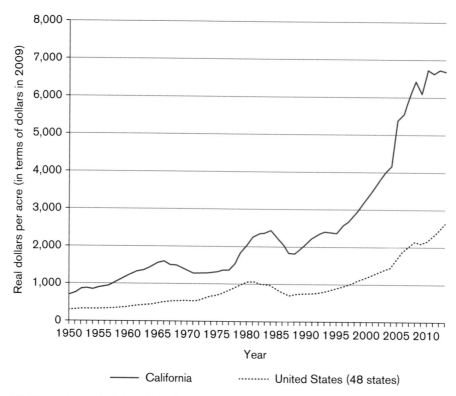

FIGURE 2.11. Average inflation-adjusted (real) value per acre of California and US farm real estate, 1950–2013. US values exclude Hawaii and Alaska, and real values have been deflated by GDP deflator from US BEA (2015).

DATA FROM: USDA NASS (n.d.) and USDA NASS (2014b).

In general during the second half of the twentieth century, the value of land for urban uses far exceeded the price of land for agricultural uses, except in some very specific premium wine growing areas of the Napa Valley. As a result, variations in land prices for agricultural uses had relatively little impact on the conversion to nonfarm uses. Moreover, changes or differences in crop value per acre (including the effect of farm subsidies) had little or no influence on conversion of land to nonfarm uses during that period (Kuminoff and Sumner, 2001). These relationships have attenuated since the mid-2000s. The contraction in home construction brought by the Great Recession lowered demand for conversion of agricultural land to housing and other forms of suburban and urban development. Over the same period, increases in tree nut and other export commodity prices have driven significant increases in California agricultural land prices; it remains to be seen what effect the drought (ongoing at the time of preparation of this book) will have on farmland values.

Public policies intended to affect farmland conversion have taken different approaches. Zoning regulations, farmland conservation easements, and related local policies such as the Marin Agricultural Land Trust have had significant effects on land conversion (Sokolow, 2006). Of particular note, the Williamson Act of 1965 was designed to enable local governments to establish contracts with private landowners in which landowners commit to restricting specific parcels of land to agricultural or related open space use. Landowners are compensated through lower property tax assessments. About 16 million acres have been enrolled in easement contracts under the act. However, the Open Space Subvention Act (OSSA), which provided the funding for these easement contracts, was suspended during fiscal year 2009–2010. Federal funding is available through the federal Farmland Protection Program with a mandated budget of $743 million nationally for 2008–2012 (USDA ERS, 2009).

2.3.2. Cost of Irrigation Water and Water Institutions

California's primary source for water is precipitation, which occurs largely in the north of the state. The diversion and conveyance of water in California is the responsibility of the Central Valley Project and the State Water Project. Much of the precipitation is stored as surface water in reservoirs or as groundwater. In a normal precipitation year, the state will receive a total of about 247 km³ (200 million acre-feet [maf]) of water, including 6–12 km³ of imports from adjacent watersheds (CA DWR, 2005). Of the total surface supply, about 60% is used directly by native vegetation, pasture, or land used for crops, evaporates, or flows to salt sinks like the Pacific Ocean, saline aquifers, and the Salton Sea. This water is mainly rain or snow that does not run off or percolates to aquifers. The remaining 40%, or about 99 km³ (80 maf), is referred to as "developed" or "dedicated" and is

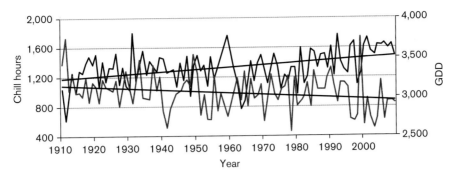

FIGURE 2.12. Historical chilling hours and growing-degree days in Yolo County, California, 1910–2010. Chill hours are in red and growing degree days in black.

SOURCE: Reprinted with permission from AIC (2011).

distributed among agricultural, urban, and environmental uses or is stored in surface water or groundwater reservoirs (CA DWR, 2005). About 42.2 km³ (34.2 maf) is used for agricultural irrigation and about 11.0 km³ (8.9 maf) is devoted to urban and industrial uses in a normal year (CA DWR, 2005).

The Department of Water Resources occasionally publishes the results of surveys on agricultural water costs (CA DWR, 2005). The complexity of water contracts makes systematic evaluation of cost trends difficult. Prices paid by farmers for irrigation water differ widely by water district and no summary measure is available to assess time trends. Variations across locations are easier to identify and the two most robust patterns are the gradient of increased prices from north and east to the south and coast, and the generally higher charges paid by urban users (AIC, 2012b). The cost of water in California is a central force in the development of both agriculture and urban areas (Hundley, 1992). There is, however, no long-term analysis of the effects of water costs and institutions on nitrogen use in agricultural production.

2.3.3. California's Climate: Trends and Variability

California's climate is a fundamental resource for agriculture and changes in climate that affect precipitation and water availability, chilling hours, and growing degree days have a large potential to change both the commodity mix and the practices of California's agriculture. In turn, climate is a central factor in both natural and anthropogenic flows of nitrogen in California. California's climate is diverse and provides appropriate growing conditions for a large number of crops. Future changes in climate, both in temperatures and precipitation, have the potential to affect agriculture in both positive and negative ways.

At a global scale, the IPCC Fifth Assessment (IPCC, 2014) found that "warming is unequivocal," including likely effects on the global water cycle, with "many of the observed changes unprecedented over decades to millennia." The most recent scenarios for climate change in the western United States show substantial uncertainty for both future temperatures and precipitation, but for each model simulation, the warming is unequivocal and large compared to historical temperature variations (Cayan et al., 2010; Kunkel et al., 2013). See Vermeulen et al. (2012) for a global review of current understanding and evidence on trends and interactions between climate change and food systems.

For some crops an increase in growing degree days or the occurrence of weather suitable for pollination may have positive impacts on agricultural production. Possible adverse effects of climate change include decreases in water availability and chilling hours, or increased occurrence of extreme events such as floods, storms, drought, heat waves, and spring frosts. As an example of the trade-offs that can occur, over the last century Yolo County has seen an 8% increase in growing-degree days, which benefits alfalfa production, and a 13% decrease in chilling hours, which can be detrimental to certain orchard crops (e.g., stone fruit) (Jackson et al., 2012a; figure 2.12).

Precipitation in the North and the Sierra Nevada mountains provides an indispensable source of water for agricultural, urban, and industrial users. Due to California's Mediterranean climate, a large fraction of the annual precipitation falls during the winter season and is subsequently stored in reservoirs and as snowpack in the Sierras. State records indicate that mean annual temperatures have increased by 0.6–1.0°C during the past century, with the largest increases observed at higher elevations (CA DWR, 2008). This warming trend have led to a 10% decline in Sierra snowpack over the same period, and a loss of 1.5 maf of snow water storage (Barnett et al., 2008; CA DWR, 2008). Changes in the timing of snowmelt have also shifted periods of peak streamflow to earlier in the spring, which has significant implications for storage infrastructure and surface water supplies in California (Purkey et al., 2007; Stewart et al., 2005a).

At present, year-to-year variability and short climate cycles create variations in weather patterns that generally exceed the long-term changes in mean temperature and precipitation that are occurring due to climate change. But despite the uncertainty regarding how climate change will affect various locations, there is a growing consensus that the impacts on California's water resources will be outside the range of past experience (Kiparsky and Gleick, 2003; Milly et al., 2008).

California has received considerable attention nationally and internationally for its Climate Action Strategy, starting with the landmark Assembly Bill 32 passed in 2006 (CARB, 2014a). However, it is generally accepted that, even if completely successful, California's actions alone cannot significantly affect the course of global climate change; instead the strategy is to demonstrate leadership in seeking solutions that others may emulate at national and regional levels. Compared to AB 32 on mitigation, the counterpart adaptation strategy launched in 2009 is in earlier stages of scoping and implementation (California Natural Resources Agency, 2009) and, as such, there is little if any evidence on likely effectiveness of the proposed measures. California's Third Climate Change Assessment was intended to provide additional information on vulnerability and adaptation options discussed in the 2009 California Adaptation Strategy (California Climate Change Center, 2012). As part of that third assessment, a team led by Louise Jackson (2012b) produced a seminal white paper on vulnerabilities and adaptation options in California agriculture, including a spatially explicit vulnerability index derived from 22 climate, crop, land use, and socioeconomic variables. This index highlighted particularly high vulnerability in the Sacramento–San Joaquin Delta, the Salinas Valley, the Merced–Fresno corridor, and the Imperial Valley. Overall, Jackson et al. (2012 found important differences across these regions in the underlying determinants of vulnerability and resilience and suggested that "future studies and responses could benefit from adopting a contextualized 'place-based' approach"; these approaches seem sensible, but while accepted, they are unproven.

The California Water Plan (CA DWR, 2014a) describes how critical challenges for water resources management in the state already appear to be affected by changing climate:

"California has undergone a warming trend over the past century . . . Summertime heat waves are increasing. Over recent decades, there has been a trend toward more rain versus snow in the total precipitation volume over the state's primary water supply watersheds, and time of runoff has shifted to earlier in the year. The water management community has invested in, and depends on, a system based on historical hydrology, but managing to historical trends will no longer work because historical hydrology no longer provides an accurate picture of future conditions."

Because of this uncertainty, the current California Water Plan (CA DWR, 2014a) calls for innovation and investment to mitigate risks of greater drought impacts, competing water demands, increasing flood risk, degraded water quality, aging infrastructure, groundwater depletion, land subsidence, and vulnerabilities to the Sacramento–San Joaquin Delta ecosystem that serves as an "essential water supply conveyance hub for more than half of the state's population and much of Central Valley Agriculture." Because most of the land of the Delta already is below sea level, this "essential hub" is especially vulnerable to the effects of continued sea level rise.

It is impossible to say with certainty that the drought that began in 2012, and which is ongoing as this assessment is being completed, is caused by changes in the state's climate. However, a long-term analysis drawing on the record of blue oak tree ring growth and other data (Griffin and Anchukaitis, 2014) concluded that while a number of other 3-year drought periods in California's history had less precipitation, the current drought is the worst in the last 1,200 years and "is driven by reduced though not unprecedented precipitation and record high temperatures." New satellite-borne sensors that monitor small changes in Earth's gravitational fields provide unprecedented evidence of massive depletion of groundwater resources in the Central Valley (Borsa et al., 2014). CA DWR (2015) estimates that historically about 38% of California's water supply came from groundwater in an "average year" (and it is not clear what an "average year" means now). During dry years groundwater use rises to 46% or more of the total; however, many individual communities rely on groundwater for up to 100% of their annual water needs (CA DWR, 2015).

Depending on the extent of climate change observed in different regions, agricultural producers will likely adapt by shifting to crops and production systems that are suitable to new growing conditions (Jackson et al., 2012a). In California, these shifts in cropping pattern and management practice will have important, albeit uncertain, impacts on nitrogen use that merit further study. Howitt's (2014) analysis of climate change scenarios to 2050 indicates that despite possible "reductions in irrigated area and net water use, California agriculture can continue to grow in revenue value and employment." If this relatively optimistic conclusion is correct, innovations in water management and agricultural practices appear to be the keys to addressing water shortages arising from climate change and other stressors.

2.3.4. Cost of Manure Used as Fertilizer

In contrast to synthetic fertilizer, manure fertilizer is not easily transported and the availability and cost of manure fertilizer for crop production depend on the proximity and size of concentrated livestock operations. Accordingly, the drivers of livestock production in California affect the use of manure application for crop production. The size and location of livestock operations, which have been affected by technological innovation and regulations, have had an effect on the availability of manure in different crop production locations. The ongoing increases in operation size and spatial concentration in the southern part of the Central Valley have resulted in larger and more concentrated manure sources (see Chapter 3).

2.3.5. Synthetic Fertilizer Prices

Nitrogen fertilizer is an essential input of agricultural production and a large literature is dedicated to analyzing the factors that affect the use of fertilizer by farmers. Variations in fertilizer prices relative to crop prices have been shown to be one of the main underlying drivers of fertilizer use. Griliches (1958) showed that the drastic decline in the price of nitrogen amendments resulting from the development and commercialization of the Haber–Bosch process in the 1920s dramatically increased the supply of fertilizer and was the main factor behind the large and widespread increase of fertilizer use in industrialized countries.

The relationship between the quantity of fertilizer applied by farmers and the price has been quantified by many authors and estimates of demand elasticities (percent change in the quantity of fertilizer used for a percent change in price, holding all other variables constant) display a wide range. Larson and Vroomen (1991, Table 3) used data from five corn growing states and found fertilizer price elasticities ranging from –0.23 to –0.85 with variations across states and across the time period covered by the data. They also found that fertilizer demands have become less responsive to own-price changes over the period 1964–1989. Denbaly and Vroomen (1993, Table 2) differentiated the long- and short-run responses of farmers to fertilizer price changes and estimated a price elasticity of –0.21 for the short run compared to –0.41 for the long run. Most of the fertilizer demand studies focus on corn growing regions and estimates for California as a whole are rare. Carman (1979, Table 2) estimated fertilizer price elasticities for the western United States and found California's elasticity of –0.204 to be lower than other states.

Nitrogen fertilizer has been traded and shipped across continents since the nineteenth century and therefore the price of fertilizer to California producers has been driven by international supply and demand essentially throughout the era of rapid development of the agriculture sector. In addition to decreasing the price of nitrogen fertilizer for growers, the development of the Haber–Bosch process in the early twentieth century coupled the cost of fertilizer production to the price of natural gas, and indirectly to the price of other energy sources (US General Accounting Office, 2003). In addition to shifts in the production costs of fertilizer, changes in the demand for fertilizer from farmers in the United States and the rest of the world result in changes in fertilizer prices for California growers. For instance, Huang (2009) found that a price spike in 2008 reflected the inability of the US fertilizer industry to quickly adjust to surging demand or sharp declines in international supply. Importantly, the increase in demand for nitrogen fertilizer by China has shaped the international trade of fertilizer in the last few decades, with China's share of world

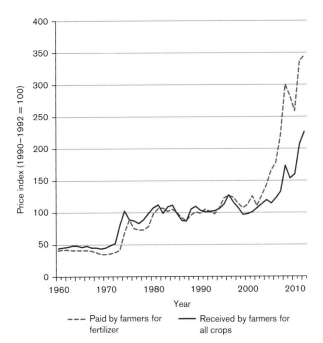

FIGURE 2.13. Producer price index for fertilizer and crops in the United States from 1960 to 2012 (1990–1992 = 100).
DATA FROM: USDA ERS (n.d., *Fertilizer Use and Price*).

fertilizer consumption growing from 11% to 34% between 1970 and 2008 (World Bank, 2015).

In the latter part of the twentieth century, variations in the price of fertilizer were comparable in timing and magnitude to variations in agricultural commodity prices. From 1960 until about 2005, price indexes for both fertilizer and crops in the United States followed similar patterns, with a dramatic rise during the 1973 oil crisis. However, during the rest of the 2000s, prices for fertilizer increased faster than crop prices (figure 2.13). Recent data continue to suggest that prices paid for fertilizers may no longer be as tightly coupled to prices received for crops (USDA ERS, n.d., "Fertilizer use and price"; USDA NASS, 2015b).

There is no federal or state policy that affects directly and significantly the price of fertilizer to California growers. In 1945, the state of California adopted Regulation 1588 which restated a preexisting exemption of the sales tax for fertilizer and seeds. A small tax of $0.0005 per dollar of fertilizer sale was established in 1990 in order to fund research efforts related to nitrate pollution in California.

In addition to fertilizer prices, several other factors influence fertilizer use. In particular, variations in crop yields or profitability, due to weather for instance, play an important role in farmers' behavior and a large literature focused on the impact of risk and variability on fertilizer use has developed (Boyer et al., 2010; Carriker, 1995; Rajsic et al., 2009).

2.3.6. Energy Prices

Social accounting matrix analysis by Roland-Holst and Zilberman (2006) identifies three distinct groups of

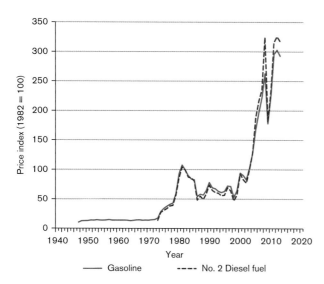

FIGURE 2.14. Index of US prices for gasoline and No. 2 diesel fuel, 1947–2013 (1982 = 100).

DATA FROM: DOL BLS (2015b).

California's agricultural products regarding vulnerability to energy prices: livestock and low-value-per-volume field crops are most vulnerable and high-value nursery products and flowers are least vulnerable, with fruit, vegetables, and poultry in between. The oil crisis of the early 1970s led to sharply higher prices for gasoline and diesel through the early 1980s (figure 2.14). From the mid-1980s until around 2003 prices did not show any particular trend despite some large fluctuations. However, between 2003 and 2012, the price for gasoline increased more than fivefold and the price for diesel increased more than sixfold. Between June and December 2014, gasoline retail prices had fallen by approximately 30% (US EIA, 2015), suggesting continuing variability and possibly increasing uncertainty regarding the future course of energy prices rather than the secular adjustment to high energy prices expected by some in the 2000s.

Relative to other states, fuel is more expensive in California because of mandated blend standards. For instance, in 2007 the California Air Resources Board adopted a new standard to set the minimum content of ethanol at 10% for gas sold in California starting in late 2009. Moreover, both the state government and the federal government collect fuel taxes on diesel and gasoline. In 2013, California's gasoline taxes ($0.719 gal^{-1}) were the highest in the country, followed by the states of New York ($0.682) and Connecticut ($0.677) (API, 2013). California's diesel taxes ($0.749 gal^{-1}) were also the highest, followed by the states of Indiana ($0.742) and New York ($0.74) (API, 2013).

2.3.7. Labor Costs and Agricultural Labor Institutions

The cost of labor is a crucial driver of agricultural production in California, in particular for the many crops that require manual thinning, weeding, and harvesting. According to the Census of Agriculture (USDA, 2009a), in 2007, California had

TABLE 2.6
Average annual multifactor productivity growth rates in California and US agriculture, 1949–2002

	1949–1960	1960–1970	1970–1980	1980–1990	1990–2002
California	1.66	2.22	2.84	1.01	1.24
United States	1.89	1.69	2.46	2.07	0.97

SOURCE: Data from Alston et al. (2010).

the highest number (about 450,000) of hired farm workers, followed by Washington and Texas with about 250,000 and 150,000 hired workers, respectively. Martin (2001) estimated that in 1999 the average monthly employment on California farms was 418,000 with large yearly variations due to seasonality. Changes in labor costs have resulted in changes in the commodity mix. For instance, Martin and Calvin (2011) show that the decline in asparagus production in California has been driven by availability of labor.

Immigration is the main driver of the availability and cost of farm labor and according to the Public Policy Institute of California (PPIC), in 2009 immigrants accounted for nearly 37% of the labor force in California, up from 11% in 1970. In California, the legal minimum wage was $8.00 hr^{-1} in 2011, which is higher than the federal minimum wage ($7.25) (DOL WHD, 2015). In 2011, Texas's minimum wage was the same as the federal minimum wage, whereas Washington was higher than California at $8.67 hr^{-1} (DOL WHD, 2015). California's minimum wage regulation is binding for some operations such as weeding and thinning, but harvest workers are often offered incentives based on harvested prices that can result in higher wages (Martin, 2001).

2.3.8. Development and Adoption of New Technologies

Innovations by individual farmers and by research and development institutions are an important driver of agricultural productivity, often described as the ratio of measures of the quantity of outputs produced to the quantity of inputs used. Because of the predominance of specialty crops in California's agriculture and because of California's unique soils and climate, both private and public research and development efforts organized through federal and state programs have been significant sources of technological change in California.

Although agricultural productivity has increased over the last several decades, the average annual productivity growth rates in California and US agriculture have declined since the 1980s and rates of productivity growth have fallen below what they were in the 1950s and 1960s (Table 2.6).

The growth in the amount of resources dedicated to agricultural research and development in the United States has shown a similar pattern. After a period of steady growth

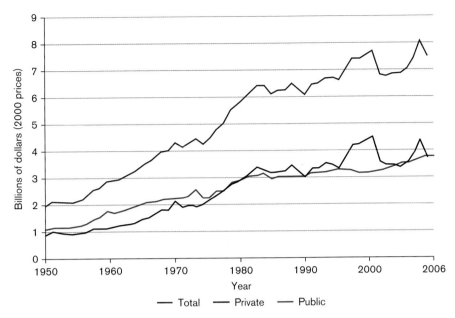

FIGURE 2.15. US agricultural research and development expenditures, 1950–2007 (2000 prices).
SOURCE: Reprinted from Alston et al. (2010). Copyright 2010. Springer. Used by permission.

from 1950 until around 1980, both public and private research and development expenditures grew much more slowly through 2007 (figure 2.15). In 2007, more than 51% of agricultural research and development was undertaken by the public sector. Universities and colleges represented about 35% of this research expenditure and federal government research laboratories another 16.7% (Alston et al., 2010). California's public research on agriculture is performed by the California Agricultural Experiment Station of the University of California, Division of Agriculture and Natural Resources (UC ANR). Cooperative Extension constitutes the ANR's main outreach program, with about 350 specialists and advisors dispersed throughout the state in 2013. The annual expenditures for both Cooperative Extension and California Agricultural Experiment Station increased between 1993 and 2007 in nominal dollars; however, in inflation-adjusted dollars both expenditures have declined slightly since 2002 (AIC, 2009).

Overall, agricultural biotechnology patenting in the United States has been increasing, and at a faster rate than patenting of other sectors (figure 2.16) (US Patent and Trademark Office, 2014). Commercial firms, followed by US nonprofits and universities, receive the majority of agricultural biotechnology patents. In 2004, California was issued more agricultural biotechnology patents than any other state. Of the 7,097 such patents issued in the United States that year, California received 1,506. Private research tends to focus on patentable innovations rather than general productivity-enhancing improvements (Alston et al., 2010).

Transfers of biotechnologies from outside of California have also played an important role in increasing California's productivity. For example, agricultural research and development in the Spanish region of Valencia have affected citrus production in California, where local research, development, and extension have contributed to adapting Spanish varieties to California conditions.

2.3.9. Research and Development Has Enhanced Productivity in California: So What?

There is great uncertainty regarding future climate, water supply, prices of energy (and hence synthetic fertilizer), and labor costs faced in California; similarly there is great uncertainty about the patterns of technological change. The point here is not to yearn for precise long-term forecasts, which are impossible, but to consider how technological change can drive adaptation in the context of climate, water supply, energy price, and labor costs and availability. Although largely speculative, Howitt's (2014) conclusion that agriculture can continue to grow in revenue value and employment is consistent with past performance of California agriculture. From this, it would follow that our focus should be on investing to increase this capacity for innovation and adaptation that underpins resilience to various input supply and price shocks.

2.4. Policies Affecting Nitrogen Flows in California

This section focuses on nitrogen-related policies that have had measurable effects on nitrogen flows over the last several decades. Chapter 8 provides the details and analysis of policy responses to changes in nitrogen flows, focusing primarily on flows associated with agriculture.

Nitrogen pollution has been a target of numerous policies and regulations for several decades in California and the

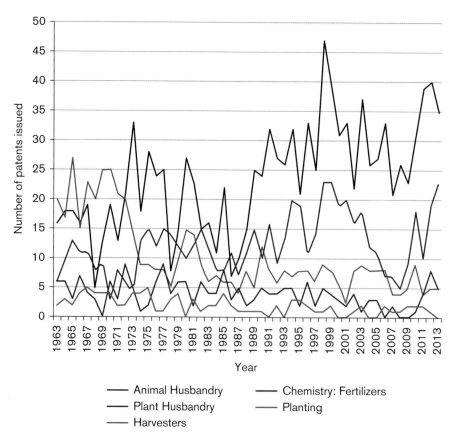

FIGURE 2.16. Issuance of patents to holders in California for selected agricultural technology classes, 1963–2013.

DATA FROM: US Patent and Trademark Office (2014).

United States. For the most part, policies have targeted the degradation of individual resources. As a result, regulations affecting different media have generally evolved independently and there is no federal, state, or local integrated nitrogen policy. The extent to which nitrogen flows have been affected by environmental policies varies widely by resource.

Of all the human activities that contribute to nitrogen pollution, the combustion of fossil fuels and the management of human and animal waste have been the most strongly affected by policies, most of which have taken the form of regulations. However, there are currently no direct regulatory restrictions or reporting requirements for nitrogen management in crop production when manure is not involved. Across activities and across resources, there has been a visible trend towards more widespread and binding regulatory policies, with economic incentives and other policy instruments having played a much smaller role so far.

Regulations for surface water, under the Clean Water Act of 1972 and its subsequent amendments, have contributed to an observed decrease in nitrogen concentrations in many but not all watersheds in California (Dubrovsky et al., 2010). Point sources such as sewage collection and treatment plants, industrial facilities, and confined animal facilities have been the main targets of surface water policies.

The Clean Air Act of 1963 initiated a nationwide effort to regulate air quality with a focus on fossil fuel combustion and has resulted in reducing or curbing concentrations of NO_x in several airsheds in California. Some of the local air districts that are responsible for the implementation and enforcement of air quality standards have targeted air pollution from farming in order to reduce concentrations of particulate matter, for instance. Yet, dairies and other confined animal facilities were exempt from regulation by the state of California until 2003. Establishing regulations for emissions of ammonia from concentrated dairy operations has received increased attention nationwide. Conservation programs of the Farm Bill have had some impact on farming practices with the distribution of subsidies encouraging the adoption of conservation practices, including manure management.

The impact of policies on nitrate leaching to groundwater has been limited. California's water quality regulations differ from federal regulations by including both surface water and groundwater objectives in the main law, the Porter-Cologne Act of 1969. However, until recently agriculture has been exempt from regulations related to groundwater through local agricultural waivers. These waivers, which affect both surface water and groundwater pollution, are in the process of being publicly revised in several administrative water regions.

Policies targeting emissions of nitrogen greenhouse gases (GHG) are at early stages of development and implementation. California's Global Warming Solutions Act, or Assembly Bill 32 (AB32), was passed in 2006. AB32 allows for the development of agricultural offset programs from livestock and crop operations but does not include California's agricultural sector in its central cap-and-trade and other measures.

Other environmental policies have likely had some local effects on the management of nitrogen pollution but there is no published analysis of their impact on nitrogen in agriculture. For example, the federal Endangered Species Act of 1973 has regulated actions that threaten the survival and the habitat of listed species, which includes the Delta Smelt. State and local programs have also been developed over the last few decades to tackle nitrogen pollution but no estimate of their impact is available. The Fertilizer Research and Education Program (FREP) was created in 1990 and implemented by the California Department of Food and Agriculture to tackle nitrate pollution from animal waste management and fertilizer use. It is funded on a tax of $0.0005 per dollar of fertilizer sales in the state and funds research and education programs.

2.4.1. Water Quality Policies

The Porter-Cologne Act is the backbone of water quality regulation and policy in California. The goal of the Porter-Cologne Act is to prevent the loss of beneficial uses of water both from surface and ground resources. It applies federal regulations of the Clean Water Act to the state and provides the framework for the actions and rulings of local water boards that are in charge of implementing quality standards. The Clean Water Act does not directly address groundwater contamination, which is regulated federally by the Safe Drinking Water Act and the Resource Conservation and Recovery Act of 1976, which regulates the disposal of hazardous waste.

2.4.1.1. SURFACE WATER REGULATIONS

The two central measures of the Clean Water Act are the definition of Total Maximum Daily Loads (TMDL) and the establishment of the National Pollutant Discharge Elimination System (NPDES) permit program. These programs have targeted nitrogen in surface water through their quality standards on dissolved oxygen, which is depleted when nitrogen pollution favors algal development. In addition, the Coastal Zone Act of 1972 and the Coastal Zone Act Reauthorization Amendments (CZARA) of 1990 are federal regulations targeting the pollution of coastal waters from nonpoint sources. In California, the coastal zone includes the entire state and the regulations require that the state submit and implement a nonpoint source program.

The impact of surface water quality regulations on livestock operations is less clear. The qualification of livestock operations as point sources, and therefore the applicability of regulations, depend on herd size and records of emissions, and vary across jurisdictions (Morse, 1995). The regulation of nitrogen pollution to surface water and groundwater is still in development. For instance, the Central Valley Water Board is currently developing a general Waste Discharge Requirements (WDR) Order to regulate dairy operations.

Crop production, which qualifies as a nonpoint source, has been the target of regulations for surface water pollution but regional exemptions, called agricultural waivers, have limited the actual implementation and effect of these regulations. The ongoing process of revision of the agricultural waivers may result in significant changes to cropping practices that affect nitrogen pollution, such as tailwater recycling and fertilizer application rate and timing.

Manure application to crop fields, which lies at the interface between livestock and crop production, is in the process of being regulated through the implementation of nutrient management plans. Although reporting by farmers of nutrient management plans is now mandatory, application rates per acre are only subject to recommendations with no enforceable standard.

2.4.1.2. GROUNDWATER REGULATIONS

The Safe Drinking Water Act is the federal regulation that affects nitrate pollution in groundwater most directly. However, the regulation determines the quality standards allowable for drinking water without a direct mandate for the degradation of the quality of aquifers. The California Department of Public Health is the state agency responsible for monitoring and enforcing quality standards for the water provided to the public by utilities and municipalities.

Drinking water standards, for nitrates as well as for other contaminants, have resulted in water providers investing in water treatment equipment as well as switching from groundwater to surface water sources.

Although the Porter-Cologne Act was designed to address both surface water and groundwater pollution, the effect of policies on nitrate leaching to groundwater in California has been limited. Policies that target both crop production and livestock operations are in the process of being developed. For instance, the current revisions of the agricultural waiver for the Central Coast water region include provisions for both surface water and groundwater (see Chapter 8).

2.4.2. Air Quality Policies

The Clean Air Act (1970) is the air counterpart of the Clean Water Act. The federal Environmental Protection Agency (EPA) establishes air quality standards and enforces their application by states and local air districts using attainment criteria on which federal funding is conditioned. The

California Air Resource Board is responsible for monitoring the regulatory activity of the 35 California air districts. Other federal regulations such as the Comprehensive Environmental Response, Compensation, and Liability Act (CERCLA or Superfund) and the Emergency Planning and Community Right-to-Know Act (EPCRA) have also targeted air quality in livestock operations and require reporting of ammonia emissions.

The Clean Air Act has contributed to large reductions in air emissions from fossil fuel combustion and improvements of air quality in regions such as the Los Angeles area. The EPA estimated that between 1970 and 1990 the costs of achieving the pollution abatement dictated by the Clean Air Act were $523 billion for the country, compared to an estimated $22 trillion in avoided health and environmental costs (EPA, 1997).

Agriculture has been, for the most part, exempt from air permitting requirements until recently. Livestock operations, some of which are subject to air quality regulations according to the EPA, were exempted from state-level air emissions permitting until 2004 and the implementation of Senate Bill 700 (Florez, 2003). The EPA designated non-attainment areas in California related to volatile organic compounds (VOCs) from agricultural operations and put the state on notice to change its regulation of livestock operations. The change in regulations has resulted in districts establishing rules for livestock operations. For instance, in June 2006 the Air Pollution Control District of the San Joaquin Valley adopted a rule mandating the adoption of conservation management practices by dairy operators (Rule 4570). These practices include dust control, manure handling and treatment, and silage management. In addition to federal regulations, policies implemented by regional air districts have also had an impact on nitrogen flows to the air. For instance, the Connelly-Areias-Chandler Rice Straw Burning Reduction Act limits the burning of rice residue in the northern region of the Central Valley.

2.4.3. Climate Change Policies

Although regulation of emissions of GHG has been the topic of policy discussions for almost two decades, AB32 is one of the first policies to set regulatory objectives for GHG emissions. Specifically, AB32 aims to reduce statewide GHG emissions to 1990 levels by 2020, and a further 80% by 2050 (CARB, 2008). Agriculture contributes roughly 6% to California's overall GHG emissions and its role in the new climate policies is minor relative to the energy, transportation, and industrial sectors (see Chapter 4).

The state's cap and trade program, which started in 2013, does not require agricultural producers to report emissions, nor does it place a cap on emissions from agriculture. Instead, California's climate change scoping plan encourages agricultural producers to mitigate emissions on a voluntary basis, with the adoption of manure digesters a main target for action (CARB, 2008). In contrast, energy producers, food processors, and others in the industrial sector face a mandatory cap on emissions, a policy that is likely to have important, albeit uncertain, economic effects on agriculture (Haden et al., 2013; Sumner and Rosen-Molina, 2010). At present, few studies have examined the breadth and magnitude of these effects on California agriculture.

The cap and trade program does allow capped industries to purchase carbon credits from mitigation projects that meet the criteria of being real, additional, permanent, quantifiable, verifiable, and enforceable (Niemeier and Rowan, 2009). Some offset protocols may involve agriculture and thus provide economic incentives to farmers who adopt practices and technologies that mitigate emissions or sequester carbon in soils or vegetation. For example, offset protocols for dairy manure digesters and rice cultivation are already under development and will soon be evaluated for inclusion in the cap and trade program (Climate Action Reserve, 2011; Sumner and Rosen-Molina, 2010). A voluntary offset protocol for nitrogen management is being developed for corn in the Midwest, but this does not currently apply to California crops (Climate Action Reserve, 2012).

While state agencies have provided a framework for these climate policies, much of the responsibility for the implementation of AB32 has been delegated to local governments. For example, AB32 and the California Environmental Quality Act (CEQA) now require local governments to develop detailed plans to mitigate climate change whenever they update their general plan (California Attorney General's Office, 2009). Local "climate action plans" generally include an inventory of 1990 and present-day emissions, specific plans to mitigate future emissions, and in some cases strategies to adapt to the impacts of climate change (Jackson et al., 2012b; Wheeler, 2008).

At present, there is a great deal of uncertainty about how these new climate policies will affect the use of N in agriculture. The use of inventory methods and GHG modeling tools that can accommodate both state and local data on agricultural emissions sources is becoming more common among agencies serving rural communities in California (Colorado State University, 2012; Haden et al., 2013). For example, Haden et al. (2013) found that prior to AB32, N_2O emissions from agriculture in Yolo County had already decreased by more than 20% since 1990, due to a combination of declining cropland area, a market-driven shift toward crops that require less N (e.g., grapes, alfalfa), and improved N management for certain crops.

Policies to preserve farmland and encourage "smart growth" in California are complementary to the overall goals of AB32, because by keeping land in agriculture, further reductions in N_2O emissions may be achieved by supporting stakeholder efforts to optimize N management through incremental adoption of recommended fertilizer regimes and improved technology. That said, considerable uncertainty remains regarding how these policies, and the associated costs to producers, will influence agricultural production both in California and elsewhere. The possibility of "leakage," where agricultural production of certain commodities is shifted to other states or countries with less stringent regulatory policies, has not been adequately studied in the context of AB32 (Peters et al., 2011). Future research in California should examine the effects of leakage from the perspective of impacts on California's agricultural economy and the overall effects on "exported" emissions to other regions (Davis and Caldeira, 2010).

2.4.4. Federal Conservation Programs

Over the last two decades, conservation programs have grown in importance and funding within the Farm Bill. Although the Conservation Reserve Program (CRP) has been a central feature of federal agricultural conservation since its establishment in 1985, programs that target conservation practices on land that is maintained in production have had a relatively greater impact in California, where land retirement remained minimal. In California, land retirement programs are of minor importance. California landowners enrolled only about 138,000 acres in the CRP. California represented only 0.4% of the national CRP acreage and the CRP in 2007 represented only 1.25% of cropland within California, compared to about 8% of the cropland nationally (USDA FSA, 2007). The CRP originally focused on soil erosion, which is less of an issue for most of California cropland. In addition, land values in California are relatively high. Accordingly, cost per acre of land retirement is high, especially for irrigated cropland, which makes up the largest share of cropland in California.

Working land programs provide subsidies and technical assistance through the Natural Resources Conservation Service to encourage adoption of conservation practices. The Environmental Quality Incentive Program has been the most important working land program in terms of scope and funding. The ability to maximize environmental benefit per dollar of subsidy was made more difficult by a restriction in the 2002 Farm Bill that eliminated the option for farmers to increase the likelihood of their project being funded by indicating a willingness to accept lower cost share percentages (USDA ERS, 2006). There is no published analysis of the impact of working land conservation programs on nitrogen flows in California.

2.4.5. Other Environmental Policies

The CEQA of 1970 requires state and local agencies to identify the environmental impacts of their actions and to avoid or mitigate those impacts, if feasible. CEQA, through its regulation of construction and extension of livestock facilities, has had an impact on the size of herds by making adjustments more costly (Deanne Meyer, pers. comm.).

There are no estimates on the net effect of CEQA on livestock-related nitrogen pollution.

The Endangered Species Act of 1973 regulates actions that may affect threatened and endangered plants and animals. The policy is administered by the EPA and the US National Oceanic and Atmospheric Administration. For example, the habitat of the Delta Smelt, listed as a threatened species, has been the target of conservation efforts, which include the improvement of water quality related to nitrogen.

2.4.6. Current N Policy Is Fragmented across Resources and Flows: So What?

In contrast to emissions from motor vehicles, no current policies exert a strong direct effect on nitrogen flows in California agriculture. One concern voiced by farmers is the proliferation of conflicting and at times perverse regulations across different issues. Expectations of some regulation of nitrogen are a source of uncertainty in the state's agricultural sector. The novelty and great uncertainty in what policy strategies will be pursued regarding nitrogen is taken up in the scenarios in Chapter 6 and the analysis of policy and institutional options in Chapter 8.

2.5. Conclusion

The tremendous economic and population growth that occurred over past decades, both throughout the world and in California, has affected nitrogen flows in California through a large number of interrelated effects. The effect of population and income increase in California on fossil fuel combustion in the state and the corresponding consequences on NO_x, NH_3, and N_2O emissions are relatively clear (see Chapter 5). In contrast, understanding the underlying drivers of nitrogen flows related to nitrogen in agriculture is more challenging because of the many connections that California agriculture has with the global economy. Economic forces and trends far from California affect both the demand for California's products and the supply of inputs such as fertilizer, fuel, or labor. These effects vary across the large spectrum of commodities grown in California, as will be discussed in Chapter 3 on direct drivers of crop choice and production technique. The strength of these economic connections also varies across crops according to specific changes in transportation costs and trade barriers. As a result, the full effect of the policies on nitrogen flows related to agriculture in California can only be estimated by carefully accounting for the impact these policies have on the behavior of California's producers, which is the focus of Chapter 8. In the past, California's producers have readily adjusted the commodity mix to changes in economic incentives and it is likely that they will continue to do so.

Direct Drivers of California's Nitrogen Cycle

Lead Authors:
T. S. ROSENSTOCK AND T. P. TOMICH

Contributing Authors:
H. LEVERENZ, D. LIPTZIN, D. MEYER, D. MUNK,
P. L. PHELAN, AND J. SIX

What Is This Chapter About?

The release of nitrogen (N) into the environment is in part a consequence of the inherent properties of the N cycle but is greatly affected by human decisions. This chapter assesses those human and natural processes that directly alter N cycling (hereafter referred to as "direct drivers"). It considers trends in on-the-ground actions that influence N use, leakages, and emissions, following examination of underlying drivers (Chapter 2) and leading to calculations of the relative magnitude of N flows in the state (Chapter 4). We document six activities that have and will continue to shape California's N cycle: (1) fertilizer use; (2) feed and manure management; (3) fossil fuel combustion; (4) industrial processes (e.g., chemicals, explosives, and plastics); (5) wastewater management; and (6) land use and management.

Stakeholder Questions

The California Nitrogen Assessment (CNA) engaged with industry groups, policy-makers, nonprofit organizations, farmers, farm advisors, scientists, and government agencies. This outreach generated more than 100 N-related questions, which were then synthesized into five overarching research areas to guide the assessment (figure 1.3). Stakeholder-generated questions addressed in this chapter include the following:

- What are the current N rate recommendations? Are current nitrogen application guidelines appropriate for present-day cropping conditions?
- How is nitrogen use efficiency (NUE) determined and what are the most efficient and inefficient production systems?

Main Messages

Everyday actions of Californians radically alter the nitrogen (N) cycle. Activities such as eating, driving, and even disposing of waste modify N stocks and flows, transferring N statewide and influencing N dynamics beyond California's border. Six actions fundamentally change N cycling in the state. Each of these drivers has intensified since 1980.

Direct drivers catalyze specific N transformations and transfers between environmental systems. There are close relationships between a direct driver and the N cycle. Some direct drivers are much more important than others for specific impacts. Fertilizer use dominates nitrate (NO_3^-) leaching and nitrous oxide (N_2O) emissions. Fuel combustion drives volatilization of nitrogen oxides (NO_x). The spatial distributions of activities create distinct regional patterns of consequences (both benefits and costs).

Fertilizer use—inorganic and organic—represents the most significant N cycle modification. Sales of chemical N fertilizers have increased considerably since World War II and have risen by at least 40% since 1970, but consumption has leveled off in the past 20 years. Increases in agricultural productivity have been even greater. N fertilizer has been critical for the growth of California's agricultural industry and rural economy. Despite progress, inorganic N fertilizer application rates (kg ha^{-1}) increased an average of 25% between 1973 and 2005. Data show the majority of California crops recover well below half of applied N, with some crops capturing as little as 30%. Similar or even lower N recovery rates are found when organic N sources are used. Differences between the NUE in research trials at plot and field scale and statewide averages suggest there may be substantial potential for improvement in fertilizer management.

Until recently, manure management decisions were made without much regard to N consequences. The breadth of techniques used, limitations in available information, and large variability among operations, especially for San Joaquin Valley dairies, make any conclusion about changes in manure management practices tentative. Surveys, however, suggest the recent adoption of manure management techniques helps manage nutrients more effectively. It is important to note that optimal manure N handling is the consequence of many processes and thus must be considered as a system.

Fuel combustion increased significantly, but emissions declined steadily since 1980. Over the past 30 years, sales of diesel and gasoline fuel, size of the vehicle fleet (both passenger cars and heavy-duty trucks), and the number of stationary sources (e.g., energy production and industry) increased measurably, often doubling. Emissions however have been controlled by aggressive technology forcing regulations. This is most evident in the declining importance of the small vehicle fleet for NO_x emissions by comparison to diesel engines.

Ammonia (NH_3) is an ingredient in a variety of industrial products, including plastics, nylons, chemical intermediaries, and explosives. However, much of its use and related impacts are poorly documented. In addition to the release of N compounds during production, the longevity of N-derived industrial products (varying from spatulas to counter tops) results in a latent pool of N in human settlements. Slow degradation means they are a long-term threat to human and environmental health. Industrial use may be as large as 55% of inorganic N fertilizer use annually.

About 77% of food N will enter wastewater collection systems and about 50% of wastewater is dispersed in the environment without N removal treatment. This includes wastewater treatment plants with limited nitrification, leakage from sewers, and wastewater infiltration systems. Recent attempts to control N pollution have increased the level of treatment practiced at municipal wastewater facilities throughout California. In 2008, nearly 50% of wastewater treatment facilities reported performing at least advanced secondary treatment and 20% performed tertiary treatment processes. On-site wastewater systems treat the wastewater of more than 3.5 million Californians, with approximately 12,000 new units installed each year. Despite relatively small potential N emissions, improperly sited or malfunctioning systems can cause N discharge hot spots.

Changes in land cover, land use, and land management fundamentally alter N cycling in ways only recently appreciated. Change can result from a shift in land cover or change in intensity of use. Urbanization has caused agriculture to relocate, often to lands more marginally suited for these systems. The net effect of urbanization and agricultural relocation/expansion has led to a 1% decrease in total agricultural land between 1972 and 2000. This has been accompanied by an intensification of use. The mix of crops produced has changed from relatively N-extensive to N-intensive species. Field crops were still grown on 53% of cropland in 2007 (largely because of the land area dedicated to alfalfa), but this is a significant decrease from 74% in 1970. Simultaneously, the dairy cow population has doubled and the broiler population has tripled, concentrating N-rich feed in California and amplifying manure N handling concerns.

3.0. Factors Controlling the N Cycle

This chapter describes the human actions and natural processes that modify California's nitrogen (N) cycle, referred to hereafter as "direct drivers" (MA, 2005b). We will first describe relative influence in terms of impact on N stocks and flows, and then trace historical trends.

3.1. Relative Influence of the Direct Drivers

Nitrogen's centrality in agriculture, transportation, and industry means that virtually every human activity, ranging from cooking dinner to waging wars, will affect N cycles, oftentimes in profound, cascading, and multiplicative ways. Population growth, development, and affluence have all contributed to a greater quantity of reactive N in the environment today, by an enormous proportion (Davidson et al., 2012). In 1860, humans created approximately 15 Tg of reactive N per year to meet energy and food demand. That amount has now increased by more than an order of magnitude (Galloway et al., 2008). Few indicators suggest these trends will reverse or even slow significantly in the foreseeable future. Indeed the opposite, continued rapid growth, seems more likely when one considers forecasts of demand for food and energy use.

Analysis of trends in reactive N creation at small spatial scales (such as California) is unavailable (Box 3.1). However, a recent analysis shows that between 2002 and 2007, reactive N creation in the United States increased approximately 4% on balance (Houlton et al., 2012). But reactive N created to enhance food production (cultivation-induced N fixation and inorganic fertilizer use) increased ~10% (from 22.8 to 24.7 Tg) and reactive N from transportation and industry decreased by 19% (from 5.9 to 4.8 Tg) (Houlton et al., 2012). Differences in the magnitude and trends in N cycling illustrate the significance of developing N budgets by activity to better understand individual direct drivers and target remedial actions (Robertson, 1982).

Human activities modify the N cycle through a variety of pathways, each with different magnitudes. For example, burning fossil fuels in transportation and industry is the principal source of reactive N compounds into the atmosphere, the largest fraction of which are nitrogen oxides (NO_x). Ammonia (NH_3) gas is also released but to a much lesser extent. Fossil fuel combustion activities create little threat to groundwater, at least prior to their deposition on downwind landscapes. In contrast, inorganic N fertilizers applied to cropland or urban areas are transported downward through the soil profile (leaching) or laterally on the

"Scale" is a critical framing concept when thinking about N cycling. Depending on the context of its use, scale can refer to two ideas.

First, scale can be used as a synonym for spatial extent. This is significant for N cycling because each process that affects the turnover, transformation, and transmission of N compounds and the consequential impacts have characteristic spatial extents for which they occur, from local to global. For example, denitrification takes place at a very small, local spatial scale within the soil complex, that of microns, but a product of denitrification and a principal concern, N_2O, has global effects. Leaching is a function of local soil texture and moisture conditions. Regardless, if the rest of the field is dry, a depression or local soil fissure may be a hot spot of leaching activity. The local scale nature of N cycling processes contrasts with the more regional and global nature of N cycling concerns. The principal N issues happen at large spatial scale—kilometers—based on the aggregate of local dynamics.

Second, scale can also be thought of as a synonym for magnitude. Here it is important as a consideration for source activities and impact. As shown in the mass balance calculations (Chapter 4) and re-reported throughout this chapter, the magnitudes of N flows from source activities differ considerably. Both spatial extent and overall magnitude are important properties to consider with regard to N flows.

soil surface (runoff), typically as dissolved nitrate (NO_3^-). Two important points flow from the propensity for certain activities to cause specific N transformations and transfers between environmental systems. First, there is a close relationship between a direct driver and specific N stocks or flows. Wildfires, for instance, liberate organic N contained in soils and biomass (N stocks) and release reactive N compounds and N_2 into the atmosphere (N flow). Second, the importance of a direct driver and likely changes in the N cycle depend on the extent of activities ("activity level"). The diversity and spatial patterns of human activities in California mean that direct drivers have differential regional impacts. Urban areas of Southern California receive a larger proportion of the reactive N input from fossil fuel combustion or wastewater treatment, while fertilizer use determines the reactive N levels in the Central Valley.

The CNA's mass balance calculations (Chapter 4) identify five direct drivers that control California N cycling: (1) fertilizer use on croplands; (2) feed and manure management; (3) fossil fuel combustion; (4) industrial processes (e.g., chemicals, explosives, and plastics); and (5) wastewater management. Statewide, fertilizer use on croplands and urban areas is the largest source of N in California, totally 32% of N imports (figure 3.1). Fossil fuel combustion contributes a significant amount of new reactive N to California each year too (25%), followed by biological nitrogen fixation (21%), and imported feed (12%). Only a few direct drivers regulate the release of reactive N into air and water resources (figure 3.1). Fossil fuel combustion dominates gaseous emissions (44%), with the vast majority of these emissions as NO_x. Meanwhile, manure handling is responsible for the majority of the NH_3 emissions, which account for 22% of total atmospheric N release. Croplands are overwhelmingly responsible for N loading into groundwater across the state (88%). Harter et al. (2012) indicate that cro-

plands contributed 96% of the NO_3^- to groundwater in the Salinas Valley and the Tulare Lake Basins, with 54% and 33% from inorganic fertilizer and manure use, respectively. By comparison to groundwater, multiple sources contribute to surface water N loading including natural lands (40%), fertilizer use (49%), and wastewater (11%). The mass balance is a static model documenting one year's (2005) N flows; it does not capture temporal dynamics. That limitation, and the understanding that individual land uses affect N cycling in vastly different ways, leads us to identify "land use, land cover, and land management" as a sixth direct driver. The following sections analyze trends in these key activities and provide context for historical changes.

3.2. Fertilizer Use on Croplands

Nitrogen availability generally limits plant productivity. Farmers respond by applying N fertilizer to soils to enhance plant growth and reproduction. Fertilizer N typically stimulates soil N cycling. Not only does the size of the soil mineral N pool increase, but also soil microbial activity increases, the pace of N transformations and soil N turnover intensify, and the risk of N emissions typically increases. The fundamental nature of the soil N cycle requires producers to apply more N than plant demand to ensure adequate nutrition (Boxes 3.2 and 3.3). If managed well, plants capture a sizeable fraction of the fertilizer. However, agriculture and lawns are inherently leaky systems and some N inevitably escapes into the environment.

Fertilizer use on croplands introduces the most significant annual amount of new reactive N from a single source into California (Chapter 4). Inorganic N fertilizer use on croplands amounts to 466 Gg N yr^{-1}. Organic N use introduces nearly an equal amount (459 Gg N yr^{-1}) through

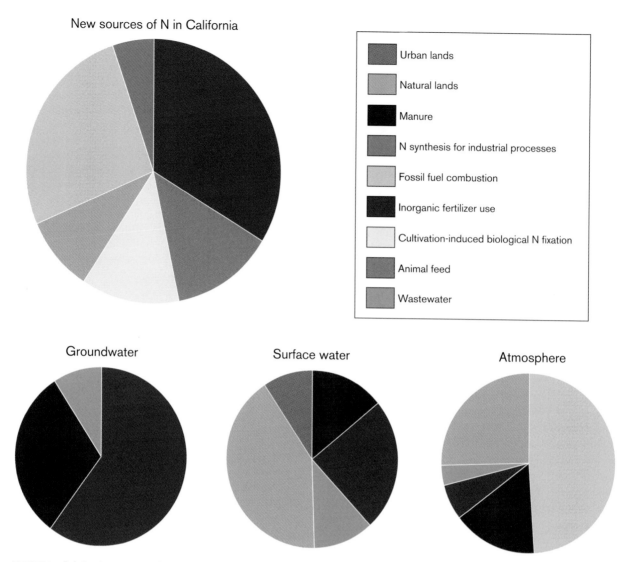

FIGURE 3.1. Relative importance of the direct drivers on California's nitrogen cycle, 2005. Values are percentages of the total and may not add up to 100% because only drivers contributing at least 5% to the total are included in charts and/or due to rounding. N-BNF and C-BNF refer to natural lands and cultivation-induced biological nitrogen fixation, respectively. It is important to note that "fertilizer use and soil management" for groundwater and surface water includes both inorganic and organic N sources (e.g., chemical fertilizers, C-BNF, and manures) used on croplands.

manure application and cultivation-induced biological nitrogen fixation (C-BNF; or cropland fixation). Fertilizer use—inorganic and organic combined—thus is responsible for slightly less than 1 Tg of reactive N and has significant effects on the dynamics of N in California.

3.2.1. Inorganic N Fertilizer Use on Farms

Inorganic N fertilizer (synthetic N fertilizer) has played a critical role in increasing agricultural productivity and food security globally. It has been suggested that the scientific discovery (e.g., Haber–Bosch process) of creating inorganic N fertilizer has resulted in more than 2 billion people alive today than would be otherwise (Erisman et al., 2008). Smil (2001) suggests inorganic fertilizer N is the basis for more than 50% of food production. Data from long-term experi-

ments suggest that between 40% and 60% of crop yields in the United States and Europe can be attributed to inorganic N fertilizer use, a slightly lesser proportion in tropical environments (Stewart et al., 2005b). The fundamental importance of fertilizer N to food security requires that any discussion of past, present, or future inorganic fertilizer use must acknowledge its benefits to society.

3.2.1.1. TRENDS IN INORGANIC N USE AND YIELDS

Sales of inorganic N fertilizer have increased 12-fold since materials became widely available after World War II. Prior to this time, inorganic N fertilizers, also known as mineral fertilizers, were derived from Chilean nitrate deposits. However, with the invention of the Haber–Bosch process in 1908, availability of inorganic fertilizer N radically changed (Eris-

BOX 3.2. BRIEF DESCRIPTION OF N CYCLING IN SOILS

N occupies various pools in the soil, including inorganic N, microbial N, and organic N, the latter comprises a broad range of carbon compounds with varying susceptibility to microbial mineralization. The vast majority of N in soils, especially in natural ecosystems, is bound within soil organic matter or stored in microbial biomass, from which it is slowly released as plant-available N over time and hence does not pose an immediate threat to the environment or humans. Each year, a fraction of this organic N reservoir is mineralized to NH_4^+. Mineralization serves an essential function for plants, and in agricultural systems supplies as much as 50% or more of the N assimilated by crops. Mineralized N is highly mobile and is readily transformed by soil microbes among different N species: organic N, ammonium (NH_4^+), ammonia (NH_3), nitrite (NO_2^-), and nitrate (NO_3^-). In the reverse process, immobilization, inorganic N is integrated into the living biomass of plants and microbes. The amount of organic matter returned to the soil, soil moisture, and management practices like tillage combine to affect soil microbial populations and activity and the rate of N storage or release. Adding inorganic N fertilizers can increase the total amount of N cycling through soils which can promote long-term fertility (Ladha et al., 2011). High inorganic N availability may promote high plant productivity, but can also be associated with large surpluses of N. This excess N can lead to environmental degradation either by percolation through the rootzone (leaching) or through volatile emissions of N gases into the atmosphere (e.g., NH_3, nitrogen oxides [NO_x], or nitrous oxide [N_2O]). Inert dinitrogen (N_2) is the gaseous emission released in the highest quantities. Though it is difficult to measure because of the relative concentrations in ambient air, N_2 to N_2O ratios in agricultural systems are an average of 1.8:1 (Schlesinger, 2009) but can be higher than 75:1.

The most important aspects that distinguish nutrient cycling in conventional agricultural soils relative to those of natural terrestrial systems are (1) conditions of nutrient saturation; (2) the decoupling of N, P, and C cycles; and (3) an inadequate synchrony and synlocation of nutrient sources and sinks (Drinkwater and Snapp, 2007). The inputs of inorganic N fertilizer generally exceed the demands of plants and the soil community, a situation made worse by the decoupling of nutrient cycles, which disrupts the primary mechanisms for inorganic N immobilization and storage. When sufficient C is present, microbes are the major channel for immobilization of inorganic

N in the soil. In most natural systems, microbial N far exceeds inorganic N, whereas the reverse is true in agricultural soils with low C inputs. Under steady-state conditions, the balance in rates of mineralization and immobilization, combined with the rapid turnover of the microbial community, leads to a low level, but stable supply of N availability to plants.

Need for Better Understanding of Natural Processes

While N cycling in natural systems is understood on a gross level, more needs to be known about the fine-scale mechanisms and processes and about the relative roles of various organisms to regulate, store, and provide feedback for nutrient retention. Comparative studies of N pools and flux between them in agricultural soils versus unmanaged native grasslands and forests are instructive in how the natural processes have been altered by different soil management schemes. In addition, as a result of the doubling of reactive N globally by human activity (Vitousek et al., 1997), natural terrestrial ecosystems have experienced chronically high levels of N deposition. Studies indicate variation among ecosystems in how quickly they reach nutrient saturation, indicating differences in their capacity for N retention. In a spruce forest subject to decades of high N deposition, Kreutzer et al. (2009) describe a dynamic system of N cycling, characterized by high rates of microbial mineralization and immobilization of N, accompanied by rapid turnover of the microbial community. This high internal flux between N pools mediated by the microbial community produces relatively high N retention while maintaining plant-available N levels sufficient to cover the entire budget of all of the trees. The potential coexistence of both high rates of ammonification and nitrification with low accumulation of ammonium and nitrate at any point in time, as demonstrated in this and other studies, provides encouragement for agricultural fertility management.

It is also worth noting that a multitude of abiotic and biotic factors (e.g., pH, temperature, organic carbon, microbial activity, soil texture, etc.) affect the N cycle in soils over a wide range of temporal scales, from as short as minutes (e.g., gaseous NH_3 volatilization) to decades (e.g., movement of NO_3 through the vadose zone to the aquifer). Spatiotemporal heterogeneity in soil N cycling is one factor contributing to the diversity of N fertilizer use and pollution potential among fields and farms.

BOX 3.3. LINKS BETWEEN THE N AND HYDROLOGIC CYCLES

Water regulates the nitrogen cycle. For example, nitrate in soils will not move toward plant roots without water (mass flow); the extent of soil moisture alters microbial activity, N transformations, and the form of gaseous emissions (nitrification and denitrification); dissolved N in solution is transported in streams and waterways (runoff); and airborne N is transported to the ground with precipitation (deposition).

Given the presence or absence of water governs N dynamics through physical and biological processes, changes in the natural hydrologic cycle or management of water resources by humans, climate change, or both will have cascading effects throughout the N

cycle at plot and larger spatial scales. For example, on-farm it may catalyze a shift to low-volume irrigation with the potential to reduce solution N losses at the threat of greater gaseous emission. But equally plausible is a reduction in agricultural area reducing total inputs. At watershed levels, altering the timing or amount of precipitation may cause erratic pulses of nutrients. It is not possible to forecast the net impacts the changes in the hydrologic cycles are yet to exert on N cycling in the state at this time because of the multitude of drivers and potential responses. However, it is important to consider the significant linkages between the two global cycles when reflecting on potential future N trajectories.

man et al., 2008). After World War II, demand for explosives—another product derived from the Haber–Bosch process and the root motivation for its development—declined, and a rapid increase in the production and distribution of inorganic fertilizer ensued. The consequence has been a massive increase in the use of N fertilizer (Galloway et al., 2008).

In California, inorganic N fertilizer sales (and presumably use) have grown at an average annual rate of 5% between 1946 and 2009 (figure 3.2). Annual sales grew at their fastest pace prior to 1980. Since that time, sales of N fertilizers have leveled off; recent annual sales of approximately 600,000 Mg of N fertilizers are not distinctly higher than sales in 1980. In the recent past (since 2005), N fertilizer sales have continuously declined. It is worth noting that California no longer fixes NH_3 at industrial scales and virtually all the inorganic N fertilizer sold in California today is imported. California's fertilizer manufacturers refine imported NH_3 into other products, such as ammonium nitrate or specialty fertilizer blends, which are then applied in California's fields.

Statewide sales data present a limited picture of inorganic N fertilizer use. Farm operators make fertilizer decisions at the field level. Decisions for an individual parcel of land determine the intensity, effectiveness, and outcomes of N use. Disaggregated knowledge of inorganic N use at this level is thus essential to understanding the cause and effect of N fertilizer use in California. Unfortunately, finer resolution N use data are practically nonexistent for California (Appendix 4.1).

A first step to identify leverage points and hot spots, and to target action is to examine N use by crop. We documented changes in N application rates, yields, and cropped area for 33 important California crops between 1973 and 2005 (Rosenstock et al., 2013). Average N fertilizer application rates (kg ha^{-1}) across the 33 crops surveyed increased 25% over this 33-year period (35% when considered on an area-weighted basis), the magnitude and direction of change being crop specific (Appendix 3.1). Application rates for a few crops increased by more than 75%. Yet, for 10 of the 33

crops examined, the average rate at which N fertilizers were applied declined. Nitrogen fertilizer use on vegetables and nut crops showed the largest increases. This is particularly significant because the area dedicated to these crops increased simultaneously with higher N application rates (figure 3.3; Appendix 3.1). Since many high-value vegetable and nut crops saw the greatest increase in fertilizer N use and typically recover a smaller percentage of that N than the field crops they replaced, cropping and N use trends suggest a greater threat for N loading to the environment.

Our estimates of fertilizer use represent a first approximation of inorganic N fertilizer use in California. However, no data exist on the variation in inorganic N applications among fields, farms, and regions. Application rates may vary by 50% to >100% depending on soils, irrigation, weather, and grower preference, even for the same crop. A 1973 survey of fertilizer use in California demonstrates this heterogeneity (Mikkelsen and Rauschkolb, 1978): among regions, average application rates varied approximately 34%.[1] The variations reported in the 1973 survey are only illustrative and cannot be assumed to reflect today's cropping conditions. Significantly better data are needed to answer basic questions on who, where, and how inorganic N fertilizer is used throughout California (Rosenstock et al., 2013).

Yields of California crops increased dramatically during the period of rapid expansion expansion of inorganic N use. For example, between 1950 and 2007, yields of almonds, processing tomatoes, and rice increased by 349%, 221%, and 136%, respectively (USDA NASS, n.d. *Historical Data*. Accessed July 1, 2012a). Most cropping systems have seen similar rates of yield increase. The relative contribution of yield increases that can be directly attributed to inorganic fertilizer N has not been systematically analyzed for California. However,

---

1. Calculated as the average of coefficient of variations (standard deviation/mean) among average reported rates for each commodity for all reported commodities.

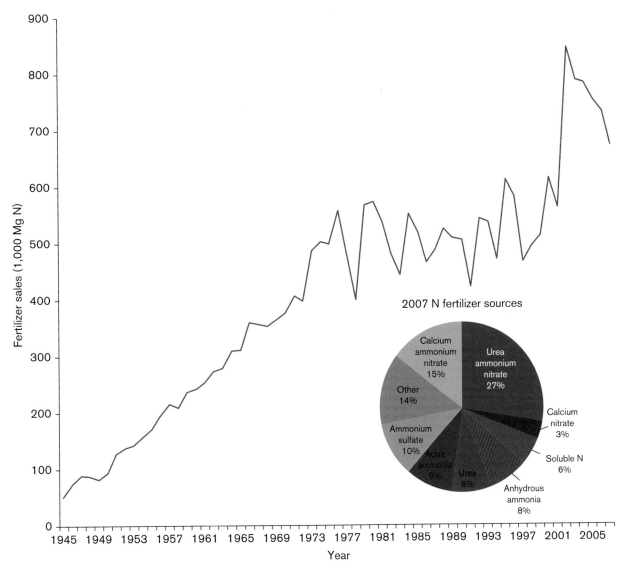

FIGURE 3.2. Synthetic nitrogen fertilizer sales in California, 1945–2007. Since their introduction after World War II, sales (and presumably use) of synthetic N fertilizers have increased by an average of 5% per year, although they have largely leveled off since the early 1980s. The large rise in fertilizer sales between 2001 and 2002 calls the reliability of these data into question.

DATA FROM: CDFA (2009).

trends of inorganic N fertilizer sales and agricultural productivity in California show increases that parallel global trends suggesting similar benefits.

3.2.1.2. INORGANIC N USE ON MAJOR CALIFORNIA CROPS

Californian farmers grow a remarkable diversity of crops on more than 4 million hectares. The total number ranges from approximately 150 to greater than 400 crops, depending on the source of information and year of interest. Despite the variety, much of California croplands are planted with only a handful of species. Fewer than 20 crops are grown on at least 1% of the state's cropland. Alfalfa, almonds, grapes, rice, wheat, and corn cover approximately 16%, 9%, 8%, 8%, 7%, and 6%, respectively, of the harvested cropland, more

than 50% of the total (USDA NASS, 2012). These values are a first approximation of inorganic fertilizer use. Manure applications to silage corn and cereal forages are significant and total N applications from manure to these crops may be similar to the other crops listed here.

Crop species require different amounts of N for growth and reproduction. Plant N requirements regularly exceed 100 kg N ha^{-1} and can be more than 250 kg ha^{-1}. Differential N recommendations among crops reflect this variation in demand (Appendix 3.2). Average application rates differ by an order of magnitude among widely cultivated species (Appendix 3.3). For example, wine grapes receive an average of less than 30 kg N ha^{-1}, while celery receives closer to 300 kg N ha^{-1}. Perennial crops only have one crop per year. Annuals often are double- or even triple-cropped. Rotating

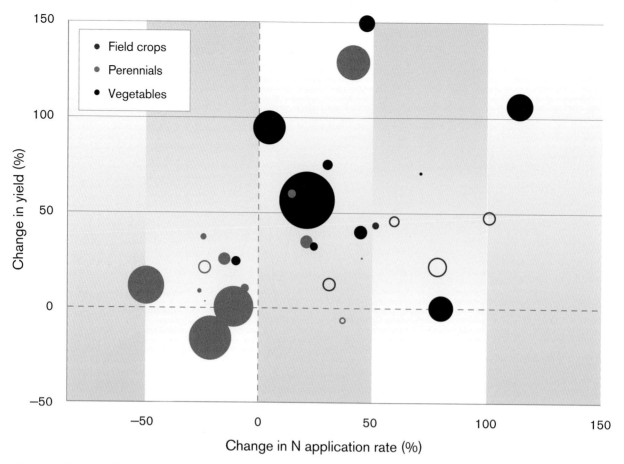

FIGURE 3.3. Changes in N application rates, yields, and cropped area for 33 crops, 1973–2005. The size of circle represents the relative magnitude of change in the area cultivated for that particular crop between 1973 and 2005. Closed circles represent increases in cropped area and open circles are declines in area between 1973 and 2005.

SOURCE: Reprinted from Rosenstock et al. (2013). Copyright © 2013. Regents of the University of California. Used by permission.

annuals on a single piece of land greatly increases N intensity. Fertilizer N use on a lettuce–broccoli rotation in Salinas may receive between 300 and 550 kg ha^{-1} yr^{-1}, far greater than either commodity alone. Considering the cropping system rather than individual crops is an important distinction for understanding the causes and effects of N use.

Substantial differences in cropped area and N fertilizer application rates suggest certain crops have a larger impact on overall N dynamics than others. Multiplying the area harvested by average fertilizer N application rates for 33 crops in California shows that only four crops—cotton, almond, rice, and wheat—account for 51% of total N applied. Thus, relatively few cropping systems have a disproportionate share of California cropland N use. Notably, nursery or greenhouse industries were excluded from these calculations because of both data limitations and the fact that ornamental horticulture production systems tend to be among the most intensive N users with use ranging from 100 to 7,000 kg ha^{-1} (Evans et al., 2007).

Over the last 35 years, California's crop mix has shifted heavily from field crops that often receive less N fertilizer to more N-intensive species (e.g., vegetables and nuts). Field crops are still grown on the majority of croplands as of 2008 (figure 3.4), but the land dedicated to field crops declined from 74% to 53% between 1970 and 2007. Fruits and vegetables are now grown on a nearly equivalent area (53% vs. 47% in 1970). The shift in crop production towards N-intensive crops is partially responsible for increases in N consumption.

3.2.2. Organic N Use on Croplands

Crop producers, at times, apply organic N in lieu of, or in addition to, inorganic N fertilizers. Commonly used organic N fertilizing materials include manures, composts, waste products, and leguminous cover crops (Gaskell et al., 2000; Hartz and Johnstone, 2006; Mikkelsen and Hartz, 2008).

Organic N use can represent either a transfer of N internally around California or an introduction of new reactive N into the state. Manures and composts are examples of the former. Cows do not create N; it simply passes through them during the conversion of feed into manure. Nitrogen in manure is derived from biological N fixation (e.g., from alfalfa), the Haber–Bosch process (e.g., when fertilizer is

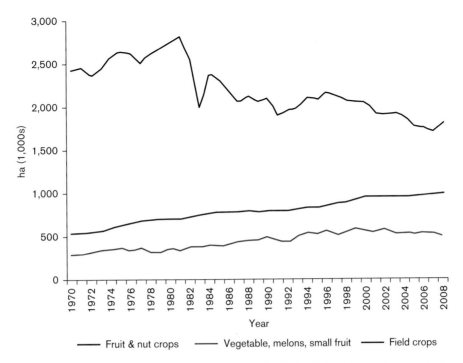

FIGURE 3.4. Change in cropland area by major crop types in California, 1970–2008. The amount of cropland dedicated to field crops has declined steadily since 1980. Today, almost 50% of cropland is used to grow horticultural commodities.

DATA FROM: USDA (2009a).

applied to feed crops), or from soil reserves. Compost represents another transfer since it is a collection of N from different waste products (e.g., food waste, manure, and urban green waste). Leguminous plants grown for green manures are the exception. They introduce new N into the biosphere by fixing atmospheric N through biological means and incorporating it into biomass and eventually soil.

Organic N use and recycling drives two significant components of California's N dynamics. Manure use redistributes slightly less than half as much N as inorganic N fertilizer applications. Alfalfa production creates 180 Gg N yr⁻¹ through biological N fixation (Chapter 4). Thus, organic N drives multiple N transfers in California's N system and its fate is important to understanding overall N dynamics.

Some evidence suggests that organic N sources typically improve soil health (Reganold et al., 2001). Additional organic matter applied with organic N is the source of many benefits to soil, including improved soil structure, hydraulic conductivity, water holding capacity, biotic activity, and nutrient retention (Drinkwater et al., 1995). It has also been suggested that organic systems reduce pollution pressure by stimulating higher rates of denitrification to N_2 (Kramer et al., 2006) and reduce leaching pressure by comparison to inorganic N sources (Drinkwater et al., 1998). However, organic N cannot be assumed to be less damaging to the environment under all conditions. Research shows that organic materials represent a significant source of reactive N to the environment, both gaseous and in solution,

because of the difficulty in managing the timing of N release from soil organic matter (Barton and Schipper, 2001; Kirchmann and Bergström, 2001).

3.2.2.1. TRENDS IN ORGANIC N USE

Unlike inorganic N fertilizer, organic N information is not publically available in most cases. Thus, it is extremely difficult to piece together a coherent account of organic N use in California. A survey conducted by Dillon et al. (1999) suggests that organic N use is common. More than 20% of approximately 800 farmers surveyed applied composts or manures in 1986. In the subsequent 10 years, the use of these N sources became 24% more prevalent. Indirect indicators further support the conclusion that organic N is increasingly demanded in California. The N fertilizer used by certified organic farms and the land dedicated to these systems has grown rapidly in recent years (Smukler et al., 2008), though it still only accounts for a small fraction of cultivated cropland in any given year (less than 4%). Between 2000 and 2005, the area of certified organic farms in California increased 31% from 59,421 to 77,963 ha (Klonsky and Richter, 2005). The most recent USDA Organic Agricultural Census reports that more than 110,000 ha were certified organic in 2008, suggesting nearly a doubling in the 8 years between 2000 and 2008 (Klonsky and Richter, 2005; USDA, 2010). According to the Organic Census, 58% of certified organic farms produced or applied organic compost and 49% applied green or animal manures in 2008 (USDA, 2010). Large increases in animal and human populations have resulted in a greater availability of N-rich manures, composts, and urban wastes destined for land application than ever before.

The distribution of organic N is not well documented. The State Water Quality Control Board (SWRCB) requires documentation of distribution of liquid manure and biosolids from dairies in the San Joaquin Valley. Dairy manure is the largest organic N source in the state and thus more information on characteristics, distribution, and attributes of use would be a major step forward. Modernizing the reporting system would increase the utility of the data collected and could reduce compliance costs for producers. Another significant gap in the current reporting system is information on the distribution and application of solid manure. When sold and transported off-farm (often to composters), manure quantities are recorded, but the manure's final location of application is not. With as much as 50% of dairy manure and 100% of poultry and beef feedlot manure applied to land offsite, assessing the significance of the transfer of N from animal systems to croplands is impossible. Additionally, the total size of the solid manure N flow may increase in the future as dairy operators are forced to send more manure solids offsite to comply with water quality regulations.

3.2.2.2. MANURE USE ON CROPLANDS

Approximately 263 Gg of manure N is collected and applied to croplands (Chapter 4). Utilizing manure N as a fertilizer is discussed in other sections, along with organic N sources more generally. This section emphasizes two issues of particular relevance to understanding manure N dynamics: material placement and geography.

Where manure N is applied, either on the surface or injected into the soil, preconditions its fate. Manure N applied to the soil surface is more likely to be volatilized. Higher rates of emission from surface applications than for incorporated manure are a result of soils being strong NH_3 sinks; injection of liquid manure and incorporation soon after broadcasting solid manure creates a boundary layer between manure N and the atmosphere. Placement of manure even 2 cm below the soil surface reduces NH_3 emissions from 25% to 37% (Sommer and Hutchings, 2001). Manure incorporation however is not a panacea. It increases soil N concentrations and can lead to higher rates of NO_3^- leaching unless additional steps are taken (Velthof et al., 2009).

Confined dairy systems in the San Joaquin Valley apply liquid manure to feed crops close to the production unit. In the most recent manure practice survey, no respondents reported injecting manure below the soil surface (Meyer et al., 2011), which suggests common practice exposes N to extensive volatilization. Because manure injection requires specialized equipment, switching practices requires costly investments.

A second consideration for land application of manure is the spatial distribution of animals. California animal production has historically been in concentrated areas (e.g., Chino basin and now the southern San Joaquin Valley), but has become more intensive. Intensification has increased herd and flock size per unit area. The result is a concentration of waste and an increased probability of overapplication. Operators become N rich and land/crop poor, putting pressure on N disposal. However, there may be enough land associated with dairies to effectively utilize manure N. Pettygrove et al. (2003) estimate that as much as 200,000 ha may be associated with dairies in the San Joaquin Valley and available to receive manure applications. Since manure N application rates are now determined by crop uptake due to the SWRCB General Order for Dairy Waste Discharge, one might expect an increase in the number of operators moving to triple crop practices (3 crops in 1 year) to increase offtake. Triple crop systems assimilate more than 600 kg ha^{-1}, which permits operators to apply 840–990 kg N ha^{-1}, making these the most N-intensive production systems in the state. In comparison, the most N-intensive cropping systems (e.g., double-cropped cool-season vegetables) typically apply inorganic fertilizer N at approximately two-thirds these rates, ~600 kg N ha^{-1}.

3.2.2.3. CULTIVATION-INDUCED BIOLOGICAL N FIXATION

A specialized and taxonomically diverse group of prokaryotes use the enzyme nitrogenase to convert atmospheric N_2 gas to NH_3. The organisms can be free-living soil and aquatic biota (e.g., *Azotobacter* or *cyanobacteria*) or form associative (e.g., *Azolla*) or symbiotic (e.g., *Rhizobium*) relationships with higher plants. Symbionts, *Rhizobium* bacteria, are the most important group of N fixers in agricultural ecosystems. Prior to the invention of the Haber–Bosch process, biological nitrogen fixation (BNF) was the primary way N moved from the atmosphere to the biosphere, and the abundance of N fixers regulated ecosystem productivity. Still, BNF is thought to contribute ~128,000 Gg N yr^{-1} globally (Galloway et al., 2004).

Biological nitrogen fixation adds approximately 335 Gg N yr^{-1} to California's terrestrial ecosystems, an amount equal to 65% of that applied as inorganic N fertilizer (Chapter 4). The majority of BNF (58%) is cultivation induced (C-BNF). That is, production of food and feed drives the planting of crops that utilize BNF to satisfy N requirements. BNF in California takes place in systems planting legumes and rice. While BNF is possible in multiple cropping systems, alfalfa dominates the total C-BNF flux (92% of total; Chapter 4) because of its productivity and extent. The relative impact on the overall N cycle in California of other legumes is minor.

Understanding how alfalfa yields and cropped areas have changed provides information on the importance of C-BNF as a direct driver. Absolute N fixation rates for California alfalfa are difficult to assess because studies have not thor-

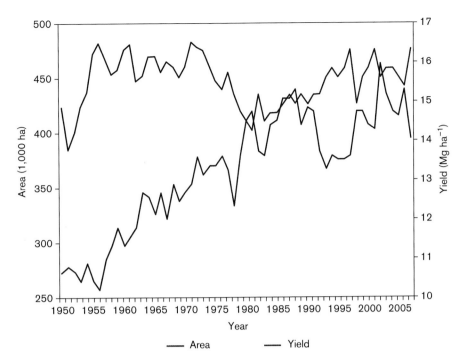

FIGURE 3.5. Cropped area and yield of alfalfa in California, 1950–2007. Data show that area has remained relatively the same but productivity has increased markedly. Because biological N fixation is correlated with dry matter production, data suggest C-BNF introduces considerably more N into California's biosphere than a half century ago.

DATA FROM: USDA (2009a).

oughly measured aboveground and belowground biomass production across the range of soils and weather conditions. But fixation rates can be inferred from yields. N fixation in alfalfa is proportional to dry matter production (Unkovich et al., 2010). Between 1950 and 2007, statewide average alfalfa yields increased 53% from 10.5 to 16.1 Mg ha^{-1}. Over the same time period, the area of cropland dedicated to alfalfa remained almost unchanged. It increased 4% from 423,000 to 440,000 ha, but averaged 432,000 ha and ranged between 368,000 and 484,000 ha across these years (figure 3.5). Assuming a direct proportionality between N fixation and yield, the yield increase and negligible areal increase suggest alfalfa transfers 44% more N from the air to the land's surface each year on a similar land base (USDA NASS, n.d. *Historical Data*. Accessed July 1, 2012a). Though a significant increase, the rise has been less pronounced than the trends seen for inorganic fertilizers and fuel combustion.

Alfalfa yields are highly regionally dependent. For example, production was more than 50% greater in the San Joaquin Valley than in the Intermountain Region in 2004 and 2005 (Summers and Putnam, 2008). Higher yields largely result from a longer growing season that increases the number of cuttings. Latitude is generally a good predictor of yield (and hence fixation). Differential yield suggests that the amount of N fixed and the importance of BNF to N cycling will be unique to each region and hence the total impact will depend on the spatial distribution of crop patterns and the location in the state.

3.2.3. Agronomic Nitrogen Use Efficiency (NUE)

Higher rates of N fertilizer application are not a problem, if fertilizer N recovery increases proportionally. Concerns about field practices arise because growers must apply more N than crops require for growth and reproduction because of inherent inefficiencies of production systems and soil N dynamics (Box 3.2). Hence, the portion of N not taken up by plant roots remains in soil after harvest, and thus is vulnerable to release as reactive N compounds, or is denitrified to inert N_2 gas. The relative proportion attributable to each fate depends heavily on soil physical and chemical properties and crop management (Appendix 3.4). Only a small fraction of the N applied beyond plant uptake ("surplus") is used in the subsequent growing seasons, often less than 10% (Ladha et al., 2005). Research demonstrates that the surplus N is particularly vulnerable to loss from the soil system as reactive N. Both NO_3^- leaching potential and the rates of gaseous N_2O emissions increase nonlinearly with increasing surplus N (Broadbent and Rauschkolb, 1977; Van Groenigen et al., 2010). Surplus N emissions may occur either during the growing season, as in the case with leaching in many irrigated systems, or following harvest when soil N levels are high. Furthermore, surplus N represents an unused resource and expenditure for the producer. Therefore, knowledge of the amount of N fertilizer applied and taken up is critical to understanding the fate of N fertilizer use.

3.2.3.1. NUE WHEN USING INORGANIC FERTILIZER

Measures of agronomic NUE[2] are ratios of plant N uptake to the amount of N fertilizer applied. NUE is one of the most often cited, and unfortunately, most often misinterpreted indicators of cropland N use. Mistakes arise because there are at least 18 different ways to calculate NUE, each quantifying slightly different components of the soil–crop system (Ladha et al., 2005). Thus, assessment of NUE needs to be executed with caution, explicitly defining the terms and knowing their limitations (Appendix 3.5).

Globally, the efficiency of inorganic fertilizer N applications ranges between 30% and 50% in the first growing season for cereal crops (Tilman et al., 2002) and less than 5% in the second growing season (Fritschi et al., 2005; Ladha et al., 2005). The NUE of California grain production systems are within this range, or even slightly higher. Recently developed management practices for rice show capacity to increase NUE even further, to >60% (Linquist et al., 2009). By comparison to field crops, fruits and vegetables tend to have lower NUE (Table 3.1). This is important because trends in crop mix show a shift to high-value horticultural commodities that are typically more technically N inefficient than the crops they replace. Practically, every high-value horticultural commodity averages NUE below 50%, some far below this value. This is significant because recovery of N is significantly lower in farmer fields than the controlled conditions NUE research is typically conducted under. Depending on the crops, low NUE may be attributed to the sensitivity of the crop to N limitation, physiological limitations, scale of production systems, poor knowledge of N demand, or application of N "insurance" against annual fluctuations in crop demand.

Partial nutrient balance (PNB) is one way to measure NUE. A PNB is equal to the amount of nutrient, in this case N, in the material exported off the field divided by the amount of nutrient applied (Dobermann, 2007; Snyder and Bruulsema, 2007). Given that PNB specifies an input–output ratio, a value near to one denotes a system where applications equal removal, a system in steady state. For PNB, greater than one indicates nutrient mining of soil resources; less than one, surplus either builds up in soils or is lost to the environment (benign or otherwise). PNB interpretation relies on the assumption that the soil N pool is in a steady state. That is, the amount of N mineralized from organic matter is equal to the amount immobilized, a zero sum. Conditional on the field location and management, the assumption of steady state may be violated, especially when considering short-term dynamics (Lund, 1982). In long-term experiments, though, the assumption may be more reasonable given that soil N concentrations are not changing rapidly in California's croplands. Resampling previously sampled agricultural soils throughout California 50–60 years later indicates an average increase in N of 0.20% (0.09–0.29%) or about 0.0036% per annum (Singer, 2001). The advantage PNB presents compared to other measures of NUE is that it can be calculated post hoc with data often available. It therefore can provide decision-makers a metric to evaluate the performance of fertility programs and a tool to evaluate changes in NUE over time at field-scale, even when NUE was not the original goal of the data collection. We first estimated PNB from data found in fertilizer response trials in California (see Appendix 7.2). Mean PNB rarely neared one, even for tightly controlled experiments, when analyzing N application rates that reflect those used in the field. These findings suggest ample room for improvement. However, it is again true that many of the studies are dated and may not reflect the sophistication of modern production or more importantly the yield levels.

The CNA also calculated PNB for 33 crops based on average yield (USDA, 2011a), N application rates (Mikkelsen and Rauschkolb, 1978; Appendix 3.1), and moisture and N content (USDA, 2012) for 1973 and 2005 to examine historical trends in NUE. Results suggest California cropping systems have become more N efficient over the 33-year period, with PNB increasing 37% on average. This was expected, as the rate of average yield increases (>50%) far outpaced that of N application rates (25%) (figure 3.3). Similar to N application rates, crops differ significantly in the magnitude and direction of their trend. An area-weighted PNB for 2005 suggests that an amount of N equivalent to 54% of statewide sales could be accounted for in crop products and by-products exported from the field. Assuming the PNB values are representative of California cropland as a whole, this statewide PNB suggests there was a surplus of almost 310,000 Mg of N sold (and presumably applied) in 2005. Though the estimate of surplus is striking, it is worth reiterating that it is impossible to reduce the amount of surplus N to zero and sustain high yielding agriculture. Nevertheless, this suggests substantial room for efficiency gains.

3.2.3.2. NUE WHEN USING ORGANIC FERTILIZER

Management of organic N is complex by comparison to inorganic sources. Organic N is bound within the soil organic matter and is not immediately plant available. It must first be mineralized into plant available forms, NH_4^+ and NO_3^-. The rate at which mineralization occurs depends on the origin of the material, N concentration, and environmental conditions (e.g., temperature and water), especially with respect to its resistance to microbial breakdown. Variable and uncontrollable rates of N release coupled with the fact that organic N generally must be applied (in the

---

2. There is a distinction between agronomic NUE and economic N efficiency. They are not interchangeable terms and care should be taken when discussing NUE. Agronomic efficiency measures the ratio of N assimilated to N applied and represents the technical potential of the system. In contrast, economic efficiency measures the rate of economic return for adding an additional unit of N and is subject to market prices of crops and inputs. When the agronomic efficiency of N application rate is at a maximum, economic efficiency is usually not. The remainder of this section refers to agronomic NUE when it discusses "NUE."

TABLE 3.1
Fertilizer nitrogen use efficiency (NUE) for select crops based on the available research conducted in California. The ^{15}N and difference methods differ in their methodology and are a direct and indirect measure of fertilizer recovery, respectively (see online see Appendix 4.1). AGB recovery efficiency relates to the N recovered in all aboveground biomass (AGB), not only the marketable crop yield

Crop	Mean N rate (kg/ha)	Mean harvest N (kg/ha)	Mean NUE by difference (%)	Mean NUE by ^{15}N yield recovery efficiency (%)[a]	Mean NUE by ^{15}N AGB recovery efficiency (%)	Source
Almond	267.8	151.9	29.4	9.7		Weinbaum et al. (1980); Weinbaum et al. (1984)
Avocado				6.7	35.0	Rosecrance et al. (2012)
Bell pepper	168.0	89.5	25.7			Hartz et al. (1993)
Cauliflower	141.8	127.7		37.2		Welch et al. (1985)
Celery	247.8					Feigin et al. (1982a)
Corn	179.6	148.9	39.1	43.5	18.8	Arjal et al. (1978); Broadbent and Carlton (1978); Doane et al. (2009); Hartz et al. (1993); Hills et al. (1983); Kong et al. (2009)
Cotton	130.7	87.9	23.1	26.3	47.0	Fritschi et al. (2004)
Garlic	280.7	202.4	30.4			Tyler et al. (1988)
Grape	50.6	143.8		7.5	30.3	Christensen et al. (1994); Hajrasuliha et al. (1998); Williams (1991)
Lettuce	130.7	56.6	20.9	19.1	19.3	Hartz et al. (2000); Jackson (2000); Welch et al. (1983, 1979)
Nectarine	190.4	97.7	17.0			Daane et al. (1995)
Peach	112.5	34.7	9.2			Niederholzer et al. (2001); Saenz et al. (1997)
Pistachio	318.0			29.6		Weinbaum et al. (1994b)
Potato	136.9	121.5	43.5	52.8		Tyler et al. (1983); Meyer and Marcum (1998); Lorenz et al. (1974)
Prune	285.4	104.4	55.0	0.9	13.5	Southwick et al. (1999); Weinbaum et al. (1994a)
Raisin	40.0	32.0	6.0			
Rice	129.2	97.4	50.0	22.8	31.7	Adviento-Borbe et al. (2013); Bird et al. (2001); Eagle et al. (2001); Linquist et al. (2009); Mikkelsen (1987); Peacock et al. (1991); Ye et al. (2014)
Strawberry	111.7	72.0	7.4			Abshahi et al. (1984); Bendixen et al. (1998); Welch et al. (1979)
Sugar beet	131.9	181.2	19.1	24.9	42.4	Hills et al. (1983, 1978)
Tomato	119.3	97.9	17.1	53.4	15.0	Broadbent et al. (1980); Doane et al. (2009)[b]; Hartz et al. (1994); Hills et al. (1983); Miller et al. (1981)
Walnut	194.6	105.8	1.4	23.3	29.9	Richardson and Meyer (1990); Weinbaum and Van Kessel (1998)
Wheat	165.3	124.6	20.8		12.8	Abshahi et al. (1984); Baghott and Puri (1979); Wuest and Cassman (1992)

a. Recovery of ^{15}N measured over one growing season/year except the following (years): almond (2), avocado (0.25), pistachio (2), and walnut (6).

b. Mean ^{15}N recovery efficiency only includes recovery of isotopically labeled synthetic fertilizer, not treatments with labeled cover crop.

case of manure) or incorporated into the soil (in the case of cover crops) prior to production makes timing soil N supply with plant N demand difficult (Pang et al., 1997). Further, because only a part of the N in organic material mineralizes in a year (e.g., <10% from manures; Hartz et al., 2000), producers using organic N sources typically apply much more N than would be required using inorganic N fertilizers, at least until new soil carbon (C) and N equilibrium are reached (Pratt and Castellanos, 1981).

High application rates and difficulty in controlling N release suggest that systems utilizing organic N have a low NUE. Indeed Crews and Peoples (2005) reviewed ^{15}N recovery in legume-based rotations and found that between 10% and 30% of N from legumes was harvested in subsequent plant tissue. Low NUE in systems using inorganic N fertilizers, however, may be of greater concern. Biologically fixed N appears to be more readily utilized by soil biota and incorporated into organic matter, increasing its retention in the rootzone (Crews and Peoples, 2005). Results from an unpublished long-term experiment in California show similar low NUE in Mediterranean climates with annual crops. When calculating the difference between N inputs and N in harvested product in an organic corn–tomato rotation, only 27% of the amount of N applied was accounted for (Reed et al., 2006). Fields fertilized with liquid dairy manure have historically had low NUE. Following a series of assumptions about source and sink attribution, Harter et al. (2002) suggest that NUE (as PNB) could be approximately 50–60%, lower than field crops fertilized with inorganic fertilizer. Overall, the NUE of systems utilizing organic N is poorly documented.

3.3. Feed and Manure Management

Animals require dietary N and amino acids (building blocks of proteins containing N) for maintenance, growth, and production. Meeting the protein demand of California's animal population (cattle, poultry, horses, and pets) requires more than 557 Gg N yr^{-1}, 75% of which is fed to dairy cattle (Chapter 4).[3] The N needed to support California's livestock economy is 15% greater than the inorganic N used to support crop production (557 vs. 466 Gg N yr^{-1}) and 53% of total cropland N (1038 Gg N yr^{-1}).[4]

Only a fraction of the N contained in feed is converted into milk, meat, or eggs. N that is not converted is excreted in manure (Kebreab et al., 2001; Powell et al., 2010). Where manure is deposited and how it is managed determine the fate of embodied N. Managed well, manure N represents a resource for farmers. Managed poorly, manure N is a serious pollution concern. The authors estimate 416 Gg N yr^{-1} is excreted, 263 Gg N yr^{-1} (63%) of which is recycled to croplands as fertilizer (see Chapter 4). The balance is released into air or water and stored in soils.

3.3.1. Trends in California Livestock Production

Since 1980, there has been a considerable increase in the livestock population of California (figure 3.6). The population of dairy cattle nearly doubled and the population of broilers tripled in only 27 years. There were more than 1.8 million dairy cattle and 266 million broilers in 2007. But not all animal populations grew over this time period. Populations of feedlot steers, cattle and calves, and non-broiler poultry species (e.g., layers and turkeys) varied over this time frame. Depending on the species, the populations of these animals in California in 2007 were roughly equal to or significantly less than the population size in 1980.

The increasing size of the animal population has a direct effect on N transferred into California's biosphere, though the absolute impact is not known. Additional animals require additional protein which requires N. N in feed crops originates from the atmosphere and is fixed via either biological (e.g., alfalfa) or industrial means (e.g., inorganic N fertilizer). A fraction originates from California and most of this is alfalfa, while the majority of other dietary needs are imported. Increases in feed demand therefore can determine cropping patterns in the state (e.g., alfalfa and silage corn) and influence those in other regions.

Growth in livestock and poultry production has helped fuel California's agricultural economy and US food security. Livestock products were worth US$ 9.8 billion in 2010, up 25% from 2009 (CDFA, 2012) and contribute nearly 30% of annual agricultural receipts in recent years. Notably, California dairy operators produce 21% of the US milk and cream, and egg producers rank fifth among states, generating 6% of US total production (CDFA, 2012). Receipts from the production of dairy, poultry, and cattle and calves are one reason why California has been the top earning agricultural state (in terms of farm receipts) in the nation for every year since 1948. California livestock support the rural economy and fill a vital niche in the US food system.

3.3.2. Dietary N, N Use Efficiency, and N Excretion

Protein nutrition has a significant impact on productivity, profitability, NUE, and sustainability of animal production systems (figure 3.7). Protein is critical to animal metabolism, and animals consuming more protein yield more milk, meat, or eggs (Kebreab et al., 2001; Nahm, 2002). Dairy cows, for example, fed a mixed ration with forages, grains, and protein supplements will generally yield more milk than a cow consuming only forages. Yields of poultry products increase in a similar fashion when fed well-balanced high-protein diets. But like fertilizer applied to the

---

3. Actual feed N demand for California livestock and poultry is greater than this amount because this estimate only accounts for the confined animal population. Protein requirements of grazing animals are not included.

4. Total cropland N is the sum of N applied from inorganic and organic (manure and C-BNF) sources.

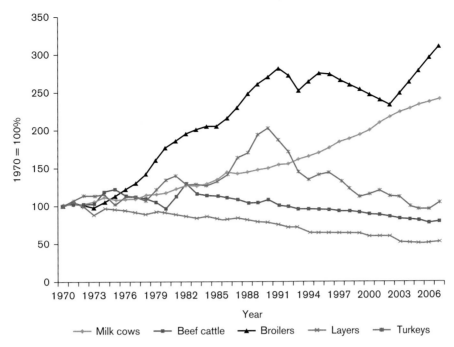

FIGURE 3.6. Change in California's animal inventory, 1970–2007. The number of milk cows and broilers has more than doubled since 1970, while some other animal populations have declined.

DATA FROM: USDA (2010); USDA NASS (n.d. *Historical Data*. Accessed January 6, 2015b).

soil, the relative increase in yields declines with increasing protein consumption due to inherent biological limits of the animal. When the physiological threshold of assimilation is reached, excess protein is excreted.

Animals are often fed more protein than necessary to obtain the greatest possible production. For example, the NRC (2001) recommends a diet containing 16.5% crude protein (CP) content for lactating dairy cows. However, milk production in some systems can be equivalent when cows are fed as little as 12% CP (VandeHaar and St-Pierre, 2006). The actual amount of CP required to meet production goals will depend on genetics and husbandry techniques unique to each environment.

Improvement in analytical techniques and investment in research has allowed formulation of diets to better meet animal nutritional needs of CP, rumen degradable/nondegradable protein, or specific limiting amino acids (Morrison and Henry, 1940; NRC, 1994, 2001). Diets can be formulated to meet minimum and/or maximum protein and/or amino acid requirements. The general objective in formulating diets is to provide the necessary nutrition for the least cost, so the minimum protein constraint is typically used because protein ingredients are usually more expensive. The possible exception is inexpensive by-product feeds. By-product feeds, such as distiller's grains, almond hulls, cottonseed, or carrot tops, may or may not increase dietary concentrations of proteins or minerals. The widespread feeding of by-products in California highlights another important point in formulating diets. The formulation of diets is constrained by the availability of raw materials, composition, and cost. A balance must be reached between what is scientifically plausible and practically feasible to achieve economic and environmental goals. The major obstacle in achieving a tight coupling between protein supply and animal requirements is cost and the resulting decline in farm profit.

When the protein and amino acid requirements are in balance with the animal's requirement, N is used more efficiently (a higher percent of the consumed N is incorporated into animal product). Partial efficiencies of N use can be calculated during each stage of production as the ratio of N converted to animal product and/or retained to N consumed by the animal (ASAE, 2005). Careful attention must be directed to the unit of time involved for each category of animal. For turkeys and broilers, total NUE is equivalent to partial NUE. For all other production animals (i.e., beef, dairy, swine, layers), total NUEs can be calculated over the life of the animal as the sum of lifetime N retained and/or converted to animal product divided by total lifetime N consumed. Partial efficiencies range from 15% to 64% depending on the species and production category (Table 3.2). Average partial efficiency of N conversion to animal product is 14.9% for feedlot steers during the 153-day feeding period, 24.4% for high producing dairy cattle, 63.7% for milk-fed calves, 34.0% for grow–finish pigs, and 35.4% for layers. Efficiencies for broilers are near 60%. Ingested N not converted to animal product or used for growth is excreted (Hristov et al., 2011; Nahm, 2002).

Diet has a profound impact on N excretion and loss. As discussed, the quantity of protein intake determines the quantity of N excreted but consumption also determines

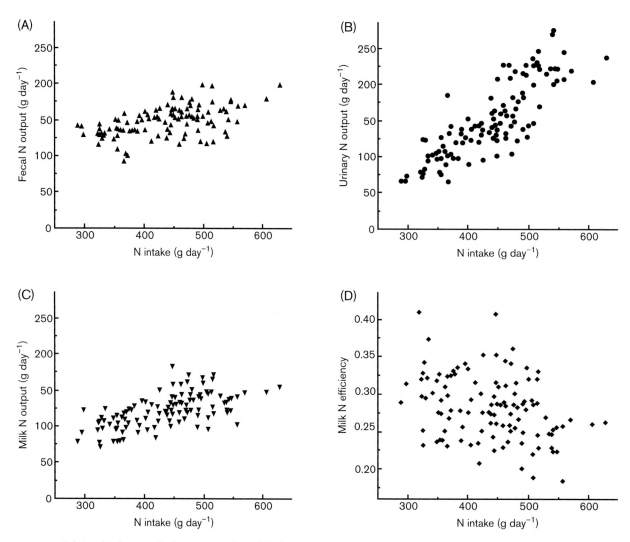

FIGURE 3.7. Relationship between feed nitrogen intake and (A) fecal nitrogen, (B) urine nitrogen, (C) milk nitrogen, and (D) milk nitrogen efficiency. As N intake increases, part of the additional N may increase milk N, but the majority is excreted as highly volatile urea in urine.

SOURCE: Reprinted from Dijkstra et al. (2011). Copyright 2011. Elsevier. Used by permission.

TABLE 3.2
Partial nitrogen utilization efficiencies for select economically important animal species

Animal category	Unit of time	kg N intake	kg N excreted	Intake excreted (%)	Partial N utilization efficiency (%)
Layers	20–80 weeks	1.04	0.67	65	35
Broiler	48 days	0.13	0.05	40	60
Lactating dairy cow	Daily	0.60	0.45	76	24
Feedlot beef cattle	153 days on feed	29.38	25.00	85	15
Milk fed calf	Daily	0.02	0.01	36	64
Growing finisher pig	120 day grow out	7.12	4.70	66	34

NOTE: Partial nitrogen utilization efficiency is calculated as PNUE = (1 kg N excreted/kg N intake)*100).
SOURCE: Data from ASAE (2005).

manure characteristics (e.g., form of N and moisture content). Manure composition, in turn, defines the probability for certain N transformations. Urea and uric acid formation and excretion increase with increased consumption of dietary N, especially when animals consume N above recommended nutritional levels. Urea N voided by cattle and uric acid voided by birds may be quickly hydrolyzed to NH_3 when urease and microbes are present, increasing the risk of NH_3 volatilization (VandeHaar and St-Pierre, 2006; Xin et al., 2011). If physical and chemical conditions are favorable, the process from excretion to volatilization takes place rapidly, in a time span ranging from hours to days. Decomposition of organic N excreted from cattle occurs at slower rates than hydrolysis of urea since organic N must be mineralized first. The greater environmental stability of organic N, by comparison to urea N, increases the feasibility of N collection and conservation. However, organic N is of lower utility as a fertilizer than inorganic urea and NH_3 because of the difficulty of predicting and controlling its release (Section 3.2.2). A conflict, thus, arises between the ability to conserve N within the animal production unit and planning for its end use as a fertilizer.

3.3.3. Manure Management

3.3.3.1. MANURE MANAGEMENT WITHIN CONFINED ANIMAL FEEDING OPERATIONS

From a rancher's point of view, the goal of manure management is to maintain a clean environment for the animal, reduce nuisance from odors, and improve animal health. From an environmental standpoint, manure management should conserve manure N until it can be recycled to cropland. Although manure treatment presents many pathways for N loss, and some emissions are inevitable, the primary loss pathway is volatile emissions of NH_3 into the atmosphere. It is estimated that between 20% and 40% of the N excreted on dairies in the San Joaquin Valley (Committee of Experts on Dairy Manure Management, 2005a) and 4–70% in poultry houses worldwide (Rotz, 2004) is emitted as NH_3. These wide ranges reflect the large impact of management and environmental conditions on emissions. Leaching of NO_3^- to groundwater may also be a concern from concentrated facilities (Cassel et al., 2005). Significantly elevated soil NO_3^- levels have been found under a dairy corral in Southern California (Chang et al., 1973), but the evidence of N accumulation under feedlots and corrals from elsewhere is mixed. Regardless, manure contains 416 Gg N yr^{-1}, of which only 263 Gg N yr^{-1} are estimated to be applied to cropland (Chapter 4). The remainder (153 Gg N yr^{-1}) contributes to air and water pollution, threatens downwind ecosystems, and represent lost nutrient resources.

Manure management practices and systems are diverse and constrained by the design of the facility. Differences between freestall and open lot dairies in the Central Valley are a good example (figure 3.8). Manure deposited in freestall barns is collected by flushing water over the concrete surfaces transferring it to a pond (lagoon) to be stored/treated as wastewater. Collection of manure in liquid form can help minimize emissions from housing, but economic considerations limit the distance it can be transported for land application (Committee of Experts on Dairy Manure Management, 2005a). In contrast, manure in open lot dairies is deposited on the soil surface where it dries. While manure resides in place, open lots are sources of NH_3 (Cassel et al., 2005). Lots are scraped and manure removed at specified intervals, typically two to four times per year. After collection, solid manure is stacked and stored prior to use.

Until recently, manure management decisions on many California dairies were made independent of N conservation or utilization. Yet, manure handling practices significantly change the form and concentration of N in manure and, therefore, it is important to understand consequences of changes in practices. Four surveys documenting California manure management practices have been published, but differences in the geographic extent and questions asked among the surveys make comparisons difficult (Mellano and Meyer, 1996; Meyer et al., 1997, 2011). Nevertheless, it appears dairy operators are adopting practices that increase their ability to manage N (Table 3.3). For example, between 1988 and 2002, the percentage of respondents that used settling basins to separate solids from liquids doubled to 66% and those that composted solid manure rose from 6% to 21% statewide. These two manure treatment options provide greater control over manure N by isolating more homogenous manure components and stabilizing N into organic matter, respectively (San Joaquin Valley Dairy Manure Technology Feasibility Assessment Panel, 2005a). In the most recent survey, more than 95% of respondents now use lagoons to store liquid manure (Meyer et al., 2011), helping to provide greater flexibility on when to apply manure. Many important details of manure management that potentially alter N dynamics on a dairy are not covered in the surveys (e.g., frequency of collection), greatly limiting the ability to determine how modifications of manure management schemes have affected California N cycling.

Manure management in poultry operations is more uniform than the dairy industry. In confined poultry production facilities, birds are under roof structures. This minimizes contamination of manure with rainwater and maintains a solid product that is manageable and transportable. The frequency of manure removal can range from once weekly to only twice yearly for California layer production systems (Hinkle and Hickle, 1999; Mullens et al., 2001), while manure is generally removed between flocks for broiler and turkey production. Dried material is then sold for animal feed, as a soil amendment, or transported to commercial processing plants for pelletization or composting. Manure characteristics (e.g., moisture content), environmental conditions (e.g., temperature and wind speed), and drying method (e.g., depth of stack) will alter NH_3 emissions in the house and during processing (Xin et al., 2011). Like dairy systems, the future of California poultry

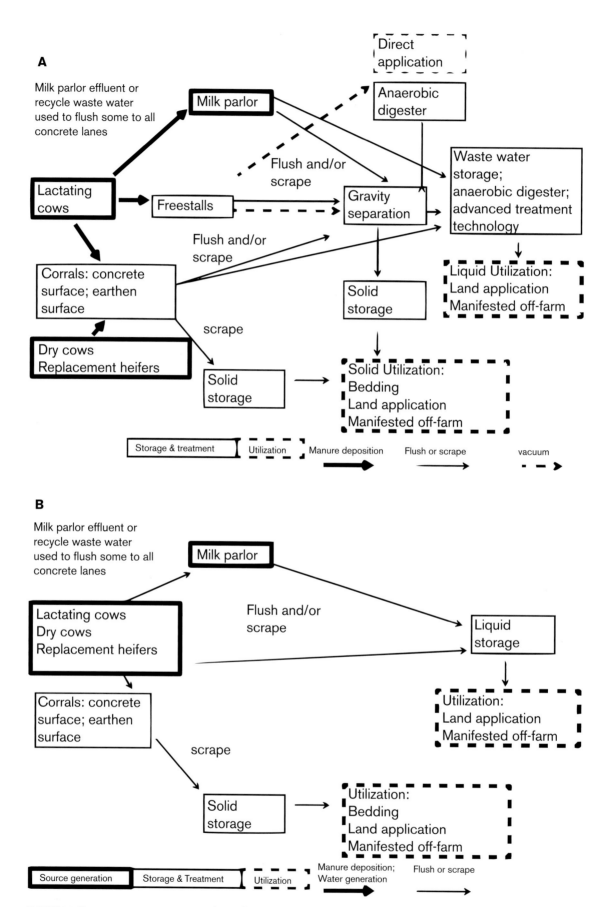

FIGURE 3.8. Common manure treatment trains on San Joaquin Valley dairies, 2010. (A) Manure flow pathway in freestall systems with or without open corrals. (B) Manure flow pathway in open corral systems. The diagrams shown here demonstrate major processes and the intricacy of manure handling on dairies. Manure management is a complex interdependent system constrained by the facility design.

SOURCE: Reprinted from Meyer et al. (2011). Copyright 2011. Elsevier. Used by permission.

TABLE 3.3
Manure management practices in California dairy production, 1988, 1994, 2002, and 2007

	Percentage of respondents				
Practice	1988	1994	2002³	2007	2007
Location of dairies	Statewide	Southern SJ Valley	Statewide¹	Glenn County	Tulare County
Number of dairies		139²	428	19	88
Housing and manure collection					
Flushed freestall⁴	61.7	77	66 [9, 23]	63	39
Manure storage ponds	67	96	99		
Solid separation		54		63	71
Settling basins	33	30	66	42	32
Mechanical separation		10	32	5	11
Gravity and mechanical combo.		15		16	27
Solids processing					
Scraped and piled	60	95		80	93
Compost	6	5	21	26	11
Utilization					
Solid	72	78.4	20	89.5	62.5
Liquid	91	70.4	48	100	100
Both			23		
Bedding		27	22	81.8	79.4
Removed from farm		6.8	3		
Sold as liquid		12.2			
Sold as solid	8	58.1	22	26.3	69.3

1. Survey did not include dairies on the North Coast region.

2. Only includes responses from written survey. An additional 45 phone surveys were conducted.

3. Animal housing in SAREP (2004) only reflects the percentage of milking cows under each system. The range for dry cows, bred heifers, calves, open heifers, and other milking livestock are shown in square brackets.

4. Flushing in 2002 refers to flushed lanes in scraped drylot and in 1994 refers to "flushing" but does not indicate housing. Even though managing N was not a primary objective until recently, manure management practices used on a dairy will affect N transformation, conservation, and loss. It is thus important to understand how they have changed over time.

SOURCE: Data from Meyer et al. (2011, 1997) and SAREP (2004).

manure management practices is uncertain. Implementation of newly defined housing systems (Proposition 2) may change manure handling practices and subsequent N dynamics on ranches.

Manure management practices are changing constantly as managers seek to comply with environmental regulations. Regulations have caused operators to evaluate and modify practices, which has undoubtedly changed N dynamics.

3.3.3.2. MANURE MANAGEMENT FOR GRAZING ANIMALS

Cattle and calves feed on natural lands and irrigated pastures before entering feedlots for fattening and finishing, being shipped out of state, or entering into the dairy supply chain. Grazing lands can be found in almost every part of the state. Depending on the season, the foothills of the Sierra Nevada, the Intermountain Region, North Coast, and Central Valley are common grazing lands, with animals being transported among them. Historically, pastures were fertilized with approximately 88 kg N ha⁻¹ (Mikkelsen and Rauschkolb, 1978) to increase productivity. Today, fertilization of pasture is rare. A more common practice for improving feed quality and protein content of pastures is to plant leguminous species, specifically clover.

NUE of grazing cattle is generally lower than that of confined animals (Powell et al., 2010). Lower NUE results from the inability of operators to assess the CP content of pastures and make adjustments to achieve the dietary balance to increase efficiency. Consequently, an even higher rate of

N is excreted in manure per unit of weight gain or product than in confined systems.

Manure excreted on pasture is not collected or stored. The distribution of deposition has a significant influence over the manure N fate. On pasture, urine and feces are deposited in heterogeneous patterns creating small hot spots of N addition. Depending on microbial activity, hoof action, soil type, plant species composition, topography, and climate, the N may be incorporated into plant roots, adsorbed to soil particles, lost atmospherically, leached, or run off (Liebig et al., 2010; Mosier et al., 1998; Oenema and Tamminga, 2005). Since the manure itself is not managed, pasture management becomes critical. Grazing patterns, stocking density, and pasture productivity will determine the ability of the environment to buffer and utilize the deposited manure N. Assuming appropriate stocking densities and pasture management, manure deposited in grazing systems tends to be relatively N neutral environmentally (Tate et al., 2005)

3.3.4. Whole Farm N Balances

Livestock production systems are complex operations with multiple codependent unit processes taking place simultaneously. Opportunities for N loss during manure processing abound. Because of interactions between treatment processes, N sustainability for livestock production systems is best assessed at the scale of the whole farm, instead of individual system components. N inputs at the farm scale include feed N, and sometimes bedding materials contain N (figure 3.8). N is exported in milk, meat, eggs, and manure (when it is transported off-site). Manure applied to croplands associated with the farm does not factor into the calculations since it is generated and applied on farm. The balance of inputs and outputs then provides a simple but imperfect tool to assess N flows of a particular farm.

Reviews of dairy production systems show significant N imbalances at the whole farm scale (Castillo, 2009; Powell et al., 2010). European dairy farms yield between 16% and 56% of the N imported in feeds and US dairy farms between 16% and 41%. On 41 dairy farms in the Western United States, an average yield of 36% of N was found. N that is not exported in agricultural products (e.g., 64% of imports in Western dairy farms) is volatilized to the atmosphere, leached to groundwater, or stored (temporarily) in soils under cropland and corrals. Whole farm N balances for California livestock production systems are poorly documented and it is not possible to draw conclusions about their relative environmental performance. However, if California systems are within the range of US and European systems, these results suggest significant room for improvement in manure management. Decreasing N imports and increasing N exports would help relieve pressure on the surrounding environment. Strategies that enhance N use such as staged feeding and surge irrigation of manure on croplands are available (Chapter 7).

3.4. Fossil Fuel Combustion

Fossil fuel combustion during transportation and industrial activities releases reactive N compounds, NO_x and NH_3, into the atmosphere. NO_x is produced in two principal ways.[5] (1) "Thermal NO_x" is created by the reaction of N and oxygen in air at high temperatures. Relative temperature and the length of time N is at high temperature regulate the rate of NO_x production. (2) "Fuel NO_x" results when N contained within fossil fuels, in particular oil and coal, is converted to NO_x during combustion. Biogenic processes that occur in soils can also produce NO_x; however, in California, 89% of NO_x, a total input of 359 Gg N year^{-1}, results from fuel combustion, making it the dominant driver of atmospheric concentration of this gas by far (Chapter 4).

Technologies used to control NO_x emissions sometimes unintentionally cause the release of NH_3. Instead of reducing NO_x to the environmentally benign N_2, catalytic converters can reduce NO_x to NH_3 when the air-to-fuel ratio is high, a common occurrence during acceleration (Baum et al., 2001; Kean et al., 2000). NH_3 is also used as a reagent to control NO_x emissions from stationary sources, specifically with selective catalytic reduction (SCR) technology. If the SCR system is not optimized (e.g., too much NH_3 in the gas stream, temperature is too low, or the catalyst has aged), NH_3 is released directly with flue gas without completing its intended reaction.

Once airborne, NO_x and NH_3 may travel long distances. NO_x can be transported from 0.010 to thousands of kilometers, while NH_3 usually deposits after short distances. One estimate indicates that nearly half of the NO_x and NH_3 produced in Los Angeles lands outside the South Coast Air Basin (Russell et al., 1993). Environmental conditions controlling the atmospheric chemistry and transport of N dictate when and where the N will land. Transport of airborne N compounds away from the source of emissions makes combustion-derived N an issue of concern beyond the location of initial emission (Ying and Kleeman, 2009) and means there is a spatial dimension to atmospheric N pollution (Durant et al., 2010; Hu et al., 2009; Karner et al., 2010; Zhu et al., 2002).

N emissions from fossil fuel combustion are an important source of air pollution and contribute to a multitude of human health concerns (Chapter 5). NO_x reacts with other pollutants in the presence of sunlight to form tropospheric (ground-level) ozone. Atmospheric NH_3 is an ingredient of particulate matter (PM), specifically particles of ammonium nitrate. Creation of N-derived PM depends on having sufficient levels of NO_x and NH_3 in the atmosphere, meaning in certain airsheds, PM reactions are NO_x limited (e.g., Southern San Joaquin Valley) and in others NH_3 is limiting (e.g., South Coast).

---

5. There is a third category of NO_x production called "prompt NO_x." It includes all NO_x produced that cannot be explained by either of the other two categories. It generally accounts for insignificant amounts by comparison to the other two mechanisms.

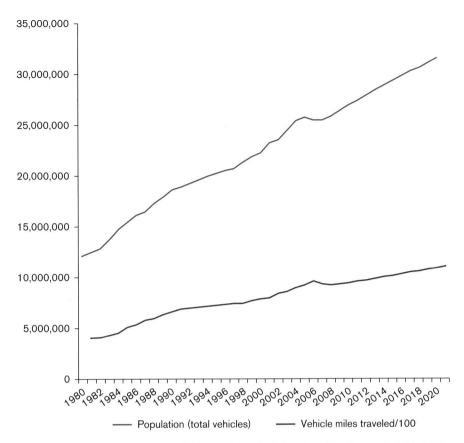

FIGURE 3.9. Vehicle inventory and total distance driven in California, with values projected to 2020. Mobile sources including on- and off-road activities are the primary source for NO$_x$ emissions (greater than 86% of the total). Despite large increases in the number of vehicles (population) and the distance traveled (VMT), there has been a significant decrease in emissions due to technological advances.

DATA FROM: CARB (n.d.).

3.4.1. Transportation

The terms "transportation" and "mobile sources" are not perfectly synonymous. Mobile sources include a wide range of on-road and off-road activities, of which transportation is a part. Vehicles used in the transportation of humans and goods (passenger cars, light and heavy-duty trucks, etc.) dominate atmospheric NO$_x$ emissions. The cumulative consequence of transportation sources far outweighs the impact of less common mobile sources such as lawnmowers and off-road recreational vehicles, despite higher emissions per quantity of fuel from these other sources. Thus we focus the discussion on transportation.

3.4.1.1. TEMPORAL AND SPATIAL TRENDS

Because N emissions from fuel combustion are correlated to fuel consumption, fuel sales data provide a starting point to understand emissions from the transportation sector. According to the California Board of Equalization, annual sales of gasoline increased 77% from 8,940 to 15,807 million liters and sales of diesel increased by 430% from 0.6 to 3.1 million liters between fiscal years ending in 1970 and

2007 (BOE, 2011). The average annual rate of change over the same time period was 2% and 5% per annum for gasoline and diesel, respectively. Sales trends demonstrate that there have been massive historical increases in the consumption of fuel for transportation in recent years, which has undoubtedly heightened the risk of additional atmospheric N loading. A recent projection suggests gasoline consumption in 2030 will be 54% higher than that in 2008 (CalTrans, 2009).

Increased fuel sales have been in part catalyzed by growth of the vehicle fleet and distance traveled per vehicle (figure 3.9). For example, the vehicle population in California increased 109% from 12.1 to 25.4 million vehicles between 1980 and 2007. The number of light-duty trucks on the road increased 212% (from 1.7 to 5.3 million vehicles), medium-heavy-duty truck population more than doubled (111%) to 0.24 million, and the number of passenger vehicles increased 68% from 7.6 to 12.8 million vehicles. In addition, vehicles were traveling further distances. In 2007, total vehicle kilometers traveled equaled 1.49 billion kilometers (CARB, n.d.). That distance was a 129% increase from 0.65 billion kilometers in the 27 years since 1980. The most significant growth in distance traveled was

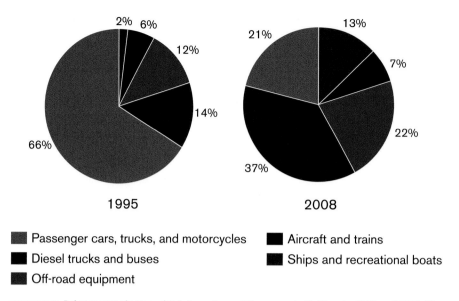

1995   2008

■ Passenger cars, trucks, and motorcycles   ■ Aircraft and trains
■ Diesel trucks and buses   ■ Ships and recreational boats
■ Off-road equipment

FIGURE 3.10. Relative contribution of NO$_x$ by major mobile sources in California, 1995 and 2008. The importance of certain sources has changed recently largely as the consequence of technology forcing policies. Regulations have yet to be implemented to control emissions from diesel engines and port activities but are currently under consideration with CARB.

DATA FROM: CARB (1999, 2010a).

for medium-heavy trucks (109% to 0.014 million) versus 77% for passenger vehicles. This trend contrasts sharply with the comparison between the populations of these two vehicle classes. Of the two, passenger vehicle population increased more rapidly. Less substantial but significant rises in the activity of larger mobile emission sources—trucks, buses, aircraft, and trains—have been demonstrated in some parts of the state as well (Corbett et al., 1999; Reid et al., 2007). Recently, ocean-going vessels have received increased attention because as much as 70% of emissions take place near ports (Corbett et al., 1999).

Transportation activities have historically been and continue to be the driving force in combustion derived NO$_x$ emissions (CARB, 2010a, 2015a). In 2008, 86% of NO$_x$ was derived from on- and off-road mobile sources statewide (CARB, 2015a). However, the relative significance of the various vehicle classes is changing. Of these, heavy-duty diesel vehicles, trucks, and buses were responsible for 37% of the mobile source emissions (or roughly 31% of the total emissions) (figure 3.10). Emissions from heavy-duty diesel vehicles are now the largest source of NO$_x$ in the state. Interstate trucks account for 14–17% of the truck population and 28–29% of the distance traveled (Lutsey, 2008). This represents a departure from previous trends (figure 3.10). As little as 16 years ago, NO$_x$ emissions resulted mostly from passenger vehicles. The change in the relative significance of NO$_x$ sources can be traced to aggressive technology forcing regulations on passenger vehicles and more lax policy for diesel engines (Sawyer et al., 2000). Rules to regulate emissions from the latter sources are currently under various stages of development and implementation by CARB.

3.4.1.2. TECHNOLOGICAL CHANGE

Human actions such as driving, however, do not by themselves determine N gas production. Emissions are the product of the activity level such as number of cold starts or distance driven and the technology such as catalytic converters or fuel being employed. These factors interact in dynamic ways to create (and control) emissions. Traffic conditions, the age of the vehicle, and gasoline composition significantly affect combustion and hence the total amount and relative proportions of compounds in emissions (Bishop et al., 2010). Technological change is the reason NO$_x$ emissions in California have been declining, despite significant increases in vehicle population and distance traveled.

Major components of vehicle design, comprising vehicle type, engine, and fuel combinations, can be thought of as an integrated system that together affects emissions and mitigation options. The technology in use is largely determined by the fuel and vehicle type. Technological changes for light-duty vehicles that run on gasoline have been the most radical. Utilization of positive crankcase ventilation systems, exhaust gas recirculation systems, and three-way catalytic converters has helped control NO$_x$ pollution. More recently, computer-controlled fuel injection systems and on-board diagnostic systems provide the engine information that helps it maintain the appropriate stoichiometric air-to-fuel point for the catalysts that convert NO$_x$. In addition to engine refinements, fuels have been reformulated to enhance engine modifications. Low sulfur concentrations—which are standard in California now—are a common feature of reformulated gasoline. Use of low sulfur gasoline is significant because sulfur ruins catalysts' effectiveness.

Technological changes for other vehicle/fuel combinations (e.g., medium- and heavy-duty vehicles) have been less extensive. For the most part, however, control technologies are similar for other gasoline-powered vehicles. Diesel engines have changed combustion chamber design, operate at lower engine speeds, and use electronic control for improved timing, among other improvements to reduce NO_x. Opportunities are also available to use exhaust gas recirculation systems and particle traps to reduce NO_x and primary PM emissions.

The importance of technological change to control N emissions from fossil fuel combustion cannot be overstated. Typically only 10% of the fleet is responsible for the majority of emissions, meaning there are a small number of high polluting vehicles on the road. High polluting vehicles are generally, but not always, older. Age of the vehicle is important because it defines the technology in use and often the condition of the technology. Catalysts and other control technologies deteriorate over the life span of a vehicle. It is for this reason that fleet turnover and renewal have been critical to past gains and will continue to underscore future N emission reductions.

3.4.2. Energy and Industry (Stationary Sources)

In California, major stationary source categories include boilers, steam generators and process heaters, utility boilers, gas turbines, internal combustion engines, cement kilns, glass melting furnaces, waste combustion, residential water heaters, and residential space heaters. Stationary sources were only responsible for approximately 11% of NO_x emissions in 2008 (CARB, 2009a). Of this, fuel combustion contributes 71%, or roughly 8% of the total NO_x inventory—335 Mg. Fuel combustion by stationary sources is therefore a relatively insignificant driver of N cycling in California today. That was not always the case, however. In 1980, stationary source fuel combustion contributed >21% of the state's NO_x, 954 tonnes. In the 20 years between 1987 and 2007, emissions were cut by nearly two-thirds. Reductions occurred despite the number of stationary sources producing NO_x increasing from 3,437 to 9,296, a 170% increase, over that time frame (CARB, 2010a).

Though there are a large number of individual NO_x producers (e.g., >9,000 in 2007), the vast majority contribute very small fractions, if any, to the total. Eighty percent of emissions were derived from 152 and 187 facilities in 1987 and 2007, respectively. The skewed distribution towards a relatively few sources improves the ability for targeted response and increases efficiency of point source control actions. Compared to agricultural facilities (>80,000 farms), the number of significant stationary sources is small.

Remedial actions have been enhanced by development and uptake of control technology. There is some evidence that technological advances that reduce N emission are becoming more prevalent (Kirchstetter et al., 1999; Yeh et al., 2005). Emissions reductions generally are the consequence of either modifying combustion conditions or capturing gases prior to release. Popp (2010) examines trends in adoption of NO_x-reducing technology at coal-fired power plants across the United States and found that between 1990 and 2002 there was a 375% increase in the adoption of combustion modification technologies, but the use of post-combustion technologies lags behind.

Power plants in California are typically not coal-fired. However, California energy demand requires import of energy from beyond state boundaries, much of which is produced from coal. In California power plants, greater market penetration of post-combustion technologies has occurred. More than 60% of the energy generated with fuel-fired gas turbines in the state applies post-combustion controls (CARB, 2004).

The example of California power plants illustrates an important concern: the potential for pollution leakage. Leakage refers to shifting the pollution burden from one entity to another, be it a location or environmental system. In this case, stringent regulatory controls coupled with high market demand have created a system where California's needs and the resulting pollution are shifted to other places.

3.5. Industrial Processes

Nitrogen is used for a variety of industrial purposes. Globally, industrial uses account for 18% of synthesized NH_3 (Yara, 2009, 2012). In the United States, estimates of non-fertilizer use range from 12% to 28% of the total consumption (produced or imported), depending on the year and the data source (Chapter 4). A recent estimate indicates that non-fertilizer N use accounted for 14% of total US NH_3 consumption in 2010 (USGS, 2012).

NH_3 fixed via the Haber–Bosch process is the starting point for N-based chemicals. The resulting NH_3 can be used as a raw material in industrial systems itself or further processed into a series of ingredients—nitric acid, ammonium nitrate, or urea (Appendix 3.6). NH_3 is primarily used in the production of ammonium salts—ammonium phosphates, ammonium nitrate, and ammonium sulfate. Ammonium salts are common fertilizers and, by comparison, have relatively few industrial uses. NH_3 does however have a role in the production of certain chemicals, especially melamine and caprolactam, that are important in the production of nylon and plastics. NH_3 can also be used to remove air toxins and reduce pollutant loads of exhaust gases from point sources burning fossil fuels.

Many industrial N uses rely on intermediate N products such as nitric acid and urea. Conversion of NH_3 to nitric acid occurs via the Ostwald process. Nitric acid is most commonly used in making explosives—e.g., ammonium nitrate and nitroglycerine. California consumed an average of about 35,000 Mg of industrial explosives and blasting agents a year (1994–2009). Nitric acid also can be used in producing primary metals, including as an extracting agent for copper and gold. Production of nitric acid is now one of the top three most common non-fertilizer uses of N in the United States (USGS, 2012).

Industrial N use is arguably the most poorly characterized, monitored, and understood parts of the N cycle. This is significant because the demand for industrial NH_3 is projected to increase. Forecasts estimate that global demand will increase by 21% between 2007 and 2013 (IFA, 2010). Market expansion will result from both increased demand for N-containing products and discoveries of new uses.

Significance of industrial N in the N cycle of California is difficult to quantify. One approach is to estimate the size of these flows from consumption. Estimates suggest per capita consumption in the United States ranges between 2 and 9 kg N $capita^{-1}$ yr^{-1}, not including N used for explosives (Domene and Ayres, 2001). Though there is a greater than fourfold difference between the minimum and maximum, the higher end of the range may be more plausible since similar levels have been found for Western Europe (Jensen et al., 2011). Assuming 7.5 kg N $capita^{-1}$, industrial N use would be responsible for approximately 283 Gg N yr^{-1}. That suggests the industrial N use would be responsible for a transfer of N into California equivalent to more than half that of inorganic N fertilizer applications.

Industrial N use is not environmentally benign, as industrial processes can be a significant source of emissions directly and over the life span of the materials created. Emissions from chemical processes may end up in air or water depending on the product. Nitric acid production released 3% of US N_2O in 1996 (Domene and Ayres, 2001). Explosives release most of the embodied N as N_2 but a fraction is NO_x and N_2O. Industrial N end products also tend to accumulate in high-density settlements, in structures or landfills. Given the longevity of many industrial N products, this pool of reactive N provides a legacy that releases N slowly into the environment. Where it is concentrated, industrial N may pose considerable long-term environmental and human health concerns.

3.6. Wastewater Management

Human consumption concentrates N in settlements and urban areas, much of which is discarded in garbage, refuse, and human excretions, creating N-enriched wastes. Wastes are then collected, processed, and discarded as part of the municipal solid waste or wastewater stream (Appendix 3.7). Spent water, in particular, contains a substantial latent pool of N due to its constituent mass of feces, urine, industrial waste, and by-products of food preparation. In California, the size of the excrement-derived wastewater N flow is approximately 174 Gg N yr^{-1} (Chapter 4). Whereas this wastewater represents only a relatively moderate sized flow of N by comparison to others (e.g., fertilizer or fuel combustion), its importance is partially derived from the fact that N removal was not a historical goal of treatment. Consequently, it was discharged directly into receiving ecosystems. Discharge to the ocean is the most common fate of wastewater N in California, with smaller amounts ending

up in biosolids, emitted as gases during treatment, applied to soils, or discharged to surface waters (Chapter 4).

Irrespective of the ultimate receptacle receiving the wastewater (freshwater, land, or marine), wastewater N presents environmental concerns. The potential for N to pollute marine systems is well known (e.g., the hypoxic zones in the Gulf of Mexico and Chesapeake Bay), though similar impacts off California's coast are less pronounced (see Chapter 5). But addition of even small concentrations of N into freshwater systems can often overwhelm them. Background N levels in aquatic systems are typically quite low. Any addition can disrupt the functioning of food webs and ecosystem health. Discharges to land are no more environmentally friendly. N not denitrified by natural soil attenuation processes elevates soil N content and increases leaching potential. Lund et al. (1976) investigated inorganic N concentrations below sludge ponds and found elevated NO_3^--N and NH_4^+-N levels at multiple depths below the wastes, indicating downward percolation of N from waste. Therefore, although the results from a study performed nearly 40 years ago may no longer be accurate, this suggests that wastewater discharge can cause acute pollution.

Technologies to remove N from wastewater are available, however. Wastewater treatment takes place in two ways. In California, it is typically processed at a centralized, regional wastewater treatment plant (also known as publicly owned treatment works). Or when sewage systems are not available to collect and convey the material to a centralized location, wastewater can be treated with on-site wastewater treatment systems (sometimes referred to as septic systems). N removal from wastewater is a time-, energy-, and money-intensive process. Discussion of the extent of wastewater treatment must consider the social, economic, and environmental context in concert (Muga and Mihelcic, 2008). Chapter 7 of this volume further discusses wastewater treatment options.

3.6.1. Publicly Owned Treatment Works (POTWs)

Centralized treatment plants process about 90% of human wastewater generated in California (Chapter 4). The amount of wastewater treated at each plant is relative to the size of the population it serves, with a typical value around 379 L $capita^{-1}$ day^{-1} depending on the degree of water conservation. Wastewater contains about 13.3 g N $capita^{-1}$ day^{-1} (Metcalf & Eddy, Inc., 2003).

3.6.1.1. WASTEWATER TREATMENT

When considering the effects of wastewater on N cycling, it is useful to start with collection systems. For a majority of the population in California, wastewater and raw sewage are transported through a system of pipes and pumps to a municipal POTW. For a variety of reasons, including cost, most conveyance systems are not maintained adequately. Aging infrastructure, poorly fitted pipes, and seasonally

TABLE 3.4
The level of treatment at California wastewater treatment plants, 1997 and 2008

Treatment level	N removal efficiency (%)	Facility treatment capacity 1996–1997 (%, N = 643)	Facilities treatment capacity 2007–2008 (%, N = 716)	Percent of total CA flow, 2007–2008
Primary	3–5	13	12	1.1
Advanced primary	10–50	9	11	19
Secondary	40–60	53	36	30
Advanced secondary		7	15	32
Tertiary	50–90	18	20	18
Onsite systems	3–5			

NOTE: Three pieces of information are important to understand: (1) increased treatment decreased N load of wastewater effluent, (2) wastewater is being treated to higher standards, and (3) traditional onsite treatment systems remove only trace amounts of N from wastewater.

SOURCE: Data from CA SWRCB (1997, 2008).

high flow can cause wastewater collection networks to leak through overflow and seepage during transit.

Sewage systems overflows (SSO) and sewage exfiltration (leakage) cause wastewater to escape into the surrounding soil and potentially reach surface waters or leach into groundwater (Wakida and Lerner, 2005). Between 1970 and 2011, there were 11,084 SSO incidents reported throughout California (CA SWRCB, 2011). Only 10% of the sewage was recovered and 84% or approximately 141 million liters reached surface waters (CA SWRCB, 2011). Overflows are most significant when the untreated wastewater enters sensitive water systems, which can affect aquatic systems and potable water supply. Common causes of SSO are infiltration and inflow of stormwater, and blockages by grease, debris, or plant material. Sewage exfiltration is more difficult to identify or quantify because it tends to occur below ground. Leakage may range anywhere from 1% to 25% of N transport (Viers et al., 2012).

Once sewage reaches the POTW, it may undergo physical, chemical, and/or biological treatment. The type and extent of wastewater treatment processes employed has a large effect on nutrient removal and the final N load of the effluent (Table 3.4). Broadly, the technologies can be grouped into primary, secondary, and tertiary treatment (Tchobanoglous et al., 2013). During primary treatment, a portion of the floating and settleable solids is removed through screening and/or sedimentation in clarifiers. Secondary treatment converts wastewater organic matter into new bacterial cells and carbon dioxide. The greatest potential to remove N from wastewater occurs during the secondary treatment processes. However, many large wastewater treatment plants do not remove nitrogen and instead control the treatment process to prevent nitrification, resulting in high-effluent ammonium concentrations. To remove N during secondary treatment, an increase in retention time and energy for aeration is needed to accomplish nitrification, followed by denitrification in anoxic zones. Thus, the removal of N requires a more intensive secondary treatment process, which is referred to as biological nutrient removal (BNR). To maintain a steady-state secondary process, microbial cells must be removed periodically. These cells, along with the primary solids, are collectively called "sludge" and are removed for further processing (see discussion of biosolids below). Tertiary treatment aims to remove any remaining suspended materials following secondary treatment using filtration. Tertiary treatment is most often performed to meet regulatory requirements for water reuse projects and does not have a significant impact on effluent N content.

It is important to remember that nitrification/denitrification transform a significant portion of wastewater N into N_2 and other nitrogenous gases. N_2 gas is the main product of these processes, with more than 90% of the N being volatilized in this form. However, N_2O is produced as a by-product of incomplete conversion by denitrifying microbes. Consequently, utilizing nitrification/denitrification increases emissions of this climate forcing gas while achieving the goal of reducing the N load in wastewater. A recent study of wastewater treatment in California shows that treatment for N removal increases N_2O production from ~0.5% of the N in influent to as much as 2%. However, the authors also recommend making comparisons to N_2O emissions from high-N wastewater subject to primary and secondary treatment only, which was not available in this study (Townsend-Small et al., 2011b).

The amount of N in effluent discharged from POTW depends on the level of treatment, be it primary, secondary, or tertiary, and the conditions of biochemical controls. The efficacy of N removal in wastewater treatment processes is related to the availability of carbon, temperature, alkalinity, use of anoxic zones, solid retention time, dissolved oxygen, and hydraulic retention time (EPA, 2008a). By using advanced secondary treatment, effluent levels can be well below 10 or even 2 mg L^{-1} NO_3^--N.

Following processing, wastewater effluent may be reused for various applications or, more commonly, discharged to surface waters or applied to land. For small POTWs, the specific effluent dispersal scheme will depend on the location of the POTW and time of year. However, nearly all large POTWs discharge to surface waters, including rivers and lakes for inland systems, and to the ocean for coastal cities. By one estimate, 49,227 Mg of solids and 5,110 million liters of effluent each day are discharged directly into the ocean (Hauser et al., 2010). Most of the ocean discharge is from the Los Angeles (38%) and San Diego (33%) regions. Many coastal wastewater facilities do not remove N prior to ocean discharge. However, inland POTWs are being scrutinized because of the realization, by the public, that wastewater effluent is being discharged into rivers and lakes that are key water supplies for downstream communities, a practice known as "unplanned indirect potable reuse" (Asano et al., 2007).

Biosolids consist of primary and secondary solids from centralized POTWs and sludge removed from septic tanks, known as septage. As a result of increasing population, the generation and use of biosolids (processed sludge) is increasing in California. In 1988, it was estimated that 339,450 dry Mg were produced, while in 2009 more than 650,000 dry Mg were generated, a 91% increase over a 20-year period. Most of the biosolids are produced at 10% of the POTWs within Region 4—Los Angeles—producing nearly 40% of the state total in 1988, 1991, and 1998 (CASA, 2009). These reports also suggest the use of biosolids is changing. In 1988, 60% of biosolids went to landfills, while in 2009 more than 61% were applied to land. While the application of biosolids to land is controversial, in part due to the past practice of combining industrial wastes with domestic and commercial sources, it does represent an important opportunity for recycling organic N back to soil systems, and thereby could also reduce the need for synthetic fertilizer.

3.6.1.2. TRENDS IN WASTEWATER N AND TREATMENT

The concentration of N in wastewater is predictable and correlated with the size of the population. The population census can therefore be used as a reasonable proxy for wastewater N, greatly enhancing our knowledge of trends in wastewater N impacts. According to the 2010 census, California is now home to more than 37 million people. Much of the growth has occurred since the middle of the last century. Ten million people lived in California in 1950, up from less than 2 million in 1900. By 2020, California's population is estimated to reach 42–48 million. Assuming direct proportionality and a constant percentage of persons serviced by POTWs, the quantity of wastewater N produced in California has increased more than twofold over 60 years. Over this period, diets have been changing, which affects N concentrations in wastewater, and population growth has largely resulted in more developed areas, which are usually connected to centralized treatment systems. The increase in population and increased protein consumption suggest that the estimate of a tripling of wastewater N processed by POTWs since 1950 is conservative.

Reports suggest California facilities are treating wastewater to the highest standard in history. Between 1997 and 2008, the percentage of facilities using advanced secondary and tertiary processing increased from 7% to 15% and 18% to 20%, respectively (Table 3.4). As described in the 2007–2008 report (CA SWRCB, 2008), nearly 80% of processed wastewater receives at least secondary treatment and 50% of the total flow potentially receives advanced secondary and tertiary treatment.

Though the trend seems to indicate enhanced N removal, it is a challenge to estimate the true impact of wastewater management on N at POTWs. Facilities report the levels at which they have the capacity to treat wastewater and the amount of flow they are capable of treating. Neither the proportion of wastewater nor the extent to which it is treated is reported. Furthermore, standard N removal relies on the biological mediated process of nitrification and denitrification, processes very sensitive to environmental conditions—e.g., carbon and oxygen availability and temperature. Because of fluctuating conditions through time, wastewater processed with the same unit process at the same facility will have variable effluent N concentrations.

3.6.2. On-site Wastewater Treatment Systems (OWTS)

Developments in remote areas and some industrial sites cannot be connected economically to sanitary sewer infrastructure. These facilities utilize OWTS, sometimes referred to as septic systems, to treat wastewater prior to discharge. (The term "septic system" refers to widespread use of septic tanks for low-maintenance primary solids removal.) Between 1970 and 1990, the percentage of California's population using OWTS declined from 12.2% to 9.8% (United States Census Bureau, 2012). Despite this relative decline, 28% more people (1.09 million) reported using septic systems in 1990 due to population growth. In 2002, it was estimated that approximately 10% of California's population, about 3.5 million people, relied on OWTS to treat wastewater and about 12,000 new OWTS are set up each year (Leverenz et al., 2002; SWRCB, 2015).

Historically, a septic tank provided the only treatment prior to land application from OWTS, usually by subsurface infiltration. Because only a small fraction of wastewater N accumulates in the sludge in septic tanks, the effectiveness of the system for the treatment of N depends on the physical, chemical, and biochemical characteristics of the soil (EPA, 2002). The basic model for soil-based N removal from septic tank effluent is adsorption of ammonium on clay particles around the dispersal system, nitrification when unsaturated conditions develop, and denitrification under saturated conditions that occur with the next hydraulic load (e.g., flush of wastewater). Thus, nitrogen removal is compromised under certain circumstances, including sandy soils, high groundwater areas, and in saturated systems.

At 10% of California's wastewater, OWTS account for only a small share of the total N cycle in California (e.g., 17.4 Gg N yr^{-1}, Chapter 4). In situations where OWTS function improperly, sewage discharges N directly into the surrounding environment. OWTS N in these areas may be a local concern (Boehm et al., 2009; Walters et al., 2011). In 2001, a survey of 47 California jurisdictions with 912,949 individual sewage systems issued 4,831 repair permits, a median of 0.5% of the operating systems (CWTRC and EPA, 2003).

Modern on-site systems have been engineered to utilize the same processes used in centralized treatment systems to convert wastewater NH_4^+ into an inert gas, nitrification and denitrification. A variety of these OWTS are available. Nitrification and denitrification can either be performed in conjunction in a single unit or in segregated units. In the single-stage process, aerobic and anoxic decomposition takes place within the same reactor. Periods of aeration alternate with periods without aeration to accomplish nitrification and denitrification. The availability of carbon (as an electron donor) is the primary limitation of N removal in single-stage treatment. The effectiveness of single-stage systems range between 40% and 65%, and the efficacy of N removal can reach 75% if effluent is recycled back into the reactor. In the two-stage unit, nitrification occurs in a separate location from denitrification. Moderating pH during the nitrification stage and providing an electron donor in the second stage are concerns with these systems. However, if operated properly, two-stage systems achieve high levels of efficiency of N removal (60–95%). Theoretically, modern OWTS can achieve high levels of effluent quality, similar to that of centralized facilities, but the vast majority do not. As with POTWs, OWTS must provide the requisite conditions to sustain biological treatment mechanisms. Under intermittent management and sewage flow, treatment conditions are typically not optimized. Realized N removal efficiency of advanced, well-maintained systems typically is only 40–60%, well below the efficiency of POTWs that treat for N (EPA, 2008a; Leverenz et al., 2002).

Between 70% and 80% of N in OWTS influent is from human excrement (Lowe et al., 2009). The remainder of the N mass is a function of consumer chemical and product use and food preparation. The isolation of waste streams with unique characteristics facilitates tailored management of the N properties of each. Source separation of wastewater is an emerging strategy in Europe for nutrient recovery from domestic sewage. However, the cost of retrofitting infrastructure, toilets, and domestic pipes is a limiting factor at this time.

Because of lack of control and other challenges associated with incidental N removal in the soil, engineered N removal systems are required in some areas. The effluent quality requirements for on-site systems are based on site-specific considerations, mostly concerned with leaching and accumulation of nitrate in groundwater. It is anticipated that groundwater quality regulations will increase use of OWTS designed for N removal (e.g., SB 885).

3.7. Land Use, Land Cover, and Land Management

Public and private entities modify land use, cover, and management practices. Each conversion implies a unique type of change to the physical characteristics of a given parcel. Land use change refers to a shift between two different classes of use (e.g., among agricultural, natural, or development). Changes in land cover denote transformations of the surface material (e.g., from forest to grassland). Perhaps the most common changes are those where land use and land cover change simultaneously (e.g., grasslands to agriculture). Land management, a less often discussed third category, does not necessarily change use or cover. It is included here because it typically alters the intensity of N fluxes and flows (e.g., increased N fertilizer use with more intensive agriculture or increased fuel use with exurban development).

Land use, cover, and management decisions affect N dynamics in at least two ways. First, they alter the magnitude and speed of N cycling because the magnitude of N inputs, the potential for specific transformations, and the likelihood of certain N loss pathways differ considerably between the original and the derivative state (Table 3.5). The last effect is particularly significant because it means that landowner choices determine not only N dynamics on the piece of land itself but also how it interacts with the wider N cycle. For example, agricultural areas tend to be sources of NO_3^- to groundwater, while urban areas tend to emit N compounds into the atmosphere. Various land uses alter the entry point of reactive N compounds into the environmental systems.

The importance of transitions among land uses, cover, and managements for N cycling and the environment has only recently become appreciated and remains poorly characterized in California. Viers et al. (2012) demonstrate the potential impact. Examining trends in land use area, crop mix, yields, and N fertilization rates since 1945, the authors' analysis indicates that conversion of natural areas into intensive agriculture of the Tulare Lake Basin has contributed to higher NO_3^- levels in the aquifer. These estimates are consistent with the hypothesis that land use change in California has the potential to increase nonpoint source pollution (Charbonneau and Kondolf, 1993). However, changes in land use, cover, or management do not necessarily lead to greater N loading to the environment. Between 1971 and 2001, there was a 31% increase in effluent volume pumped into oceans in Southern California as a result of development, yet mass emissions of NH_4^+ decreased 18% (Lyon and Stein, 2008). Improved management at large POTWs mitigated development's impact. Historical land use, cover, and management shifts have caused massive changes to N input, exports, and storage in California's landscape. The net effect depends on factors that interact in ways that are difficult to predict. Quantification of major land use, cover, and management trends is a first step in

TABLE 3.5
Relative magnitude of N flows on different land uses

	Land use class				
	Urban core	Suburban–exurban	Crop agriculture	Animal agriculture	Natural areas
Nitrogen sources					
Deposition	+++	++	++		++
Fertilizer (all sources)	++	++	+++	+	
Food	+++	++			
Feed	+	+		+++	
Nitrogen exports					
Food			+++	+++	
Wastewater	+++	++			
Manure				+++	
NH_3	+	+	+	+++	
NO_x	+	+	++	++	
N_2O	+	+	+++	++	
NO_3^-	+	++	+++	++	
Flow control process					
Biological activity	+	++	+++	+++	+++
Human engineering	+++	++	+	++	
Soil infiltration	+	++	+++	+++	

NOTE: Symbols are relative both within row and within column (on a per unit area basis).

SOURCE: Expert opinion.

understanding the potential consequences of changes in California's landscape.

3.7.1. Developed Areas

Developed areas of California have been expanding over the past 40 years. Between 1973 and 2000, developed areas increased their land base by 37.5% and now account for 4.2% of California's total area (Table 3.6). Over the same time period, regions experienced a variety of development patterns. Development declined by 5.0% in the East Cascades and Foothills. In the Southern California Mountains, it increased 44.8%. Population density has risen with the expansion of developed areas but growth rates are variable depending on the city. The number of people per square kilometer in Fresno and Redding rose by 187% and 382%, respectively, between 1970 and 2010. Larger cities grew less rapidly. The Sacramento population rose 87% and South San Francisco grew 41% over the same time period (United States Census Bureau, 2013). Not surprisingly, data show that California has become more urban and populous in

the last 40 years. A question relevant for this discussion becomes, what was lost as a result?

Expansion of developed areas came at the expense of agricultural and natural areas. Reconstructions from satellite imagery between 1973 and 2000 show that 3,884 km² of agricultural land, grasslands, and shrubland have been developed (Sleeter et al., 2010). The relative proportion of the converted land shifted over time. Development was largely on agricultural land between 1973–1980 and 1992–2000, with 697 and 470 km² converted, respectively (Sleeter et al., 2010). Conversion of agricultural land to development has reached double-digit growth rates in some regions since the early 1980s (CDC LRP, 2007). During the two intervening periods, development occurred more on grasslands and shrublands than agricultural lands with 448 km² (1980–1986) and 1,037 km² (1986–1992) converted.

Growth of developed areas radically modifies the N cycle. Development increases N imports from food, fertilizer, and fuels, creating N hot spots. Urban expansion replaces plant cover, often those of agricultural or natural lands, with a

TABLE 3.6
Land-use change throughout California (%), 1973–2000

California

Year	Developed	Forest	Grassland/shrub	Agriculture	Mechanically disturbed	Non-mechanically disturbed
1973–1980	9.2	−1.0	−0.5	0.2	−29.4	151.9
1980–1986	6.6	0.3	−0.4	0.0	57.8	−51.0
1986–1992	11.0	−1.3	−0.3	−1.9	98.9	3.5
1992–2000	6.4	−2.1	−1.2	0.7	−25.8	457.4
1973–2000	37.5	−4.1	−2.4	−1.0	64.3	611.7

Coast Range

Year	Developed	Forest	Grassland/shrub	Agriculture	Mechanically disturbed	Non-mechanically disturbed
1973–1980	2.6	−0.6	−0.8	−0.2	24.9	0.0
1980–1986	2.3	76.0	6.5	4.4	2.2	0.0
1986–1992	4.1	74.5	6.8	4.3	3.5	0.0
1992–2000	3.5	2.7	2.9	−1.6	−29.5	−100.0
1973–2000	13.1	−7.0	31.6	−5.3	170.0	0.0

Sierra Nevada

Year	Developed	Forest	Grassland/shrub	Agriculture	Mechanically disturbed	Non-mechanically disturbed
1973–1980	0.0	−0.4	3.0	0.0	−72.6	34.9
1980–1986	0.0	−0.2	0.7	0.0	191.5	−99.6
1986–1992	2.5	−0.8	−0.7	0.0	169.1	25922.2
1992–2000	3.6	−3.5	3.5	0.0	−47.8	949.4
1973–2000	6.1	−4.9	6.7	0.0	12.2	1449.6

Chaparral and Oak Woodlands

Year	Developed	Forest	Grassland/shrub	Agriculture	Mechanically disturbed	Non-mechanically disturbed
1973–1980	10.0	−3.8	−1.7	−2.1	−56.1	519.7
1980–1986	5.9	2.9	0.3	−2.1	107.7	−62.3
1986–1992	7.7	−0.1	0.1	−3.3	68.4	−47.6
1992–2000	6.2	−5.8	−3.3	−1.1	−46.4	898.4
1973–2000	33.1	−6.9	−4.6	−8.3	−17.6	1122.2

Southern California Mountains

Year	Developed	Forest	Grassland/shrub	Agriculture	Mechanically disturbed	Non-mechanically disturbed
1973–1980	12.7	1.0	0.2	−1.8	−88.3	−71.2
1980–1986	9.7	−2.3	−1.2	0.6	117.5	453.6
1986–1992	10.7	0.0	1.5	−3.1	116.4	−75.4
1992–2000	5.7	1.8	−1.6	−0.7	11.6	110.2
1973–2000	44.8	0.4	−1.1	−4.8	−38.8	−17.8

East Cascades and Foothills

Year	Developed	Forest	Grassland/shrub	Agriculture	Mechanically disturbed	Non-mechanically disturbed
1973–1980	−22.6	−0.9	1.8	0.7	46.7	−100.0
1980–1986	3.6	0.2	−0.2	−0.2	5.6	0.0
1986–1992	8.0	−1.3	−0.5	1.1	93.9	
1992–2000	9.7	−1.5	1.5	1.0	−3.1	−100.0
1973–2000	−5.0	−3.5	2.7	2.5	191.2	−100.0

Mojave Basin and Range

Year	Developed	Forest	Grassland/shrub	Agriculture	Mechanically disturbed	Non-mechanically disturbed
1973–1980	12.5	0.0	−0.2	−8.9	−9.0	0.0
1980–1986	12.2	−3.5	−0.5	6.1	59.7	0.0
1986–1992	36.8	0.0	−1.9	−2.2	194.2	
1992–2000	3.6	0.0	0.0	−6.4	−6.1	−100.0
1973–2000	79.0	−3.4	−2.6	11.5	301.9	

Klamath Mountains

Year	Developed	Forest	Grassland/shrub	Agriculture	Mechanically disturbed	Non-mechanically disturbed
1973–1980	3.5	−0.3	7.4	0.0	−64.4	−38.8
1980–1986	2.4	0.5	−4.0	0.2	61.4	−96.9
1986–1992	2.3	−0.8	−1.9	0.4	41.6	32654.5
1992–2000	3.5	−0.1	4.2	0.6	−43.4	−20.6
1973–2000	12.2	−0.7	5.4	1.2	−53.9	386.6

(continued)

TABLE 3.6 (continued)

Central Valley

Year	Developed	Forest	Grassland/shrub	Agriculture	Mechanically disturbed	Non-mechanically disturbed
1973–1980	9.9	−5.7	−8.1	1.0	75.8	0.0
1980–1986	5.5	−0.7	−5.6	0.6	−17.5	0.0
1986–1992	8.0	−0.7	3.8	−1.7	48.0	0.0
1992–2000	9.8	−2.1	−11.4	1.2	106.3	0.0
1973–2000	37.7	−8.9	−20.2	1.1	342.7	0.0

Sonoran Basin and Range

Year	Developed	Forest	Grassland/shrub	Agriculture	Mechanically disturbed	Non-mechanically disturbed
1973–1980	4.1	0.0	−0.1		650.0	0.0
1980–1986	35.3	−9.3	0.0	0.0	−70.0	0.0
1986–1992	0.5	10.2	0.0	0.0	23.8	0.0
1992–2000	1.9	0.0	0.0	0.0	25.6	0.0
1973–2000	44.3	0.0	−0.2		250.0	0.0

NOTE: Statewide, the land dedicated to agriculture has declined only slightly, 1%, while developed area has increased 38%. The rate of conversion and specific conversions among land uses is region specific. In short, develop refers to land covered with built structures and impervious surfaces; forests have greater than 10% tree cover; grassland/shrubs have at least 10% of grasses, forbs, or shrubs; agriculture includes croplands and confined livestock areas; mechanically disturbed are transition areas such as clear-cuts or human-induced changes; nonmechanical disturbed are transition areas caused by natural phenomenon such as fire, wind, or flood. See original source for descriptions of each land cover class.

SOURCE: Reprinted from Sleeter et al. (2010).

Cities and their populations significantly influence N cycling (Grimm et al., 2000; Pickett et al., 2008). Transferring food, fiber, fuel, and industrial materials from the surrounding landscape into more densely inhabited settlements causes a large influx of N to concentrate in cities. Once imported, fundamental components of the built environment—roads, buildings, waste handling facilities, and engineered drainage—have a profound impact on how N is used, processed, and transported (Kaye et al., 2006; Kennedy et al., 2007). Infrastructure advertently and inadvertently changes N dynamics in cities, transforming it (e.g., wastewater treatment plants), storing it (e.g., landfills), and shifting its location (e.g., impervious surface). Which environmental system ultimately receives the previously imported urban-N and the N composition depends on policies, processing, and disposal activities (Bernhardt et al., 2008). For example, approximately two-thirds of N in California wastewater is dumped into the Pacific Ocean from coastal cities, while inland urban areas generally treat wastewater N due to regulations limiting land and freshwater N disposal (Section 3.6).

Understanding the impact of cities on N cycling in California is desperately needed. Consequences of N use range from freshwater pollution in drainages from lawn fertilizer to species endangerment due to wastewater discharge. A systematic examination of city-N cycling for a diverse range of cities is clearly warranted to create ideas on how to mitigate N transfers and pollution because the impacts of cities on N dynamics can be counterintuitive. For example, one might imagine that high-density growth would decrease vehicle miles traveled and reduce NO_x as a result. However, the opposite seems to be true.

Evidence from two California cities shows there is no relationship between urban planning and vehicle miles traveled, demonstrating a paradox (Melia et al., 2011).

Though not formally codified, the current and historical importance of California cities to the state's N cycle is apparent. Today, the vast majority of Californians live in urban areas. According to the 2010 US Census, 36.4 million people lived in urban areas in the state, more than 97% of the total population and approximately double the urban population in 1970 (United States Census Bureau, 2013). Thus, it stands that changes in the N cycle resulting from activities used to support the livelihoods of most Californians can be attributed directly (for example, with fossil fuel emissions from the small vehicle fleet), or indirectly (as with food production), to cities. Food production, in particular, demonstrates the power of cities to affect N cycling in distant regions. A large fraction of food consumed in the state is imported from beyond the state's borders, despite the net food balance being relatively small and positive. Assuming population geography and N dynamics continues along the same trajectory as in the past 10 years (i.e., business as usual), the impact of urban areas will continue to grow. Urban population grew 10% between 2000 and 2010 (from 33.1 to 36.4 million people), while the rural population increased 6% (from 796,198 to 845,229) (United States Census Bureau, 2013). With an estimated population of 50 million people living in California in 2050 and almost 49 million of them living in urban areas, demand to support their everyday activities and reduce the harm of the N influx will be enormous. Indeed it may well be nearly 50% greater than apparent today.

built environment (Box 3.4). Natural hydrologic and soil processes are altered or arrested. The extent of impervious surfaces and drainage increases, though the magnitude depends on the type of development—high-density, suburban, or exurban. Expansion of engineered structures results in efficient collection and conveyance of N around the landscape. N accumulated on pavement moves in stormwater runoff, trimmed grass becomes green waste, and waste discarded by human becomes sewage or trash. All eventually are deposited and stored within the urban areas (e.g., landfill) or exported beyond its boundaries (e.g., into the Pacific Ocean or local streams in California). The high concentration of N in wastes has the tendency to saturate and overwhelm the receiving environment's buffering capacity and can cause local and regional environmental contamination (Groffman et al., 2004).

3.7.2. Agriculture

Relocation and intensification are two dominant processes shaping California agriculture. Agricultural relocation significantly affects N cycling since N flows and turnover in agricultural systems are generally much larger than that in natural areas.

When faced with urban encroachment, farm operators have transferred their operations to new locations in new regions. Displacements of dairy and citrus producers from the Chino Basin and Los Angeles to the San Joaquin Valley are two examples from the 1970s. More recently, vineyards have been spreading into the foothills of the Sierra Nevada or the North Coast, replacing grasslands and oak woodlands. Merenlender (2000) shows that more than 4,500 ha of grapes were planted in 7 years (1990–1997) in Sonoma

County alone (almost 25% of the total). Farmers often take the opportunity to change management practices by updating technology when they move (Hart, 2001).

Relocation has brought only a small decline in the agricultural land base despite urban encroachment. Estimates based on USDA Agricultural Census Data and remote sensing suggest that there has only been about 1% reduction in agricultural area statewide since the early 1970s (Hart, 2003; Sleeter et al., 2010). The statewide balance may be deceiving, however. Some regions have lost most of their agricultural heritage to development. Others, such as the Imperial Valley, have seen considerable growth in agriculture. Agriculture has generally moved from prime locations with high-quality agricultural soils and water access to more marginal lands. According to the CDC LRP (2007), average annual rates of decline of "prime farmland" and "farmland of statewide importance" in their surveyed regions were 21% and 9%, respectively, between 1984 and 2006. Shifting production to farmland of lesser quality may have negative but also counterintuitive effects on N cycling processes. Marginal lands typically are steeper and have thin, erodible soils, and may require more N fertilizer. The combination of these factors would likely increase the potential for N loading to the surrounding environment. Since at least 1993, this indirect consequence of agricultural relocation has been recognized in California (Charbonneau and Kondolf, 1993), but this hypothesis is difficult to test. Interactions between the environment and management practices make generalization difficult, though N loss is probable without significant adjustments in management.

Conversely, agricultural intensification (a change in land management and sometimes cover) presents one of the clearest effects on N cycling. The most obvious result of agricultural intensification is increased N fertilizer use. California croplands are becoming more N intensive; an average of 25% more N fertilizer was applied per crop per acre in 2005 versus 1973 (Section 3.2.1). For the most part, this increased N use has brought increases in yield (Section 3.2.3). Moreover, average application rates differ by an order of magnitude among widely cultivated species. For example, wine grapes receive an average of less than 30 kg N ha^{-1}, while celery receives closer to 300 kg N ha^{-1}. Plant N uptake regularly exceeds 100 kg N ha^{-1} and can be as high as 250 kg N ha^{-1}. Because of the difference in plant N demand, changes in crop mix will alter total statewide crop N use. Over the last 35 years, California's crop mix has shifted heavily from field crops that often receive less N fertilizer to more N-intensive species, e.g., vegetables and nuts. As of 2008, field crops are still grown on the majority of croplands (figure 3.4), but the land area dedicated to field crops declined from 74% to 53% between 1970 and 2007. Fruits and vegetables are now grown on a nearly equivalent amount of land (53% vs. 47%). The shift in crop production towards N-intensive crops is at least partially responsible for greater N consumption in the state. Animal production has become more intensive too, with significant implications for the N cycle. As discussed, animals require N-rich feed and excrete N-rich manures (Section 3.3). Demand for animal feeds is responsible for a greater amount of N entering California's terrestrial biosphere than fertilizer used on crops and lawns, when summing N fixed by biological and synthetic means. Not only does feed production dictate N dynamics in the state (e.g., alfalfa and field corn), but it also influences N cycling in other regions of the United States. Approximately one-third of the N fed to California animals is grown elsewhere. By changing feed demand (and increasing dependence on off-farm feeds), animal production in California indirectly contributes to N fertilizer use concerns in other regions including the Mississippi River Basin. Larger animal populations create more N-rich waste, although the relationships are not proportionate to the number of animals due to changes in NUEs over time.

3.7.3. Other Land Uses: Forestry, Wetlands, Grasslands, and Shrublands

N cycling in natural lands is at a much lower magnitude than that of intensive agricultural production. However, because of the extent of natural areas, the total effects are significant.

Forests, grasslands, and shrublands accumulate and emit N compounds. Many naturally occurring and exotic plant species in these areas of California have the capacity to form symbiotic relationships and biologically fix nitrogen, in much the same way as in croplands; however, not all BNF results from symbiotic relationships. Free-living N fixers also are common and it is estimated that approximately 10% of statewide BNF in natural lands may result via this mechanism (Chapter 4). The actual amount of fixation, symbiotic or free-living, is sensitive to soil N availability. Hence, with increasing rates of atmospheric N deposition, N fixation in many areas may be being suppressed, lowering the total influence of this mechanism.

Simultaneous to N being added to the system, N is lost through gaseous and solution emissions. Land cover change processes in natural lands can have acute impact on N cycling. Wildfires are an important example of this in California. During combustion, N contained in the biomass and litter is released to the atmosphere (Sugihara et al., 2006). Airborne N can either be redeposited on the landscape or transported away from the site with air currents, depending on environmental conditions. Incomplete combustion of materials will result in some N remaining in the partially burned biomass. If the fire burns hot enough, N contained in soil organic matter can be volatilized in gaseous N forms as well (Neary et al., 1999). Wildfires change relationships between soil C and N. Lower soil C:N ratios that follow wildfires stimulate N mineralization, causing N to be converted from organic to inorganic forms and released into the soil where it is predisposed for loss. It can either be transported off-site as NH_4^+ by soil erosion or leach downward through the soil profile after it is transformed to NO_3^-.

The degree of N loss is related to a wildfire's intensity. When wildfires burn at high temperatures, e.g., between 400°C and 500°C, 75–100% of N is lost; at cooler temperatures, e.g., less than 200°C, only small amounts of N are lost (DeBano et al., 1979; Wohlgemuth et al., 2006). The relationship between temperature and N loss is partially the consequence of more complete and rapid combustion of aboveground biomass. The amount of N contained in the biomass (and the latent potential to be released) depends on plant species and density. For a mixed-conifer forest, Nakamura (1996) estimates that approximately 10% of the total system N (706 kg ha^{-1}) is contained in the biomass. To put this in perspective, complete loss of this N would be more than an order of magnitude greater than soil N emissions from the most intensive cropping systems (assuming 10% gas losses and 600 kg N ha^{-1}). Or put another way, the impact on air quality of 1 ha severely burned is greater than 10 ha of the most intensive crop use. Wildfire intensity is also correlated with fuel load, fuel type (e.g., shrubs, litter, trees, logging slash, fallen woody material), and the vertical and horizontal continuity of fuels. Fuel loads in California have been increasing due to periodic droughts, fire suppression, and, in some cases, invasive species. Increasing annual precipitation in some areas of the central and northern California mountains may also be leading to more fine fuels growth. Together, these factors make the probability of ignition and fire spread more likely and increase the potential intensity of the fire.

Recently the area burned by wildfire in California has increased. Research conducted as part of the 2010 Forest and Range Assessment Program (FRAP, 2010) indicates that between 1950 and 2008, the area burned by wildfires averaged 128,000 ha yr^{-1} but ranged between 12,400 and 548,000 ha yr^{-1}, a 44-fold difference. Even with high annual variation, recent trends (1990–2008) indicate the extent of wildfires is increasing statewide. Evidence from the Sierra Nevada, Cascades, and Klamath Mountains supports this conclusion and shows considerable increases in mean area burned since the beginning of the 1980s (Miller et al., 2008, 2012). The 3 years that had the largest area burned all took place in the last decade (2003, 2007, and 2008). And the trend will likely not abate. Modeling efforts agree that fire activity and intensity are likely to increase over the next 50–100 years (Hayhoe et al., 2004; Lenihan et al., 2003, 2007; Miller and Urban, 1999). Past wildfire, however, has not been equally distributed across ecosystems. Shrubland wildfires have always been the most common, but there has been an exponential increase in burning in conifer forests since the turn of the century (figure 3.6). The increased extent and future projections of wildfires suggest this driver has and will continue to exert pressure on air and water resources.

3.8. Universal Historical Increases but Future Uncertainty

In this chapter, we introduced the six activities and processes that drive N cycling processes in California and traced historical trends in activity levels. Data clearly show that the intensity of the activities regulating N cycling in California has increased. The consequence of universal intensification has undoubtedly been a greater perturbation of California's N cycle and more total N released in the environment, on balance. But the impacts are uneven. Certain N emissions have been tempered dramatically, despite increased use (e.g., NO_x emissions from fuel combustion). Others such as NO_3^- losses from croplands have seen contrasting trends. Despite the likelihood of continued increases in activity levels well into the future, impacts are highly uncertain. Currently ongoing technological and policy discussion will undoubtedly change the trajectory of their future impact. Technological and policy responses that address critical control points of these direct drivers are discussed in Chapters 7 and 8, respectively.

CHAPTER FOUR

A California Nitrogen Mass Balance for 2005

Lead Authors:

D. LIPTZIN AND R. DAHLGREN

Contributing Author:

T. HARTER

What Is This Chapter About?

A mass balance of nitrogen inputs and outputs for California was calculated for the year 2005. This scientifically rigorous accounting method tracks the size of nitrogen flows which allows us to understand which sectors are the major users of nitrogen and which contribute most to the nitrogen in the air, water, and ecosystems of California. New reactive nitrogen enters California largely in the form of fertilizer, imported animal feed, and fossil fuel combustion. While some of that nitrogen contributes to productive agriculture, excess nitrogen from those sources contributes to groundwater contamination and air pollutants in the form of ammonia, nitric oxides, and nitrous oxide. In addition to statewide calculations, the magnitude of nitrogen flows was examined for eight subsystems: cropland; livestock; urban land; people and pets; natural land; atmosphere; surface water; and groundwater. Understanding the major nitrogen contributors will help policy-makers and nitrogen users, like farmers, prioritize efforts to improve nitrogen use.

Stakeholder Questions

The California Nitrogen Assessment engaged with industry groups, policy-makers, nonprofit organizations, farmers, farm advisors, scientists, and government agencies. This outreach generated more than 100 nitrogen-related questions which were then synthesized into five overarching research areas to guide the assessment (figure 1.3). Stakeholder-generated questions addressed in this chapter include the following:

- What are the relative contributions of different sectors to N cycling in California?
- What are the relative amounts of different forms of reactive nitrogen in air and water?

- Are measurements of gaseous losses and water contamination accurate?

Main Messages

Synthetic fertilizer is the largest statewide import (519 Gg N yr^{-1}) of nitrogen (N) in California. The predominant fate of this fertilizer is cropland including cultivated agriculture (422 Gg N yr^{-1}) and environmental horticulture (44 Gg N yr^{-1}). However, moderate amounts of synthetic fertilizer are also used on urban land for turfgrass (53 Gg N yr^{-1}).

Excretion of manure is the second largest N flow (416 Gg N yr^{-1}) in California. The predominant (72%) source of this N is dairy production, with minor contributions from beef, poultry, and horses. A large fraction (35%) of this manure is volatilized as ammonia (NH_3) from livestock facilities (97 Gg N yr^{-1}) and after cropland application (45 Gg N yr^{-1}). However, there is limited evidence for rates of ammonia volatilization from manure. While liquid dairy manure must be applied very locally (within a few kilometers of the source), the solid manure from dairies and other concentrated animal feeding operations can be composted to varying degrees and transported much longer distances (>100 km). However, because of the increased regulation of dairies in the Central Valley (see Chapter 8), it will soon be possible to determine what fraction of dairy manure is used on the dairy farm compared to what is exported based on the nutrient management plans produced for each dairy.

Synthetically fixed N dominates the N flows to cropland. Synthetic fertilizer (466 Gg N yr^{-1}) is the largest flow of N to cropland, but a large fraction of N applied in manure and irrigation water to cropland is also originally fixed synthetically. On average, we estimated that 69% of the N added annually to cropland statewide is derived from synthetic fixation.

The biological N fixation that occurs on natural land (139 Gg N yr⁻¹) has become completely overshadowed by the reactive N related to human activity in California. While this flow was once the major source of new reactive (i.e., biologically available) N to California, it now accounts for less than 10% of new imports at the statewide level. The areal rate (8 kg N ha⁻¹ yr⁻¹) representing the sum of all N inputs to natural lands, including N deposition, is an order of magnitude lower than either urban or cropland.

The synthetic fixation of chemicals for uses other than fertilizer is a moderate (71 Gg N yr⁻¹) N flow. These chemicals include everyday household products such as nylon, polyurethane, and acrylonitrile butadiene styrene (ABS) plastic. These compounds have been tracked to some degree at the national level (e.g., Domene and Ayres, 2001), but the true breadth and depth of their production, use, and disposal is poorly established.

Urban land is accumulating N. Lawn fertilizer, organic waste disposed in landfills, pet waste, fiber (i.e., wood products), and non-fertilizer synthetic chemicals are all accumulating in the soils (75 Gg N yr⁻¹), landfills (68 Gg N yr⁻¹), and other built areas associated with urban land (122 Gg N yr⁻¹).

Nitrogen exports to the ocean (39 Gg N yr⁻¹) from California rivers account for less than 3% of statewide N imports. In part, this low rate of export is due to the fact that a major (45%) fraction of the land in California occurs in closed basins with no surface water drainage to the ocean. While concentrations of nitrate in some rivers can be quite high, the total volume of water reaching the ocean is quite low.

Direct sewage export of N to the ocean (82 Gg N yr⁻¹) is more than double the N in the discharge of all rivers in the state combined. Because of the predominantly coastal population, the majority of wastewater is piped several miles out to the ocean. A growing number of facilities (>100) in California appear to be using some form of N removal treatment prior to discharge.

Nitrous oxide (N₂O) production is a moderate (38 Gg N yr⁻¹) export pathway for N. Human activities produce 70% of the emissions of this greenhouse gas, while the remainder is released from natural land. Agriculture (cropland soils and manure management) was a large fraction (32%) of N₂O emissions in the state.

Ammonia is not tracked as closely as other gaseous N emissions because it is not currently regulated in the state. While acute exposures to NH₃ are rare, both human health and ecosystem health are potentially threatened by the increasing regional emissions and deposition of NH₃. However, rigorous methods for inventorying emissions related to human activities as well as natural soil emissions are currently lacking.

Atmospheric N deposition rates in parts of California are among the highest in the country, with the N deposited predominantly as dry deposition. The Community Multiscale Air Quality (CMAQ) model predicts that 66% of the deposition is oxidized N and 82% of the total deposition is dry deposition not associated with precipitation events. In urban areas and the adjacent natural ecosystems of southern California, deposition rates can exceed 30 kg N ha⁻¹ yr⁻¹, but deposition is, on average, 5 kg N ha⁻¹ yr⁻¹ statewide.

The atmospheric N emitted as NOₓ or NH₃ in California is largely exported via the atmosphere downwind (i.e., east) from California. Approximately 65% of the NOₓ and 73% of the NH₃ emitted in California is not redeposited within state boundaries, making California a major source of atmospheric N pollution. Further, atmospheric exports of N are more than 20 times higher than riverine N exports.

Leaching from cropland (333 Gg N yr⁻¹) was the predominant (88%) input of N to groundwater. It appears that N is rapidly accumulating in groundwater with only half of the annual N inputs extracted in irrigation and drinking water wells or removed by denitrification in the aquifer. On the whole, groundwater is still relatively clean, with a median concentration ~2 mg N L⁻¹ throughout the state. However, there are many wells in California that already have nitrate concentrations above the Maximum Contaminant Level (10 mg NO₃⁻-N L⁻¹). Because of the time lag associated with groundwater transport (decades to millennia), the current N contamination in wells is from past activities and current N flows to groundwater will have impacts far into the future.

The amount of evidence and level of agreement varies between N flows. The most important sources of uncertainty in the mass balance calculations are for major flows with either limited evidence or low agreement or both. Based on these criteria, biological N fixation on cropland and natural land, the fate of manure, denitrification in groundwater, and the storage terms are the most important sources of uncertainty.

In many ways, the N flows in California are similar to other parts of the world. In a comparison with other comprehensive mass balances—the Netherlands, the United States, Korea, China, Europe, and Phoenix—California stands out in its low surface water exports and high N storage, primarily in groundwater and urban land. Further, when compared to these other regions of varying size, California has a relatively low N use on both a per capita and, especially, on a per hectare basis.

4.0. Using a Mass Balance Approach to Quantify Nitrogen Flows in California

Human activities, including agriculture and urban development, have led to dramatic increases in biologically available or reactive nitrogen (N). As such, the anthropogenic alteration of the N cycle is emerging as one of the greatest challenges to the health and vitality of California's people, ecosystems, and agricultural economy. Input of N to terrestrial ecosystems has more than doubled in the past century due to nitrogen fixation associated with food production and energy consumption (Galloway, 1998). This mobilization of anthropogenic N has been connected with increased N loading to aquatic ecosystems, emissions of nitrous oxide (a greenhouse gas), and associated ecosystem and human-

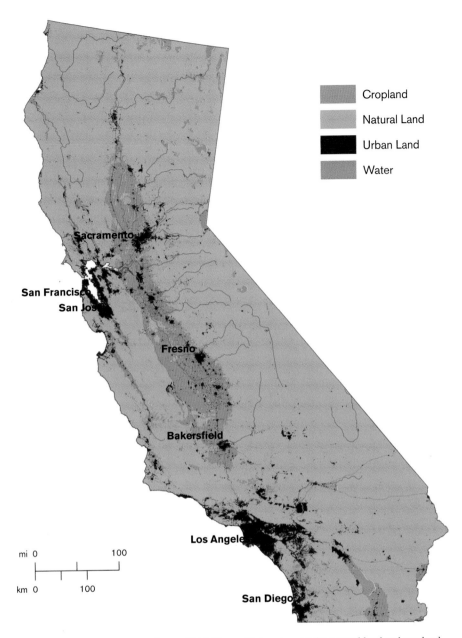

MAP 4.1. Land cover map of California, 2005. The multiple categories for natural land and cropland were lumped for display purposes (data from Hollander, 2010; map by R. Murphy).

and 2008 were used. The N flows were calculated by compiling the necessary data from both peer-reviewed and non-peer-reviewed literature, government databases, and in some cases expert opinion. When possible, we calculated multiple independent estimates of the N flows during this time period. A quantitative measure of uncertainty is reported in Section 4.1 as part of the estimates of N flows.

The concurrent goals of this mass balance were (1) to quantify current statewide N flows and (2) to evaluate the scientific uncertainty in the magnitude of N flows. The assessment provides a summary of the statewide N imports and exports and the N flows in the eight subsystems. Both the absolute and relative sizes of the N flows were grouped

into categories to help highlight which flows are particularly important (Box 4.2). A detailed description of data sources and calculations used in the mass balance are available in the Appendices 4.1 and 4.2. The spatial and temporal variability of important stocks and flows of N will be addressed in detail in the Ecosystem Services and Human Well-Being chapter (Chapter 5).

Uncertainty in the mass balance is addressed in this chapter as well as in the Data Tables. The discussion in this chapter largely focuses on comparing multiple independent estimates of the same N flow. In the Data Tables, we concentrate on the uncertainties associated with individual data sources and methodologies. Following the model of the

BOX 4.1. LANGUAGE USED TO CATEGORIZE DIFFERENT N FLOWS

Import	Flows of nitrogen entering the state
Export	Flows of nitrogen leaving the state
Input	Flows of nitrogen entering a subsystem from another subsystem within the state
Output	Flows of nitrogen leaving a subsystem to another subsystem within the state
Storage	Nitrogen that remains (i.e., stored) within a subsystem

BOX 4.2. LANGUAGE FOR DESCRIBING ABSOLUTE AND RELATIVE N FLOWS*

Absolute flow (Gg N yr⁻¹)	*Flow category*	*Relative flow (%)*	*Fraction category*
<1	Insignificant	1–10	Small
1–25	Minor	10–25	Medium
25–100	Moderate	25–50	Large
>100	Major	>50	Predominant

*We adopted a specific vocabulary for categorizing the absolute and relative magnitudes of N flows in the mass balance chapter. A gigagram (Gg) is equivalent to 1 million kilograms.

Intergovernmental Panel on Climate Change, we use reserved words to quantify the level of scientific agreement and the amount of evidence (Box 4.1; Data Tables). The uncertainties associated with each N flow depicted in figure 4.1 are presented both in figure 4.2 and in the various tables showing the state-level and subsystem mass balances, and in the Data Tables.

4.1. Statewide and Subsystem N Mass Balances

This section describes the magnitude of the N flows at the statewide level as well as the eight subsystems examined in the mass balance: cropland, livestock, urban land, people and pets, natural land, atmosphere, surface water, and groundwater. For the statewide flows of agricultural products, we report net flows in the cases of food, feed, and fiber, and not the transport of individual commodities. We calculate the net flow as the difference between production and consumption. Based on our results, feed and fiber represent statewide imports of N and food represents a statewide export of N. At the statewide level, the California atmosphere was considered internal to the system with advection resulting in N import to and export from the atmosphere.

4.1.1. Statewide N Flows

There were six moderate to major statewide imports of N to California: synthetic N fixation, fossil fuel combustion, biological N fixation, atmospheric imports (i.e., advection of N), feed, and fiber in the form of wood products (figure 4.1). Products created from synthetic N fixation by industrial processes, typically by the Haber–Bosch process, represent the largest statewide import (590 Gg N yr^{-1}) and a large (36%) fraction of the new statewide imports (Table 4.1a; figure 4.3a). Of this synthetically fixed N, the predominant (88%) form was fertilizer. However, the mixture of other chemicals (e.g., nylon, polyurethane, ABS plastic) created from synthetically fixed NH_3 also represented a moderate (71 Gg N yr^{-1}) N flow. Fossil fuel combustion was the second largest import (404 Gg N yr^{-1}) of N to California, with NO_x the predominant (89%) form. This flow represents N emissions to the atmosphere and is not equivalent to atmospheric N deposition in California (Section 4.1.7). Biological N fixation was also a major statewide N import (335 Gg N yr^{-1}) with more occurring on the 400,000 ha of alfalfa compared to the 33 million hectares of natural land. While there was medium evidence for this flow, there was low agreement. The import of livestock feed and fiber in the form of wood and wood products to meet the demand in California represented major (200 Gg N yr^{-1}) and moderate (40 Gg N yr^{-1}) statewide imports of N, respectively.

To satisfy the mass balance assumption, statewide N exports and storage were defined to be equivalent to N imports at the statewide level. Atmospheric exports of N gases and particulate matter were estimated based on the assumption of no N storage in the atmosphere. All nitrous oxide (N_2O) and nitrogen gas (N_2) emitted was assumed to be exported from California while the export of NO_x and NH_3 was calculated as the difference of emissions and deposition to the land subsystems. The atmospheric N export (NO_x, NH_3, N_2O, and N_2) accounted for the predominant (86%) fraction of the N exports from California (figure 4.3b, Table 4.1b). More NO_x (270 Gg N yr^{-1}) than NH_3 (206 Gg N) was exported. Nitrous oxide was a moderate (38 Gg N yr^{-1}) statewide export of N, while N_2 represented a major statewide export (204 Gg N yr^{-1}). This total includes groundwater denitrification even though the N_2 produced may not reach the atmosphere for several decades until the groundwater is discharged at the surface. Including groundwater denitrification, the inert N_2 emissions account for 29% of the gaseous N export from the state. While most of the NO_x export was related to the N import related to fossil fuel combustion, the export of the other gaseous forms represents N that was transformed within the state. For example, a large fraction of the NH_3 derives from manure, which previously was feed,

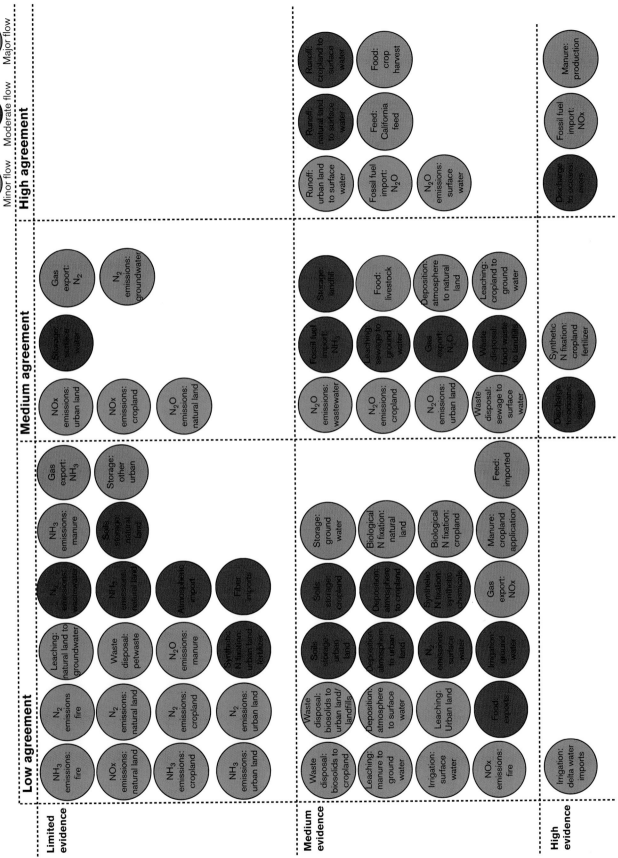

FIGURE 4.2. Measuring uncertainty in the California nitrogen mass balance. This figure reflects the amount of evidence and level of agreement for the various flows of nitrogen covered in the mass balance. Flows represent inputs and outputs as well as transfers of nitrogen within California. For a more complete catalog of evidence and agreement among data sources, see Tables 4.1, 4.2, 4.6, 4.8, 4.9, 4.11, 4.12, 4.14, 4.16, and Appendices 4.1 and 4.2.

TABLE 4.1A
TABLE 4.1A

California statewide nitrogen mass balance for 2005: imports

Nitrogen flow	Methods section*	Flow direction	Flow (Gg N yr⁻¹)	Evidence	Agreement
Fossil fuel combustion					
NO_x	4.2.1	Import	359	High	High
NH_3	4.2.1	Import	36	Medium	Medium
N_2O	4.2.1	Import	9	Medium	High
Atmospheric import	4.2.1	Import	40	Limited	Low
Biological N fixation					
Natural lands	4.2.3	Import	139	Medium	Low
Cropland	4.2.3	Import	196	Medium	Low
Synthetic N fixation					
Fertilizer	4.2.4	Import	519	High	Medium
Chemicals	4.2.4	Import	71	Medium	Low
Feed	4.2.5	Import	200	Medium	Low
Fiber	4.2.5	Import	40	Limited	Low
Delta water imports	4.2.9	Import	8	High	Low

*See Appendix 4.2.

TABLE 4.1B

California statewide nitrogen mass balance for 2005: exports and storage

Nitrogen flow	Methods section*	Flow direction	Flow (Gg N yr⁻¹)	Evidence	Agreement
Food	4.2.5	Export	79	Medium	Low
Gas export					
NO_x	4.2.2	Export	270	Medium	Low
NH_3	4.2.2	Export	206	Limited	Low
N_2O	4.2.2	Export	38	Medium	Medium
N_2	4.2.2	Export	204	Limited	Medium
Discharge to ocean					
River	4.2.9	Export	39	High	High
Sewage	4.2.7	Export	82	High	Medium
Storage					
Groundwater	4.2.10	Storage	258	Limited	Medium
Other storage	4.2.11	Storage	443	Medium	Medium

*See Appendix 4.2.

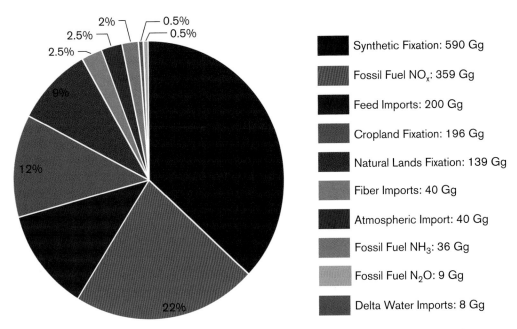

Synthetic Fixation: 590 Gg

Fossil Fuel NO_x: 359 Gg

Feed Imports: 200 Gg

Cropland Fixation: 196 Gg

Natural Lands Fixation: 139 Gg

Fiber Imports: 40 Gg

Atmospheric Import: 40 Gg

Fossil Fuel NH_3: 36 Gg

Fossil Fuel N_2O: 9 Gg

Delta Water Imports: 8 Gg

FIGURE 4.3a. Statewide nitrogen imports to California in 2005 (1,617 Gg N yr^{-1}). Synthetic fixation is the largest single import of N to California, contributing 37% of the total. Fossil fuel combustion adds N in the form of NO_x (22%), NH_3 (2%), and N_2O (0.5%) to the atmosphere. Biological N fixation is an N input in both cropland (12%) and natural land (9%). The net import of agricultural products is a source of N in the form of feed (12%) and fiber (2.5%). Minor sources of N import are (0.5%) water pumped by the Central Valley Project and State Water Project in Tracy because it occurs in the Delta downstream of the river gauges and (2.5%) import of reactive N gases in the atmosphere from across the Pacific Ocean.

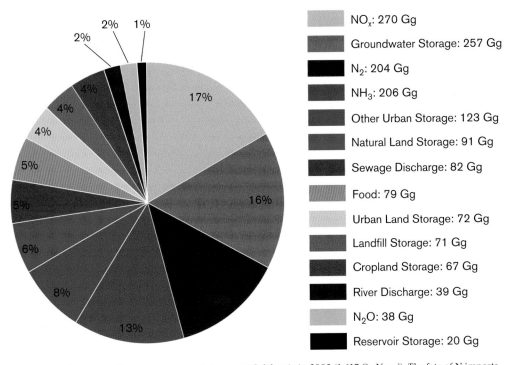

NO_x: 270 Gg

Groundwater Storage: 257 Gg

N_2: 204 Gg

NH_3: 206 Gg

Other Urban Storage: 123 Gg

Natural Land Storage: 91 Gg

Sewage Discharge: 82 Gg

Food: 79 Gg

Urban Land Storage: 72 Gg

Landfill Storage: 71 Gg

Cropland Storage: 67 Gg

River Discharge: 39 Gg

N_2O: 38 Gg

Reservoir Storage: 20 Gg

FIGURE 4.3b. Statewide nitrogen exports, and storage in California in 2005 (1,617 Gg N yr^{-1}). The fate of N imports to California is almost divided between exports (56%) and storage (44%). Atmospheric export is the dominant fate with NO_x (17%), NH_3 (13%), N_2O (2%), and N_2 (13%) accounting for almost half of the N imports. Nitrogen is exported to the ocean in much smaller amounts from rivers (2%) and as sewage (5%). The net food balance contributes 5% of the N export. Groundwater (16%) is the single largest fate as storage with the various other forms of storage in soils and urban environments combining to account for 28% of the total N import.

which in turn may have been imported to the state. California was a net exporter of food. That is, the total production of N in food was 79 Gg N yr^{-1} greater than the estimated consumption of N in food. The gross flow of food is likely significantly higher with many fresh fruits and vegetables as well as dairy products transported out of the state. Moderate statewide exports of N to the ocean occurred in both rivers (39 Gg N yr^{-1}) and direct sewage discharge (82 Gg N yr^{-1}).

A large (43%) fraction of the N imports were stored in some form in California (701 Gg N yr^{-1}). Accumulation of N in groundwater was estimated to be 258 Gg N yr^{-1}, with the input predominantly from cropland. Storage in the soils or vegetation of the three land subsystems was estimated to be 230 Gg N yr^{-1}. Within the urban subsystem, there was N storage associated with landfills (71 Gg N yr^{-1}), but a major (122 Gg N yr^{-1}) source of storage was related to the buildup of synthetic chemicals and wood products in structures and long-lived household items like nylon carpets, electronic equipment, and lumber. Finally, storage in surface water bodies (i.e., lakes and reservoirs) was 20 Gg N yr^{-1}. We assumed no storage in the atmosphere subsystem.

There are some examples of measured increases in N storage in California, but there is more evidence related to carbon storage. Agricultural soils in California (DeClerck and Singer, 2003) and turfgrass soils (Raciti et al., 2011) have been shown to accumulate both carbon (C) and N. Ornamental lawns in southern California were found to accumulate 1,400 kg C ha^{-1} yr^{-1} for more than three decades after lawn establishment (Townsend-Small and Czimczik, 2010). Assuming a soil C:N ratio of 10, this would represent 140 kg N ha^{-1} yr^{-1}, similar to the results of N accumulation reported by Raciti et al. (2011) for Maryland. In other contexts, storage of N can be inferred from measurements of carbon storage. For example, the increasing acreage of perennial crops in California (Kroodsma and Field, 2006) results in net uptake of carbon by ecosystems in California (Potter, 2010). The disposal of organic materials like wood products and food waste in landfills results in 10% of the total dry mass of solid waste sequestered in the form of C (Staley and Barlaz, 2009). Depending on the chemical environment in the landfill and the C:N ratio of the materials, varying amounts of N would be accumulating as well. With these multiple avenues for C sequestration, it is very likely that N storage would be increasing as well in these settings. Some of these storage pools (soils and vegetation) have an asymptotic capacity for N uptake which may be saturated within years or decades. However, the disposal of waste in landfills and the use of long-lived wood and synthetic materials can potentially keep increasing over time. The high capacity for retention of N in surface water bodies is well established especially in reservoirs (e.g., Harrison et al., 2008), but the fraction buried in sediments versus the fraction denitrified is not well known.

Nitrogen flows can also be tracked through the land-based subsystems: cropland, urban land, and natural land. Because of the N flows among subsystems, the total sum of N inputs across all subsystems was greater than the statewide N imports. For example, manure N was an input to the cropland subsystem, but not an import to the state as it was considered a transformation of existing N at the scale of California. Agriculture, including cropland and livestock, dominated the N inputs in California (figure 4.4). Cropland had greater N inputs than urban land and natural land combined. Similarly, livestock feed was more than double the amount of human and pet food. The two biggest inputs to the three land subsystems were synthetic N fertilizer (to cropland and urban land) and manure (to cropland). Less than half of the N inputs to cropland and one quarter of the N inputs to livestock were converted into food or feed (figure 4.4). More than a third of cropland N inputs were leached to groundwater and a similar fraction (40%) of livestock N inputs was emitted as ammonia. Other gaseous N emissions from cropland and the other land subsystems were only minor N flows. Human food was largely converted to sewage with the exception of food waste that was disposed of in landfills. While natural land and cropland were estimated to store small fractions of their N inputs, the predominant fate of N inputs to urban land was storage in soils, landfills, or as long-lived synthetic materials or wood.

4.1.2. Cropland N Flows

Cropland covers only 4.9 million of the 40.8 million hectares in California, but accounts for a disproportionate amount of the N flows (Table 4.2; figure 4.5). A total of 1,027 Gg N yr^{-1} was added to cropland, resulting in an average areal N input to cropland of 250 kg N ha^{-1} yr^{-1}.

4.1.2.1. CROPLAND N IMPORTS AND INPUTS

The use of synthetic fertilizer on cropland represented the largest flow of N in California (figure 4.5; Table 4.2). The 2002–2007 average statewide synthetic N fertilizer sales were 762 Gg N yr^{-1}. However, it is unclear why there was nearly a 50% increase in sales from 2001–2002 or similarly a 50% increase from the 1980–2001 mean to 2002–2007 mean fertilizer sales (Box 4.3). There was no significant linear change ($p = 0.28$) in fertilizer sales over the 1980–2001 period. We believe that the mean from this period, 519 Gg N yr^{-1}, provides a more realistic estimate of statewide fertilizer sales than the 2002–2007 mean. The fraction of fertilizer sales applied to cropland was calculated as the difference between turfgrass use (53 Gg N yr^{-1}; see Section 4.1.4) and total fertilizer sales. Synthetic fertilizer use was therefore a major flow of N (466 Gg N yr^{-1}), representing a large (45%) fraction of total N flows to cropland soils. Manure application was also a major (263 Gg N yr^{-1}) N input to cropland (see Section 4.1.3). A large uncertainty is related to the partitioning of manure between gaseous NH_3 losses and application to cropland (figure 4.2). In our accounting methodology we only considered synthetic N applied as fertilizer as an N import (i.e., new input of N) in the budget calculations at the statewide scale. However, many of the other sources of

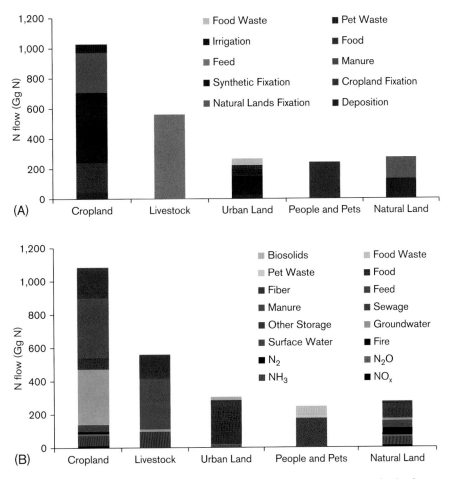

FIGURE 4.4. Summary of nitrogen imports/inputs (A) and exports/outputs/storage (B) for the three California land subsystems in 2005. The flows to and from the livestock subsystem (i.e., feed) and the people/pets subsystem (i.e., food) are shown for comparative purposes, but these subsystems are calculated independently from the land subsystems.

N to cropland (e.g., manure, irrigation, atmospheric deposition) also originally derive in part from synthetic fertilizer applied to cropland. We assumed that half of the biosolids produced in the state were applied to cropland soils.

Synthetic fertilizer applied to cropland can also be estimated based on the crop-specific fertilization rates and harvested acreages. For the 2002–2007 period, cultivated crops were estimated to receive 539 Gg N yr^{-1}. This value would be expected to be higher than the synthetic fertilizer sales data for cropland if any manure was used as fertilizer. A total of 263 Gg N yr^{-1} of manure was estimated to be applied to cropland. If 73 Gg N yr^{-1} of manure was used instead of synthetic fertilizer, then the two estimates would agree perfectly.

Synthetic fertilizer use in environmental horticulture was calculated separately because it relied on different sources of data. There were 7,100 ha of sod, 6,200 ha of floriculture, and 13,100 ha of open grown nursery stock which were estimated to receive 44 Gg N yr^{-1}.

Biological N fixation was also a major (196 Gg N yr^{-1}) flow to cropland and almost entirely associated with alfalfa (Table 4.3). We were not aware of any N fixation rates for alfalfa measured in California, where productivity, and thus N fixation, is much higher than the Midwestern states where data have been collected. While there was variability associated with the productivity–N fixation relationship, the biggest source of uncertainty in the estimate of N fixation is the amount of fixed N belowground.

Two moderate N flows to cropland are atmospheric deposition and N applied in irrigation water. The total atmospheric deposition of N to cropland was estimated at 43 Gg N yr^{-1} based on the results of the CMAQ model (Table 4.2). The mean N deposition rate for cropland, 8.7 kg N ha^{-1} yr^{-1}, was higher than the state average of 5.0 kg N ha^{-1} yr^{-1}. Irrigation water provided a similar quantity of N (59 Gg N yr^{-1}) to cropland statewide as N deposition. Surface water was withdrawn at a rate of 2.6 × 10^{13} L yr^{-1} for irrigation use in California in 2000 (Hutson et al., 2004). In 2000, a total of 0.6 × 10^{13} L yr^{-1} was pumped from the Sacramento–San Joaquin Delta (the Delta) at Tracy for the Delta Mendota Canal and the California Aqueduct (Governor's Delta Vision Blue Ribbon Task Force, 2008). Because the Delta pumps are located downstream of the location of river

TABLE 4.2
California cropland nitrogen mass balance in 2005

Nitrogen flow	Methods section*	Flow direction	Flow (Gg N yr⁻¹)	Evidence	Agreement
Biological N fixation	4.2.3	Import	196	Medium	Low
Deposition	4.2.2	Input	43	Medium	Low
Synthetic fertilizer	4.2.4	Import	466	High	Medium
Manure application	4.2.6	Input	307	Medium	Low
Biosolids	4.2.7	Input	11	Medium	Low
Irrigation					
Groundwater	4.2.9	Input	33	Medium	Low
Surface water	4.2.9	Input	18	Medium	Low
Delta	4.2.9	Import	8	High	Low
Gas emissions					
NO	4.2.8	Output	12	Limited	Medium
NH$_3$	4.2.8	Output	60	Limited	Low
N$_2$O	4.2.8	Output	10	Medium	Medium
N$_2$	4.2.8	Output	17	Limited	Low
Feed	4.2.5	Output	357	Medium	High
Fiber	4.2.5	Output	1	Medium	High
Food	4.2.5	Output	185	Medium	High
Runoff	4.2.9	Output	41	Medium	High
Leaching	4.2.10	Output	333	Medium	Medium
Soils	4.2.11	Storage	65	Medium	Low

NOTE: All flows were calculated independently except soil storage which was calculated by difference. However, there is independent evidence suggesting increases in cropland soil storage. This term may also include storage in perennial crops.

*See Appendix 4.2.

gauges (which we consider to be the boundary of the study area), this pumping resulted in the return of 8 Gg N yr⁻¹ to the state. The remaining surface water withdrawals for irrigation, calculated as the difference between total surface water use and Delta pumping, provided another 18 Gg N yr⁻¹ to cropland. Groundwater nitrate (NO_3^-) concentrations (0.572 mg NO_3^--N L⁻¹) were even higher than the N concentration in the water pumped from the Delta. However, only 1.3×10^{13} L yr⁻¹ were pumped from groundwater in 2000 for irrigation, resulting in a total of 33 Gg N yr⁻¹.

4.1.2.2. CROPLAND N OUTPUTS AND STORAGE

Harvesting crops was a major flow of N and the largest N output from the cropland subsystem. The top 20 crops in terms of harvested N are shown in Table 4.4. For 2002–2007, total harvest of food crops was 185 Gg N yr⁻¹ and feed crops was 357 Gg N yr⁻¹ (Table 4.2). Cotton lint was the only fiber crop grown on California cropland (timber was considered harvested from natural land), with only 1 Gg N yr⁻¹ harvested. The production of nursery and floriculture crops was

14 Gg N yr⁻¹. While there is transport of this nursery material in and out of California, we estimate that CA produces 14% of the national total and would use 12% based on its population resulting in no net flow of nursery material.

The total production showed minimal variability over this time period with less than a 10% difference between the lowest (2002) and highest (2003) quantity of N harvested. The two sources of crop acreages, the county Agricultural Commissioners and National Agricultural Statistics Service (NASS) annual surveys, were highly correlated ($r > 0.95$) for the common crops that are reported by both agencies. The largest source of uncertainty in the crop calculations is in the conversion of production to the N content of the biomass. The USDA crop nutrient tool is a compilation of data from several decades ago, but no more recent database exists. The potential for large errors are greatest for the forage crops where the whole plant is harvested and for the vegetables with high water content.

Gases were emitted from cropland soils as a result of both physical and biological processes. Ammonia volatilization is a physical process based on the temperature and pH-

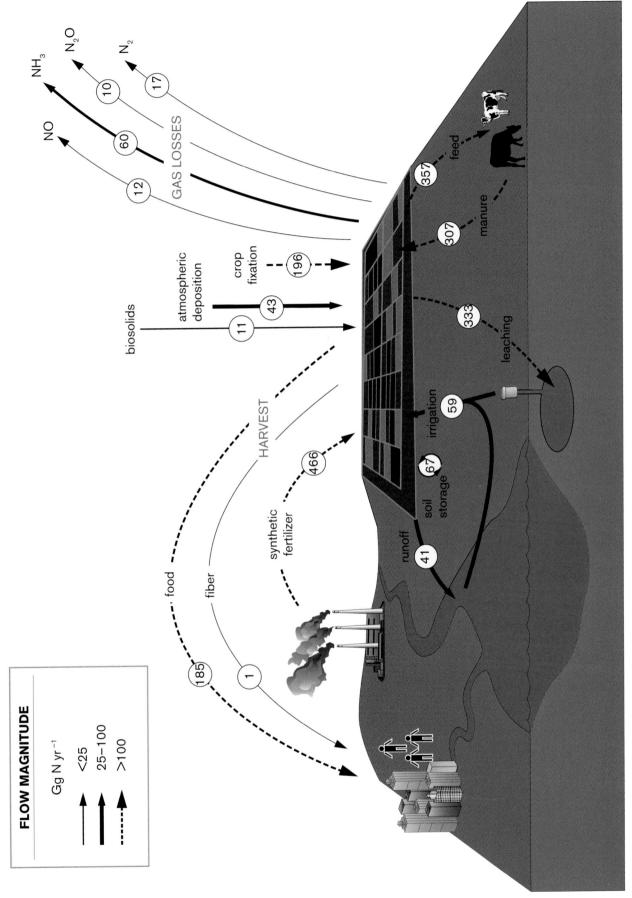

FIGURE 4.5. Flows of nitrogen in California cropland in 2005. The circled values indicate the absolute magnitude of the flow in Gg N yr⁻¹ with arrow thickness specifying the relative magnitude of the flow. Storage terms are indicated with arrows on the circled values.

BOX 4.3. THE PROBLEM OF FERTILIZER ACCOUNTING

Nitrogen (N) fertilizer use is virtually certain to be the largest single flow of N in California. In general, fertilizer sales data are assumed to be the best proxy for fertilizer use. In California, fertilizer sales are tracked by the California Department of Food and Agriculture (CDFA). The values are calculated twice per year and compiled and published in semiannual Tonnage Reports (CDFA, 2013). The data are reported by licensed fertilizer dealers to track sales of fertilizer to unlicensed buyers. This system is designed to prevent double counting. Currently the licensed dealers fill out a "tonnage report for commercial fertilizers" form for every county where they have sales indicating the tonnage of all types of fertilizer. There are 17 listed grades of N fertilizer (e.g., anhydrous ammonia [NH$_3$], ammonium nitrate [NH$_4$NO$_3$], urea [CO(NH$_2$)$_2$]) along with an "other" category without a grade listed. A similar list is used for phosphorus and potassium fertilizers as well as other agricultural minerals. The tonnage of each fertilizing material sold in each county by each licensed dealer for farm use is thus recorded. Farm use is defined as "Commercial Use, Farm, Golf Courses, Professional Landscaping, Not Home and Garden." Nonfarm use sales of "Registered Specialty Fertilizer/Packaged Ag Minerals for Home and Garden Use" is recorded by the ton as well, but only subdivided into tons of dry bulk, dry packaged, liquid bulk, and liquid packaged with no grade associated with the sales. The tonnage reports are compiled by a third party that enters the data manually to convert it into a digital format. The raw data are then processed by Joe Slater, a professor at the University of Missouri, Columbia, into the final tonnage reports.

A few possibilities exist to explain the large jump of reported N fertilizer sales in 2002. First, it is possible that there was a change in methodology at this time. We have been unable to get confirmation from CDFA regarding any changes in the reporting form or data processing. Secondly, it is possible that the conversion of tons of fertilizing material into tons of nutrients is a large source of error, as this process requires that the grade of fertilizer (i.e., the nutrient content) be

known. For example, the grade of anhydrous ammonia is 82-0-0, indicating 82% N, 0% P, and 0% K; however, the form does not require a breakdown of the nutrient content for "other" materials. For example in 2002, "other" farm-use fertilizing materials were reported at 395,115 tons. If the N grade was 10, the tonnage would be 39,115 tons, whereas with an N grade of 30, this would represent more than 100,000 tons. At this point, it is not clear what grade was assumed by CDFA for this calculation and what grade the materials actually were. Further, while the 2002 "other" tonnage was almost double the 2001 value, by 2003 the amount was back to within 10% of the 2001 value, suggesting that the grade of "other" fertilizing materials was not the major source of error. Third, it is possible that double reporting could be happening. It is unclear why the sales of anhydrous ammonia increased from a few thousand tons before 2002 to over 100,000 tons starting in 2002. The Calamco ammonia depot in Stockton, where ships bring in synthesized ammonia from around the world, is located in San Joaquin County. This facility sells anhydrous ammonia, aqua ammonia (solution of ammonia in water (NH$_4^+$)(OH$^-$)), and AN-20 (20% ammonium nitrate solution). This facility also supplies J.R. Simplot with ammonia for their N fertilizer production plant. Thus, it is possible that the anhydrous ammonia is being reported sold as well as the other products created from the anhydrous ammonia. Finally, the reporting system for nonfarm fertilizer is problematic for estimating total N tonnage and partitioning fertilizer into farm use and nonfarm use. This is not likely related to the high sales since 2002, as virtually no fertilizer is reported as nonfarm use. One problem with the nonfarm reporting is that there is no grade reported for any materials. One growing source of revenue is potting mixes amended with nutrients. While the tonnage of these materials is required to be reported as they are considered fertilizers, they tend to have an N grade of less than 2. Thus, it is unclear once again how the conversion from tonnage of materials to tonnage of nutrients is calculated.

dependent equilibration of gaseous NH$_3$ and dissolved ammonium (NH$_4^+$) in the soil. Based on an emission factor of 3.2% for the various synthetic fertilizers in California, as well as emissions from land applied manure, NH$_3$ outputs were a moderate flow (60 Gg N yr^{-1}). The other gas outputs are associated with the microbial processes of nitrification and denitrification. Based on the average of all sources of data (Table 4.5), nitric oxide (NO) and N$_2$O outputs were also minor flows (12 and 10 Gg N yr^{-1}, respectively; Table 4.2). Using the limited number of published literature estimates from California cropland soils, the median NO and N$_2$O

fluxes were 1.9 kg NO-N ha^{-1} yr^{-1} and 2.9 kg N$_2$O-N ha^{-1} yr^{-1}, respectively (see Appendix 4.2). These rates are considerably higher than the global median for NO (0.9 kg NO-N ha^{-1} yr^{-1}) and N$_2$O (1.4 kg N$_2$O-N ha^{-1} yr^{-1}) from the largest global compilation of gaseous emissions from cropland soils (Stehfest and Bouwman, 2006). The total emissions of N$_2$O calculated from the California areal rates and cropland area were 14 Gg N yr^{-1}. This value is similar to the estimate of 9 Gg N yr^{-1} using the emissions factor approach. Emissions of nitrogen (N$_2$) gas from soils from denitrification were also a minor flow (17 Gg N yr^{-1}), estimated using a fixed N$_2$:N$_2$O ratio of

TABLE 4.3
Biological nitrogen fixation for agricultural crops in California in 2005

Crop	Acreage (1,000 ha)	Fixation rate (kg N ha⁻¹)	Fixed N (Gg N)	Fixation rate reference
Alfalfa	457	393	180	Unkovich et al. (2010)
Dry beans[1]	83	40	3	Smil (1999)
Fresh beans[2]	11	40	0.4	Smil (1999)
Rice	226	25	6	Smil (1999)
Pasture (clover)	434	15	7	Smil (1999)
Total			196	

1. Includes all dry beans including dry lima beans.

2. Includes snap beans and green lima beans.

TABLE 4.4
Harvested N by crop. The production (Gg N yr⁻¹) and acreage (ha) and N yield (kg N ha⁻¹ yr⁻¹) of the top 20 crops in terms of harvested N

Crop	Harvested N (Gg N yr⁻¹)	Production (%)	Harvested acreage (ha)	Percentage of acreage	N yield (kg N ha⁻¹ yr⁻¹)
Alfalfa hay	187	39.0	457467	12.5	410
Corn silage	41	8.6	179382	4.9	230
Haylage, non-alfalfa	28	5.8	161289	4.4	172
Wheat	27	5.6	205956	5.6	131
Rice	26	5.4	226499	6.2	114
Cotton[1]	23	4.9	264060	7.2	89
Almonds[2]	23	4.7	254527	7.0	89
Tomatoes, processing	16	3.4	125420	3.4	131
Corn grain	9	1.9	56737	1.6	161
Walnuts	8	1.6	94943	2.6	83
Lettuce	7	1.5	99584	2.7	73
Sudan hay	7	1.4	34201	0.9	197
Small grain hay	6	1.3	86360	2.4	73
Grapes	6	1.2	335890	9.2	17
Broccoli	5	1.0	48070	1.3	98
Pistachios	5	1.0	43963	1.2	106
Oranges	4	0.7	78441	2.1	46
Sugar beets	4	0.7	18617	0.5	190
Potatoes	3	0.6	16912	0.5	173
Carrots	3	0.6	22159	0.6	121
Other crops	44	9.2	848788	23.2	52
Total	481	100	3659264	100	132

1. Includes lint and seed.

2. Includes kernels and hulls.

TABLE 4.5
Sources of data for biome-specific NO and N$_2$O fluxes

Source	NO (Gg N yr^{-1}) Cropland	NO (Gg N yr^{-1}) Natural land	N$_2$O (Gg N yr^{-1}) Cropland	N$_2$O (Gg N yr^{-1}) Natural land	Type of data	Spatial extent
Dalal and Allen (2008)				19	Field	Global
Davidson and Kingerlee (1997)	18	8.9			Field	Global
Li et al. (1996)			6.9		Model	California
Potter et al. (1996)	7.4	12		7.9	Model	Global
Stehfest and Bouwman (2006)	4.9	13	5.9	11	Field	Global
Emissions factor			9		Field	Global
California literature	9		14		Field	California
Average estimate	12	11	10	13		

NOTE: Biome-specific NO and N$_2$O fluxes were calculated as the average of several published sources for cropland and natural lands. For the published studies with areal rates by biome (cropland, desert, coniferous forest, hardwood forest, grassland, shrubland), we used the biome areas from California Augmented Multisource Landcover Map. For the emissions factor approach, we assumed 1% of fertilizer (both synthetic and manure) was converted to N$_2$O, like the California Air Resources Board. However, we also included a background cropland emission of 1 kg N$_2$O-N ha^{-1} based on Stehfest and Bouwman (2006) to calculate total, not just anthropogenic N$_2$O emissions. We also compiled published estimates of NO and N$_2$O for California cropland. In the case of N$_2$O we used the median flux across all crops and management practices, while for NO we calculated the mean of the daily flux estimates for the crops measured by Matson et al. (1997).

1.66. Because of the high variability in N$_2$:N$_2$O ratios and the high reported rates measured in California in the 1970s, we estimated a lower and upper bound for the N$_2$:N$_2$O as 1.25 and 2.31 as the mean ± 2 SE of the Schlesinger (2009) dataset. Taking into account the uncertainty, the range of N$_2$ emissions would be 13–23 Gg N yr^{-1}.

Dissolved outputs of N to surface water from cropland were estimated based on a predicted N yield (14 kg N ha^{-1} yr^{-1}). As only 2.9 million hectares of California cropland was located in watersheds with surface water drainage, outputs of N to surface water (i.e., runoff) from cropland was a moderate N flow (41 Gg N yr^{-1}). Kratzer et al. (2011) reported similar N yields for the Central Valley sub-watersheds with the highest fraction of agricultural land, the Orestimba Creek watershed (17.9 kg N ha^{-1}) and the portion of the San Joaquin River near Patterson (16.3 kg N ha^{-1}).

Leaching below the rooting zone was a major flow (333 Gg N yr^{-1}) of N from cropland. This value is the average of two approaches which differ considerably in magnitude (see Appendix 4.2). Multiplying recharge volume by the median concentration of NO$_3^-$ from published studies in California estimating leaching below the rooting zone, cropland leaching was estimated to be 395 Gg N yr^{-1}. In contrast, using the median of the fraction of applied fertilizer (synthetic + manure) that leaches from published studies in California predicted only 272 Gg N yr^{-1} leached. Thus, the level of agreement on the magnitude of N leaching is low. Conditions in the vadose zone in California are not conducive to denitrification (Green et al., 2008a). Therefore, this leached nitrate would be predicted to reach the groundwater table. Like many other fluxes, there was high spatial and temporal variability. However, while it is relatively simple to measure NO$_3^-$ concentrations in leachate, it is more difficult to estimate N load as it also requires an estimate of the recharge volume. One recent estimate of leaching that actually calculated the areal rate of N loading in recharge was nearly 100 kg N ha^{-1} yr^{-1} in an almond orchard near the Merced River (Green et al., 2008b). Based on our statewide total N load and cropland area, we estimated that the average areal rate of cropland N loading would be 68 kg N ha^{-1} yr^{-1}.

Based on the difference between inputs and outputs, soil storage was calculated as 65 Gg N yr^{-1}. There was limited evidence for storage of N in cropland soils in California. Based on the repeated sampling of agricultural soils throughout California, on average, N content in cropland soils increased from 0.09% to 0.29% in the upper 25 cm (DeClerck and Singer, 2003). Assuming no change in bulk density over the 55-year period between samples, cropland soils would accumulate 1 kg N ha^{-1} yr^{-1} for a total of 5 Gg N yr^{-1} statewide. This suggests that the estimate of storage by difference is too high. If we used the estimate of soil storage and calculate leaching by difference, the N flow would be 395 Gg N yr^{-1}, equivalent to the higher of the two estimates for leaching based on recharge volume and concentration. Based on data from Post and Mann (1990), soils will accumulate carbon when carbon concentrations are less than 1% in the top 15 cm, assuming an average soil bulk density of 1 g cm^{-3}. Many of the agricultural areas in the state, with the notable exception of the Delta, were established in areas with relatively low organic matter soils. Therefore, increases in soil N would be expected as well. However, these increases in soil N are not linear over time, with the highest

TABLE 4.6
California livestock nitrogen mass balance in 2005

Nitrogen flow	Methods section*	Flow direction	Flow (Gg N yr^{-1})	Evidence	Agreement
Feed					
California feed	4.2.5	Input	357	Medium	High
Imported feed	4.2.5	Import	200	Limited	Low
Manure	4.2.6	Output	416	High	High
Food	4.2.5	Output	141	Medium	Medium

*See Appendix 4.2.

NOTE: The total amount of feed was calculated as the sum of manure production based on livestock population and the amount of animal food products. Imported feed was calculated as the difference between feed crops harvested in California and the total amount of feed.

TABLE 4.7
Confined livestock populations and manure and animal food products in California in 2005

Class	Inventory (1,000 head)	Annual sales (1,000 head)	Total N requirement (Gg N)	Manure production (Gg N)	Calculated food produced (Gg N)	Food produced as feed and manure (Gg N)
Dairy cow	1,715		351	266	85	86
Dairy heifer	772		42	33		
Dairy calf	772		25	18		
Layers	21,115		12	6	6	7
Beef steer	644		43	32	18	11
Horses	876		32	35		
Turkeys		6,327	4	7	4	−3
Broilers		270,480	35	19	15	15
Pigs		303	3	1	1	2

NOTE: The total N requirement and manure production are population-based estimates based on inventory or sales data. The calculated food produced column is the independent estimate of food N based on the tonnage of animal food products and their N content. For comparison "Food produced as feed and manure" is calculated as the difference between the N requirement and manure production.

increases expected soon after land conversion and saturating after a certain time (e.g., Garten et al., 2011).

4.1.3. Livestock N Flows

The N flows for the livestock subsystem assumed that all of the livestock in the state (with the exception of beef cows and all calves) were on feed.

4.1.3.1. LIVESTOCK FEED

The majority of crop production (357 Gg N yr^{-1}) in California was harvested to feed livestock (Table 4.2). However, this production must be supplemented with another 200 Gg N yr^{-1} of feed imported from out of the state. Corn grain from the Midwest is a major source of livestock feed. The waybill

samples from the Surface Transportation Board suggest that over 300 million bushels of corn arrive in California annually on trains originating in Nebraska and Iowa (USDT, 2013). This feed supply was converted into 141 Gg N yr^{-1} of food and 416 Gg N yr^{-1} of manure (Table 4.6). Dairy cows and replacement stock dominated the demand for livestock feed and manure production, but beef cattle, poultry, and horses contributed a significant fraction as well (Table 4.7).

4.1.3.2. LIVESTOCK MANURE

The majority of the N in livestock feed is excreted. Livestock manure is potentially a nutrient resource, but concentrated quantities can pose a waste disposal problem (Table 4.8). The fraction of manure that is volatilized as NH_3 depends on the type of livestock. Using the US Environmental Protection

TABLE 4.8
Fate of manure nitrogen from confined livestock in California in 2005

Nitrogen flow	Methods section*	Flow direction	Flow (Gg N yr^{-1})	Evidence	Agreement
Manure production	4.2.6	Input	416	High	High
Gas emissions					
NH_3	4.2.6	Output	97	Medium	Low
N_2O	4.2.6	Output	2	Medium	Low
Leaching	4.2.6	Output	10	Medium	Low
Cropland	4.2.6	Output	307	Medium	Low

*See Appendix 4.2

NOTE: Manure production was calculated based on livestock populations as were ammonia (NH_3) emissions. Nitrous oxide (N_2O) emissions were direct emissions from manure prior to land application from the California Air Resources Board greenhouse gas inventory. All manure except dairy manure was assumed to be utilized as a solid. Leaching was calculated based on the fraction leached from dairy facilities reported in van der Schans et al. (2009).

Agency (EPA) emission factors and our estimates of excretion, we calculated an emission of 97 Gg NH_3-N yr^{-1} from livestock facilities. When combined with manure-associated emissions from cropland (45 Gg NH_3-N yr^{-1}), this N flow is almost identical to the reported tonnage of NH_3 in EPA (2004a); however, this is far larger than the value of 69 Gg N yr^{-1} in the 2005 EPA National Emissions Inventory for California (EPA, 2008b). While there is a high amount of evidence for the amount of manure excreted by livestock, there is limited evidence for the fate of that manure. There are relatively few data measuring NH_3 emissions for the management practices and climate specific to California. Therefore, the emissions of NH_3 and the land application of manure are speculative (figure 4.2). However, as the residence time of NH_3 in the atmosphere is relatively short, this reduced N may essentially be land applied from atmospheric deposition downwind of dairy facilities. The underestimate of modeled compared to measured N deposition in the ecosystems on the west slope of the Sierra Nevada could result from an underestimate of NH_3 emissions in the model.

We assumed that all non-volatilized manure was applied to cropland. Dairy manure is unique because it occurs in both solid and liquid form and its disposal is now regulated in the Central Valley. With the 2007 General Order from the Central Valley Regional Water Board, there should soon be information available on the amount of dairy manure applied on the dairy facility and the amount transferred off the dairy. The crops that receive manure in any form and the amount of manure applied are not well established and whether the manure is used more as an organic amendment or a source of nutrients is not clear.

4.1.4. Urban Land N Flows

Urban land covers 2.3 million hectares or 6% of the state. Nitrogen flows of 284 Gg N yr^{-1} correspond to an areal input of 124 kg N ha^{-1}, with much of the N remaining in the soils, structures, and landfills in the urban system (figure 4.6).

4.1.4.1. URBAN LAND N IMPORTS AND INPUTS

Atmospheric N deposition was relatively high in urban areas (11 kg N ha^{-1} yr^{-1}) adding 25 Gg N yr^{-1} (Table 4.9). Synthetically fixed N was a major (124 Gg N yr^{-1}) flow of N and accounted for a large (44%) fraction of the N flow to the urban land subsystem. Synthetic fertilizer use, predominantly for residential, commercial, and recreational turfgrass, was an import of 53 Gg N yr^{-1}. Other synthetic N-containing chemicals, such as resins, plastics (in particular ABS), polyurethane, and nylon, were an input of 71 Gg N yr^{-1} that largely remains in urban landscapes. Wood and wood products (i.e., fiber), though relatively low N content materials, still contribute 51 Gg N yr^{-1}. Finally, a variety of materials such as retail and consumer food waste (54 Gg N yr^{-1}), pet waste (16 Gg N yr^{-1}), and biosolids (11 Gg N yr^{-1}) were added to the urban subsystem (soils or landfills).

4.1.4.2. URBAN LAND N OUTPUTS AND STORAGE

The estimated outputs of N from urban land are relatively minor. Gaseous outputs in all forms, from the fraction of urban areas covered by turfgrass, amounted to only 7 Gg N yr^{-1}. Few data exist on gas fluxes from turfgrass in California. For N_2O, Townsend-Small et al. (2011a) found that the turfgrass direct emission factor ranged from 0.6% to 2.3% of fertilizer inputs, similar to the emission factor for cropland. The literature-based N yield in surface water runoff (5.6 kg N ha^{-1} yr^{-1}) was higher than for natural land areas, resulting in urban runoff being a minor (10 Gg N yr^{-1}) output similar in magnitude to gas outputs.

The vast majority of N entering urban land remains there in some form. Storage occurs in soils (72 Gg N yr^{-1}), landfills

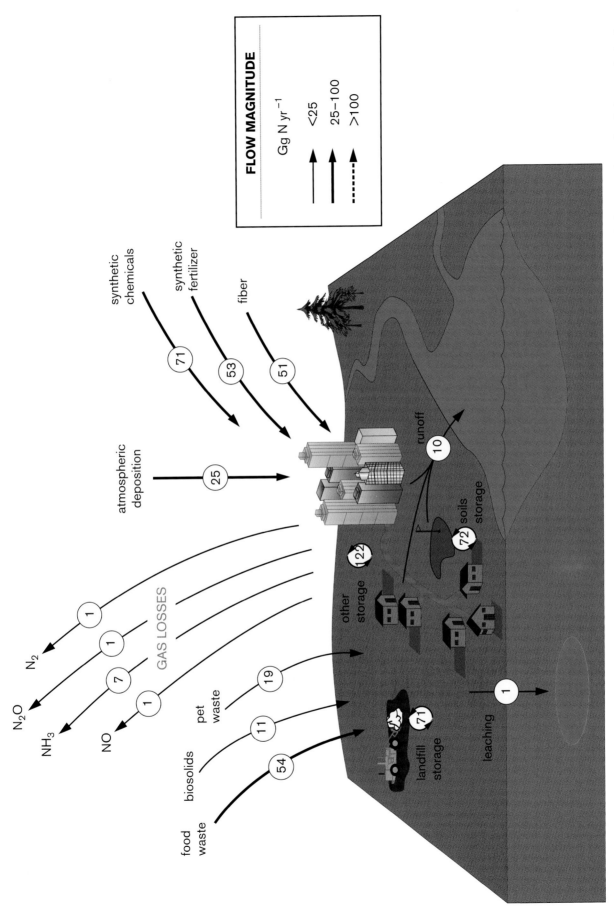

FIGURE 4.6. Flows of nitrogen in California urban land in 2005. The circled values indicate the absolute magnitude of the flow in Gg N yr⁻¹ with arrow thickness specifying the relative magnitude of the flow. Storage terms are indicated with arrows on the circled values.

TABLE 4.9
California urban land nitrogen mass balance in 2005

Nitrogen flow	Methods section*	Flow direction	Flow (Gg N yr^{-1})	Evidence	Agreement
Deposition	4.2.2	Input	25	Medium	Low
Synthetic fertilizer	4.2.4	Import	53	Limited	Low
Synthetic chemicals	4.2.4	Import	71	Medium	Low
Fiber	4.2.5	Import	51	Limited	Low
Food waste to landfill	4.2.7	Input	54	Medium	Medium
Pet waste	4.2.7	Input	16	Limited	Low
Biosolids	4.2.7	Input	11	Medium	Low
Gas emissions					
NO	4.2.8	Output	1	Limited	Medium
NH_3	4.2.8	Output	7	Limited	Low
N_2O	4.2.8	Output	1	Medium	Medium
N_2	4.2.8	Output	1	Limited	Low
Runoff	4.2.9	Output	10	Medium	High
Leaching	4.2.10	Output	1	Medium	Low
Landfill	4.2.7	Storage	71	Medium	Medium
Soils	4.2.11	Storage	72	Medium	Low
Other	4.2.11	Storage	122	Limited	Low

*See Appendix 4.2

NOTE: All terms except soil storage and other storage were calculated independently. Soil storage was calculated as the difference between the inputs of deposition, synthetic fertilizer, and dog waste, and the outputs of gases and runoff to surface water. This storage term may also include storage in perennial vegetation in urban landscapes. Other storage includes the materials that cannot be tracked to landfills. This includes synthetic chemicals and some fiber products.

(71 Gg N yr^{-1}), and the built environment (122 Gg N yr^{-1}). While there are some data related to N storage in landfills, there is limited evidence for most other forms of storage. Turfgrass soils are well known for their capacity to accumulate N in soils for decades (e.g., Raciti et al., 2011), but there are no data for California. Synthetically fixed N in forms other than fertilizer often is used for long-lived components of structures or is disposed of in landfills along with N from food and yard waste like grass clippings. For example, polyurethane resins and nylon carpets will remain in buildings for years to decades. A major use of ABS plastic is for the housing of electronic equipment and in cars. There is no quantitative information on the ultimate fate of these synthetic N-containing chemicals. While plastic disposal to landfills is tracked, there is no information on what fraction of that plastic is ABS. There is also a growing recycling capability for these compounds as technologies for separating materials have improved. Much of the synthetic N and organic N in urban land is eventually disposed of in landfills. Of the known sources to landfills, food waste is the predominant (64%) source of N, but yard waste (e.g., prunings, stumps, leaves, and grass; 14%) and wood products (e.g., lumber; 13%) comprise a medium fraction of the landfill nitrogen disposal (Table 4.10). In the same way that the inputs and outputs of the livestock subsystem were quantified apart from the cropland subsystem, the household (food and waste) subsystem was considered separately from the urban land subsystem. Therefore, food was only considered part of the urban subsystem if it was disposed of in a landfill.

4.1.5. Household N Flows

We assumed that the food supply for humans and their household pets (dogs and cats only) consisted of the same materials.

4.1.5.1. HUMAN FOOD

On average, there was 6.4 kg N yr^{-1} in food available per person in the United States according to the USDA Economic Research Service (USDA NASS, 2012) for 2002–2007. Therefore, with a population of 35.6 million people, a total of 228 Gg N yr^{-1} of food was available for California's human population. Assuming a demographic-based food consumption of 4.9 kg N yr^{-1} per capita (Baker et al., 2001), a statewide total of 174 Gg N was consumed, leaving 54 Gg N, or 23%, as food waste. This is close to the 27% food waste reported by Kantor (1997). With a

TABLE 4.10
Sources of nitrogen to landfills in California in 2005

Material	Tonnes	Moisture (%)	N (%)	Gg N
Paper	7,678,172	20	0.1	6.1
Lumber	3,528,376	15	0.1	3.0
Prunings	836,687	15	0.1	0.7
Stumps	108,867	15	0.1	0.1
Food	5,322,138	74	3.2	44.3
Leaves and grass	1,541,838			9.0
Manure	33,187	72	1.6	0.1
Cat waste				3
Biosolids				11
Total				68

NOTE: With the exception of cat waste and biosolids, which are based on the mass balance, the tonnage of materials sent to the landfill is based on CIWMB (2005). All moisture and N contents are from Rynk (1992) with the exception of food waste (Zhang et al., 2007). The category including leaves and grass was assumed to be equally composed of these two materials. Only food waste and cat waste are considered a new input to urban land, while the other organic materials were already considered part of the urban landscape.

total production of 185 Gg N yr^{-1} of food crops and 141 Gg N yr^{-1} of animal products, we estimated that there was a net export of 79 Gg N yr^{-1} from California.

4.1.5.2. HUMAN WASTE

The analysis of the fate of food was based on three decision points. First, 25% of the available food was not consumed by people, but was disposed of in landfills, while the other 174 Gg N yr^{-1} was excreted and became sewage. Secondly, ~10% of households in California use on-site waste treatment (i.e., septic) for waste disposal instead of centralized wastewater treatment. Based on the literature, we assumed that 9% of septic N would be removed as biosolids, but there is limited evidence for the fate of the other 91%. It is very likely that some N from septic systems is taken up by vegetation near the leach fields or quickly reaches surface water bodies; however, we assumed that all of this N would reach groundwater to maximize the potential impact of septic systems on groundwater N. Finally, the N entering wastewater treatment plants (WWTPs) can be disposed of in liquid form (effluent), solid form (biosolids), or gaseous form (predominantly denitrification to N$_2$).

Because the population of California tends to live along the coast, the predominant (61%) fate of wastewater influent is discharged into the Pacific Ocean (82 Gg N yr^{-1}) (Table 4.11). This includes the discharge from the Sacramento regional and the Stockton regional wastewater treatment plant (WWTP). Even though they discharge into the Sacramento River and San Joaquin River, respectively, their effluent is discharged downstream of the US Geological Survey (USGS) gauges where N concentrations are measured. Only a small amount (12 Gg N yr^{-1}) of wastewater N was discharged into other surface water bodies of California from WWTPs. Discharge of treated wastewater to land (11 Gg N yr^{-1}) that largely leaches to groundwater was a small (9%) fraction of wastewater based on the sum of N from facilities without an NPDES permit but with a "NON 15" land discharge permit from the State Water Resources Control Board. The statewide production of biosolids is estimated to be 22 Gg N yr^{-1}, which we assumed was equally split between application to cropland and use as alternative daily cover at landfills. A fraction of the sewage is converted to gas during wastewater treatment; facilities with advanced secondary or tertiary treatment convert approximately two-thirds of the total N into gaseous forms by denitrification. A small (2 Gg N yr^{-1}) amount of N$_2$O is produced during treatment, but N removal by advanced wastewater treatment produces largely N$_2$. Based on the assumption that half of the N load is converted to gaseous forms by advanced treatment, 16 Gg N yr^{-1} would be emitted from wastewater facilities. If 2 Gg N yr^{-1} were in the form of N$_2$O based on the greenhouse gas inventory, then 14 Gg N yr^{-1} would be emitted as N$_2$.

However, calculating all of the outputs independently results in a discrepancy of 15 Gg N yr^{-1} between sewage input (174 Gg N yr^{-1}) and output pathways (159 Gg N yr^{-1}). This discrepancy could be explained by several potential errors. First, the empirical relationship of effluent N and WWTP design flow is based on NH$_3$ discharge and not total N discharge. Many, but not all, of the WWTPs in the state are required to monitor NH$_3$ concentrations monthly, but the data in most cases are only publicly available in paper form at the regional Water Quality Control Board offices. Further, the other dissolved N forms (NO$_3^-$) and organic N are rarely monitored because the predominant form of

TABLE 4.11
Fate of nitrogen in human food in California in 2005

Nitrogen flow	Methods section*	Flow direction	Flow (Gg N yr^{-1})	Evidence	Agreement
Excretion	4.2.7	Input	174	Medium	Low
Biosolids	4.2.7	Output	22	Medium	Low
Gas emissions					
N_2O	4.2.7	Output	2	Medium	Medium
N_2	4.2.7	Output	29	Limited	Low
Surface water	4.2.7	Output	12	Medium	Medium
Leaching					
Septic	4.2.7	Output	16	Limited	Low
Natural land	4.2.7	Output	11	Limited	Low
Ocean	4.2.7	Output	82	High	Medium

*See Appendix 4.2

NOTE: This table does not include the 54 Gg N yr^{-1} of food waste that ends up in landfills or the 79 Gg N yr^{-1} of food exported from California. The difference between inputs and estimated outputs was accounted for as N_2 loss. For comparison, we estimated that N_2 emission associated with N removal in wastewater facilities was only 14 Gg N yr^{-1}.

discharged N is NH_3 unless the facility uses advanced treatment to remove N. Secondly, we may underestimate the N content of biosolids. The literature values vary widely, but the N content of biosolids in California is not monitored. A third possibility is that there are emissions of N_2 in facilities without advanced wastewater treatment. Finally, the missing N might never have reached the WWTPs. That is, 15 Gg N yr^{-1}, or ~10% of the N in human waste, could be leaking out of sewer pipes into groundwater during the collection process. While the magnitude of N leaking from sewer pipes is difficult to measure directly, the presence of leaky sewer pipes in urban areas is well documented (e.g., Groffman et al., 2004). For the purposes of the mass balance we assumed that the missing N was in the form of N_2, resulting in 29 Gg N yr^{-1} as N_2 instead of the 14 Gg N yr^{-1} calculated based on the amount of N denitrified (Table 4.11).

4.1.5.3. HOUSEHOLD PETS

With 7.0 million dogs and 8.8 million cats in the state, 19 Gg N yr^{-1} of food N was needed to feed household pets. Assuming household pets and humans eat from the same food supply, total food demand was 242 Gg N yr^{-1}. The predominant fate of pet waste was urban soils (12 Gg N yr^{-1}) with some cat waste (3 Gg N yr^{-1}) disposed of in landfills and a minor (4 Gg N yr^{-1}) flow of N emitted as NH_3.

4.1.6. Natural Land N Flows

Natural land covers 33 million hectares, or more than 80% of the area of the state. Total N inputs of 271 Gg N yr^{-1} resulted in an average areal input of 8 kg N ha^{-1} yr^{-1} (figure 4.7).

4.1.6.1. NATURAL LAND N INPUTS

The input from atmospheric N deposition was 132 Gg N yr^{-1} for natural land as reported in Fenn et al. (2010) (Figure 4.7, Table 4.12). This value is based on results from the CMAQ model but modified for several ecosystems that have higher measured than modeled N deposition rates. However, the spatial distribution of N deposition measurements is too sparse statewide to rigorously evaluate the model's results.

Based on the biome-specific approach, biological N fixation in natural land ranged from 139 to 411 Gg N yr^{-1} for an average areal fixation rate of 4–13 kg N ha^{-1} yr^{-1}, depending on the value of relative cover of N fixing species. This total includes nonsymbiotic fixation which is estimated to produce 10% of biologically fixed N. A second approach using the empirical relationship predicting N fixation from modeled evapotranspiration (ET) found statewide natural land N fixation ranged from 59 to 381 Gg N yr^{-1} for an average rate of 2–12 kg N ha^{-1} yr^{-1}. Finally, with the mass balance approach (i.e., outputs minus inputs assuming no change in storage), the statewide N fixation on natural land was estimated at 53 Gg N yr^{-1} or 1.6 kg N ha^{-1} yr^{-1}. The overall range of these values translates into 30–75% of the new reactive inputs of N to natural land and 4–23% of the inputs statewide (Tables 4.1a and 4.12).

The estimates of natural land N fixation are speculative. One problem with using the compilation of data to estimate N fixation is that the data may not be representative of the landscape as a whole. That is, measurements are likely made in areas where N fixation is higher. For example, the N fixation value of 16 kg N ha^{-1} yr^{-1} for forests is likely to be an overestimate for California since there is relatively little area

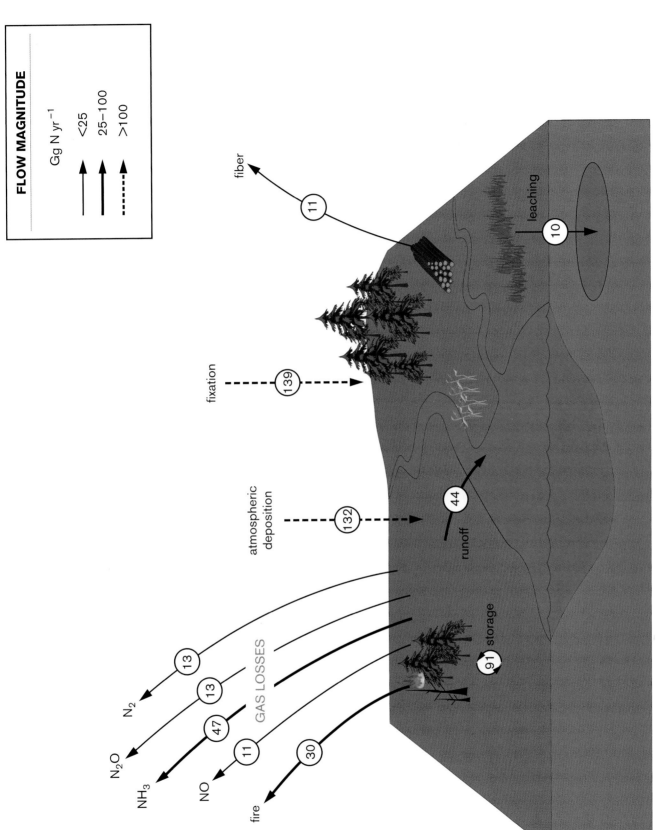

FIGURE 4.7. Flows of nitrogen in California natural land in 2005. The circled values indicate the absolute magnitude of the flow in Gg N yr⁻¹ with arrow thickness specifying the relative magnitude of the flow. Storage terms are indicated with arrows on the circled values. To distinguish it from other gaseous emissions, there is a separate arrow for wildland forest fires, representing the total amount of N volatilized (predominantly N_2).

TABLE 4.12
Californian natural land nitrogen mass balance in 2005

Nitrogen flow	Methods section*	Flow direction	Flow (Gg N yr⁻¹)	Evidence	Agreement
Biological N fixation	4.2.3	Import	139	Medium	Low
Deposition	4.2.2	Input	132	Medium	Medium
Gas emissions					
NO	4.2.8	Output	11	Limited	Low
NH₃	4.2.8	Output	47	Limited	Low
N₂O	4.2.8	Output	13	Limited	Medium
N₂	4.2.8	Output	13	Limited	Low
Fire	4.2.8	Output	30	Limited	Low
Runoff	4.2.9	Output	44	Medium	High
Leaching	4.2.10	Output	10	Limited	Low
Fiber	4.2.5	Output	11	Limited	Low
Storage	4.2.11	Storage	91	Limited	Low

*See Appendix 4.2.

NOTE: Storage was estimated as the difference between inputs and outputs and could occur in soils or vegetation.

that has high cover of N-fixing species. In addition, many biomes in the state have relatively few N-fixing species with medium to high fixation rates present at all. Further, as atmospheric N deposition has increased by an order of magnitude from 0.5 to 5 kg N ha⁻¹ yr⁻¹ over the last century, there may have been a corresponding decrease in N fixation with increasing N availability. This could be due to changes in the amount of N fixed by N-fixing species or the decreased cover of N fixing species (Suding et al., 2005). On the other hand, there are increasing numbers of invasive N-fixing species which are likely expanding their areal extent. Therefore, we feel that the low-end estimate of 139 Gg N yr⁻¹, based on the biome-specific rates, would be the most appropriate value for statewide natural land N fixation.

4.1.6.2. NATURAL LAND N OUTPUTS AND STORAGE

The largest N output from natural land soils is in gaseous forms. The biome-specific rates of gaseous emissions and biome areas result in the output of 11 Gg NO-N yr⁻¹, 47 Gg NH₃-N yr⁻¹, 13 Gg N₂O-N yr⁻¹, and 13 Gg N₂-N yr⁻¹. While the biome level rates of N₂O and NO are averages of multiple datasets often based on many published papers, it is difficult to discern how well they represent California ecosystems. For example, abiotic NO emissions are possible in desert regions, where the surface temperature can reach over 50°C (McCalley and Sparks, 2009).

Wildfire produces another 30 Gg N yr⁻¹ of gaseous N emissions. The area burned annually is monitored carefully by the California Department of Forestry and Fire Protection. However, the amount and form of N released by fire is more difficult to discern because it varies depending on the

amount of biomass and the burn characteristics. The 2005 EPA National Emissions Inventory reported 2 Gg N yr⁻¹ emitted as NOₓ and 2 Gg N yr⁻¹ emitted as NH₃ related to natural land fires for California. Insignificant (<1 Gg N yr⁻¹) amounts of N₂O were also emitted. Thus, by difference N₂ must be the dominant N form in wildfire emissions. Nitrogen volatilization from fires is considerably larger than the harvest of timber (11 Gg N yr⁻¹) from natural land for wood products.

Runoff to surface water accounts for 44 Gg N yr⁻¹ output from natural land soils based on an export coefficient of 2.4 kg N ha⁻¹ yr⁻¹. However, based on the California-specific data in Kratzer et al. (2011), we estimated the export coefficient to be only 1.3 kg N ha⁻¹ yr⁻¹. A large part of this difference may be associated with the managed hydrology in California. Significant fractions of the Sacramento and San Joaquin watersheds, especially the natural land, are located upstream of dams. Surface water bodies, especially reservoirs, can retain large amounts of N (Harrison et al., 2008). In closed basins, dissolved constituents cannot be transported to the ocean via surface water, but can only be leached through the soil to groundwater. In desert regions of the southwest with a deep water table, the estimated flux of 0.6 kg N ha⁻¹ yr⁻¹ would result in 10 Gg N yr⁻¹ leaching to groundwater. This annual rate is based on the NO₃⁻ stock of subsoil horizons that has accumulated over millennia. This subsurface inorganic N storage can be considerably larger than the surface soil organic N pool.

The mass balance calculations indicate that storage is a moderate N flow (91 Gg N yr⁻¹) in natural land. There are three possible explanations. First, our estimate of N inputs may be too high, especially for N fixation. Secondly, our estimate of N outputs may be too low, especially gaseous emis-

TABLE 4.13
Atmospheric nitrogen balance for California in 2005

N flow	Methods section*	Flow direction	NO_x (Gg N yr^{-1})	NH_3 (Gg N yr^{-1})	N_2O (Gg N yr^{-1})	N_2 (Gg N yr^{-1})	Total (Gg N yr^{-1})
Fossil fuel combustion	4.2.1	Import	359	36	9		404
Soil	4.2.8	Input	24	110	24	31	188
Manure	4.2.6	Input		97	2		99
Upwind sources	4.2.1	Import	20	20			40
Wastewater	4.2.7	Input			2	29	31
Pet waste	4.2.7	Input		4			4
Fire	4.2.8	Input	3	3		24	30
Surface water	4.2.9	Input			2	34	36
Groundwater	4.2.10	Input				91	91
Deposition	4.2.2	Output	135	67			202
Export	4.2.2	Export	270	203	38	204	716

*See Appendix 4.2.

NOTE: Only the fossil fuel combustion and the upwind sources of N were new statewide inputs of N. All N_2 and N_2O emitted were assumed to be an output from the state. For NO_x and NH_3, export beyond the state boundary of these gases was calculated as the difference between emissions and deposition. Because of reactions in the atmosphere, a significant fraction of the export of oxidized and reduced N has been converted to chemical forms (e.g., nitric acid, ammonium nitrate particles, peroxyacetyl nitrate) other than NO_x and NH_3. These oxidized and reduced forms are often summarized as NO_y and NH_x.

sions. Finally, N may be accumulating in vegetation and soils in California. The estimated storage term, while large with respect to the annual mass balance, was small in terms of the soil N pool. Assuming that the top 10 cm of soil in natural land is 0.1% N with a bulk density of 1 g cm^{-3}, the addition of the calculated annual change in N storage averaged across all natural lands represents an increase of 0.25% in the size of the soil N stock. That is, the top 10 cm would increase from 100 to 100.25 g N m^{-2}. This increase would be difficult to detect analytically, and even more so considering that the top 10 cm of soil only contains a fraction of the total soil N pool.

4.1.7. Atmosphere N Flows

The atmosphere is 78% N_2 gas; an essentially unlimited supply of N. At the scale of California, we assumed the atmospheric stock of N_2 is not changing, but we did estimate the export of fixed N as N_2 related to denitrification. For the atmosphere subsystem N mass balance, we estimated (1) how much reactive N was added to the portion of the atmosphere above the state, (2) deposition from the atmosphere to the land surface, and (3) export from the state (with all N_2O and N_2 emissions considered N exports because of their long atmospheric residence times). Overall, California is a large source of reactive N to the atmosphere with the majority of the N exported beyond the political boundaries of the state via atmospheric transport (Table 4.13).

4.1.7.1. ATMOSPHERE N IMPORTS AND INPUTS

Fossil fuel combustion is the major (40%) source of N to the atmosphere and NO_x is the predominant (89%) form of fossil fuel N generated. In total, 359 Gg NO_x-N yr^{-1}, 36 Gg NH_3-N yr^{-1}, and 9 Gg N_2O-N yr^{-1} were emitted during fossil fuel combustion (Table 4.13).

Soils and manure were also large sources of N to the atmosphere and are discussed in more detail in previous sections. Soils were the second largest contributor of N to the atmosphere with 24 Gg NO-N yr^{-1}, 110 Gg NH_3-N yr^{-1}, 24 Gg N_2O-N yr^{-1}, and 31 Gg N_2-N yr^{-1}. These emissions encompass all land cover types, as well as emissions from the land application of manure. Direct emissions from manure management on livestock facilities and after land application were a moderate (97 Gg N yr^{-1}) flow and a major (36%) source of NH_3 to the atmosphere. Dairy manure was the predominant (80%) source of NH_3 emissions from manure management. Manure management on livestock facilities was also a small (2 Gg N yr^{-1}) source of N_2O.

Wildfires, wastewater treatment, and surface water were all moderate N flows of similar magnitude to the atmosphere (30–36 Gg N yr^{-1}). For these three sources, unlike soils or fossil fuel combustion, N_2 is the dominant form of emissions.

A fraction of the reactive N in the atmosphere originates from areas upwind of California. Based on the atmospheric deposition rates generated by the CMAQ model in areas off

TABLE 4.14
California surface water nitrogen mass balance in 2005

Nitrogen flow	Methods section*	Flow direction	Flow (Gg N yr⁻¹)	Evidence	Agreement
Runoff to rivers					
Natural land	4.2.9	Input	44	Medium	High
Cropland	4.2.9	Input	41	Medium	High
Urban land	4.2.9	Input	10	Medium	High
Sewage	4.2.7	Input	12	Medium	Medium
Deposition	4.2.2	Input	2	Medium	Low
Irrigation	4.2.9	Output	18	Medium	Low
Gas emissions					
N_2O	4.2.9	Output	2	Medium	High
N_2	4.2.9	Output	28	Medium	Low
Ocean	4.2.9	Export	39	High	High
Lake/reservoir storage	4.2.9	Storage	20	Limited	Medium

*See Appendix 4.2.

NOTE: The reservoir storage term was calculated by difference. An independent estimate of N storage in lake and reservoir sediments was 14 Gg N yr⁻¹.

the coast of California, the current background deposition rate is 1 kg N ha⁻¹ yr⁻¹, split evenly between oxidized and reduced N. This rate does not represent the preindustrial N deposition rate because it includes anthropogenic N from other regions of the world, particularly Asia. This deposition rate applied for the whole state would result in 40 Gg N yr⁻¹ deposited in California even in the absence of any N emissions to the atmosphere in California. This background N deposition is considered an N import to California's atmosphere because it originates beyond the political boundaries of the state. We cannot estimate how much reactive N enters California's atmosphere from outside California and passes through the state without being deposited.

4.1.7.2. ATMOSPHERE N EXPORTS AND OUTPUTS

We assumed that there was no N storage possible in the atmosphere. Therefore, NO_x and NH_3 emissions had to be redeposited in California or exported downwind from the state. In addition, all of the N_2O and N_2 emitted were assumed to be exported. For both oxidized (33%) and reduced (25%) forms of N, less than half of the emissions were redeposited in the state. Oxidized N emissions (NO_x) were four times higher than reduced N emissions (NH_3), while oxidized deposition was only double that of reduced deposition, highlighting that a greater fraction of oxidized emissions are exported. The emitted N compounds can be exported in more stable forms after transformation to compounds like ammonium nitrate particles, nitric acid, or various organic N compounds.

4.1.8. Surface Water N Flows

Surface water drainage differs in California for several reasons. First, more than 40% of the state has no surface water drainage to the ocean. The watersheds in the Mojave Desert, Great Basin, Carrizo Plain, and Tulare Basin were assumed to have no external drainage. Secondly, almost every major river in the state is dammed and water is transferred among river basins. Finally, the timing and amount of nutrient inputs to surface water may differ from other parts of the United States because of the Mediterranean climate (Ahearn et al., 2004; Sobota et al., 2009).

4.1.8.1. SURFACE WATER N INPUTS

We estimated that the N input to rivers from runoff of the three land cover types was 95 Gg N yr⁻¹ with an additional loading of 12 Gg N yr⁻¹ from WWTPs (Tables 4.14 and 4.15). Nonpoint sources in natural land (44 Gg N yr⁻¹), cropland (41 Gg N yr⁻¹), and urban land (10 Gg N yr⁻¹) dominated the N inputs based on the export coefficients for these three land cover types. A small amount of deposition (2 Gg N yr⁻¹) fell directly on water bodies in the state.

4.1.8.2. SURFACE WATER N EXPORTS, OUTPUTS, AND STORAGE

Of the N entering rivers, less than half (39 Gg N yr⁻¹) reached the ocean (Table 4.15). Nitrogen dissolved in irrigation water withdrawals accounted for 18 Gg N yr⁻¹ of the

TABLE 4.15
Estimated annual N discharge to the ocean by watershed for California

Watershed	N loading to rivers based on export coefficients (Gg N yr⁻¹)		Estimated N discharge to ocean (Gg N yr⁻¹)		Measured N discharge to ocean by watershed (Gg N yr⁻¹)			Best estimate of discharge to ocean
	Wickham et al. (2008)	Central Valley watersheds	Wickham et al. (2008)	Central Valley watersheds	Sobota et al. (2009)	Schaefer et al. (2009)	Kratzer et al. (2011)	
Bay Delta	8.1	6.9	3.3	3.5				3.3
Central Coast	4.1	3.1	2.3	2.3				2.3
Colorado	2.9	2.0	1.9	1.8				1.9
Cuyama	0.9	0.6	0.9	0.9				0.9
Delta Rivers	4.0	3.7	2.2	2.5	0.2			0.2
Eel	2.5	1.6	1.7	1.5		2.7		2.7
Klamath	7.4	5.0	3.1	2.9		4.6		4.6
North Coast	2.9	1.8	1.8	1.6				1.8
Oregon	0.1	0.1	0.3	0.3				0.3
Pajaro	1.4	1.2	1.2	1.3		1.4		1.4
Russian	1.6	1.4	1.3	1.4		1.1		1.1
Sacramento	28.8	24.7	6.8	7.0	7.8	7.1	6.9	7.3
Salinas	3.4	2.9	2.0	2.1		0.9		0.9
San Joaquin	13.8	12.8	4.5	4.9	2.6		4.9	3.7
Santa Ana	2.4	1.8	1.7	1.7		2.0	1.6	1.8
Santa Clara	1.4	1.0	1.2	1.2				1.2
South Coast	10.8	8.3	3.9	3.9				3.9
California total	96	79	40	41				39

NOTE: For watersheds that were drained by rivers that reach the ocean, we used literature estimates of N loads at the furthest downstream gauge. The three sources of data were Sobota et al. (2009), Schaefer et al. (2009) and Kratzer et al. (2011) with the data representative of the years 2000–2003, 1992, and 2000–2004, respectively. In watersheds where there were no literature values for N discharge, we first calculated the estimated N loading to the watershed based on the export coefficients. We used export coefficients for cropland, urban land, and natural land from two sources: (1) values reported in Wickham et al. (2008) and (2) values calculated for the Central Valley from Kratzer et al. (2011) and multiplied these values by the area of each land cover. We compared the predicted values of annual N loading based on export coefficients to the measured values for the eight watersheds available in the literature. Based on the log–log regression (R^2 = 0.71) of predicted against measured data, we adjusted the predicted N loading to the watershed from the export coefficients in Wickham et al. (2008) to estimate the N discharged to the ocean for the watersheds. To simplify these calculations, we lumped the small (<1,000 km²) coastal watersheds into four basins: (1) the north coast, from the Oregon border to San Francisco Bay, (2) the San Francisco Bay/Delta downstream of the USGS gauges at Vernalis on the San Joaquin River and Freeport on the Sacramento River, (3) the central coast from San Francisco Bay to the Santa Clara River, and (4) the south coast from the Santa Clara river south to the Mexican border. The Oregon watershed includes the N loading from tributaries of the Rogue River that flow from California into Oregon.

output from the surface water subsystem. We estimated denitrification to N_2 from rivers, lakes, and reservoirs to be 30 Gg N yr⁻¹ and production of N_2O to be 2 Gg N yr⁻¹. By difference, we calculate storage in surface water bodies as 20 Gg N yr⁻¹ (Table 4.14). The independent measures of N in surface water storage were similar. First, using the sedimentation rate and N concentration of sediments, we estimated 65 Gg N yr⁻¹ buried in sediments. Based on Harrison et al. (2008), N retention was 8 Gg N yr⁻¹ in lakes and 57 Gg N yr⁻¹ in reservoirs. The denitrification estimate of 30 Gg N yr⁻¹ means that 35 Gg N yr⁻¹ would be accumulating in sedi-

ments. The dominance of reservoirs for N retention is consistent with the results of Harrison et al. (2008) that found that reservoirs retained N at rates 10 times higher than lakes.

4.1.9. Groundwater N Flows

Groundwater N flows are rarely quantified directly, but we estimated their magnitude at the statewide level as a function of recharge or withdrawal volume and N concentration.

TABLE 4.16
California groundwater nitrogen flows in 2005

Nitrogen flow	Methods section*	Flow direction	Flow (Gg N yr⁻¹)	Evidence	Agreement
Soils leaching					
Cropland	4.2.10	Input	333	Medium	High
Urban land	4.2.10	Input	1	Medium	Low
Natural land	4.2.10	Input	10	Limited	Low
Manure leaching	4.2.6	Input	10	Medium	Low
Sewage leaching	4.2.7	Input	27	Medium	Medium
Irrigation	4.2.10	Output	33	Medium	Low
Denitrification	4.2.10	Output	91	Limited	Medium
Storage	4.2.10	Storage	258	Medium	Low

*See Appendix 4.2.

NOTE: We assumed no net transport of N between surface water and groundwater. Storage of N in groundwater was calculated as the difference between inputs and outputs.

4.1.9.1. GROUNDWATER N INPUTS

Leaching to groundwater was a major (380 Gg N yr⁻¹) flow of N (Table 4.16). Almost 90% of the N flow to the groundwater leached from cropland soils (333 Gg N yr⁻¹). Small fluxes of N were related to leaching from manure in dairy facilities (10 Gg N yr⁻¹), natural land in areas with no surface drainage (10 Gg N yr⁻¹), and discharge of treated wastewater (27 Gg N yr⁻¹). The latter was a combination of septic systems (16 Gg N yr⁻¹) and treatment plants that dispose of treated wastewater on land (11 Gg N yr⁻¹). The estimate for septic systems is likely an overestimate of inputs to groundwater as we assumed that all of the N, with the exception of the biosolids, would reach the groundwater, but even if 50% of the septic N had some other fate, the impact on total groundwater N inputs would be minimal.

4.1.9.2. GROUNDWATER N OUTPUTS AND STORAGE

Groundwater pumping for irrigation removed 33 Gg N yr⁻¹, with water containing, on average, 0.572 mg NO₃⁻-N L⁻¹. Denitrification produced 91 Gg N yr⁻¹ as N₂ in 2005, but this flow is tentatively agreed by most (Box 4.4). The three estimates ranged from 26 Gg N yr⁻¹, using a fixed rate of denitrification, to 85 Gg N yr⁻¹, using historical estimates of N loading and a fixed half-life of N, to 162 Gg N yr⁻¹, using a fixed ratio of denitrification to N inputs based on current inputs. Taking into account the irrigation withdrawals and denitrification of historical nitrate in groundwater, almost 70% of the annual groundwater inputs for 2005 would contribute to an increase in groundwater N storage of 258 Gg N yr⁻¹ (Table 4.16). This assumes no net exchange of N with surface waters because the bidirectional flow is close to zero and the N concentrations in groundwater and surface water

are similar. We assumed that groundwater denitrification produces solely N₂ and not N₂O. However, this N₂ would not actually be returned to the atmosphere until the groundwater discharges to surface waters which could take decades to millennia.

4.2. Synthesis

Calculating nitrogen mass balances has been occurring for decades. The first global N budget was published by Delwiche in 1970 with the first watershed N budget published by Bormann et al. (1977) for Hubbard Brook. The general approach has largely remained the same ever since, but more types of fluxes (especially in urban areas) have been incorporated. The published N mass balances vary in their spatial extent (from watersheds <100 km² to the entire planet) and the types of boundaries (political vs. watershed, just agriculture vs. the whole landscape).

One common approach developed in the 1990s is termed Net Anthropogenic Nitrogen Inputs or NANI (Howarth et al., 1996; Jordan and Weller, 1996). The NANI approach estimates imports of new reactive N from atmospheric deposition (total or just oxidized, which is more likely to be from fossil fuel combustion instead of recycled N), net food and feed imports, crop biological N fixation, and synthetic N fertilizer. Typically the goal is to calculate the fraction of N imports that are accounted for in surface water exports. One advantage of this approach is that the imports are standardized so it is easy to compare across watersheds. A recent synthesis by Howarth et al. (2012) suggests that on average one quarter of the N imports were exported from watersheds globally. However, the watersheds included are largely temperate with moderate precipitation. The fraction of surface water exports has been suggested to be much

BOX 4.4. DENITRIFICATION IN GROUNDWATER

Denitrification is the process that converts nitrate (NO_3^-) to inert nitrogen (N_2) gas through a series of chemical reactions. It is typically a biological process in which microorganisms, such as bacteria, respire NO_3^- instead of oxygen to meet their metabolic needs. Denitrification can occur when three conditions are met: nitrate is present, oxygen concentrations are low, and a source of electrons (e.g., energy) is available. Denitrifying organisms are ubiquitous in soils and sediments, as well as surface water and groundwater environments; these organisms can also be harnessed to remove nitrate from high-nitrogen (N) waters such as in wastewater treatment plants and agricultural runoff. Denitrification is a key transformation in the N cycle as it is the dominant process that converts reactive N back to atmospheric N_2. As such, it reduces risks of excess N on human health and the environment (Moran et al., 2011).

In most environments, there are methods, albeit expensive and requiring specialized equipment, for measuring denitrification rates in situ. Part of the difficulty is that the product of denitrification, N_2, comprises almost 80% of the atmosphere. Therefore, it is impossible to detect the small flux of N_2 from surficial environments where atmospheric air is present. Because of these methodological issues, denitrification is often quantified by difference in mass balance studies because of the difficulties in measuring it directly. In groundwater, denitrification is typically detected by chemical signatures and dissolved excess N_2 gas left behind by the process. Analysis of the isotopes of nitrogen and oxygen in groundwater NO_3^- can indicate whether denitrification is occurring but may also reflect a signature of the original source (e.g., manure vs. fertilizer) of the NO_3^-. Quantifying groundwater denitrification rates typically involves measuring excess N_2. Because groundwater is isolated from the atmosphere, the N_2 produced by denitrifica-tion remains dissolved. This "excess N_2" can be measured and the amount of NO_3^- originally dissolved in the water can be determined.

Denitrification occurs in groundwater when nitrate-rich water recharged from the surface reaches portions of the aquifer with low dioxygen (O_2). In addition, either organic carbon leached from the surface or reduced minerals like sulfides need to be present in the sediments or rocks of the aquifer as a source of energy. In some aquifers, conditions are such that denitrification can convert a significant amount of the nitrate to N_2, while in others high O_2 or a limited supply of energy precludes the complete conversion of NO_3^-. Data on denitrification in groundwater are particularly difficult to obtain and much more research is needed on the subject (see, e.g., Böhlke and Denver, 1995; Browne and Guldan, 2005; Fogg et al., 1998). It is tentatively agreed by most that denitrification rates are relatively low in the major groundwater basins in the Central Valley, especially the shallow aquifers. The few studies that have been conducted found that the aquifers in California do not typically have the combination of conditions that would be conducive for the removal of all NO_3^- by denitrification (Green et al., 2008b; Landon et al., 2011; Moran et al., 2011). Harter et al. (2012) suggest that it is practical and sensible to conclude that most NO_3^- in California aquifers used for irrigation and municipal supplies is unlikely to be denitrified. The most prominent exception, perhaps, are denitrifying conditions found in the vicinity of the major streams and near valley troughs that have accumulated lake and marshy sediments with significant organic matter (Landon et al., 2011; Moran et al., 2011). Additional studies using excess nitrogen/argon (N_2/Ar) ratios, natural N isotopes, and mass balance calculations could further our understanding of the spatial and temporal variability in denitrification in California's groundwater.

---

lower in arid areas, but relatively few arid watersheds have been examined (Caraco and Cole, 2001). Further, this approach neglects several potentially important flows: fiber (particularly wood products) and synthetic chemicals like plastics. Including these would not affect the magnitude of surface water exports, but they would decrease the fraction of surface water exports.

In addition to the NANI method, several other approaches have been utilized (Table 4.17). These range from studies that focus solely on agricultural regions, intermediate studies that include more imports and exports than NANI, and comprehensive studies that include all significant N flows. The spatial extent of these studies can be watersheds or political boundaries ranging from regions (e.g., the San Joaquin Valley) to states (e.g., Wisconsin) to countries (e.g., the United States) to continents and Earth (Galloway et al., 2004). The comprehensive mass balances have largely been attempted in the last decades. Because these comprehensive studies include measurements of all N flows, they can also be standardized by area, by population, or as fractions of the total flow.

We standardized five of the published comprehensive mass balances: the Netherlands and Europe (Leip et al., 2011), the Guangzhou region of China (Gu et al., 2009), the Phoenix region (Baker et al., 2001), and South Korea (Kim et al., 2008), and compared them to the California N mass balance (Figures 4.8 and 4.9). These five areas differ in size, climate, and land cover, but the comparison among them can provide useful insight into N dynamics. In some cases, this required adjustments to the boundaries of the study,

TABLE 4.17
References for other nitrogen mass balance studies

Author	Year	Type	Spatial extent
Antikainen et al.	2005	Comprehensive	Country
Baisre	2006	NANI	Country
Baker et al.	2001	Comprehensive	Region
Bormann et al.	1977	NANI	Watershed
Boyer et al.	2002	NANI	Watersheds
Carey et al.	2001	NANI	Watersheds
Castro et al.	2003	NANI	Watersheds
David and Gentry	2000	NANI	State
Delwiche	1970	Intermediate	Global
EPA SAB	2011	Intermediate	Country
Galloway et al.	2004	Intermediate	Global/continents
Goolsby et al.	1999	Intermediate	Watershed
Gu et al.	2009	Comprehensive	Region
Han et al.	2011	NANI	Region
Han and Allen	2008	NANI	Watersheds
Howarth et al.	1996	NANI	Watersheds
Howarth et al.	2012	NANI	Watersheds
Janzen et al.	2003	Agriculture	Country
Jordan and Weller	1996	NANI	Watersheds
Keeney	1979	Intermediate	State
Kim et al.	2008	Comprehensive	Country
Leip et al.	2011	Comprehensive	Country
Messer and Brezonik	1983	Agriculture	State
Miller and Smith	1976	Agriculture	Region
NRC	1972	Intermediate	Country
OECD	2001	Agriculture	Country
Parfitt et al.	2006	Intermediate	Country
Prasad et al.	2004	Agriculture	Country
Quynh et al.	2005	Intermediate	Watersheds
Robertson and Rosswall	1986	Intermediate	Countries
Salo et al.	2008	Agriculture	Country
Schaefer and Alber	2007	NANI	Watersheds
Schaefer et al.	2009	NANI	Watersheds
Sobota et al.	2009	NANI	Watersheds
Söderlund and Svensson	1976	Intermediate	Global
Valiela and Bowen	2002	NANI	Watersheds
Van Drecht et al.	2003	Intermediate	Global
Velmurugan et al.	2008	Intermediate	Country

NOTE: Types of mass balances include the Net Anthropogenic Nitrogen Inputs (NANI) approach described by Jordan and Weller (1996), comprehensive approaches that include all N flows, and intermediate approaches that examine only a subset of the landscape (e.g., agriculture) or a subset of the flows across the entire landscape.

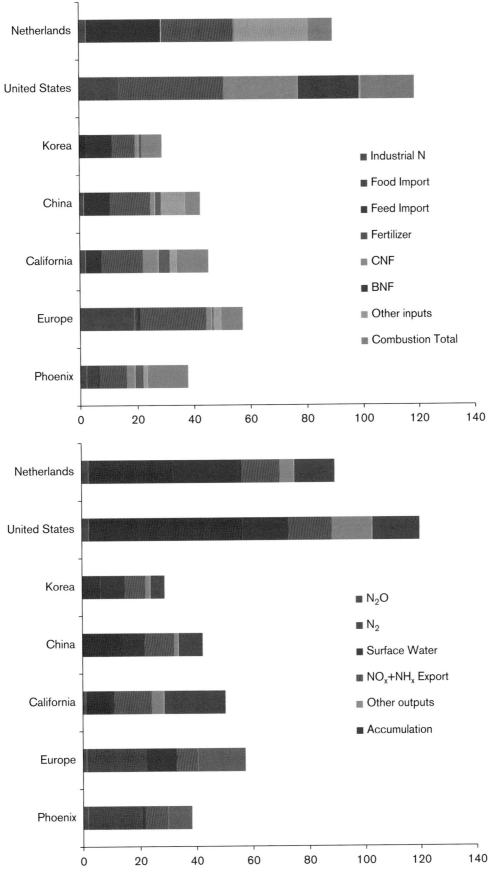

FIGURE 4.8. (top) N imports and (bottom) exports/storage per capita (kg N person^{-1} yr^{-1}). Comparison of N flows on a per capita basis for the California N Assessment (California) to six representative comprehensive N mass balance studies at various spatial scales around the world.

SOURCE: Data for the Netherlands and Europe are from Leip et al. (2011), the United States from EPA SAB (2011), China from Gu et al. (2009), Korea from Kim et al. (2008), and Phoenix from Baker et al. (2001).

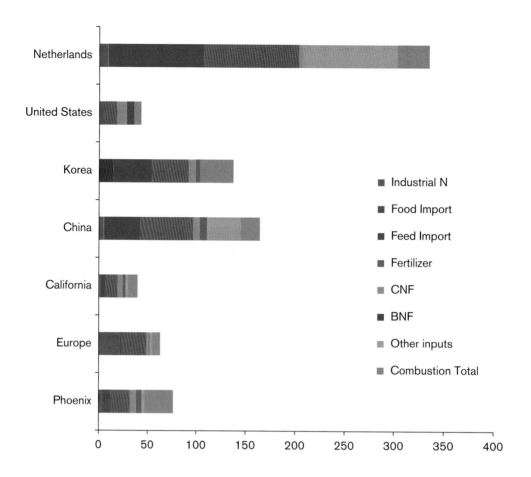

Legend (top):
- Industrial N
- Food Import
- Feed Import
- Fertilizer
- CNF
- BNF
- Other inputs
- Combustion Total

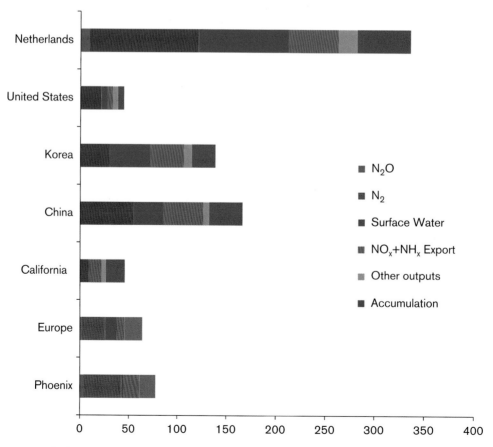

Legend (bottom):
- N₂O
- N₂
- Surface Water
- NOₓ+NHₓ Export
- Other outputs
- Accumulation

FIGURE 4.9. (top) N imports and (bottom) exports/storage per unit area (kg N ha⁻¹ yr⁻¹). Comparison of N flows on an areal basis for the California N Assessment (California) to six representative comprehensive N mass balance studies at various spatial scales around the world.

SOURCE: Data for the Netherlands and Europe are from Leip et al. (2011), the United States from EPA SAB (2011), China from Gu et al. (2009), Korea from Kim et al. (2008), and Phoenix from Baker et al. (2001).

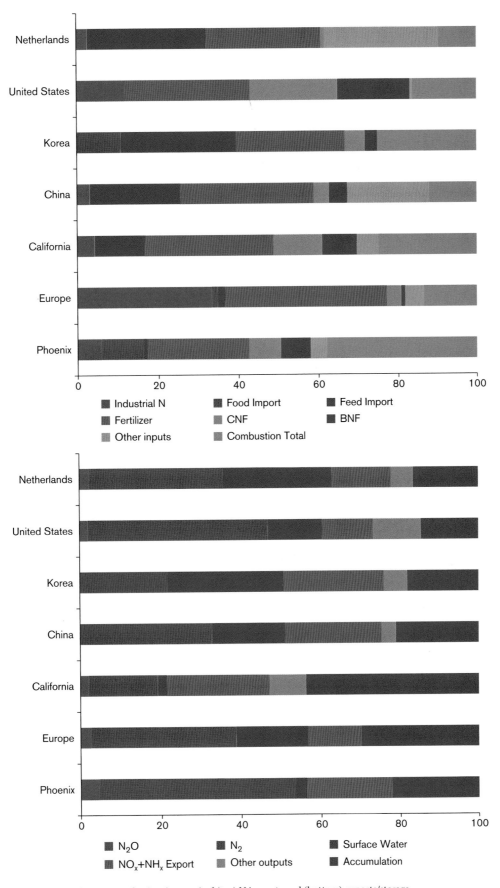

FIGURE 4.10. Relative contribution (percent) of (top) N imports and (bottom) exports/storage.

SOURCE: Data for the Netherlands and Europe are from Leip et al. (2011), the United States from EPA SAB (2011), China from Gu et al. (2009), Korea from Kim et al. (2008), and Phoenix from Baker et al. (2001).

but the flows included were the same to compare across the different areas. For example, in Europe the coastal regions were included in the study area, whereas the boundary in California was the coastline. In a few cases, flows were not available. For example, in both China and South Korea, there was no estimate of N_2O distinct from total denitrification and in South Korea there was no estimate of synthetic N chemicals. In some cases, storage or N accumulation was not explicitly calculated, but we estimated it by difference between imports and exports. Based on the data available in these studies, the population, and the area of the study region, we calculated N imports and exports per unit area and per capita. We also calculated N flows as a percentage of the total.

When compared to other regions of various sizes, California has a relatively low N use both on a per capita and, especially, on a per hectare basis (figures 4.8 and 4.9). The United States has by far the largest per capita imports of N (118 kg N person^{-1} yr^{-1}). Similarly, the Netherlands has by far the largest imports of N on an areal basis (334 kg N ha^{-1} yr^{-1}).

Synthetic fertilizer is the largest N import in all studies, with the exception of the Netherlands where there was slightly more N feed imported. With the exception of the United States as a whole, all studies reported feed import, often as a large fraction of the total N imports (figure 4.10). Similarly, only the United States as a whole has a medium (22%) fraction of imports from crop N fixation with most studies reporting less than 10%. Food import is less common and typically only a small fraction of new imports. The N import from fossil fuel combustion ranges from 10% in the Netherlands to 38% in Phoenix.

Denitrification or N_2 production was the largest export of N except in California and South Korea. In both these studies, export of $NO_x + NH_3$ was larger. In South Korea, surface water exports were the largest export flow. Surface water export ranged from 2.2% in California to 29% in South Korea. Low surface water export (2.7%) was also observed in Phoenix, corroborating the phenomenon of low fractional export in arid areas. While most studies reported less than 20% of export in surface water, these values cannot be directly compared to the NANI approach because the imports are more inclusive. Nitrogen accumulation was reported or inferred from all the studies. The highest storage fraction (37%) occurred in California with other studies ranging from 14% in the United States to 30% in Europe.

In many ways, the N flows in California are similar to other parts of the world. The two ways that it stands out are the low surface water exports and the high N storage, primarily in groundwater and urban land.

CHAPTER FIVE

Ecosystem Services and Human Well-Being

Authors:
V. R. HADEN, D. LIPTZIN, T. S. ROSENSTOCK, J. VANDERSLICE,
S. BRODT, B. L. YEO, R. DAHLGREN, K. SCOW, J. RIDDELL,
G. FEENSTRA, A. OLIVER, K. THOMAS, D. KANTER, AND T. P. TOMICH

What Is This Chapter About?

Changes in nitrogen (N) levels in soils, air, and water affect the benefits people derive from ecosystems. These benefits, known as ecosystem services, fall into the four categories of *provisioning, regulating, cultural,* and *supporting* services. In this chapter we examine ecosystem services that are known to be affected by nitrogen levels and management activities, with a focus on those that are relevant to California. The five sections of this chapter address the central role of N in food production and agriculture (Section 5.1); how N affects the ecosystem goods of clean drinking water and clean air (Sections 5.2 and 5.3); the regulating service that N provides in maintaining a stable climate (Section 5.4); and the cultural and spiritual values that N affects, most notably how excess N alters biodiversity in terrestrial and aquatic ecosystems, changing the way humans interact with and enjoy nature (Section 5.5).

Stakeholder Questions

The California Nitrogen Assessment engaged with industry groups, policy-makers, nonprofit organizations, farmers, farm advisors, scientists, and government agencies. This outreach generated more than 100 N-related questions, which were then synthesized into five overarching research areas to guide the assessment (figure 1.3). Stakeholder-generated questions addressed in this chapter include the following:

- What is the state of knowledge on how nitrogen influences air and water quality and impacts human health?
- What is the cost of N management—to growers and to society in terms of public health costs, and costs related to environmental contamination?

5.0. Introduction

This chapter outlines the impacts that changes in the nitrogen (N) cycle have on the environment and human well-being. On the one hand, perturbation of the N cycle facilitates greater production of food (e.g., crops and livestock) and fiber, greatly benefiting the economy of California and the health of people worldwide. On the other hand, excessive reactive N in the environment from agricultural and urban activities is polluting the soils, air, and water, and is linked to environmental damage including acidification, invasive species, particulate matter and ground-level ozone formation, depletion of the stratospheric ozone layer, climate change, endangered species decline, eutrophication, and changes in the composition of terrestrial and aquatic biotic communities. Changes in N levels in these resource stocks affect the goods and services Californians derive from their surroundings, such as clean drinking water, clean air, and recreational activities.

In this assessment, we examined how changes in ecosystem services affect human well-being, including food security, human health, and a healthy environment. Ecosystem services include provisioning, regulating, and cultural services that directly affect people, as well as supporting services needed to maintain other services (MA, 2005a). Provisioning services are the products obtained from ecosystems (e.g., food, fuel, clean water, clean air). Regulating services are the benefits obtained from regulation of ecosystem processes (e.g., climate regulation). Cultural services are nonmaterial benefits obtained from ecosystems through spiritual enrichment, recreation, and aesthetic experiences (e.g., swimming, fishing, wildlife viewing, and ceremonial uses of particular plant and animal species). Supporting services, processes needed for production of all other ecosystem services, include soil formation and regeneration of atmospheric oxygen. Supporting services differ from other services because their impacts on people are often indirect or occur over a long time period whereas changes in the other categories have relatively direct and short-term impacts. Building on Compton et al. (2011), we examined ecosystem services known to be affected by nitrogen levels and management activities (Table 5.0.1), and refined this list to focus on those that are relevant in the California context. Trends and impacts on the environment and human health in California are synthesized within this framework, providing an assessment of the qualitative effects of nitrogen on ecosystem services and processes.

This chapter is divided into five main sections. Section 5.1 describes the central role of N in food production (crop and animal) and other agricultural products in California and includes temporal trends in their patterns. Furthermore, it details the direct and indirect effects California agricultural production has on the economy of California, as well as the important role California agriculture has in the food system of the United States and worldwide. Sections 5.2 and 5.3 discuss how N affects ecosystem goods—clean drinking water and clean air, respectively. Section 5.2 shows the spatial and temporal trends of nitrate concentration levels in groundwater in California, and explains the human health consequences of drinking water contaminated with high levels of nitrate. Section 5.3 explains how N affects air quality, illustrates trends in air quality, and details the different human health consequences of exposure to nitrogen dioxide (NO_2), ozone (O_3), and particulate matter (PM). Section 5.4 details the regulating service that N provides in maintaining a stable climate, including roles in formation of greenhouse gases (GHGs) and contribution to global warming as well as cooling. Section 5.5 discusses how N affects cultural and spiritual values of human society, including how humans interact with and enjoy nature.

5.1. Healthy Food and Other Agricultural Products

Main Messages

Production of California livestock and agricultural crops has increased since 1980, accompanied by greater N fertilizer application. Between 1980 and 2007, production of vegetables and melons, fruits and nuts increased 128% and 17%, respectively, reflecting shifts in the diet composition of the US population. To meet increasing demands for animal protein, feed crops was also one of the highest crop production categories, almost tripling over this period. Correspondingly, livestock production was on an increasing trend, with the average annual milk cow and heifer population doubling.

While N is indispensable in increasing the production of agricultural systems, much of the N applied is lost to the environment, resulting in a variety of impacts on atmospheric, terrestrial, and aquatic ecosystems. The difference between the tonnes of N fertilizer applied and N harvested is on a decreasing trend for cotton since 1980. However, the estimated amount of N that is not taken up by crops is on a slightly increasing trend for vegetables, fruits, and nuts. This corresponds to the amount of fertilizer applied by crop, with estimated application rates on many vegetable, fruit, and nut crops having increased in recent decades, at the same time as the total acreage for these crops has also increased.

California's agricultural sector is important to the state's economy and also contributes significantly to the provision of food security for the United States and globally. California's agricultural economy is the largest in the United States with over $37.5 billion in earnings in 2010, producing 21% of the nation's dairy commodities and more than 50% of the fruits and vegetables. The state is also the largest producer of ornamental horticultural goods in the United States with $2.3 billion in wholesale sales and $235 million in retail sales in 2009.

Nitrogen is an essential component of food. As a building block of proteins, DNA, and chlorophyll, N is a critical requirement for the growth and development of plants and animals (Marschner, 1995). In most agricultural systems globally, and in virtually all the agricultural systems in California, N is often the most limiting nutrient (Vitousek et al., 1997). Hence, application of fertilizer N or the importation of high protein feeds results in greater food production (Hartz and Bottoms, 2010; Kebreab et al., 2001; Letey et al., 1979; Oenema et al., 2008). Amendment of cropland with synthetic and organic N fertilizers and supplementation of animal diets with N-rich feedstock is a common practice across California (see Chapter 3).

Quantification of the impact of supplemental N on agricultural productivity and human well-being is confounded by the complexities of agricultural production systems. Despite the multiple interacting factors, however, it is well established[1] that there has been a substantial, globally positive effect of N on food production. A synthesis of global long-term studies found widespread use of synthetic N fertilizers responsible for at least 60% of agricultural production (Stewart et al., 2005b). Galloway et al. (2008) suggest that nearly 2 billion people are alive today because of synthetic fertilizer, while another study estimates that N fertilizers support an additional 27% of the world's population than would have been possible otherwise (Erisman et al., 2008).

5.1.1. Role of Nitrogen in Agricultural Production

The contribution of N to California agriculture has not been systematically analyzed. Long-term research trials such as the Century Experiment at University of California

---

1. Throughout the assessment, "reserve wording" was used to quantify areas of uncertainty in the available data and level of scientific agreement (see Box 1.6).

TABLE 5.0.1

Ecosystem services affected by increased N in the environment

Type	Ecosystem service	Beneficial or adverse impact	Mechanism of impact	N-related cause	Source
Provisioning	Production of food and materials	+	Increased production and nutritional quality of food crops	N fertilizer increases crop growth	Synthetic and organic N fertilizer
		+	Increased production of building materials and fiber for clothing or paper	N fertilizer increases crop growth	Synthetic and organic N fertilizer
		−	Soil acidification, nutrient imbalances and altered species composition	Acid deposition	Fossil fuel combustion, agriculture
	Fuel Production	−/+	Increased N inputs required for some biofuel crops can affect other services	N fertilizer increases crop growth	Synthetic and organic N fertilizer
		+	Increased use of fossil fuels to improve human health and well-being across the globe[1]	Increase energy availability	Fossil fuel combustion
Supporting and regulating	Drinking water	−	Increased nitrate concentrations lead to blue-baby syndrome, certain cancers	Nitrate into water	Agriculture
		−	Increased acidification and mobility of heavy metals and aluminum	Acid deposition	Fossil fuel combustion, and agriculture
	Clean Air	−	NO_x-driven increases in ozone and particulates exacerbate respiratory and cardiac conditions.	NO_x into air; $PM_{2.5}$, O_3 and related toxins	NO_x and NH_3/NH_4 from fossil fuel combustion, and agriculture
		−	Increased allergenic pollen production	Pollen production	Crops with airborne pollen
		−	Stimulation of ozone formation, which in turn can reduce agricultural and wood production and act as a greenhouse gas	Ozone and acid deposition	Fossil fuel combustion
	Visibility	−	Increased NO_x and NH_3 in air stimulates formation of particulates, smog, and regional haze	Fine particulate matter	NO_x and NH_3/NH_4 from fossil fuel combustion and agriculture
	Climate regulation	+/−	Variable and system-dependent impacts on net CO_2 exchange	N deposition	Fossil fuel combustion, agriculture
		−	Stimulation of N_2O production, a powerful greenhouse gas	N_2O into air	Agriculture, animal manure management, sewage treatment, fossil fuel combustion

(continued)

TABLE 5.0.1 (continued)

Type	Ecosystem service	Beneficial or adverse impact	Mechanism of impact	N-related cause	Source
	UV regulation	−	Increased N_2O release, which has strong-ozone-depleting potential	N_2O into air	Agriculture, animal manure management, sewage treatment, fossil fuel combustion
Cultural	Swimming	−	Stimulation of harmful algal blooms that release neurotoxins (interaction with phosphorus)	Excess nutrient loading, eutrophication, variable freshwater runoff	Fossil fuel combustion, agriculture
		−	Increased vector-borne diseases such as West Nile virus, malaria, and cholera	Excess nutrient loading, eutrophication, variable freshwater runoff	Fossil fuel combustion, agriculture
	Fishing	+	Increased fish production and catch for some very N-limited coastal waters	Nutrient loading, N deposition	Fossil fuel combustion, agriculture
		−	Increased hypoxia and harmful algal blooms in coastal zones, closing fish and shellfish harvests	Excess nutrient loading, eutrophication, variable freshwater runoff	Fossil fuel combustion, agriculture
		−	Reduced number and species of recreational fisheries from acidification and eutrophication	Atmospheric deposition of HNO_3, NH_3, and ammonium compounds	Fossil fuel combustion, agriculture
	Hiking	−	Altered biodiversity, health, and stability of natural ecosystems	N deposition	Fossil fuel combustion, agriculture
	Biodiversity	−	Altered biodiversity, food webs, habitat and species composition of natural ecosystems	N deposition	Fossil fuel combustion, agriculture
	Other	−	Damage to buildings and structures from acids	Acid deposition	Fossil fuel combustion, agriculture
		+/−	Long range trans-boundary N transport and associated effects (both negative and positive)	N deposition	Fossil fuel combustion, agriculture

1. This impact is not addressed in Chapter 5. Please refer to Section 3.4 for a discussion of fuel combustion as a direct driver.
NOTE: Positive and negative impacts of N on various environmental and human health services are indicated using a plus or a minus.
SOURCE: Adapted from Compton et al. (2011) and EPA (2012).

Davis' Russell Ranch Sustainable Agriculture Facility (http://asi.ucdavis.edu/rr) measure all agricultural inputs and outputs, including N, in different management systems; this database will soon provide robust N budgets in different agricultural systems. At the state scale, commercial sales of synthetic N fertilizer have increased 12-fold over the past 60 years (1946–2006), with the greatest increase between 1950 and 1980 (see Chapter 3, figure 3.2). Over the past 5 years, more than 600,000 tonnes of synthetic N fertilizer have been sold annually in California (Alexander and Smith, 1990; CDFA, 2009) with concomitant increases in yields of almost all agricultural commodities.

While it is well established from historical data that a positive relationship exists between increasing N in agricultural systems and productivity, how much of these yield increases is due to N, per se, is difficult to determine. For many crops, the relative N application rates per hectare have not increased significantly over this time period (see Section 3.1.1), whereas water infrastructure, pest management, genetics, etc., have undergone significant innovation simultaneously and are also responsible for yield increase (Johnston and McCalla, 2004). In animal systems, it is well established that increasing N fed to cattle results in greater quantities of meat and milk production (Kebreab et al., 2001; Oenema et al., 2008; Powell et al., 2010). Dairy cows are now fed more N in absolute terms (not in percentage of intake) than 30 years ago and the overall N efficiency (% N to milk production) has increased (Section 3.2.1). In contrast, though poultry production has grown in the state, growth is not due to increases in the amount of N being fed to animals but is attributable to changes in other production practices (see Section 3.2.1).

5.1.1.1. TRENDS IN INDICATORS OF CROP PRODUCTION

FOOD AND FEED CROPS

While overall crop production (harvested yield × cropping area), N applied (N rate × cropping area), and N harvested (crop production × % N in harvested portion) have generally increased in California over the past several decades, the magnitude and direction of these trends differ considerably among major crop categories (See Appendix 5.1.1 for a list of crops in each category). For example, significant increases in statewide production were observed for vegetable and melons, other feed crops, and to a lesser extent fruit and nut crops between 1980 and 2007 (figure 5.1.1). In particular, between 1950 and 2007, yields of almonds and processing tomatoes increased by 368% and 221%, respectively (figure 5.1.2) (USDA NASS, 2010).

In contrast, from 1980–2007 there was a decline in statewide production of "other food crops," a category that includes foods high in carbohydrates such as grains (rice, wheat), pulses (dry beans, peanuts), and root crops (potatoes, sweet potatoes, sugar beets). Over the same period,

production of alfalfa, cotton, and seed crops remained fairly constant.

Statewide trends in overall N applied and N harvested are driven mostly by changes in cropping area for the major crop categories, and to a certain extent by shifts in area among the dominant crops within each category. Due to the paucity of year-to-year data on crop-specific changes in N rate, for the purposes of this assessment N rates for specific crops within a category were held constant over time. Using this approach, mean N rates for a crop category were found to change over time if significant shifts in the relative area of each crop within a given category occur. As such, the increase in applied and harvested N for feed crops, and fruits and nuts (and the decrease in applied and harvested N in other food crops) mostly reflect the corresponding changes in cropping area for each category (figures 5.1.3 and 5.1.4).

The difference between N applied and N harvested provides a useful approximation of how much N is lost to the environment from various crop categories (figure 5.1.5). Based on these calculations, surplus N lost to the environment from both cotton and some food crops has declined due mostly to decreases in acreage farmed rather than to improvements in N recovery efficiency. In contrast, increasing losses of N to the environment have occurred from fruits and nuts, vegetables and melons, and other field crops since 1980. These increases in N lost are likely explained by a shift to crops with higher fertilization rates relative to the amount of N in their harvested portions which, in turn, would result in lower apparent N use efficiency (NUE; see Chapter 3, Section 3.2). As mentioned previously, NUE is calculated as the amount of N removed from the field per unit of N applied. Estimates of NUE are available for more than 20 individual crops (see Chapter 3; figure 3.3), but additional research is needed to establish long-term trends in NUE for California crops.

5.1.1.2. TRENDS IN INDICATORS OF LIVESTOCK PRODUCTION

Livestock production in California has increased significantly since 1980 (see Section 3.8.3 and figure 3.8). For example, the average annual milk cow and heifers population has doubled from 1980 to 2007, increasing from 896,000 in 1980 to 1.8 million in 2007 (figure 5.1.6). Nationally, the production of animal products has become more efficient, with more animal products produced with fewer animals (EPA SAB, 2011). In California, annual milk production per cow has increased from 15,153 to 22,440 lb milk per cow between 1980 and 2007 due, in part, to far larger amounts of N being fed to dairy cows (figure 5.1.6). It should be noted that the increase in total N intake is mostly a function of each animal consuming more feed rather than a significant increase in the fraction of N in the feed.

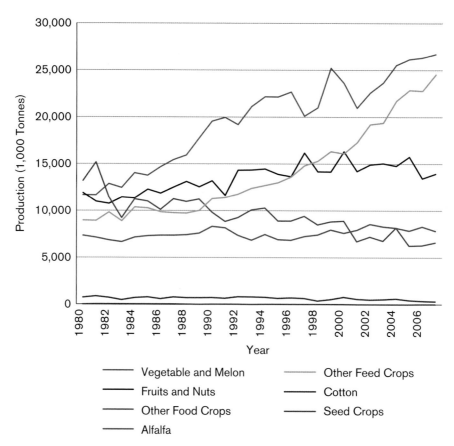

FIGURE 5.1.1. Production of major crops in California, 1980–2007. The groupings of crop categories mostly follow the categorization of the *California Agricultural Statistics, Crop Year 2010* published by the USDA, NASS, California Field Office. For the list of crops that typically fall under the "Field Crops" category, we have further divided this into several categories taking into consideration the type of nutrition and function a specific type of crop provides. "Alfalfa" is an N-fixing crop and is therefore omitted. "Seed crops" are not directly harvested for human consumption and hence has its own category. "Other food crops" consists of crops that typically provide carbohydrates for human nutrition, and "Other feed crops" are crops that are typically used for livestock production. For further details on the specific crops in each crop category see Appendix 5.1.1.

DATA FROM: USDA NASS (2012).

Dairy cattle partition N intake into milk or manure and urine, and research shows that about 20–40% of the N intake is excreted as milk, while about 60–80% of the N intake is excreted as manure and urine (Chase, 2011). Improvements in dairy cow diets can increase partitioning of N into milk production, e.g., by providing lower levels of crude protein (Chase, 2011). Though the production of animal products has become more efficient over time, more N is needed to produce the same amount of animal protein as plant protein, reducing the overall system-wide efficiency (Box 5.1.1) (Mosier et al., 2001). Animal production systems vary in their efficiency (see figure SPM.6 in Sutton et al., 2011a); for example, poultry production has a lower N footprint per kilogram of food (higher feed N recovery efficiency) than does beef production (Sutton et al., 2011a).

The statewide mass balance presented in Chapter 4 suggests that livestock production is an important driver of N

imported into the state and contributes significantly to the N flows between many of the major subsystems (e.g., cropland, groundwater, atmosphere, etc.). Feed crops accounted for almost two-thirds of the 543 Gg N harvested from cropland. Alfalfa, which obtains a significant fraction of its N from biological fixation, supplied almost 40% of the N harvested statewide. Even considering the large amount of feed crops grown in the state, there is still a need to import 200 Gg N to meet the dietary needs of livestock. The 537 Gg N in livestock feed is converted to 141 Gg N in food products and 416 Gg N in manure. Some of this manure is volatilized or leaches directly from the livestock facilities, while 307 Gg N from manure is applied to cropland, almost 30% of the total N inputs to cropland. From the mass balance approach, we cannot determine the particular fate of manure applied to cropland. However, based on the modeling results in van der Schans et al.

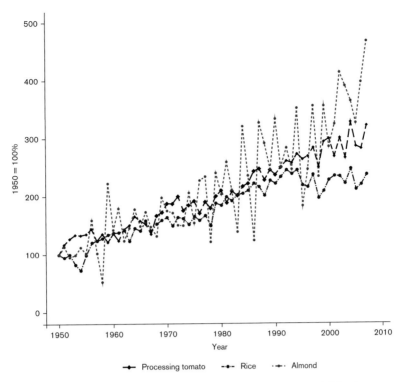

FIGURE 5.1.2. Yield increase of processing tomato, rice, and almonds in California, 1950–2007.

DATA FROM: USDA (2009a).

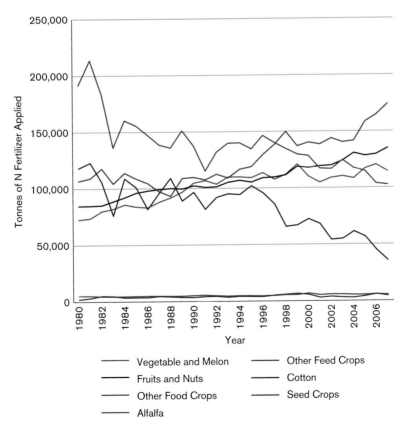

FIGURE 5.1.3. Tonnes of N fertilizer applied to various crop categories in California, 1980–2007. The change in N applied by crop category over time was calculated by multiplying the acreage of land devoted to each crop type by an average N fertilizer rate for each crop type.

DATA FROM: USDA NASS (2012); UC Davis Agricultural and Resource Economics (n.d.); USDA NASS (n.d., *Surveys*).

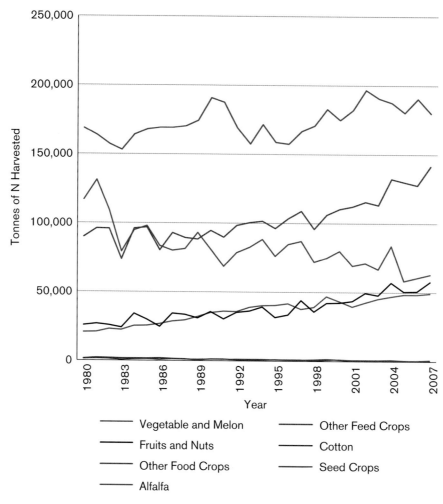

FIGURE 5.1.4. Tonnes of N harvested in California crops, 1980–2007. Tonnes of N harvested for each crop category is calculated by tonnes of crop produced × %Dry Matter × %N. The crop categorization used here is consistent with the crop groupings used in Figure 5.1.2.

DATA FROM: USDA NASS (2012); UC Davis Agricultural and Resource Economics (n.d.); USDA NASS (n.d., *Surveys*).

(2009), a large fraction of the N applied as dairy manure would likely leach from cropland soils to groundwater. Results of the mass balance also indicate that livestock systems are important sources of gaseous N emissions, accounting for approximately 53% of NH_3 and 5% of N_2O emitted in California each year.

5.1.2. Human Well-Being and Agricultural Production

5.1.2.1. FOOD AND HEALTH

Since California produces much of the nation's fruits, vegetables, and nuts (see Section 5.1.3.2), this section will review the nutritional implications of these products. According to multiple cohort studies (Dauchet et al., 2006; He et al., 2007), fruit and vegetable consumption is positively associated with reduced risk of leading causes of death including stroke and coronary heart disease. Additionally, since more than one-third of children and

two-thirds of adults are overweight or obese, the seventh edition of the *Dietary Guidelines for Americans* places even more emphasis on increasing fruits and vegetables in the diet (e.g., it advises that half the plate should be fruits and vegetables) (USDA, 2011b). The recommended amount is 2½ cups of vegetables and 2 cups of fruit per day—contributing folate, magnesium, potassium, dietary fiber, and vitamins A, C and K—which moderate evidence suggests protects against some forms of heart disease and cancer. Data from the Behavioral Risk Factor Surveillance System (BRFSS) show that only about one-third of adults consume fruit two or more times per day and only about a quarter of adults consume vegetables three or more times per day, far short of the national target (Grimm et al., 2010). Eaters are also recommended to choose a variety of plant-based protein foods including unsalted nuts and seeds. Some evidence suggests that some tree nuts (walnuts, almonds, and pistachios) reduce risk factors for heart disease as long as

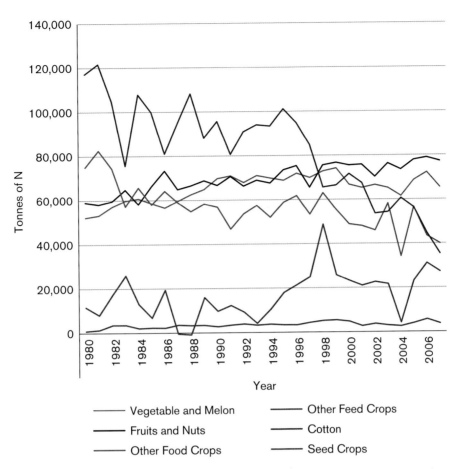

FIGURE 5.1.5. California estimated difference in tonnes of N fertilizer applied and N harvested, 1980–2007. The estimated difference between total tonnes of N applied (crop area × average N application rate) and tonnes of N harvested provide a rough approximation of how much of the N applied is lost to the environment. The crop categorization used here is consistent with the crop groupings used in Figure 5.1.2. Alfalfa is omitted since it is an N-fixing crop and very little synthetic N is applied.

DATA FROM: USDA NASS (2012); UC Davis Agricultural an d Resource Economics (n.d.); USDA NASS (n.d., *Surveys*).

they are consumed as part of a balanced diet and within calorie limitations (Kris-Etherton et al., 2008; O'Neil et al., 2011).

California also produces a significant amount of dairy products (see Section 5.1.3.2). Milk products contribute significantly to calcium, vitamin D (if fortified), and potassium in the diet. Adequate milk product intake is linked to bone health, especially in children and adolescents, and reduced risk of cardiovascular disease, type II diabetes, and lower blood pressure in adults.

5.1.2.2. N MANAGEMENT AND FOOD QUALITY: THE TRADE-OFF BETWEEN QUANTITY AND QUALITY

While fertilization of crops has increased crop yield, higher yields that result from nutrient application (not always N) tend to be inversely related with concentrations of vitamins and minerals in plant tissues (Jarrell and Beverly, 1981), as has been found in grains and berries (Davis, 2009). A

decrease in nitrate (NO_3^-) due to a decrease in N fertilizer use has been shown to increase the vitamin C content in fruits and leafy vegetables (Mozafar, 1996).

The effect of using organic versus inorganic sources of N on the mineral composition of food is debated (Lairon, 2010). It is provisionally agreed by most that food produced within organic farming systems is packed more densely with minerals and thus contains greater nutrition (Benbrook et al., 2008; Magkos et al., 2003; Rembialkowska, 2007; Williams, 2002), while others find the opposite (Bourn and Prescott, 2002; Dangour et al., 2009). Benbrook et al. (2008) found that in 61% of 236 paired comparisons, organic foods were nutritionally superior, while only 37% of the comparisons favored conventional foods. In contrast, Dangour et al. (2009) examined the same question and found that for eight nutrients and other nutritionally relevant substances ranging from nitrogen to copper, there was no difference between organic and conventionally produced food. Conventional crops contained

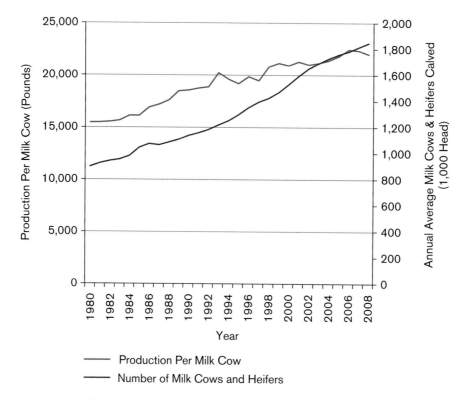

FIGURE 5.1.6. California population of milk cows and heifers and production per milk cow, 1980–2007. California has gotten more efficient in milk production over time as the production of milk per cow has increased from 15,000 pounds per milk cow in 1980 to about 22,500 pounds per milk cow in 2007.

DATA FROM: USDA NASS (n.d., *Historical Data*. Accessed July 1, 2012a).

higher levels of nitrogen, while organic crops contained higher levels of phosphorus and titratable acidity. Part of the difference in findings may result from which studies were selected for inclusion or the methods of analyzing comparisons.

The debate over nutritional quality of organic versus conventionally grown crops has also focused on California crops. The flavonoid content was greater in organic than in conventionally managed processing tomatoes and increased in the organic system after N inputs in the form of manure were reduced (Mitchell et al., 2007). Another study demonstrated that California strawberries when grown organically have higher antioxidant activity (Reganold et al., 2010). While more recent reviews survey a wider range of data, much of the results suggest that there is no definitive answer on how organic and inorganic N will affect nutrient composition of plants because of the confounding factors in the production systems.

Regardless of the source of N, it is generally accepted that the quality of California crops is sensitive to the amount of N applied, with negative consequences for crop production if too much or too little N is available. Negative effects include increased pest pressures, harvest and postharvest issues, and a lack of marketable yield (Daane et al., 1995; Hartz et al., 2005; Linquist et al., 2008). Crop sensitivity is

largely a function of growth habit, plant tissues and postharvest storage conditions, and market pressure.

5.1.3. Economic Benefit of Agricultural Production

The use of N in agroecosystems has enabled California to sustain and increase crop as well as livestock production since World War II, which has contributed tremendously to the economic well-being of Californians and the rest of the United States. California is one of the leading agricultural producers in the world and plays an important role in ensuring food security within the United States and internationally. California's agricultural economy is the largest in terms of cash receipts ($37.5 billion in 2010) in the United States and nearly twice as big as that of the third largest agricultural producing state of Texas ($19.9 billion in 2010) (USDA NASS, 2011a). California produces 21% of the nation's dairy commodities and more than 50% of the nation's fruits and vegetables. The United States is the world's largest producer of almonds, strawberries, and dairy products, where California's share of the US production for these three top commodities is 100%, 61%, and 19%, respectively. More detail about the importance of food production to California's economy and society, and to the US and global food system, is provided in online Appendix 5.1.2.

BOX 5.1.1. ANIMAL PRODUCTION REQUIRES MORE N

The agricultural sector has been identified as the largest driver of change in the nitrogen cycle on earth over the past few decades (Howarth, 2004). This is because nitrogen inputs serve human needs especially in agricultural production. Worldwide, N fertilizer accounts for about 40% of the increase in per capita food production in the past 50 years (Mosier et al., 2001). In addition to the increase in fertilizer N use, animal protein consumption in both developed and developing countries is also on the rise (Mosier et al., 2002).

The increase in worldwide demand for animal protein has led to significant changes in livestock and crop production that has contributed to increases in N loss. First, intensification of meat production increases the pressure of increasing N fertilizer into food production. This is because more N is needed to produce the same amount of animal protein as plant protein (Mosier et al., 2002). For example, Bleken and Bakken (1997) found that 3 g N must be supplied to soil to produce wheat flour containing approximately 6.3 g of protein, whereas a total of 21 g N must be supplied to soil to produce the same amount of animal protein. Further, when considering the efficiency of the whole system, estimates suggest that 4–11 units of feed N are

required to create 1 unit of animal protein (EPA SAB, 2011). The increased N requirements result from compounded inefficiencies as N is transferred through the supply chain. Tracing the N back in the food chain, Galloway and Cowling (2002) estimate that only 4% of N applied to corn is eventually consumed in beef. Although other animal production systems are typically more efficient than beef cattle on feed, this example highlights the systemic N inefficiencies when producing animal protein for human consumption.

Second, it has been observed that another contribution to the increase in N losses is due to the decoupling of livestock and crop production (Mosier et al., 2002). As a result, instead of treating animal manure as a plant nutrient source it is simply treated as a waste. This might have contributed to the increased use of synthetic N fertilizer in agricultural production.

Third, in addition to the decoupling of livestock and crop production, the level of animal production has exceeded that of crop production and this pattern is observed especially in China (Mosier et al., 2002). The excess manure N that is not used for crop production typically enters into aquatic systems, contributing to myriad ecological consequences.

5.2. Clean Drinking Water

Main Messages

The concentration of nitrate in California's surface water bodies seldom exceeds the federal maximum contaminate level (10 mg nitrate-N l⁻¹). As such, the use of surface water sources for drinking is generally considered low risk.

Nitrate levels in groundwater have increased over the past several decades, and in some parts of the state now exceed federal drinking water standards. This trend is likely to continue due to the time lag between the loss of nitrogen (N) to the environment and its accumulation in aquifers.

People in agricultural areas, particularly those with domestic wells, are more likely to be exposed to high levels of nitrate in their drinking water than those in urban and suburban areas. Groundwater from wells in the Tulare Lake Basin (TLB) and Salinas Valley (SV) regularly exceed the federal MCL and an estimated 8.0–9.4% of residents (212,500–250,000 people) in these areas are "highly susceptible" to exposure to water in excess of 10 mg nitrate-N L⁻¹.

For most adults, the amount of nitrate and nitrite consumed via foods is much greater than the amount consumed through drinking water. Infants given water or foods high in nitrate can develop "blue-baby syndrome," a potentially fatal condition where their blood cannot transport oxygen.

The International Agency for Research on Cancer concluded that nitrate and nitrite are "probably carcinogenic to humans." Nitrate

and nitrite can form nitrosamines, which are suspected to cause cancer. Consumption of nitrate and nitrite from all drinking water and food sources such as preserved meats is associated with stomach cancer in some studies.

Nitrate and nitrite can have positive effects on the body. In some patients, they are used to treat high blood pressure and reduce the risk of stroke.

Costs of treating nitrate-contaminated drinking water can pose a significant financial burden on low-income households and the public and community water systems that serve disadvantaged communities. While state-wide estimates of the cost to address nitrate in public and community water systems are needed, recent studies suggest that an increase in public and private funding on the order of $17–34 million per year over many decades will be needed to implement required nitrate mitigation projects for water systems in the TLB and SV.

5.2.1. Trends in Indicators of Water Quality

5.2.1.1. MAXIMUM CONTAMINANT LEVELS (MCL) FOR NITRATE AND NITRITE IN DRINKING WATER

This section describes the chemical and physical processes that affect N in California's drinking water, and discusses the spatial and temporal patterns of N in surface water and groundwater resources, as well as the human health and economic impacts. Drinking water in California is supplied

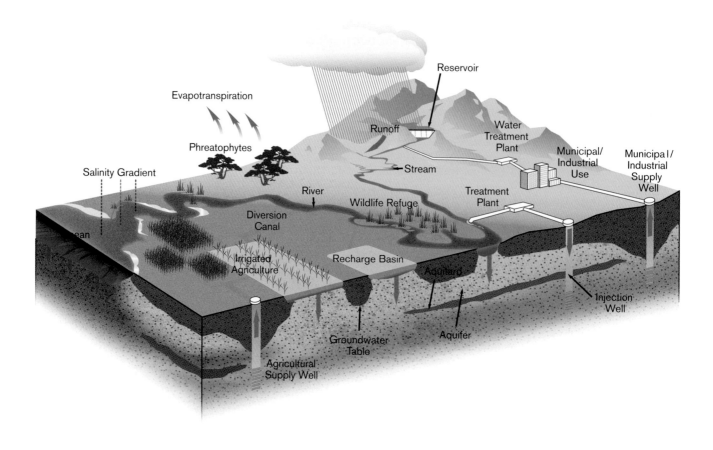

FIGURE 5.2.1. Connectivity and utilization of surface water and groundwater resources.
SOURCE: Reprinted from CA DWR (2014b).

by both surface water and groundwater, with approximately 40% of the population in part relying on groundwater as a source for their drinking water (CA DWR, 2003). Drinking water is protected by regulating both water sources and water suppliers. California treats surface water and groundwater separately although they are physically linked (figure 5.2.1). In general, the United States Environmental Protection Agency (US EPA) regulates surface water under the Clean Water Act, while the State Water Resources Control Boards implement federal regulations. In contrast, groundwater quality is regulated at the state and local levels. Regardless of the source, the US EPA under the authority of the Safe Drinking Water Act has set MCL of 10 mg nitrate-N L^{-1} and 1 mg nitrite-N L^{-1} for public drinking water supplies (EPA, 2009a). "Self-supplied water systems" (domestic wells serving 1–2 households), "local small water systems" (systems serving 2–4 households), and "state small water systems" (systems serving 5–14 households) are not subject to this water quality regulation (see online Glossary). Note that the units used by the US EPA are based on the mass of N in the nitrate or nitrite (e.g., mg nitrate-N L^{-1}), whereas European standards and guidelines are based on the mass of the nitrate molecule (e.g., mg NO_3^- L^{-1}). As such, 10 mg nitrate-N L^{-1} is equivalent to 45 mg

NO_3^- L^{-1} and 1 mg nitrite-N L^{-1} is equivalent to 3.3 mg NO_2^- L^{-1}.

5.2.1.2. THE CHEMICAL AND PHYSICAL BASIS OF NITROGEN IN DRINKING WATER

Sources of N in surface water and groundwater include weathering of bedrock, mineralization of organic N in soil, atmospheric deposition of N, N fertilizers, livestock waste, septic systems, and wastewater treatment plants (see mass balance in Chapter 4 for relative magnitudes). Ammonium (NH_4^+) and nitrate (NO_3^-) are the most abundant forms of reactive N that impact the quality of surface water and groundwater resources in California. Since ammonium is positively charged, it tends to adsorb to negatively charged soil particles and is thus not easily leached from the soil. However, under aerobic conditions ammonium is rapidly oxidized by microbes first to nitrite (NO_2^-) and then to nitrate through the process of nitrification. Nitrate is stable under aerobic conditions and highly mobile due to its negative charge and solubility in water. Hence, nitrate is generally the dominant form of N in both surface water and groundwater. Although nitrite is chemically unstable and usually prone to oxidation to nitrate, it sometimes

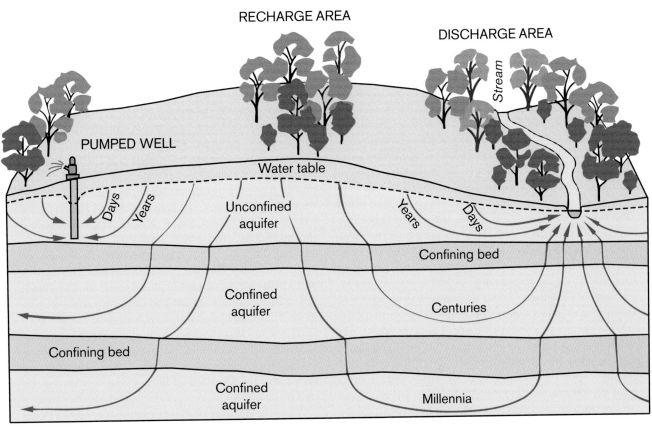

FIGURE 5.2.2. Spatial and temporal scales of groundwater movement. Groundwater flow paths vary greatly in length, depth, and travel time from points of recharge to points of discharge in the groundwater system.

SOURCE: Reprinted from Winter et al. (1998).

accumulates in surface water and ground water. Denitrification, which converts nitrate to gaseous nitrogen (N_2) under anaerobic conditions, is the main pathway that removes N from surface water and groundwater; however, if groundwater is high in oxygen and/or low in carbon, it takes decades to attenuate high groundwater nitrate loads (Green et al., 2008b). Nitrous oxide (N_2O), a potent greenhouse gas, is an intermediate product of both nitrification and denitrification. Estimates of amounts released to the atmosphere depend on environmental conditions and have a high degree of uncertainty.

Since the US EPA has established MCL for nitrate and nitrite in drinking water, in this section we will focus primarily on these two forms of N. Elevated levels of dissolved ammonia (the gaseous form of ammonium which dissolves in water) (EPA, 1999a) are sometimes found downstream of wastewater treatment sites, septic systems, and agricultural sites receiving fertilizer (Lehman et al., 2004; Parker et al., 2012). Though the US EPA has issued criteria standards for acute and chronic toxicity of juvenile fish species to dissolved ammonia, it is not considered a direct human health concern in drinking water because it rarely occurs at high concentrations (World Health Organization, 1996). Ammonia in excess of 0.165 mg ammonia-N L^{-1} (0.2 mg NH_3 L^{-1}) has the potential to significantly reduce the efficacy of chlorine disinfection of drinking water supplies (World Health Organization, 1996). Organic N in surface water draining wetlands and sometimes agricultural soils also contributes to N contamination of drinking water (Díaz et al., 2008; Hedin et al., 1995; van Kessel et al., 2009) but is not regulated and rarely monitored in drinking water.

5.2.1.3. NITROGEN IN SURFACE WATER: SPATIAL AND TEMPORAL TRENDS

Surface water originates either as runoff from land surfaces or as shallow groundwater that emerges at the surface. Runoff containing N originates from agricultural fields and other nonpoint sources, as well as from wastewater discharged from point sources, and depends on land cover types and land management practices. Loadings of N to surface water can range from less than 1 kg N ha^{-1} yr^{-1} on natural lands to approximately 20 kg N ha^{-1} yr^{-1} on agricultural lands (Beaulac and Reckhow, 1982; Jordan et al., 1997a, 1997b). Sources of N in groundwater are both natural and anthropogenic, and, in some cases, N has traveled many kilometers over thousands of years (figure 5.2.2).

In California, it is well established that drinking water drawn from the vast majority of surface water sources has relatively low concentrations of nitrate (figure 5.2.3). A large system of reservoirs, canals, and other water conveyances has been developed to move surface water from the Sierra Nevada mountains, the Sacramento/San Joaquin Delta (the Delta), and the Colorado River to the main urban areas along the coast. For example, the water supply for San Francisco and many East Bay cities is piped directly from the Hetch Hetchy Reservoir System on the upper Tuolumne River where nitrate concentrations are negligible (BAWSCA, 2012). Virtually all of the surface water bodies in the Central Valley have median nitrate concentrations well below the EPA drinking water MCL (EPA, 2006a). The concentration of nitrate in the surface water pumped from the Sacramento–San Joaquin Delta to Southern California for public drinking water supplies is less than 1 mg nitrate-N L^{-1} (4.45 mg NO_3^- L^{-1}) (EPA, 2006a; Foe et al., 2010). Nitrate concentrations in California's Lower Colorado River are also well below the MCL and appear to be improving as nutrient levels in the Upper Colorado River (e.g., Lake Powell and Lake Mead) have declined since the 1960s (Paulson and Baker, 1980). While less common, several studies also show that nitrate concentrations in some of the state's smaller rivers and sloughs (e.g., Pajaro River, Mud Slough) are sometimes above the regulatory limit (EPA, 2006a; Ruehl et al., 2007). There is considerable evidence that, for example, the San Joaquin River is affected by nitrate from anthropogenic sources (Pellerin et al., 2009) as is the Salinas River (Moran et al., 2011). In addition to the degradation of drinking water supplies, high nitrate concentrations in surface waters are linked to eutrophication and other ecological problems (see Section 5.5).

5.2.1.4. NITROGEN IN GROUNDWATER: SPATIAL TRENDS

In contrast to surface water supplies, nitrate contamination of groundwater is becoming a widespread problem in various parts of California (Harter, 2009; Harter et al., 2012; Map 5.2.1). While this problem is well established and broadly observed (figure 5.2.3), the occurrence of nitrate in groundwater can vary considerably in three-dimensional space and is influenced by a region's hydrologic features, soil type, and land-use patterns. Nitrate enters groundwater primarily via leaching, which transports excess N from the soil surface through soil pore spaces in the vadose zone until it reaches the water table. Major sources of nitrate entering groundwater are fertilizers and livestock manures, which are applied in excess of a crop's requirements. For some sources of N (e.g., dairy lagoons and septic systems), there is little opportunity for plant uptake because the nitrate never interacts with the rooting zone of crops or other vegetation. The importance of agriculture as a major source of N is demonstrated by the fact that nitrate concentrations in monitoring wells located in agricultural areas are often well above background nitrate

levels (2.0 mg nitrate-N L^{-1}; 9 mg NO_3^- L^{-1}; Boyle et al., 2012).

Several studies commissioned by the California State Water Resources Control Board have examined the current spatial patterns of groundwater nitrate in various parts of the state. Much of the recent work has focused on important agricultural regions in the TLB and SV (Table 5.2.1). This study area accounts for approximately 40% of the state's cropland, 50% of the state's livestock, and 7% of the human population (Boyle et al., 2012). Between 2000 and 2009, public supply wells in the TLB and SV regions had median nitrate concentrations of 5.2 and 4.7 mg nitrate-N L^{-1}, respectively (equivalent to 23 and 21 mg NO_3^- L^{-1}), with approximately 10% of samples exceeding the maximum contaminate level (10 mg nitrate-N L^{-1}; 44.5 mg NO_3^- L^{-1}). In several groundwater subbasins of Fresno and Tulare Counties, datasets consisting exclusively of domestic wells had exceedance rates of 30–45% (Boyle et al., 2012). Wells used for domestic and irrigation purposes often have higher concentrations than public supply wells due to their shallow depth and their proximity to agricultural land uses. In contrast, the deeper confined aquifers in the western and central TLB and the northern subbasin of the SV also tend to have relatively low nitrate concentrations.

While the Boyle et al. (2012) study was confined to the TLB and SV regions, other studies indicate that wells exceeding the drinking water MCL are also found in other parts of the state. Data from the State's Groundwater Ambient Monitoring and Assessment (GAMA) program (Belitz et al., 2003; CA EPA Water Resources Control Board, 2015), which monitors thousands of wells throughout the state, indicate that the drinking water standard is often exceeded in parts of the San Joaquin, Sacramento, and Santa Ana basins (Harter 2009; Map 5.2.1). Recent efforts by Dubrovsky et al. (2010) to predict groundwater nitrate levels by projecting observed temporal trends provide estimates that are largely consistent with the work of Harter et al. (2012), Boyle et al. (2012), and Anning et al. (2012), and suggest further that shallow groundwater resources in the Imperial Valley are also above the MCL.

5.2.1.5. NITROGEN IN GROUNDWATER: HISTORIC TRENDS AND FUTURE PROJECTIONS

Recent studies conducted in California all agree that groundwater nitrate levels have increased over the past several decades (figure 5.2.3), particularly in major agricultural regions (Boyle et al., 2012; Burow et al., 2008b; Honeycutt et al., 2012). Since the 1970s, average nitrate concentrations in public supply wells in the TLB and SV have increased by approximately 0.061 and 0.120 mg nitrate-N L^{-1} yr^{-1} (0.27 and 0.53 mg NO_3^- L^{-1} yr^{-1}), respectively (Honeycutt et al., 2012). Likewise, nitrate levels in the eastern San Joaquin Valley more than doubled between 1950 and 2000, with concentrations approaching twice the federal MCL in

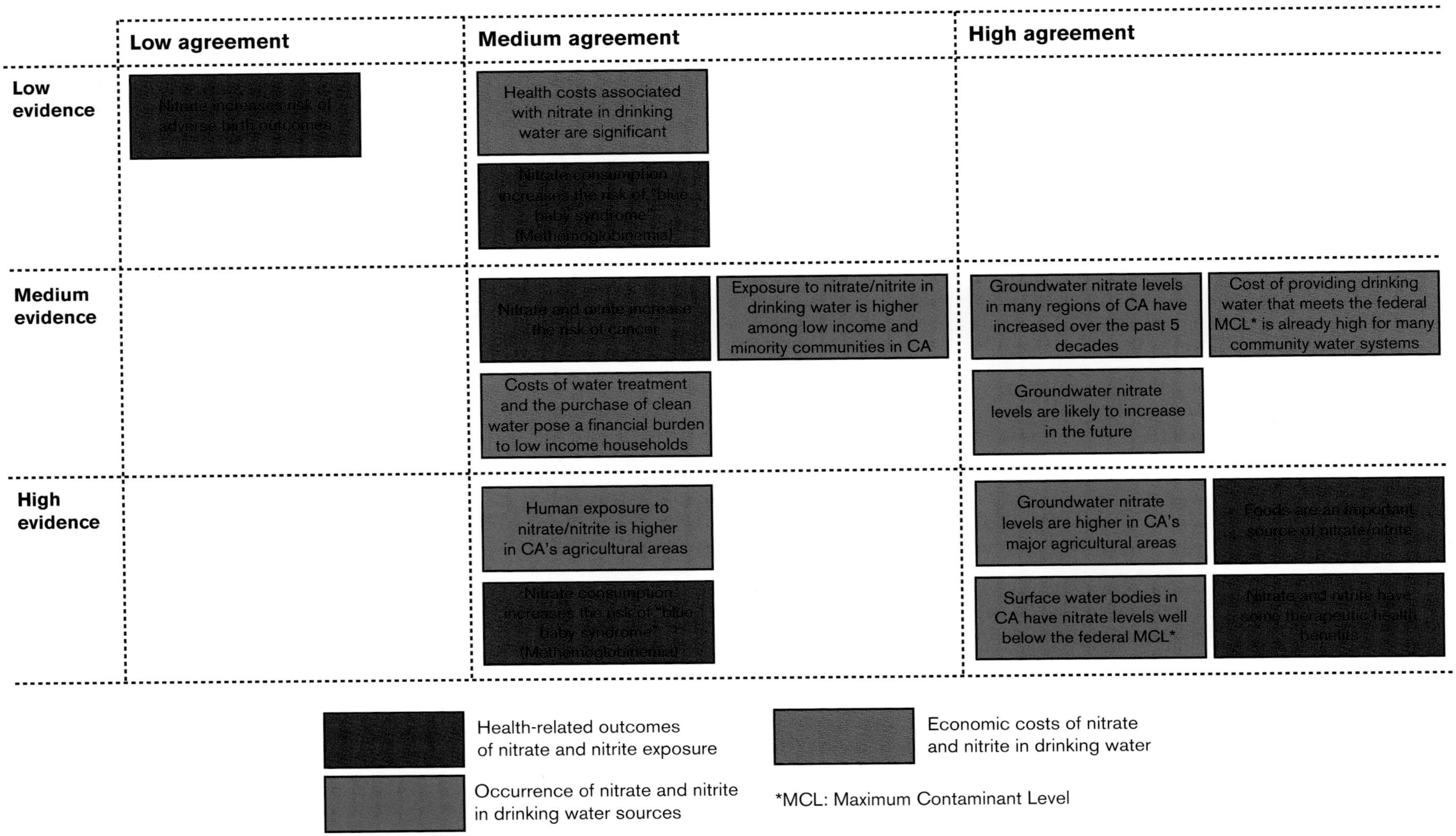

FIGURE 5.2.3. Measuring uncertainty in nitrogen's impact on human well-being. The figure shows the amount of evidence and level of agreement on various aspects of nitrogen impacts on human well-being through contamination of drinking water resources.

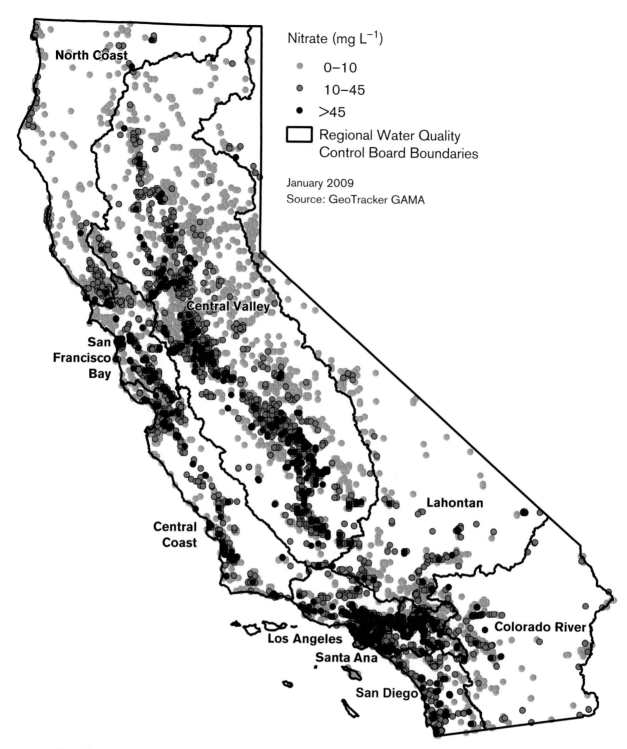

Nitrate (mg L^{-1})

- 0–10
- 10–45
- >45

☐ Regional Water Quality
  Control Board Boundaries

January 2009
Source: GeoTracker GAMA

North Coast

Central Valley

San
Francisco
Bay

Central
Coast

Lahontan

Los Angeles

Santa Ana

San Diego

Colorado River

MAP 5.2.1. Groundwater nitrate concentrations measured in wells throughout California (data from CA SWRCB, 2015; map by M. Tobias).

groundwater in the vicinity of Fresno and Modesto (Burow et al., 2007, 2008a, 2012; figure 5.2.4).

Using the CASTING database, which includes data from thousands of public supply, monitoring, domestic and irrigation wells, Boyle et al. (2012) estimated an average increase in nitrate concentration of 0.08 mg nitrate-N L^{-1} yr^{-1} (0.34 mg NO$_3^-$ L^{-1} yr^{-1}) between 1950 and 2010 across

all well types, and found the proportion of wells testing higher than background and above the MCL also increased over the past six decades (figure 5.2.5). Despite clear evidence of significant increases in groundwater nitrate, historic trends are difficult to decipher because of the paucity of samples prior to 1990 (mostly public supply wells), and large increases in the number of domestic and irrigation

TABLE 5.2.1

Data sources with the total number of samples recorded, total number of sampled wells, location of wells, type of wells, and for the last decade (2000–2010) in the Tulare Lake Basin and Salinas Valley: number of wells measured, median nitrate concentration, and percentage of MCL exceedance for the Tulare Lake Basin and the Salinas Valley

Data source[1]	Well type[2]	Total # samples	Tulare Lake Basin (2000–2010)[3]			Salinas Valley (2000–2010)[3]		
			# of wells	Median (mg NO_3 L^{-1})	% > MCL	# of wells	Median (mg NO_3 L^{-1})	% > MCL
CDPH	PS	62,153	1,769	12	6	327	8	5
CVRWB dairy	D, I, M	11,300	6,459	22	31			
DPR	D	814	71	40	45			
DWR	I	44	28	1	0			
ENVMON	M	2,601	357		52	180	27	44
Fresno Co.	D	369	349	18	15			
GAMA	D	141	141	38	43			
Kern Co.	D, I	3,825	361	5	7			
Monterey Co., Report	I, M	1,018				98	14	36
Monterey Co., Geospatial	LS	1,574				431	18	15
Monterey Co., Scanned	LS	5,674				427	17	14
NWIS	Misc.	2,151	76	35	36	4	0	0
Tulare Co.	D	444	438	22	27			
Westlands Water Distr.	I	77	31	4	0			

1. Data source: CDPH: public supply well database; CVRWB Dairy: Central Valley RWB Dairy General Order; DWR data reports from the 1960–1970s, 1985; ENVMON: State Water Board Geotracker environmental monitoring wells with nitrate data (does not include data from the CVRWB dairy dataset); EPA: STORET dataset; Fresno County: Public Health Department; GAMA: State Water Board domestic well survey; Kern County: Water Agency; Monterey County, Reports: data published in reports by MCWRA; Monterey County, Geospatial: Health Department geospatial database; Monterey County, Scanned: Health Department scanned paper records; NWIS: USGS National Water Information System; Tulare County: Health and Human Services; Westlands Water District: district dataset. Some smaller datasets are not listed. Individual wells that are known to be monitored by multiple sources are here associated only with the data source reporting the first water quality record.

2. D = domestic wells; I = irrigation wells; LS = local small system wells; M = monitoring wells; PS = public supply wells.

3. Median and MCL exceedance percentages were computed based on the annual mean nitrate concentration at each well for which data were available.

SOURCE: Data from Boyle et al. (2012).

well samples at dairies beginning in 2007. Since public supply wells tend to be deeper and have somewhat lower nitrate concentrations than the shallow wells used for domestic and agricultural purposes, this change in data sources would tend to exaggerate the increasing trend particularly after 2007 (figure 5.2.5). Deactivation and abandonment of public supply wells with water quality problems also makes assessing temporal trends in groundwater nitrate a challenge. Taking this into account, the overall trend is still an increase in nitrate concentrations prior to 2007, albeit at a more gradual slope.

Because of the time lag required for applied N to reach groundwater, nitrate concentrations are likely to increase in the coming decades even if robust measures to minimize contamination are implemented (Harter, 2009). While groundwater nitrate concentrations are likely to continue their upward trend, very few studies in California have

been conducted to project how rapidly nitrate levels may increase under various future land-use and groundwater protection scenarios (figure 5.2.3). Building on their extensive dataset of wells in the TLB, Boyle et al. (2012) have sought to address this knowledge gap by developing a process-based transport simulation model for nonpoint sources of nitrate across six hydrologic subbasins (Kings, Westside, Tule, Kaweah, Tulare Lake, Kern subbasins) within the TLB. The nitrate transport model projections exhibit significant spatial and temporal uncertainty due to inherent variability in N loading (i.e., N losses to the environment) across different land-use and source types (e.g., agricultural crops, septic systems, manure lagoons). But while the model may not forecast future groundwater nitrate levels with a high degree of accuracy, it remains a useful tool for evaluating trends and the impact of alternative land-use management scenarios.

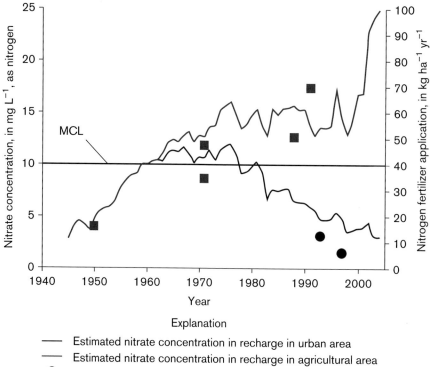

Explanation

— Estimated nitrate concentration in recharge in urban area
— Estimated nitrate concentration in recharge in agricultural area
● Monitoring well in urban area
■ Monitoring well in agricultural area

FIGURE 5.2.4. Estimated concentrations of nitrate in recharge and observed concentrations of groundwater nitrate in monitoring wells in Modesto, California, 1945–2005. Observed concentrations of nitrate from groundwater sampled in 2003–2005 (Modesto) are plotted against corresponding interpreted ages from age-dating tracers. Estimated concentrations of nitrate from nitrogen fertilizer applications represent 50% of the nitrogen fertilizer applications divided by the area of fertilized land, dissolved in 0.4 m yr^{-1} of recharge in Modesto, with the assumption that 50% of N applied as fertilizer reached the water table.

SOURCE: Reprinted from Burow et al. (2008a).

Using output from an N loading algorithm developed by Viers et al. (2012), Boyle et al. (2012) modeled four N loading scenarios that consider how changes in land use and N management from 1945 until 2050 may impact groundwater nitrate concentrations. Scenarios A and D assume that shifts in land use and improved N management will decrease N loading after 1990, while scenarios B and C assume that land-use patterns and N management result in progressively higher N loading rates in the future (figure 5.2.6). Under each of the scenarios, the transport model projected increasing nitrate concentrations for all the subbasins in all the groundwater subbasins in the TLB region. For the Westside, Kaweah, and Tule subbasins, mean nitrate concentrations are all projected to exceed the drinking water MCL between 2005 and 2030 for both the A and C scenarios. By contrast, mean nitrate concentrations in the Kings, Tulare Lake, and Kern subbasins are not projected to reach the MCL threshold prior to 2050 (figure 5.2.7). These results suggest that even with focused efforts to reduce N loading from nonpoint sources, nitrate contamination of groundwater resources is likely to become an increasingly intractable problem in certain regions. While the TLB and SV groundwater basins are

likely to be among the most impaired in California, more studies are needed to evaluate and monitor the many other groundwater basins throughout the state.

5.2.2. Human Exposure

5.2.2.1. CONSUMPTION OF NITRATE/NITRITE IN DRINKING WATER AND FOOD

As discussed above, nitrate and nitrite are the primary N species present in drinking water. In food, amino acids and proteins are the main N form, though it is also well established that nitrate and nitrite are often present in significant quantities (figure 5.2.3). Nitrate levels can be very high in leafy green vegetables, carrots, and silver beets, sometimes exceeding 677.4 mg nitrate-N kg^{-1} of vegetable (3,000 mg NO$_3^-$ kg^{-1}) (Correia et al., 2010; European Food Safety Authority, 2008; Jaworska, 2005; Matallana González et al., 2010; Tamme et al., 2010). The amount of nitrate depends primarily on the type of crop, but is also influenced by the amount of fertilizer applied, environmental conditions, type of processing, and storage time (Anjana and Iqbal, 2007; Chung et al., 2004; Prasad and Chetty, 2008). Nitrates

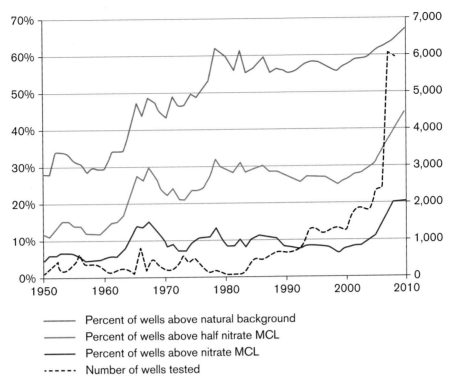

FIGURE 5.2.5. Five-year moving average of the percentage of Salinas Valley and Tulare Lake Basin wells with average annual NO_3 levels > 9 mg L^{-1} (background), 22.5 mg L^{-1} (1/2 MCL) and 45 mg L^{-1} (MCL), 1950–2010. Prior to 1990, most wells sampled were public supply wells. There are only a small number of samples available for the 1950–1970 period. Thus, trends during this period should be interpreted with care. In 2007, Central Valley dairies began testing domestic and irrigation wells which greatly increased the number of samples available for analysis.

SOURCE: Reprinted from Boyle et al. (2012).

are used in processing and preserving meats, and can be found in concentrations greater than 22 mg nitrate-N kg^{-1} of food (100 mg NO_3^- kg^{-1}). Sodium nitrite is also commonly used to cure meats and meat products such as ham, bacon, and sausage. In these foods, nitrite is present at much higher levels than in drinking water, with some cured meats having average levels of 1.5–3.0 mg nitrite-N kg^{-1} food (5–10 mg NO_2^- kg^{-1}) (European Food Safety Authority, 2008).

In drinking water, nitrate and nitrite are more of a problem when groundwater is the main source of drinking water rather than surface water. While consumption of foods, such as vegetables and processed meats are typically the main source of nitrate for most adults, in areas with high groundwater nitrate levels, drinking water can also be a significant means of exposure. Studies suggest that for those who consume drinking water well below the regulatory limit, only 7–11% of total nitrate intake comes from drinking water (IARC, 2014). However, when water sources have nitrate levels close to the regulatory limit (10 mg nitrate-N L^{-1}; 44.5 mg NO_3^- L^{-1}), as much as 50–70% of total nitrate intake may come from drinking water (Correia et al., 2010; European Food Safety Authority, 2010; Griesenbeck et al., 2010; IARC, 2014).

Nitrate in drinking water is a much more important exposure route for young infants if they are fed tap water or

foods made with tap water. This is particularly important since infants under the age of six months are most susceptible to the harmful effects of nitrate. However, there is little information about nitrate exposure levels among young infants (European Food Safety Authority, 2010). In Romania, nitrate in drinking water, given in the form of tea, was the major source of nitrate exposure (Zeman et al., 2002). A study by VanDerslice (2009) in rural Washington State found that less than 2% of the infants less than six months of age consumed any vegetables containing significant amounts of nitrate, leaving drinking water as the main source of dietary nitrate. In this same study, approximately 10% of self-supplied households using private wells and 4.7% of households served by small water systems had nitrate levels over the federal MCL. Still for over half of the sample, total intake of nitrate was quite small at less than 0.5 mg nitrate-N kg^{-1} body weight (2.2 mg NO_3^- kg^{-1}) (VanDerslice, 2009).

5.2.2.2. EXPOSURE PATTERNS IN CALIFORNIA

The US Geological Survey estimates that 7% of California residents rely on self-supplied water systems or small water systems serving fewer than 15 households, with the remaining 93% being supplied by public and community water

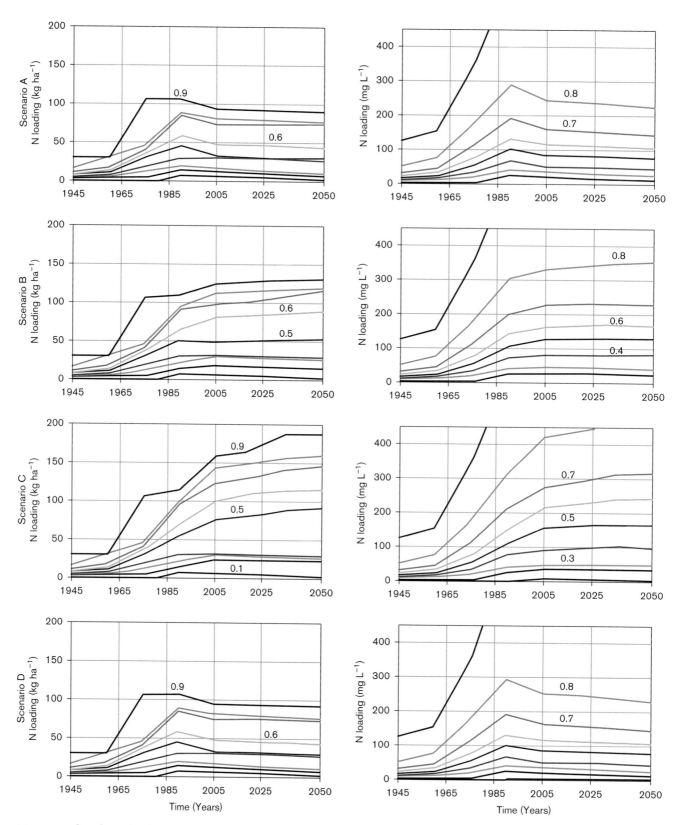

FIGURE 5.2.6. Cumulative distribution of N loading per year for model scenarios A, B, C, and D. The left panels correspond to the N loading output algorithm of Viers et al. (2012) expressed in kg N ha⁻¹ and mg NO₃ L⁻¹ in groundwater which are used as input to the nonpoint source assessment tool (NPSAT) simulation model. Scenarios A and D assume declines in N loading, while B and C assume increased N loading over time.

SOURCE: Reprinted from Boyle et al. (2012).

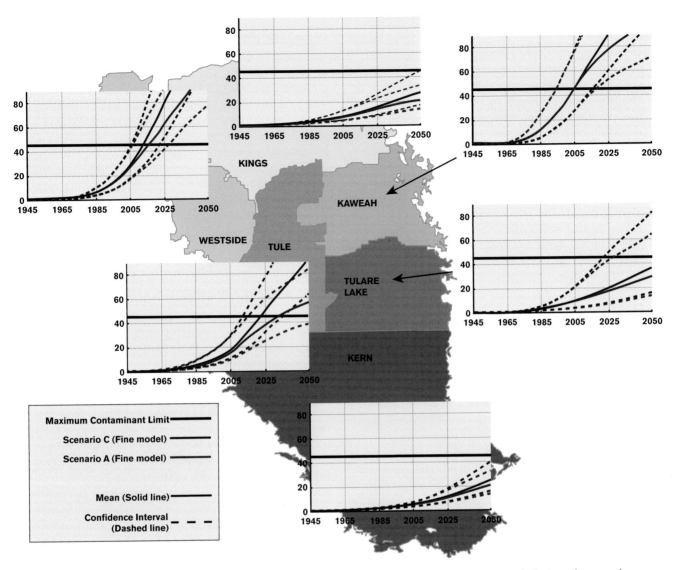

FIGURE 5.2.7. Historic and projected change in groundwater nitrate concentrations for six regions in the Tulare Lake Basin study area under model scenarios A and C. Scenario A assumes decreased N loading as nitrate over time, while scenario C assumes an increase in N loading.

SOURCE: Reprinted from Boyle et al. (2012).

systems (Kenny et al., 2009; see online Glossary). Under the Safe Drinking Water Act, public and community water systems are required to regularly test for a wide range of contaminants. Based on data from the US EPA, 21 of the 3,049 active public and community water systems in California violated the federal MCL for nitrate during 2010 (EPA, 2014). These systems served an estimated 99,162 people, of which 92,158 were from one city water system that had a single nitrate MCL violation. Of the 21 systems that violated the nitrate MCL, 14 were located in Tulare County, with a combined service population of 5,458 residents. Overall, 0.3% of those served by public and community water systems in California are potentially exposed to nitrate levels greater than the MCL.

It is more difficult to assess the potential exposure facing self-supplied water systems that have individual wells or small water systems (see online Glossary). Self-supplied and small water systems are required to test only after the well is drilled or at the time the property is sold. These results are not compiled centrally. Using data compiled from the California State Water Resources Control Board's GAMA Geotracker system, 9.8% of over 16,000 self-supplied wells tested had at least one value greater than the drinking water MCL, and 5.8% had an average level greater than this value (see Appendix 4.1). Almost 30% of the wells had maximum nitrate levels greater than 3 mg nitrate-N L^{-1} (13.3 mg NO_3^- L^{-1}), indicating human impacts on the level of nitrate. These proportions varied across the state. These results should be thought of as general indicators, as the wells in this database included many types of wells, some of which were drilled specifically to characterize areas thought to have high nitrate levels.

Honeycutt et al. (2012) examined potential exposure to drinking water in the TLB and SV by classifying the susceptibility of various private and public water systems (and the 2.65 million people they serve) based on both a qualitative definition of a water system's vulnerability (e.g., size of system, regulatory oversight, etc.) and groundwater nitrate concentrations between 2006 and 2010 (figure 5.2.8). An estimated 8.0–9.4% of the population in the study area (or 212,500–250,000 people) had "high susceptibility" to nitrate exposure through drinking water that exceeded the federal MCL. Application of a similar approach at the state level would be useful in evaluating the extent of exposure in other parts of the state. As a start, service maps of California drinking water systems have now been completed for 90% of the California population who have public drinking water systems by the California Environmental Health Tracking Program (Wong et al., 2015).

5.2.2.3. DISPARITIES IN EXPOSURE TO NITRATE/NITRITE IN CALIFORNIA

Several studies have looked at whether minority or low-income residents in California receive poorer quality drinking water than the rest of the state. However, without knowing precisely which areas, and thus which people, each system serves, assessing social disparities in water quality is very difficult. The Environmental Justice Coalition for Water conducted a county-level analysis and found that counties with the highest number of drinking water violations had a higher proportion of people of Latino ethnicity than counties with the lowest number of violations (42% vs. 16%) (Deen et al., 2005). There were smaller disparities related to income; 17% of those living in counties with the highest number of violations were living below the poverty line as compared to 12% of those in counties with the fewest violations. Another study found the proportion of residents who were Latino and the proportion who rented were significantly associated with community water system wells that had higher levels of nitrate (Balazs and Ray, 2009) and concluded this was evidence of disparity in water quality levels based on ethnicity and poverty status.

Reports of predominantly low-income Latino communities in the San Joaquin Valley served by community water systems with elevated levels of nitrate (Firestone et al., 2006) found that 74% (*n* = 29) of the 44 community water systems in California that violated the nitrate MCL in 2007 were located in this region (Moore and Matalon, 2011). US EPA data indicated that though the number of systems with nitrate MCL violations dropped from 39 to 21 between 2007 and 2010, 76% of the systems in violation (*n* = 16) were still in the San Joaquin Valley (EPA SAB, 2011).

Honeycutt et al. (2012) also examined the extent to which water systems serving disadvantaged communities in the TLB and SV exceeded the federal nitrate MCL between 2006 and 2010. In this study, "disadvantaged" and "severely disadvantaged" communities were defined as those having a median household income in 2000 below $37,994 and $28,496, respectively, which is equivalent to 80% and 60% of the statewide median household income of $47,493. Of the 328 community water systems, 51 exceeded the nitrate MCL, and 40 of the systems in violation were located in severely disadvantaged or disadvantaged communities that served approximately 379,000 people. While the studies assessed suggest that minorities and low-income populations may face higher exposures to nitrate in drinking water, more detailed studies that link individuals to their specific water systems, or which actually test the levels of nitrate in their water, are needed to gain a better understanding of the disparities in water quality throughout California.

5.2.3. Human Health Effects of Nitrate/Nitrite

5.2.3.1. ADVERSE AND BENEFICIAL EFFECTS

The consumption of nitrate and nitrite can have both adverse and beneficial effects on human health. Foods and drinking water containing high levels of nitrate and nitrite are thought to be related to three types of health problems: methemoglobinemia, adverse birth outcomes, and cancer. Studies examining these health risks are reviewed below. In addition, a small number of studies tentatively agree that increasing levels of nitrate concentration in drinking water is associated with increasing symptoms of subclinical thyroid disorders, such as hypothyroidism (Aschebrook-Kilfoy et al., 2012), although one of the studies found no such association (Ward et al., 2010). However, nitrate and nitrite and other nitrogen-containing compounds are also used as therapeutic agents to lower blood pressure, and to reduce aggregation of platelets, and nitric oxide is an important signaling molecule to regulate cellular functions (figure 5.2.3).

5.2.3.2. INTERPRETING EPIDEMIOLOGICAL EVIDENCE

Epidemiologic studies considered here are of two types: case control and cohort. In case-control studies, people with a disease are identified, and are compared in terms of exposure to similar people who do not have the disease. Optimally these people are randomly selected from the same population of the cases. In cohort studies, a group of initially disease-free people are observed over time, and are categorized by their level of exposure. The proportions of people that develop the disease are compared across exposure groups.

As with many environmental epidemiologic studies, there are conflicting results, with some studies showing associations and others not showing any associations. Results are influenced by the study design. Studies that are small are less apt to find a relationship when there is one. How exposure to nitrates and/or nitrites in drinking water and foods is measured can affect study results. Poor estimates of these exposures will often lead to results that show

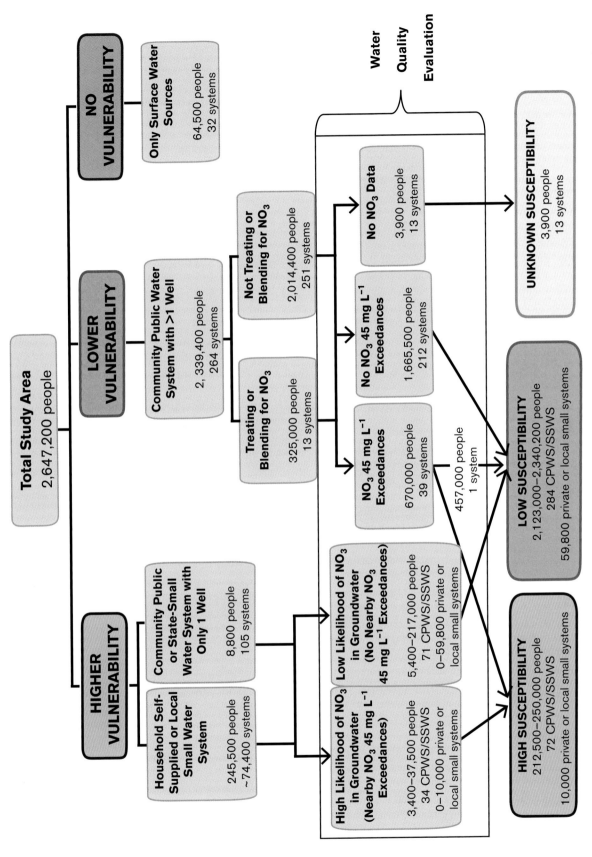

FIGURE 5.2.8. Classification of susceptible water systems and populations based on estimated vulnerability and water quality data in the Tulare Lake Basin and Salinas Valley study area.

SOURCE: Reprinted from Honeycutt et al. (2012, p.2).

a weak or no effect, even if an association is present. In the studies reviewed, exposure was estimated in several ways. In many studies, the type of water system serving the person's house (public vs. private, groundwater vs. surface water) or the nitrate level (high vs. low) defined exposure, regardless of whether the person drank tap water or how much they drank. This can mask an effect of exposure. Only a handful of studies asked about, or actually observed, water consumption patterns. Exposure is particularly difficult to determine in studies of cancer, where the important exposures occurred several years before the cancer developed. Whether the exposure is due to nitrate or nitrite in food or water is also important as the risk may be different due to chemical reactions of nitrite with certain molecules in some foods (see Section 5.2.3.5).

It is quite possible that water supplies high in nitrate/nitrite may also have high concentrations of pesticides and other contaminants given the likelihood that these inputs are often applied together in intensively managed cropping systems. Thus, it is possible that health effects attributed to nitrate/nitrite may actually be due to a different contaminant. These limitations need to be kept in mind when assessing the evidence from epidemiologic studies as we do below.

5.2.3.3. METHEMOGLOBINEMIA (BLUE-BABY SYNDROME)

Methemoglobinemia, or 'blue-baby syndrome,' is a condition where infants become hypoxic and turn "bluish," due to a lack of oxygen to the tissues. It is generally accepted that methemoglobinemia is caused when nitrite in the blood converts normal hemoglobin (which carries oxygen) to methemoglobin, a form that does not carry oxygen (McCarty, 1981; Wright et al., 1999). High levels of methemoglobin can lead to symptoms such as lethargy, dizziness, coma, and even death. Nitrate in drinking water or food is converted by stomach bacteria to nitrite, which is then absorbed into the blood stream, where it triggers conversion of hemoglobin to methemoglobin (figure 5.2.3).

This syndrome is rare in adults who can rapidly convert methemoglobin back to normal hemoglobin, whereas this process is slower in infants under six months old (Jaffé, 1993). In addition, infants have a form of hemoglobin (fetal) that is more susceptible to nitrites (McCarty, 1981; Modell and Darlison, 2008) and a higher gastric pH that supports higher rates of conversion of nitrate to nitrite than is the case for adults.

Direct effects of ingested nitrate are difficult to determine due to the complex physiological processes involving nitrate and nitrite. For example, infants with diarrhea or other intestinal inflammation have had methemoglobinemia, even without any exposure to nitrates (Hanukoglu and Danon, 1996; Hegesh and Shiloah, 1982; Lebby et al., 1993; Pollack and Pollack, 1994). Subsequent studies have demonstrated that bacterial infections and inflammation of the bowel leads to production of nitric oxide (NO), which can also produce methemoglobin (Gupta et al., 1998; Levine et

al., 1998; Tannenbaum et al., 1978; Wagner et al., 1984; Witthöft et al., 1998). Furthermore, when nitrite or nitric oxide reacts with hemoglobin to form methemoglobin, nitrate is produced (Lundberg et al., 2004). Some of the nitrate in the blood is gathered by the salivary glands and excreted into the mouth, where normal bacteria convert some portion to nitrite. This nitrite and the remaining nitrate are then swallowed, creating a complex cycle.

In the few studies linking nitrate exposure and methemoglobinemia (Fewtrell, 2004), Shuval and Gruener (1972) found no differences in mean methemoglobin levels between infants exposed to tap water with high nitrate (11.3–20.3 mg nitrate-N L^{-1}; 50–90 mg NO_3^- L^{-1}) and low nitrate (<1.13 mg nitrate-N L^{-1}; <5 mg NO_3^- L^{-1}) concentrations. However, tap water accounted for only a small proportion of the infants' nitrate exposure. A similar study found little difference in methemoglobin levels (0.75% vs. 1.22%) comparing a dose of 1–19 mg NO_3^- in the 2 hours prior to sampling to 20–50 mg NO_3^-. Neither was above what is considered to be a normal level of methemoglobin (3%) (Craun et al., 1981). Knobeloch et al. (2000) found two cases of clinical methemoglobinemia among infants who had consumed formula prepared with well water containing nitrate-nitrogen at 22.9 and 27.4 mg L^{-1} (likelihood of coliforms was thought to be low). In a study of effects of nitrate, total coliforms and *E. coli* on methemoglobinemia in 800 infants, the median nitrate-nitrogen concentration was 1.5 mg L^{-1} and 8.6% of the observations were above 10 mg L^{-1} (VanDerslice, 2009). Approximately 25% of infants in the study also consumed water that tested positive for total coliforms. Small but statistically significant relationships were found between the amount of nitrate and total coliform ingested and elevated methemoglobin levels in infants though none of the infants exhibited clear physical symptoms of methemoglobinemia (VanDerslice, 2009). A case-control study in Romania found a significant relationship between nitrate intake in diet and drinking water and methemoglobinemia episodes, and a weaker association between diarrhea and methemoglobinemia episodes (Zeman et al., 2002). Average nitrate consumption level was high among the cases whose water was >10 ppm NO_3^--N (>22.6 mg nitrate-N kg^{-1} body weight day^{-1}; >100 mg NO_3^- kg^{-1} body weight day^{-1}). Elevated methemoglobin levels were detected in children up to 8 years of age drinking well water and piped water. Children consuming well water with greater than 10 mg L^{-1} nitrate-N (>44.5 mg NO_3^--N L^{-1}) were 1.6 times more likely to have elevated methemoglobin levels (>2% methemoglobin) than those consuming the piped supply (average = 2.99 mg NO_3^--N L^{-1}) (Sadeq et al., 2008). The number of children with elevated methemoglobin increased with age, peaking at 6 years of age. There was no evidence that food consumption was related to the increased prevalence of elevated methemoglobin.

The current MCL for nitrate is based primarily on two studies that concluded that methemoglobinemia occurs when an infant is exposed to greater than 10 mg nitrate-N L^{-1} (>44.5 mg NO_3^- L^{-1}) in drinking water (Bosch et al., 1950;

EPA, 1990; Walton, 1951). As our understanding of nitrate metabolism in the body increases, some researchers questioned the role of nitrate-contaminated water as a cause of methemoglobinemia and suggested instead that bowel inflammation due to gastrointestinal infection may be a source for nitric oxide formation (Avery, 1999; Avery and L'hirondel, 2003). Groundwater contaminated with nitrate is also more likely to be contaminated with enteric pathogens, so the apparent relationship between well water nitrate levels and methemoglobinemia may be due to microbiological contamination. Studies of VanDerslice (2009) and Zeeman (2002) observed effects of nitrate intake while controlling for diarrheal disease, and there have been carefully documented cases of methemoglobinemia where drinking water was above the nitrate MCL and there was no exposure to bacteriological pathogens in the drinking water, nor evidence of diarrheal disease (Knobeloch et al., 2000). This ambiguity has called into question whether the MCL should be set at a higher level (Avery, 1999). The widespread nature of the problem, the high cost of treatment to remove nitrate or to develop alternative water sources, the lack of observed cases of methemoglobinemia, and the knowledge that the condition can be fatal combine to make this a complicated issue.

5.2.3.4. BIRTH OUTCOMES AND BIRTH DEFECTS

It has been suggested but remains unproven that exposure to nitrate may affect birth outcomes (figure 5.2.3). The National Academy of Science (McCarty, 1981) suggested that a reduction in blood oxygen levels from the creation of methemoglobin might impact the development of the fetus, and there is some evidence of this in animal studies (Fan et al., 1987). In two studies, Tabacova et al. (1998, 1997) found that maternal methemoglobin levels were associated with the risk of pregnancy complications (preterm birth, low birth weight, fetal distress, premature labor), and that the methemoglobin level in cord blood was strongly associated with the methemoglobin level in the mother's blood. While these studies did not link methemoglobin levels to drinking water nitrate levels, Bukowski et al. (2001) found that women who lived in areas with median drinking water nitrate levels over 3 mg nitrate-N L^{-1} (>13.3 mg NO_3^- L^{-1}) were twice as likely to have a low birth weight baby as women exposed to very low nitrate levels (<1.3 mg nitrate-N L^{-1}; <5.8 mg NO_3^- L^{-1}).

There are also concerns that exposure to nitrates/nitrites during pregnancy can increase the risk of spontaneous abortion. Studies examining higher versus lower methemoglobin levels during pregnancy (Skrivan, 1971), women who consumed water with variable nitrate levels (Aschengrau et al., 1989), and communities with high versus low nitrate levels (Gelperin et al., 1975) found no adverse effects. A large hospital-based case-control study in Massachusetts found no association between nitrate levels in drinking water and stillbirth or congenital anomalies, but a weak association with neonatal deaths (Aschengrau et al., 1993).

Nitrate exposure has been related to birth defects of the central nervous system (CNS). Women were more than three times as likely to deliver a baby with a defect of the CNS when exposed to drinking water (from groundwater) containing more than 15 mg nitrate-N L^{-1} (66.7 mg NO_3^- L^{-1}), compared to lower concentrations (Dorsch et al., 1984). In a case-control study of CNS congenital malformations (Arbuckle et al., 1988), an increased risk was observed when nitrate was more than 26 mg NO_3^--N L^{-1} but only in well water. Women who drank water with more than 3.5 mg nitrate-N L^{-1} (15.6 mg NO_3^- L^{-1}) and took medications that could produce nitrosamines were 14 times more likely to have a baby with a neural tube defect (NTD) (Brender et al., 2004). A California study found that in populations exposed to drinking water with greater than 10 mg nitrate-N L^{-1}, there was no association with NTD incidence, but a four times greater risk of anencephaly, a specific type of NTD (Croen et al., 2001). A survey of more than 70,000 infants found a modest increased risk for those whose mothers had groundwater as their drinking water source and thus more likely to be exposed to nitrate (Cedergren et al., 2002). A recent study of 60 congenital anomalies and 1,635 controls assessed nitrate levels in both municipal water systems and estimated nitrate levels in private wells using geostatistical interpolation (Holtby et al., 2014). There was a statistically significant increase in the risk of any anomaly associated with average nitrate between 1 and 5.56 mg NO_3^- L^{-1} and a similar effect, though not significant, for concentrations above 5.56 mg NO_3^- L^{-1}.

5.2.3.5. CANCER

Nitrate and nitrite are not thought to cause cancer directly, but are precursors of N-nitroso compounds (NOCs) that are known animal carcinogens (IARC, 2014). NOCs are a family of compounds formed through a reaction of nitrite with specific molecules that are found in amino acids, the building blocks of proteins, and other compounds. NOCs are formed in meat products cured using nitrite, as well as from reactions that take place in the human body. Thus, the level of NOCs in the body depends on the amount of nitrate and nitrite consumed, as well as the amount of NOCs ingested in foods (IARC, 2014). Some drugs, including aspirin and antihistamines, can also react with nitrite to form NOCs (Brambilla, 1985).

The relationship between the ingestion of nitrate/nitrite, foods containing NOCs, and the level of NOCs in the body is very complex as many biochemical processes are involved. While increases in nitrate or nitrite exposure may increase the level of NOCs, this also depends on the availability of foods or drugs that contain amines. Vitamin C and other antioxidants can reduce the formation of NOCs (IARC, 2014). Investigations linking specific NOCs in urine with diet or with nitrate in drinking water found that subjects with high intakes of nitrate, or chemicals which can form NOCs, excreted higher levels of the NOCs; taking vitamin C significantly reduced these levels (Mirvish et al., 1998; Vermeer et al., 1998, 1999).

Some NOCs are very reactive and can damage or cause mutations in DNA. Administration of NOCs to rodents induces tumors in the bladder, liver, nose, mouth, esophagus, kidney, pancreas, lymph, stomach, and the nervous system (IARC, 2014), in many cases only when the animals are given both nitrate/nitrite and a source of amines (Borzsonyi et al., 1976; Bryan et al., 2012; Greenblatt et al., 1971; Greenblatt and Mirvish, 1973; IARC, 2014; Lijinsky, 1984; Pliss and Frolov, 1991; Shank and Newberne, 1976). Vitamin C and other antioxidants have been shown to reduce the formation of tumors for a given exposure to NOCs (IARC, 2014). The evidence linking exposure to nitrite and nitrate with cancer in humans is not as clear as in rodents and these studies are reviewed below.

BLADDER AND KIDNEY CANCER

Bladder and kidney cancer incidence has been linked to nitrate exposure. In a cohort study of 20,000 women, those exposed to more than 2.5 mg nitrate-N L^{-1} (11.1 mg NO$_3^-$ L^{-1}) were three times as likely to have been diagnosed with bladder cancer (Weyer et al., 2001). However, there was no relationship with total nitrate (food + water) intake. Nitrite and nitrite plus nitrate in meats were associated with bladder cancer (29% increase) in a study of more than 300,000 people (Ferrucci et al., 2010). A smaller study found bladder cancer increased threefold, but only among men of Japanese descent, with consumption of high levels of nitrite (Wilkens et al., 1996). A case-control study found a relationship between number of years of consuming water above 5 mg NO$_3^-$-N L^{-1} and renal cell carcinoma, but only among individuals with high red meat consumption or below-average vitamin C consumption (Ward et al., 2007) and this was confirmed by recent cohort studies finding associations between nitrate, nitrite, NOCs, and renal cell carcinoma (Daniel et al., 2012a; Dellavalle et al., 2013). Four studies, however, found the risk of bladder cancer had no relationship with the total amount of nitrate intake from foods and water (Knekt et al., 1999; van Loon et al., 1998; Ward et al., 2003; Zeegers et al., 2006).

CANCER OF THE STOMACH AND ESOPHAGUS

Studies looking at cancers of the stomach and esophagus have examined the effects of nitrate separately from nitrite. Of 13 case-control studies that examined nitrate, only one (Boeing et al., 1991) of the three studies that assessed nitrate in drinking water found an association (Rademacher et al., 1992; Yang et al., 1998). The remaining 11 studies estimated total nitrate intake from foods and drinks; only two found a significant association (IARC, 2014; Jakszyn and Gonzalez, 2006; Rogers et al., 1995; Ward et al., 2008).

More than 20 studies investigating the relationship between stomach cancers and the amount of nitrite, or nitrite plus nitrate in the diet (IARC, 2014) found significant associations with the consumption of smoked foods, preserved fish or preserved vegetables, meats or processed meats (IARC, 2014). Relationships of cancer with total nitrite or NOC intake were less evident. Consumption of high levels of nitrite and low levels of antioxidants was also associated with stomach cancer (Bruning-Fann and Kaneene, 1993; IARC, 2014; Jakszyn and Gonzalez, 2006). Ward et al. (2008) also observed a nonsignificant association of stomach cancer incidence with nitrate from meats, and a significant association with nitrate from plant-derived foods. Two recent large cohort studies, however, did not observe associations of nitrite or nitrate with stomach cancer (Cross et al., 2011; Loh et al., 2011). In a recent review, Bryan et al. (2012) observed a lack of evidence in recent cohort studies, but concluded that the associations previously found were primarily for nitrite exposure among people with low vitamin C intake.

Fewer studies have looked at cancer of the esophagus (Berretta et al., 2012). Navarro Silvera et al. (2011) observed an association of cancer of the esophagus with red meat consumption, while Cross et al. (2011) did not. A meta-analysis of studies examining the role of a 'Western diet," including higher red meat consumption, found no association, leaving us to conclude that an association between dietary levels of nitrate and cancers of the esophagus is suggested but unproven to date (Liu et al., 2014).

BREAST AND GENITAL CANCERS

A long-term study measuring relationships between women's cancer incidences and drinking water nitrate in Iowa found nitrate levels were linked to an increased risk of ovarian cancer, an inverse relationship with uterine cancer, and no association with breast cancer (Weyer et al., 2001). Another study comparing diets of women with versus without breast cancer found no association with nitrite or nitrate intake, and a twofold increase in risk for women with a higher intake of nitrate relative to their intake of folate (Yang et al., 2010). However, in a follow-up study, nitrate intake from food or water was not found to be associated with breast cancer overall, and, unexpectedly, only associated among women with high folate intake (Inoue-Choi et al., 2012). Another study found no association of nitrate in diet and endometrial cancer (Barbone et al., 1993). No consistent association between drinking water nitrate levels and breast cancer was found in a Cape Cod study (Brody et al., 2006), while another study found an association with testicular cancer, but only for men in urban areas, casting doubt on whether the association is truly with nitrates (Møller, 1997).

BRAIN CANCERS

There have been over 20 studies on the association of nitrate and/or nitrite in foods and drinking water with the risk of various brain cancers; almost all have been ecologic or case-control studies (IARC, 2014). For adults, only 1 out of 11 case-control studies of brain cancer found any link of nitrate or nitrite exposure and the development of cancer. Ward et al. (2005) found an elevated risk from exposure to nitrate in food derived from plants but not from meat or from drinking water. A review of relationships between dietary intake of NOCs and adult glioma found no association, nor any protective effect of vitamin C (Dubrow et al., 2010).

There is greater evidence of associations between childhood brain tumors and nitrate/nitrite intake. Consumption of nitrite by women during pregnancy was significantly related to brain tumors in their offspring (Preston-Martin et al., 1996). The single study to look at nitrite levels in drinking water (based on actual water samples) of women from their residence during pregnancy did find an association with the risk of brain cancer of their subsequent child (Mueller et al., 2004). A case-control study of childhood deaths due to brain tumors found that concentrations greater than 0.3 mg NO_3^--N L^{-1} were associated with a significant increase in risk (Weng et al., 2011).

RECTAL AND COLON CANCER

A study of colon cancer incidence found that individuals were twice as likely to develop colon cancer while consuming low vitamin C diets when exposed to drinking water nitrate above rather than below the MCL (De Roos et al., 2003). They also observed a positive relationship between total nitrite intake and cancers of the colon and rectum (De Roos et al., 2003). This was in contrast to an earlier study finding no association of nitrate in drinking water and colon cancer, and an inverse association with rectal cancer (Weyer et al., 2001). A study in Wisconsin that included sampling of individual wells found that people exposed to drinking water above the MCL were 2.9 times more likely to get colon cancer (McElroy et al., 2008). Investigation of deaths due to colon cancer in Taiwan found the effect of nitrate in drinking water depended on magnesium concentration (Chiu et al., 2010), and a cohort study of over 70,000 women found a relationship between nitrate intake and colorectal cancers, but only among women with lower vitamin C intake (Dellavalle et al., 2014).

LEUKEMIA AND LYMPHOMA

Leukemia refers to cancers that occur in the blood or bone marrow, while lymphomas are cancers of the lymph system. Of the four studies that examined the risks of nitrate or nitrite exposure, two studies (Ward et al., 1996, 2006) observed an association. Consumption of nitrate in drinking water but not food was associated with non-Hodgkin lymphoma (Ward et al., 1996). In a later study, high dietary nitrite intake was associated with greater risk for non-Hodgkin lymphoma, dietary nitrate intake was inversely associated with higher risk, while drinking water nitrate showed no association (Ward et al., 2006).

PANCREATIC CANCER

None of the four studies assessing nitrate exposure and pancreatic cancer found any evidence of higher risks associated with intake of nitrate in water or diet (Baghurst et al., 1991; Coss et al., 2004; Howe et al., 1990; Weyer et al., 2001); however, one study observed increased risk associated with nitrite intake from meats (Coss et al., 2004).

SUMMARY

There are mixed results on studies of nitrate and nitrite consumption and the risk of cancer. The strongest evidence relates to exposures from nitrites in foods, and a few studies have observed much greater risk among people with low vitamin C intake. The International Agency for Research on Cancer has concluded that "ingested nitrate or nitrite under conditions that result in endogenous nitrosation [formation of nitrosamines in the body] is '*Probably carcinogenic to humans*'" (IARC 2014; figure 5.2.3). Neither the US EPA nor Health Canada has classified nitrate or nitrite in terms of carcinogenicity (EPA, 2014; Health Canada, 2013).

5.2.3.6. HEALTH BENEFITS OF INGESTED NITRATE/NITRITE

Nitrogen-containing compounds are used as therapeutic agents (Butler and Feelisch, 2008). In clinical studies, nitrite (resulting from conversion of nitrate in the body) can lower blood pressure and reduce aggregation of platelets (Gilchrist et al., 2011; Lundberg et al., 2006, 2011; McKnight et al., 1999; Webb et al., 2008). Platelet aggregation can lead to the formation of clots in the circulatory system, a risk factor for stroke. Nitrate can increase oxygen delivery to oxygen-starved tissues, and help protect against injuries resulting from a heart attack (Tang et al., 2011). While still controversial, several scientists have begun to question whether the beneficial aspects of nitrate/nitrite in foods outweigh the health risks and costs of addressing nitrate contamination in water supplies (Gilchrist et al., 2011; Hord et al., 2009; Katan, 2009; Kevil and Lefer, 2011; Powlson et al., 2008). The economic aspects are discussed in Section 5.2.4.

5.2.3.7. RESEARCH NEEDS

Several studies show evidence that nitrate or nitrite in drinking water and/or foods is an important factor in the development of "blue-baby syndrome." Nitrate, nitrite, and/or nitrosamines in foods are "probably carcinogenic to humans" (IARC, 2014), with consistent evidence that total exposure, and exposure via meats, is related to stomach cancer. Exposure to nitrate has been consistently associated with neural tube defects. However, given the complexities of these exposure–disease relationships and the state of knowledge regarding these and other potential risks, it is very difficult to determine whether current regulatory limits are adequate to protect public health or possibly more stringent than is necessary (Powlson et al., 2008; Ward et al., 2003, 2008). Key areas needing more investigation include birth defects, stillbirth, and spontaneous abortion resulting from maternal exposure to nitrate and/or nitrite during pregnancy, and impacts on thyroid disease and cancer (Ward et al., 2010). Even for "blue-baby syndrome," which has been linked to nitrate in drinking water for over 50 years, controversy remains about how much of a risk nitrate in drinking water poses and whether factors other than nitrate are the actual cause of this potentially fatal disease.

The levels of exposure to nitrate are not well known, particularly differences in exposure related to income, race, and/or ethnicity. Such research is limited by a lack of data describing which specific areas are served by which public water system (VanDerslice, 2011). In addition, water quality data for private wells are not centrally collected, making it nearly impossible to create a comprehensive picture of exposure among those using private wells.

It is often very difficult to determine whether a disease is caused by an environmental exposure, and almost impossible to precisely know the relationship between the level of exposure and the risk of that disease for individuals of different ages, racial backgrounds, and states of health. Decisions about allowable levels of nitrate and nitrite, unfortunately, have to be based on the best information possible.

5.2.4. Economic Costs of N in Drinking Water

5.2.4.1. COSTS ASSOCIATED WITH HUMAN WELL-BEING

Peer-reviewed studies evaluating the economic costs of nitrate and nitrite in drinking water are limited with few focused on California. Available data include (1) health costs associated with consumption of nitrate-contaminated drinking water, (2) household costs associated with strategies to reduce nitrate concentrations through "point of entry" and "point of use" treatment systems, (3) household costs associated with avoiding consumption of contaminated water sources through purchasing clean water (e.g., bottled or trucked water) or drilling new wells, and (4) costs associated with strategies to reduce nitrate concentrations through treatment, blending, and consolidation of water systems. Here we examine the economic costs associated with N in drinking water and focus on understanding how these cost categories may influence human well-being.

5.2.4.2. HEALTH COSTS

Accurate estimates of health costs are extremely scarce and limited by the uncertainty present in the epidemiological studies that link drinking water nitrate/nitrite to the health outcomes discussed above (van Grinsven et al., 2010). Health damages resulting from colon cancer related to nitrate in drinking water in 11 EU countries were estimated between €0.1 and €2.4 kg^{-1} N (i.e., $0.14 and $3.38 kg^{-1} N based on 2010 currency exchange rates) leached to groundwater supplies (van Grinsven et al., 2010). Several other studies have specu-

lated that health costs from the consumption of nitrate in drinking water are likely to be considerable (Compton et al., 2011; Hanley, 1990; Innes and Cory, 2001), but with the exception of van Grinsven et al. (2010), no other studies have attempted to estimate the actual costs associated with specific health outcomes (figure 5.2.3).

5.2.4.3. HOUSEHOLD COSTS

Cost studies dealing with treatment and avoidance of nitrate-/ nitrite-contaminated drinking water for single households with domestic wells are becoming increasingly common in the recent literature, and several studies have recently been conducted in California. Lewandowski et al. (2008) conducted a survey of households in Minnesota that estimated the cost of three treatment systems (reverse osmosis, distillation, and anion exchange) and two avoidance strategies (bottled water and drilling new wells). In California, the Pacific Institute carried out a similar survey of households in the San Joaquin Valley that estimated the costs for reverse osmosis systems and bottled water (Moore and Matalon, 2011). Focusing on California's TLB and SV, Jensen et al. (2012) and Honeycutt et al. (2012) present the most comprehensive analysis of the household costs for various nitrate treatment and avoidance options to date. Across all four studies, single-household reverse osmosis systems required larger upfront costs than bottled water, but were generally the lowest cost alternative when initial capital and service costs were amortized over the lifetime of the system (Table 5.2.2). Since low-income households are less able to afford the upfront costs of water treatment systems, these studies suggest that they may end up paying a large fraction of their household income to purchase relatively expensive bottled water and filtration systems or else continue consuming untreated water. For example, in one San Joaquin community (Beverly Grand), households spent 4.4% of their median income on vended and bottled water, filters, and tap water service, almost three times the 1.5% affordability threshold established by the US EPA (Moore and Matalon, 2011). These results, which are provisionally agreed upon by most, demonstrate how the cost of different household treatment options may have significant implications for both the health and economic well-being of low-income households (figure 5.2.3).

5.2.4.4. COSTS TO PUBLIC AND COMMUNITY WATER SYSTEMS

The need for projects to address nitrate contamination of groundwater in the San Joaquin Valley (and throughout the state) far surpasses available public resources. Many communities have found that funds raised through local bond measures and/or fees levied on water users are often insufficient to pay for the water quality improvements required to meet federal drinking water standards (Moore and Matalon, 2011). In California, public funds to address water quality are available from both the California Department of Public Health (CDPH) and US Department of Agriculture

TABLE 5.2.2

Estimated capital, service, and annualized costs for single self-supplied households using alternative nitrate treatment and avoidance strategies

Strategies	Capital costs[1]	Service costs[2]	Annualized costs[3]	Location, year	Reference
Treatment Strategies	$ household[-1]				
Reverse osmosis	855	87	130	Minnesota, 2006	Lewandowski et al. (2008)
	100–300	80–150	93–221	California, 2010	Moore and Matalon (2011)
	406–1,200	190–200	250–360	California, 2010	Honeycutt et al. (2012)
	330–1,430	110–330	NR	Idaho, 2007	Jensen et al. (2012)
Distillation	961	NR	NR	Minnesota, 2006	Lewandowski et al. (2008)
	275–1,650	440–550	NR	Idaho, 2007	Jensen et al. (2012)
Ion exchange	1,600	NR	NR	Minnesota, 2006	Lewandowski et al. (2008)
	660–2,425	NR	NR	Idaho, 2007	Jensen et al. (2012)
Avoidance strategies	$ household[-1]				
Bottled water	NA	190	190	Minnesota, 2006	Lewandowski et al. (2008)
	NA	380	380	California, 2010	Moore and Matalon (2011)
	NA	1,260	1,260	California, 2010	Honeycutt et al. (2012)
Trucked water	NR (storage)	950	950	California, 2010	Honeycutt et al. (2012)
Drill deeper well[4]	50–200 ft[-1]	60	860–3,300	California, 2010	Honeycutt et al. (2012)
Drill new well[5]	7,200	NR	144	Minnesota, 2006	Lewandowski et al. (2008)
	25,000–40,000	60	2,100–3,300	California, 2010	Honeycutt et al. (2012)

1. NA indicates not applicable.

2. NR indicates data not reported.

3. See Lewandowski et al. (2008), Moore and Matalon (2011), and Honeycutt et al. (2012) for assumptions, equations, discount rates, and amortization periods used to calculate the annualized costs for various strategies.

4. Honeycutt et al. (2012) assume that an existing well is deepened from 300 to 500 ft and has a pump efficiency of 0.60 and 0.15 kWh.

5. Honeycutt et al. (2012) assume a new 300 ft well with a pump efficiency of 0.60 and 0.15 kWh.

(USDA). In the San Joaquin Valley alone, approximately 100 projects ($62 million total) were proposed to address solely nitrate and a total of $150 million for projects that included nitrate among other water quality concerns (Moore and Matalon, 2011); 16 projects were funded at a cost of $21 million over 5 years. Proposed projects included strategies such as drilling new wells, treating contaminated water, blending contaminated water with cleaner water from new sources, and consolidating available clean water sources.

Many water systems in violation of federal MCL for nitrate are located in disadvantaged communities, raising questions regarding how to cover costs of reducing nitrate contamination. Honeycutt et al. (2012) suggest that an additional $17–34 million per year over many decades may be required to ensure safe drinking water for the 85 water systems in TLB and SV that already exceed the MCL for nitrate. Similar studies are needed to estimate costs required to address other regions (figure 5.2.3). Given state and federal budget constraints, water quality projects serving vulnerable communities are increasingly likely to be post-

poned, scaled back, or abandoned (Moore and Matalon, 2011). As such, some have argued that allocation of public funds to address contamination of water supplies should include a more robust assessment of the environmental and social justice concerns raised by vulnerable communities who are likely to pay a disproportionate share of the human health and remediation costs (Firestone, 2009; Firestone et al., 2006; Moore and Matalon, 2011; VanDerslice, 2011).

5.3. Clean Air

Main Messages

Nitrogen is a component of, or aids in the formation of, five known air pollutants including NO_x, NH_3, O_3, $PM_{2.5}$, and PM_{10}. Air pollutants have important impacts on the economy, the environment, and human health, and thus are regulated by state and federal agencies.

Major emissions sources include the combustion of fossil fuels in the transportation, energy generation, and industrial sectors, as well as *agricultural fertilizers and livestock*. Higher NO_x concentrations tend to be measured in and around California's urban areas and originate mostly from transportation and industrial sectors. Concentrations of ground level O_3, which is formed from emissions of NO_x and volatile organic compounds (VOCs), are highest during the summer months in the South Coast, Bay Area, and Central Valley regions. The majority of NH_3 emissions come from livestock waste and N fertilizers, thus concentrations of NH_3 tend to be higher in the southern part of California's Central Valley.

Levels of $PM_{2.5}$ and PM_{10} are highest in the south San Joaquin Valley and South Coast regions. In the San Joaquin Valley, where livestock activities occur, NH_3 is the dominant constituent of secondary particulate matter. In the urban areas of the South Coast, compounds formed from NO_x make up a larger fraction of the particulate matter.

Air quality regulations and technological innovations have led to significant declines in NO_x, O_3, $PM_{2.5}$, and PM_{10} over the past four decades. However, much of the state still has air quality that fails to meet one or more of the standards set by national and state agencies to protect human health.

There are important racial disparities in exposure to air pollutants. In the South Coast basin and San Joaquin Valley Air Basin (SJVAB), a larger percentage of the Black and Hispanic populations relative to White and other races are exposed to $PM_{2.5}$ concentrations that are above the national ambient air quality standard (NAAQS) (35 µg m^{-3}).

Air pollutants are associated with many health problems. These include difficulty breathing, reduced lung function, asthma, respiratory infections, chronic obstructive pulmonary disease (COPD), cardiovascular disease (CVD), overall deaths, and deaths due to specific respiratory and cardiac causes. In California, over 12,000 premature deaths per year from cardiopulmonary disease and ischemic heart disease are attributed to elevated $PM_{2.5}$ levels. Studies suggest that the health damages in California associated with poor air quality are on the order of tens of billions of dollars per year.

Air pollution, particularly O_3, has adverse effects on crop growth. Yield losses ranging from 1% to 33%, depending on the sensitivity of the crop and level of exposure, can reduce revenues for agricultural producers and increase food costs for consumers. The overall economic impact of O_3 on agricultural production in California is estimated to be on the order of hundreds of millions of dollars per year.

5.3.0. Introduction

This section examines how nitrogen influences air quality throughout California. While various forms of reactive N such as nitrogen oxides (NO + NO_2 are together referred to as NO_x) and ammonia (NH_3) are naturally occurring components of the earth's atmosphere, anthropogenic activities have significantly increased their ambient concentrations. Moreover, these forms of N in the air have important impacts on environmental quality and human health. For example, emissions of NO_x and NH_3 directly and indirectly influence the formation of ozone and particulate matter which are criteria air pollutants and thus regulated by the US EPA under the US Clean Air Act, the California Air Resources Board, and various regional air districts (Table 5.3.1; EPA, 2012). Because of their high reactivity, NO_x and NH_3 interact with other chemical constituents in the atmosphere to create a range of harmful secondary chemicals, with tropospheric O_3, nitric acid (HNO_3), ammonium particulates (e.g., NH_4NO_3, $(NH_4)_2SO_4$), and peroxyacetyl nitrates (PANs) among the most important (figure 5.3.1). While these secondary chemicals can all be components of smog, the relative abundance of NH_3 and SO_2 in the air will determine if these constituents are important components of smog in a given location within California. In addition to an examination of the primary and secondary compounds resulting from nitrogen emissions, we also review the available literature on the health and economic impacts of human exposure to these air pollutants, with a particular focus on vulnerable regions and populations in California.

5.3.1. Relationship between Nitrogen and Air Pollutants

5.3.1.1. EMISSIONS OF NO_x AND NH_3

The primary forms of N that influence air quality are NO_x and NH_3. NO_x is a general term used to refer to nitric oxide (NO) and nitrogen dioxide (NO_2). Though fossil fuel combustion is the main anthropogenic source of NO (figures 5.3.1 and 5.3.2), it is also emitted by natural sources such as soils, wildfires, and lightning. When released into the atmosphere, NO is rapidly oxidized to NO_2; thus, the formation of both gases are jointly referred to as NO_x emissions and the concentrations are reported in units of NO_2. Since high levels of NO_2 can have negative effects on human health and well-being, the US EPA has defined NO_2 as a primary criteria pollutant, and set an NAAQS of 0.053 ppm, averaged annually (Table 5.3.1; EPA, 2012). The California ambient air quality standard (CAAQS) set by the California Air Resources Board is even more stringent than the US EPA, at an annual average of 0.030 ppm NO_2 (Table 5.3.1; CARB, 2012).

Emissions of NH_3 originate from both anthropogenic (e.g., livestock, N fertilizers, fossil fuel combustion) and biogenic sources (e.g., soils and vegetation). In California, agricultural sources account for more than 77% of anthropogenic NH_3 emissions, with approximately 66% attributed to livestock and 12% to N fertilizers (Benjamin, 2000; Chapter 4). Fossil fuel combustion from mobile sources is responsible for about 13% of total statewide NH_3 emissions and is a dominant source of NH_3 in the air above many urban areas (Bishop et al., 2010; Nowak et al., 2012; Chapter 4). Gaseous NH_3 is in chemical equilibrium with ionized ammonium (NH_4^+), and the amount of NH_3 volatilized into the atmosphere depends on environmental factors such as tempera-

TABLE 5.3.1

National and California Ambient Air Quality Standards (AAQS)

Pollutant	Primary/secondary	Averaging time	National AAQS	California AAQS	Form[1]
Nitrogen dioxide (NO_2)	Primary	1 hour	0.100 ppm	0.18 ppm	98th percentile, averaged over 3 years
	Primary and secondary	Annual	0.053 ppm	0.030 ppm	Annual mean
Ozone (O_3)	Primary and secondary	8 hours	0.075 ppm	0.070 ppm	Annual fourth-highest daily maximum 8-hour concentration, averaged over 3 years
		1 hour	*	0.09ppm	N/A
Fine particulate matter ($PM_{2.5}$)	Primary and secondary	Annual	15 µg m⁻³	12 µg m⁻³	Annual mean, averaged over 3 years
		24 hours	35 µg m⁻³	**	98th percentile, averaged over 3 years
Inhalable particulate matter (PM_{10})	Primary and secondary	Annual	***	20 µg m⁻³	N/A
		24 hours	150 µg m⁻³	50 µg m⁻³	Not to be exceeded more than once per year on average over 3 years

1. The "Form" column is taken from the US EPA National Ambient Air Quality Standards (NAAQS).

*There is no separate 1-hour O_3 national standard.

**There is no separate 24-hour $PM_{2.5}$ standard in California, though the US EPA promulgated a 24-hour $PM_{2.5}$ ambient air quality standard of 35 µg m⁻³.

***There is no separate annual PM_{10} national standard.

SOURCE: Adapted from CARB (2012) and EPA (2012).

ture, pH, and NH_4^+ concentrations in a given substrate (e.g., soil, water, manure, fertilizer). Though NH_3 is not regulated by the US EPA as a primary criteria pollutant, it is an important precursor in formation of fine particulate matter (see below). In occupations where workers are at risk of being exposed to localized NH_3 concentrations much higher than ambient levels, the Occupational Safety and Health Administration (OSHA) has established a permissible 8-hour exposure limit of 50 ppm (EPA, 1989). This occupational exposure limit applies to industrial facilities and includes concentrated animal feeding operations (e.g., poultry houses, swine facilities, dairies, feedlots) which tend to have high levels of volatilized NH_3 from urine and manure, as well as meat packing and food processing plants that use NH_3 for refrigeration (Donham et al., 2000, 2002).

5.3.1.2. FORMATION, BUILDUP AND DECAY OF TROPOSPHERIC O_3

Tropospheric O_3 is formed from NO_x, carbon monoxide (CO), and VOCs in sunlight-driven reactions (figure 5.3.1). Oxidized NO produces NO_2, which then undergoes rapid photochemical decay to reform NO and atomic oxygen (O).

High concentrations of oxygen gas (O_2) in the troposphere allow atomic O and O_2 gas to rapidly combine to form O_3 (Seinfeld and Pandis, 1998). In the absence of VOCs, O_3 will oxidize NO back to NO_2 and thus restart the cycle with no net gain of O_3. However, when VOC molecules are present, they break down to form hydroxyl and peroxyl radicals that oxidize NO more rapidly than O_3, thus resulting in a buildup of O_3 (Seinfeld and Pandis, 1998). Tropospheric O_3 is ultimately broken down by ultraviolet light or through oxidation reactions with plant and animal tissue or other components of the land surface. National and California air quality standards for O_3 are 0.075 and 0.070 ppm, respectively, based on the annual fourth highest daily maximum 8-hour concentration averaged over 3 years (Table 5.3.1; CARB, 2012; EPA, 2012).

5.3.1.3. SOURCES AND FORMATION OF PARTICULATE MATTER

Particulate matter is one of the least understood components of atmospheric pollution, mainly due to the large variation in the source and chemical composition of aerosolized particles (Solomon et al., 2007). National and state

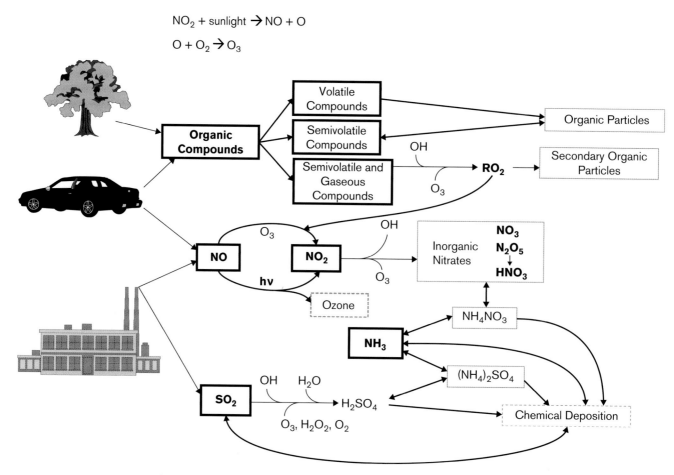

$$NO_2 + \text{sunlight} \rightarrow NO + O$$

$$O + O_2 \rightarrow O_3$$

FIGURE 5.3.1. Source and chemical links between ozone and PM formation. Major precursors are shown in boxes with thick sides. Secondary particle components are shown in boxes with thin solid sides. Mobile sources (cars, trucks, and off-road vehicles) and plants are major sources of VOCs, and mobile sources and electricity-generating units are dominant sources of NO_x, but myriad smaller sources also contribute. Trace species, such as OH, are crucial to the formation of ozone, sulfate, nitrate, and organic-carbon particulate matter. Ozone also leads to the oxidation of SO_2 and NO_2. Biological activity and fertilizer use dominate ammonia (NH_3) emissions.

SOURCE: Reprinted from NRC (2008). Copyright © National Academy of Sciences.

regulatory agencies classify particulate matter by the diameter of the particles, with fine particulate matter ($PM_{2.5}$) being 2.5 μm or less, and inhalable coarse particulate matter (PM_{10}) being 10 μm or less. National and state ambient PM standards are listed in Table 5.3.1 for both a 24-hour and annual average exposure (CARB, 2012; EPA, 2012).

While $PM_{2.5}$ can be formed directly from the combustion of fossil fuels and from various organic or inorganic materials (dust, soot, smoke, pollen, spores, etc.), secondary chemical reactions that involve NH_3, NO_x, VOCs, and sulfur dioxide (SO_2) occur in the atmosphere and are also an important mechanism of formation (figures 5.3.1 and 5.3.2) (Krupa, 2003). In agricultural areas, ammonium salts make up a large fraction of $PM_{2.5}$ through conversion of gaseous NH_3 to solid NH_4^+ via reaction with atmospheric acids (i.e., H_2SO_4, HNO_3) (Krupa, 2003). Atmospheric HNO_3 concentrations are influenced by NO_x levels; thus, both NH_3 and NO_x play a contributing role in the formation of fine particulate matter.

PM_{10} particles are commonly associated with fugitive dust arising from agricultural and forestry activities, vehicles traveling on paved and unpaved roads, construction activities, and wind erosion (Chow et al., 2003). Other important components of PM_{10} are the ash and smoke from managed burns or wildfires. PM_{10} can also be formed directly from fuel combustion by the industrial and transportation sectors or through the secondary chemical reactions described above.

5.3.1.4. ACID PRECIPITATION AND FOG

Air pollutants containing N may also impact human health, plant growth, and the environment, by increasing the acidification of precipitation in its various forms (e.g., rain, snow, fog). The occurrence of acidic fog (or the "fog–smog–fog" cycle) in California's urban, rural, and natural areas has received a moderate amount of attention in the research literature since the acidity (i.e., the H^+ concentration) in fog is typically 10–100 times greater than typical acid rain

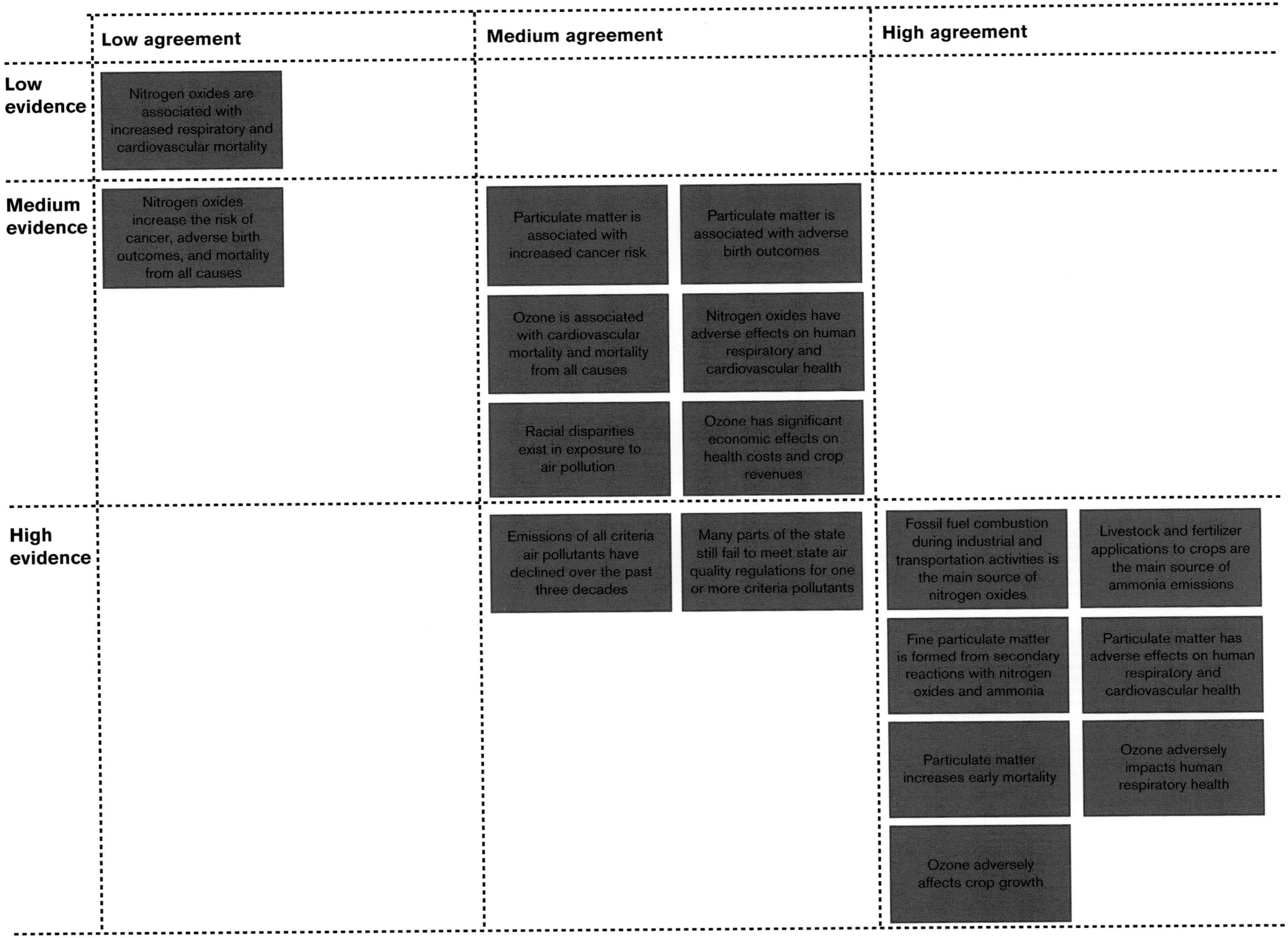

FIGURE 5.3.2. Types of uncertainty in nitrogen's impact on air quality and human health in California. This figure reflects the amount of evidence and level of scientific agreement for the effects of various nitrogen-related air pollutants.

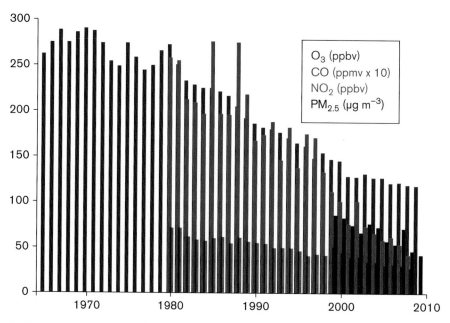

FIGURE 5.3.3. Air quality trends in the Los Angeles urban area of California. As per national standards, the O_3 data (8-hour average) are 3-year averages of the fourth highest annual maxima, the CO data (8-hour average) are annual maxima, the NO_2 data are annual averages, and the $PM_{2.5}$ data (24-hour average) are annual 98th percentiles. Data are derived from monitoring stations in the Southern California Air Basin region.

SOURCE: Reprinted from Parrish et al. (2011). Copyright 2011. Elsevier. Used by permission.

events (Brewer et al., 1983; Munger et al., 1983; Temple et al., 1987; Waldman et al., 1982). Acid precipitation and fog can be caused by the interaction between atmospheric water droplets and various air pollutants including gaseous NO_2 and SO_2, as well as other aerosols and particulates. The presence of reduced forms of N (e.g., NH_3 and NH_4^+) in fine particulates can also increase the concentration of various amino compounds in atmospheric water droplets that can have a net acidic, neutral, or basic effect depending on the compound (Zhang and Anastasio, 2003).

5.3.2. Spatial and Temporal Trends in Air Pollutants

While state regulatory agencies measure atmospheric concentrations of NO_x, O_3, and particulate matter using a series of surface monitoring stations distributed throughout the state, NH_3 levels are not systematically monitored. Placement and distribution of monitoring stations tend to be concentrated in regions with high emissions; thus, more data are available for air basins near California's major urban centers. Remote sensing techniques are increasingly being used to fill in known spatial gaps in air quality data. A collaboration among the National Oceanic and Atmospheric Administration (NOAA), CARB, and the California Energy Commission (CEC) is using heavily instrumented aircraft to periodically measure NO_x, NH_3, and a wide variety of secondary pollutants throughout California (Ryerson et al., 2013). Ground level concentrations of NO_x, NH_3, O_3, and PM are also detectable by satellites which, combined

with surface data and meteorological models, give a more complete assessment of spatial trends (Clarisse et al., 2009, 2010; Gupta et al., 2006; Hidy et al., 2009). A limitation of remotely sensed data, however, is that they lack the continuous temporal resolution needed to calculate average concentrations for the particular time periods required for compliance with national and state ambient air quality standards (Table 5.3.1; Hidy et al., 2009).

Since transportation and industrial emissions are major sources, higher concentrations of NO_x are detected in and around urban areas. Spatial data derived from interpolated surface measurements and satellite images both show that concentrations of NO_x (reported in units of NO_2) are highest in the Los Angeles region and to a lesser extent near San Francisco and Sacramento (Kar et al., 2010; Russell et al., 2010). These data also indicate that NO_x levels in San Diego, the Imperial Valley, and several cities in the Central Valley (e.g., Fresno, Bakersfield) are notably higher than less populated regions of the state.

Despite rising vehicle use and population, levels of NO_x across California have declined at a relatively constant rate over the last several decades based on emissions inventories (Cox et al., 2009; McDonald et al., 2012; Millstein and Harley, 2010), surface measurements (figure 5.3.3; Ban-Weiss et al., 2008; LaFranchi et al., 2011; Parrish et al., 2011; Russell et al., 2010), and satellite observations (Kim et al., 2009; Russell et al., 2010). Between 2005 and 2008, Russell et al. (2010) observed an annual decline in NO_x of 9% per year for Los Angeles, San Francisco, and Sacramento (figure 5.3.4)

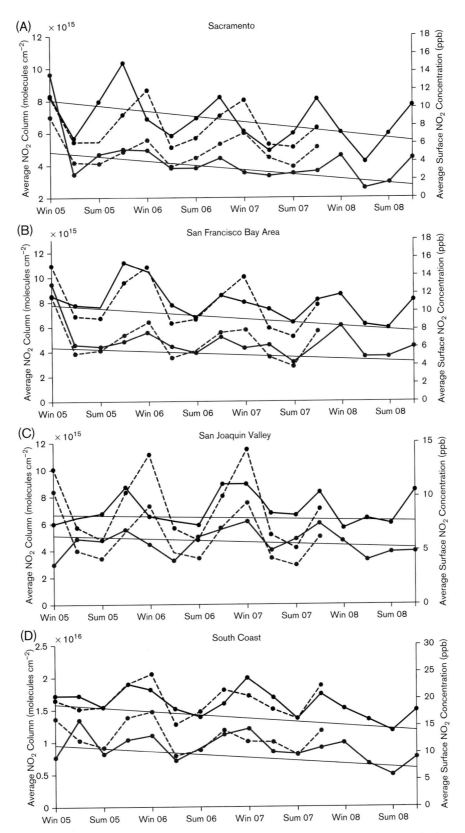

FIGURE 5.3.4. Trends in NO$_2$ concentrations through California (2005–2008). Average tropospheric NO$_2$ column concentrations (molecules cm^{-2}) from a satellite O$_3$ monitoring instrument (OMI) (solid) and surface measurements from the CARB (dashed) for weekdays (blue) and weekends (red). Study areas include (A) Sacramento County, (B) the San Francisco Bay Area, (C) the San Joaquin Valley, and (D) the South Coast regions of California. Summer trends are shown by the solid black lines. Note that figures are on different scales to make seasonal cycles visible.

SOURCE: Reprinted with permission from Russell et al. (2010). Copyright 2010. American Chemical Society.

and 4% per year in the San Joaquin Valley for Fresno and Bakersfield. A combination of regulatory policies (e.g., vehicle emissions standards) and technological innovations (e.g., catalytic converters, cleaner fuels, fuel efficient engines) are largely responsible for the declining levels of NO_x and overall improvements in California's air quality (McDonald et al., 2012; Parrish et al., 2011; Warneke et al., 2012).

Atmospheric NH_3 concentrations tend to be higher in rural areas with intensive agriculture since the majority of anthropogenic NH_3 emissions come from livestock production and fertilizer application (figure 5.3.2; Benjamin, 2000; Clarisse et al., 2009; Nowak et al., 2012). In California, NH_3 levels are highest in the San Joaquin Valley where the state's dairy and poultry industries are concentrated (Benjamin, 2000; Clarisse et al., 2009, 2010). This region also coincides with California's highest levels of groundwater nitrate (Harter and Lund, 2012). Clarisse et al. (2009) found that the San Joaquin Valley had the highest annual daily average NH_3 values (>3 mg m^{-2}) of any agricultural region in the world. Though atmospheric concentrations of NH_4^+ are generally lower in urban areas, the relative amount of NH_3 from vehicle emissions is often larger than from agricultural sources. For example, during 2010 in the South Coast Air Basin (SoCAB), automobiles emitted 62 metric tons of NH_3 per day, while dairy facilities emitted 33 metric tons of NH_3 per day (Nowak et al., 2012). Other than coarse NH_3 emission inventories conducted at rather infrequent intervals, NH_3 levels are not routinely monitored by the state, so available data are limited.

While NO_x and NH_3 levels are heavily dependent on the source of primary emissions, formation of secondary pollutants such as O_3 and particulate matter are more closely tied to the topography, meteorological conditions, and atmospheric constituents. High O_3 levels most frequently occur during California's summer "O_3 season" when high solar radiation facilitates the decay of NO_2 and subsequent formation of O_3 (Kaduwela, 2007). Concentrations of O_3 also tend to be higher downwind of urban areas because of the time lag between the primary emissions of NO_x and the secondary photochemical reactions that produce O_3 (Altshuler et al., 1995; Pusede and Cohen, 2012). Paradoxically, O_3 concentrations can be higher on weekends, especially in urban areas, when NO_x concentrations are lower (Altshuler et al., 1995). The weekend effect is likely a function of the relative concentrations of VOC to NO_x (Murphy et al., 2006, 2007). In California, a distinct weekend effect occurs in the Los Angeles, South Coast, and San Francisco air basins (Kaduwela, 2007).

In the past four decades, 8-hour average O_3 levels (3-year averages of the fourth highest annual maxima) in the Los Angeles area have declined by more than 50%, down from nearly 300 ppb in the 1970s to just over 100 ppb in 2010 (figure 5.3.3; Parrish et al., 2011). While improvements in O_3 have also been observed in the Central Valley over the same period, O_3 levels for the San Joaquin Valley remain higher than in California's coastal cities (CAPCOA, 2012). Recent satellite data indicate that the San Joaquin Valley and South Coast regions also have the highest levels of

$PM_{2.5}$ (>14 µg m^{-3}) in the state. Similar to trends in O_3, $PM_{2.5}$ levels have declined more rapidly in the South Coast than SJVAB (CAPCOA, 2012).

Throughout California, the pH of precipitation (pH = 5.2–6.2) tends to be less acidic than in other US states located in the industrial centers of the Northeast and Midwest (pH = 4.3–4.9) (NOAA, 2006). However, California's frequent fog events may still pose problems given fogwater's tendency to have higher concentrations of acidity than rainwater. For instance, several studies conducted in the 1980s recorded median pH levels of 3.3 in fog, 3.6 in mist and 4.49 in rain at various urban and rural sites in southern California, with some pH values for fog and mist reaching as low as 2.15 (Brewer et al., 1983; Munger et al., 1983; Waldman et al., 1982). Given the gradual decline in NO_x emissions observed in California and other states over recent decades, the problem of acid precipitation has become less severe (Burns et al., 2011; Parrish et al., 2011).

5.3.3. Patterns of Exposure to Air Pollutants in California

With well-established reductions in NO_x, O_3, $PM_{2.5}$, and PM_{10} documented throughout much of the state, considerable progress has been made to improve California's air quality over the past several decades (Parrish et al., 2011). But while these improvements may highlight the efficacy of certain regulatory policies and technological advances, a number of significant air quality problems persist. For example, the majority of California's counties are still designated as "non-attainment" areas based on CAAQS for O_3, $PM_{2.5}$, and PM_{10}, thus highlighting the health risks that remain for much of the state's population (figure 5.3.2 and Map 5.3.1). According to Hall et al. (2008a), air pollution levels in the SoCAB and the SJVAB remain among the worst in the United States, and during peak periods many other urban areas still reach O_3 and $PM_{2.5}$ concentrations roughly double the acceptable federal limit for vulnerable populations (Hall et al., 2008a; Parrish et al., 2011).

With the exception of Los Angeles, all regions of California are currently in compliance with the state and NAAQS for NO_2 (figure 5.3.1; Table 5.3.1; CARB, 2011; EPA, 2010). In certain high vehicle traffic areas within the SoCAB, NO_2 levels can sometimes exceed air quality standards for short periods (NAAQS = 0.18 ppm for a 1-hour average). For example, in 2009 NO_2 levels exceeding the national standard for the 1-hour average were recorded during 3 days in Los Angeles and during 1 day in both San Bernardino and Imperial Counties (CARB, 2011). Los Angeles is the only area of the state still designated as being in "non-attainment" according to more stringent California standards (CARB, 2011).

In contrast, the US EPA has classified 31 counties in the South Coast, Central Valley, and Bay Area as federal non-attainment areas for O_3, with the South Coast basin and the SJVAB further designated "extreme non-attainment areas" (CARB, 2011; Hall et al., 2008a, 2008b). From 2005 to 2007,

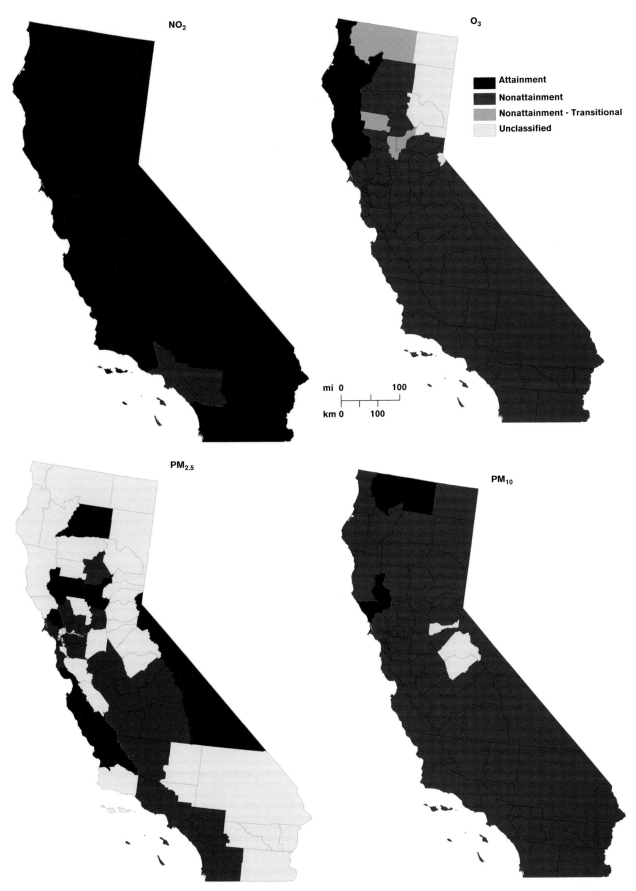

MAP 5.3.1. Air quality attainment status in California for NO_2, O_3, PM_{10}, and $PM_{2.5}$ (2007–2009). Attainment is based on California Ambient Air Quality Standards (CAAQS) (data from EPA, 2015; maps by R. Murphy).

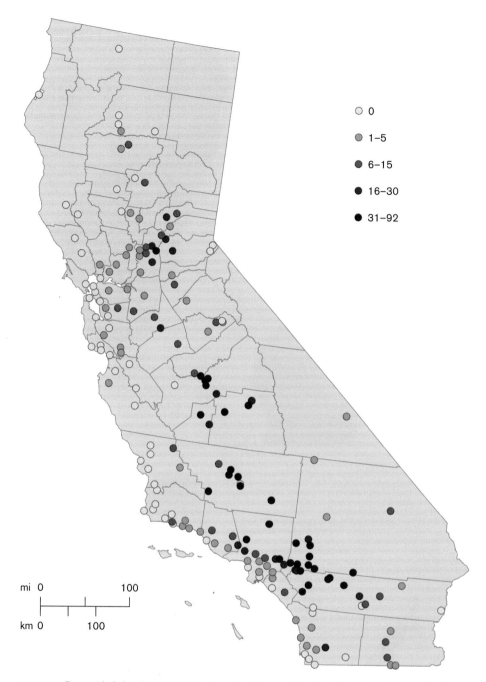

MAP 5.3.2. Days with O$_3$ levels in California above the 8-hour national ambient air quality standard (NAAQS) in 2009. The color of the monitoring stations (depicted by the dots) correspond to the number of days above the national O$_3$ standard (data from EPA, 2015; map by R. Murphy).

O$_3$ levels for large parts of Kern, Tulare, San Bernardino, and Riverside Counties exceeded the NAAQS 8-hour maximum standard (>0.075 ppm) more than 50 days a year (Hall et al., 2008a). In 2009, eight counties throughout the state had more than 30 days where the 8-hour O$_3$ level exceeded the national standard (Map 5.3.2). While Los Angeles County had fewer non-attainment days per year than counties in the San Joaquin Valley, due to its high population density it had the highest number of person–days per year (104.97

million) in which people were exposed to unhealthy levels of O$_3$ (Hall et al., 2008a).

Many areas in the South Coast, Central Valley, and Bay Area are classified as being in non-attainment for PM$_{2.5}$, and in many cases PM$_{10}$, under both state and national standards (Map 5.3.1; CARB, 2011; EPA, 2010). Between 2005 and 2007, 100% of the population in Madera, Fresno, Kings, Tulare, and Kern Counties and 75% of Los Angeles County were exposed to annual average PM$_{2.5}$ concentrations above the NAAQS (15

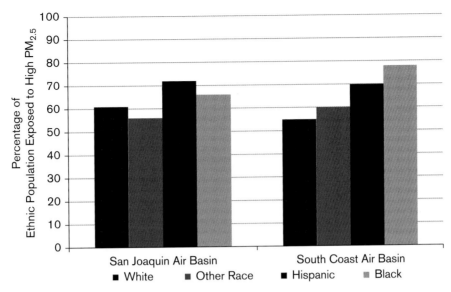

FIGURE 5.3.5. Percent of ethnic populations (White, Other Race, Hispanic, Black) exposed to average annual PM$_{2.5}$ in excess of the NAAQS (>15 μg m^{-3}). Data are from the San Joaquin and South Coast Air Basins.

DATA FROM: Hall et al. (2008a, p.200).

μg m^{-3}) (Hall et al., 2008b). The majority of people living in the Bay Area and the Sacramento Valley also appear to be exposed to average annual PM$_{2.5}$ levels above the NAAQS (CARB, 2011). While progress in reducing PM$_{2.5}$ has been made, annual average PM$_{2.5}$ concentrations in the San Joaquin Valley basin and SoCAB must still fall by approximately 30% to meet the federal standards (Hall et al., 2008b).

5.3.4. Disparities in Exposure to Air Pollutants

Since low-income and minority neighborhoods, particularly those in urban areas, are more likely to be adjacent to large roads, heavy industry, and other pollution sources, certain socioeconomic classes and ethnic groups are likely to be at greater risk of exposure to various air pollutants (Hall et al., 2006; Marshall, 2008; Marshall et al., 2006; Morello-Frosch and Lopez, 2006; Su et al., 2009) and high racial segregation in some metropolitan areas leads to greater racial disparities in exposure (Morello-Frosch and Lopez, 2006). In California, Hall et al. (2008b) observed that geographic differences in ethnicity among residents within and between different counties in Southern California result in different O$_3$ exposure frequencies. From 2005 to 2007, Hispanics in the San Joaquin Valley and Whites in the SoCAB were more frequently exposed than other racial groups to O$_3$ levels above the 8-hour NAAQS limit (Hall et al., 2006, 2008b).

Hall et al. (2008b) found even more prominent disparities in exposure to PM$_{2.5}$, with approximately 55% of White, 60% of other race, 70% of Hispanic, and 78% of Black residents in the SoCAB exposed to average annual PM$_{2.5}$ concentrations above the 15 μg m^{-3} NAAQS threshold. Similar results were also found in the SJVAB, with an estimated 61% of White, 56% of other race, 72% of Hispanic, and 66% of

Black residents exposed to annual PM$_{2.5}$ concentrations above the average annual PM$_{2.5}$ NAAQS (figure 5.3.5). While studies examining racial disparities and air pollution are not available for California's other regional air basins, the patterns of exposure among minorities in highly segregated communities are likely to occur elsewhere in the state as well (Morello-Frosch and Lopez, 2006).

5.3.5. Human Well-Being and Air Quality

The poor air quality caused by high levels of NO$_2$, O$_3$, PM$_{2.5}$, and PM$_{10}$ is known to have a number of well-established impacts on human health. There is strong evidence linking these pollutants with health problems such as difficulty breathing, reduced lung function, asthma, respiratory infections, chronic obstructive pulmonary disease, cardiovascular disease, deaths due to specific respiratory and cardiac causes, and overall deaths (Table 5.3.2; figure 5.3.2). Studies also show, albeit with less certainty, that these pollutants are related to other problems such as lung cancer, low birth weight (LBW) babies, or preterm births (Table 5.3.2; figure 5.3.2). These pollutants also have adverse effects on human well-being by increasing school absences as well as number of restricted activity days (RADs) and minor restricted activity days (MRADs) (Hall et al., 2010). The incidence of air pollution-related health impacts occurs with relatively high frequency in California, with the residents of counties in the South Coast and San Joaquin Air Basins facing high levels of exposure to various pollutants (Table 5.3.3; Hall et al., 2006, 2008b). Below, the epidemiological evidence related to poor air quality is examined with the goal of understanding relationships between various health problems and specific pollutants.

TABLE 5.3.2
Strength of evidence relating exposure to air pollutants to specific health problems

Health Outcome	NO_x	O_3	PM
Respiratory symptoms	++	+++	+++
Lung function	++	+++	+++
Asthma	++	+++	+++
COPD	+	++	+++
Respiratory infections	++	+	++
Respiratory mortality	@	+++	+++
Cardiovascular disease	++	+	+++
Cardiovascular mortality	@	++	+++
Birth outcomes	+	+	++
Cancer	+	@	++
Mortality, all causes	+	++	+++

NOTE: +++ = well established.

++ = provisionally agreed by most.

+ = speculative.

@ = suggested but unproven.

5.3.5.1. INTERPRETING EPIDEMIOLOGICAL EVIDENCE

Results of epidemiologic studies focusing on the effects of air pollutants on human populations vary considerably due to differences in study design, populations studied, and interactions among other related factors. The relationships between pollutant exposures and health outcomes all have uncertainty, which is determined based on the consistency of findings across studies. This is particularly true for rigorous studies that provide a plausible physiologic mechanism for the health impact. In some cases, an estimate of the magnitude of the effect is derived from the results of the best studies. Environmental regulations are then based, in part, on these observed effects.

Several hundred epidemiological studies in the United States have linked levels of ambient air pollutants to health problems. There have been fewer studies examining the health effects of NO_2 and NH_3 relative to those involving O_3 and particulate matter. As such, there is stronger evidence linking O_3 and particulate matter to adverse health outcomes. In this section, we present an overview of the evidence, highlighting the most significant studies, and those focusing on populations in California.

5.3.5.2. EVIDENCE OF THE IMPACTS OF NO_2, O_3, AND PARTICULATE MATTER EXPOSURE ON RESPIRATORY HEALTH

Asthma is one of the leading illnesses among children and adults, affecting 25 million Americans or approximately 8.4% of adults and 9% of children (CDC, 2008; CDC and US Department of Health and Human Services, 2009). This is nearly a 10% increase since 2001 (NCHS, 2011). Asthma can have a major impact on one's quality of life. This impact is reflected in the fact that in 2008 over 10 million school days and over 14 million work days were missed due to asthma, nationwide (NCHS, 2011).

Asthma is a condition where the muscles surrounding the bronchial airways contract, and an inflammatory response leads to the secretion of thick mucus into the airway. This makes it very difficult to breathe (NHLBI, 2007). A person experiencing an asthma attack will have trouble breathing, and will wheeze because of the narrowed airways. Such asthma attacks, or exacerbations, are brought on by a trigger, some factor that starts the inflammatory process. It can be cold air, an allergen, or a gas that irritates the airway. Asthmatics have developed a hyperresponsiveness, where the inflammatory process responds immediately to a small trigger, one that does not affect a non-asthmatic.

To better understand the effect of NO_2, O_3, and particulate matter exposure on the development and exacerbation of asthma, studies have examined the relationship between exposure levels to these pollutants and each of the steps of the disease process: increasing the inflammatory response, developing hyperreactivity, lung function, asthma exacerbations, visits to the hospital or emergency room for asthma, and deaths.

AIRWAY INFLAMMATION

When cells lining the trachea experience a physical or chemical insult, it can trigger an inflammatory response, that is, a biological response designed to protect the body from infectious agents. NO_2 and O_3 are strong oxidants. When inhaled, they can react with the membranes of the

TABLE 5.3.3

Estimated incidences of O_3- and $PM_{2.5}$-related adverse health effects in the San Joaquin Valley and South Coast Air Basins by county in 2008

	O_3-related adverse health effects by county					
	Respiratory hospital admissions (all ages)	Asthma attacks asthmatic population	Emergency room visits	Days of school absences	Minor restricted activity days	Mortality
San Joaquin Valley Air Basin						
Fresno	46	5,670	17	43,980	42,970	3
Kern	41	4,640	13	37,810	34,620	3
Kings	5	890	3	6,050	7,580	0
Madera	6	780	2	5,500	6,320	0
Merced	8	1,090	3	8,530	8,070	0
San Joaquin	17	2,290	7	13,100	17,170	0
Stanislaus	16	2,100	7	13,500	15,190	1
Tulare	24	2,940	8	23,040	21,830	2
South Coast Air Basin						
Los Angeles	380	59,100	150	653,300	483,840	12
Orange	87	17,010	45	184,500	142,380	3
Riverside	185	22,480	55	125,840	164,470	15
San Bernardino	173	22,380	55	144,690	170,720	11

	$PM_{2.5}$-related adverse health effects by county						
	Premature and postneonatal mortality	Respiratory symptoms and bronchitis	Nonfatal heart attacks	Respiratory and cardio hospital admissions	Children's asthma ER visits	Minor restricted activity days	Work loss days
San Joaquin Valley Air Basin							
Fresno	212	104,215	156	80	119	103,770	18,500
Kern	183	81,228	119	53	93	80,170	14,280
Kings	29	15,207	27	10	17	18,770	3,340
Madera	33	14,235	24	13	16	16,020	2,850
Merced	38	24,269	33	14	28	21,840	3,880
San Joaquin	110	46,908	78	43	54	49,360	8,740
Stanislaus	99	43,814	70	39	50	45,660	8,120
Tulare	110	54,678	77	37	63	50,750	9,030
South Coast Air Basin							
Los Angeles	1,727	1,000,440	1,960	903	1,175	1,224,600	241,690
Orange	411	233,310	485	175	275	300,010	59,100
Riverside	461	217,570	370	220	255	224,780	44,500
San Bernardino	412	260,480	418	187	305	266,830	52,850

SOURCE: Adapted from Hall et al. (2008b).

cells lining the trachea. This activates macrophages, a type of white blood cell, present in the lung tissue. This leads to the release of chemicals that change the membranes of blood vessels along the lung, to allow leukocytes (white blood cells) into the lung. They release other chemicals (interleukins, proteases, and oxidative species) to limit cell injury. This is called an inflammatory response.

A number of studies have examined whether exposure to these pollutants leads to specific aspects of the inflammatory response by observing the levels of specific chemicals in lung fluids. While some studies have shown evidence of inflammation in response to NO_2 exposure, particularly in children, others have not. Two comprehensive reviews, in fact, came to contradictory conclusions (EPA, 2008c; Hesterberg et al., 2009). Many laboratory studies have found evidence of inflammation after exposure to O_3 and particulate matter (Alexis et al., 2010; Dahl et al., 2007; EPA, 2006b, 2009b; Mudway and Kelly, 2004). Exposure to O_3 over several days elicited the greatest inflammatory response. Biochemical markers of inflammation occur after a single exposure to O_3 (Alexis et al., 2010; EPA, 2006b). Short-term exposure to $PM_{2.5}$ was related to higher levels of a biomarker of inflammation among asthmatic children (Delfino et al., 2006; Liu et al., 2009; Mar et al., 2005) and older adults (Adamkiewicz et al., 2004; Adar et al., 2007; Jansen et al., 2005). A cohort study of over 2,000 children found higher levels of inflammation with higher annual average $PM_{2.5}$ levels (Dales et al., 2008).

HYPERRESPONSIVENESS

Hyperresponsiveness refers to the tendency to initiate a rapid, intense constriction of the smooth muscles around the bronchi in response to an allergen. This response is typical of asthmatics and not very common among people who do not have asthma (Cockcroft and Davis, 2006). The severity of asthma, risk of exacerbations, and impact on lung function are all clearly related to the degree of hyperresponsiveness (Murray and Morrison, 1986; Xuan et al., 2000). Children who demonstrate hyperresponsiveness are at higher risk of developing asthma and not fully developing their lung capacity and function (Postma and Boezen, 2004; Xuan et al., 2000). Exposure to environmental irritants, including NO_2, O_3, and particulate matter and cigarette smoke, may induce inflammation, and as a result, indirectly impact lung function by increasing the lung's responsiveness.

Several studies have investigated whether prior exposure to air pollutants increases the level of severity of an asthma exacerbation (in terms of lung function) resulting from subsequent exposure to a trigger (such as an allergen, cold air, etc.). Evidence from animal and clinical studies shows that O_3 increases a person's response, whether they have asthma or not, to a respiratory stimulus (Holz et al., 2002; Jörres et al., 1996; Kehrl et al., 1999), but these have been at levels well above air quality standards and greater than almost all observed levels. Particulate matter, in contrast, has not been associated with increased airway hyperresponsiveness

(EPA, 2009b). Results of studies of NO_2 varied, with some of the studies demonstrating an effect and others not demonstrating any effect. However, for studies that did show evidence of NO_2-induced hyperresponsiveness, effects occurred at much lower levels of NO_2 exposure than were associated with inflammation (EPA, 2008d). An analysis of several studies by EPA indicates that exposure to NO_2 at ambient concentrations is related to airway hyperresponsiveness for mild asthmatics (EPA, 2008d) and may help explain the link between these exposures and number of hospital admissions for asthma (EPA, 2009b).

LUNG FUNCTION

Long-term exposures to air pollutants can permanently reduce the capacity of the lung. The impact of exposure to air pollutants is assessed by conducting baseline spirometry (measuring volume of air inhaled and exhaled, and velocity of exhaled air), and repeated measures at different levels of exposure (Barreiro and Perillo, 2004). Studies of school children found decreased lung function with increasing ambient NO_2 levels (Hoek and Brunekreef, 1994; Linn et al., 1996; Moshammer et al., 2006; Timonen et al., 2002). For example, Moshammer et al. (2006) found that, among children, an increase in NO_2 concentration of 20 ppb was associated with a 4% reduction in the total amount of air capacity of the lung (i.e., forced expiratory volume). Two studies of adult populations did find significant associations between NO_2 levels and lung function using spirometry among a population of never-smokers (Schindler et al., 2001) and patients with COPD (Silkoff et al., 2005). Studies that measured impacts of NO_2 or particulate matter on lung function report inconsistent results. Some studies found that six hours of exposure to O_3, even at levels found in some cities, can decrease lung function, although in many studies volunteers returned to normal in a matter of hours (EPA, 2006b).

Over 30 epidemiologic studies have examined the effects of short- and long-term particulate matter exposures on lung function and many of these have seen lung function adversely affected by exposure. Gauderman et al. (2002) found such a relationship in the Child Health Study (see below). A similar number of studies have examined the effects of O_3 exposure both in clinical studies and in epidemiologic studies (EPA, 2006b). Associations were seen in many studies, particularly among asthmatics, workers outside, and in older populations.

Normally as the lungs of a child develop, these measures of lung function increase as well. A number of studies have followed children over time (3–10 years) with annual exams and have found that growth in lung function (e.g., the amount of air inhaled) is lower in children living in areas with higher NO_2 levels (Gauderman et al., 2004; Oftedal et al., 2008; Rojas-Martinez et al., 2007). As other pollutants tended to vary in the same way as NO_2, the studies could not show that the effects were directly associated with NO_2, but only with increases in pollution levels.

RESPIRATORY SYMPTOMS

Researchers have investigated air pollutant concentrations in ambient air as well as in the indoor environment, and the prevalence of respiratory symptoms, both among asthmatics and those without asthma. Assessments of the effects of NO_2 in ambient air include three large studies, each of which focused on children (Mortimer et al., 2002; Schildcrout et al., 2006; Schwartz et al., 1994). In each of these studies, children in several cities in the United States were followed over time (from two months to a year of data collection), and their respiratory symptoms (e.g., coughing, wheezing, and/or shortness of breath), and/or use of asthma rescue medications were recorded. All three studies found significant associations between NO_2 levels and the frequency of respiratory symptoms, and the effects were greater among asthmatics. The strongest associations were for NO_2 levels averaged over the previous 2–6 days. As NO_2 levels tend to vary with other air pollutants that also induce inflammation and impact respiratory symptoms (e.g., O_3 and $PM_{2.5}$), the researchers used statistical techniques to account for the contributions of these pollutants. In most cases, there was still an effect of NO_2. These studies provide some of the most convincing evidence that NO_2 levels in ambient air are associated with increases in respiratory symptoms, particularly among asthmatics. Combined together, these studies show that a 20 ppb increase in average NO_2 levels is associated with a 14% increase in the risk of experiencing adverse respiratory symptoms (EPA, 2008d).

A number of other studies were similar in design, but used children in only one location. This reduced the variability in NO_2 levels, and in the relationship of NO_2 with other factors that potentially affected respiratory problems. Studies in Paris (Just et al., 2002), Sydney and Perth, Australia (Jalaludin et al., 2004; Rodriguez et al., 2007), the United Kingdom (Ward et al., 2002), and the Netherlands (Boezen et al., 1999) all found associations between ambient NO_2 levels and cough or other respiratory symptoms in children.

A number of recent studies that followed people over time assessed the effects of particulate matter. As pollution levels dropped between 1993 and 2000, Bayer-Oglesby et al. (2005) observed a drop in the incidence of chronic cough, bronchitis, and colds among children. The incidence of cough, phlegm, and wheezing dropped among adults in Switzerland between 1991 and 2000, as PM levels dropped (Schindler et al., 2009). Several other studies have found similar results. Short-term exposures to particulate matter have not had strong associations. Recently, Weinmayr et al. (2009) reviewed 36 studies that looked at asthma symptoms, cough, and peak expiratory flow (measure of lung function). Overall, there was good evidence of PM_{10} being associated with these symptoms, and less convincing evidence of associations with NO_2. In many studies, acute exposures to O_3 are related to respiratory symptoms among people with asthma, but there are also several studies where there is no effect (EPA, 2006b). Effects are greater with multiple days rather than single days of exposure (Escamilla-Nuñez et al., 2008; Romieu et al., 2006).

Some research has assessed the relationship between indoor levels of NO_2 and respiratory symptoms. The advantages of this study design are that NO_2 levels can be accurately assessed through direct monitoring, and that exposure is almost exclusively to NO_2 as other air pollutants do not tend to occur in the indoor environment. The most convincing study randomly selected schools to replace their unvented heaters with heaters that were vented (Pilotto et al., 2004). This change led to a substantial reduction in indoor NO_2 levels. The children in the intervention schools had significantly fewer episodes of asthma attacks, difficulty breathing, and tightness of the chest during the day. Two observational studies (Belanger et al., 2006; Kattan et al., 2007) found that the risk of poor respiratory symptoms (i.e., wheeze or cough) was 50% greater for increases of 20 ppb of NO_2. McConnell et al. (2006) found respiratory symptoms to be associated with NO_2, but that the effect of NO_2 exposure was greater for children who had a dog. Several other studies, all with a large number of subjects, did not find an association.

ASTHMA

Directly assessing the relationship between the proportion of a population with asthma (prevalence) and the number of new cases of asthma in a given time period (incidence) with air pollutant levels typically requires a large study population that is followed over time. One of the best studies of this type is the Child Health Survey, a study of children in several communities in southern California. In one part of this study, NO_2 levels were measured in children's homes. The mean NO_2 level was associated with a history of asthma and the amount of asthma medications used (Gauderman et al., 2005). Each 20 ppb increase in NO_2 levels was associated with an eightfold increase in asthma prevalence (Gauderman et al., 2004). Millstein et al (2004) found that while NO_2 was not related to the use of asthma medications, other air pollutants were, and the effect was greater among kids who spent more time outside. New onset asthma was reportedly associated with outside exercise, especially where O_3 levels were high, and with estimated O_3 levels near homes and schools (McConnell et al., 2002, 2010). O_3 level was also demonstrated to be highly related to new onset asthma among genetically susceptible children (Islam et al., 2008, 2009). Islam et al. (2007) also found that the protective effect of good lung function against new onset asthma was reduced if the child lived in a high $PM_{2.5}$ community.

Many other studies have established strong associations between the onset of asthma and O_3 and $PM_{2.5}$ levels (EPA, 2006b, 2008d, 2009b). For example, a study in the Netherlands among young children (0–4 years) found an association between NO_2 and $PM_{2.5}$ levels and ever being diagnosed with asthma (Brauer et al., 2007). A nationwide study including over 30,000 children found that relatively small increases in O_3 levels were associated with developing asthma, or having an asthma attack (Akinbami et al., 2010). Other studies have looked at exposure to traffic (based on distance to major roads) and found associations with asthma, without looking at specific pollutants.

Some investigators have used data about hospital admissions to specifically study asthma admissions. These cases of asthma are usually the most serious cases. Overall, the results of these studies are mixed and provide some indication of a relationship between daily mean NO_2 levels and asthma admissions, with stronger evidence for effects on children. Grineski et al. (2010) demonstrated that the risk of being admitted to a hospital for asthma when NO_2 levels were elevated was higher for minority children and children without health insurance.

There have been many studies of particulate matter and asthma admissions using different statistical methods and comparing admissions to particulate matter levels at different times in the past (e.g., the day before, 3 days before, etc.). Most of the studies showed some effects, with somewhat more consistent results in studies of older people.

While there have been fewer studies of O_3, this pollutant has been consistently found to be related to asthma admissions, although study results vary. In New York, chronic O_3 exposure was significantly associated with asthma severe enough to require hospitalization for children 1–6 years of age (Lin et al., 2008). An analysis of 6 years of intensive care unit (ICU) and non-ICU admissions for asthma found associations with O_3 among children 6–18 years of age (Silverman and Ito, 2010). A linear concentration–response relationship was observed, even at levels below the regulatory levels. A recent study in California found that annual average O_3, $PM_{2.5}$, and PM_{10} levels were each associated with asthma-related hospitalizations or emergency room visits (Meng et al., 2010).

CHRONIC OBSTRUCTIVE PULMONARY DISEASE (COPD)

COPD refers to a serious lung condition that makes breathing very difficult. It consists of two diseases: chronic bronchitis and emphysema. It is the result of long-term exposure to agents that impact the lungs such as smoking and air pollutants. These exposures actually affect the physical structure of the lung. COPD is the fourth most common cause of death in the United States, with over 12 million diagnosed cases (NHLBI, 2010). Most of the evidence indicates that exposure to air pollutants can lead to emergency department (ED) visits, hospitalizations, and deaths among people with COPD; there is less evidence that exposure actually leads to the development of COPD (Ko and Hui, 2010).

Only a handful of studies have looked at NO_2 and ED or hospital admissions for COPD, and the results were inconclusive. In Los Angeles, NO_2 levels were associated with COPD in older adults (Moolgavkar, 2003), as well as among adults over 30 (Linn et al., 2000). Large studies in Canada and Finland

examined the relationship of several pollutants and ED visits for specific causes. O_3 levels were associated with asthma and COPD (Halonen et al., 2010), while particulate matter was associated with asthma, particularly during the warm season (Stieb et al., 2009). Many studies also have documented associations between particulate matter and COPD, ED visits, and hospitalizations (Chen et al., 2004; Dominici et al., 2006; Medina-Ramón et al., 2006; Peel et al., 2005).

RESPIRATORY INFECTIONS

There is some evidence that short-term exposures to NO_2 can affect the body's natural defenses against viral or bacterial respiratory infections, increasing the risk of respiratory infections. Two studies of groups of children followed over time observed associations between ambient NO_2 and PM exposures and ear, nose and throat infections (Brauer et al., 2002, 2007), and ear infections (Brauer et al., 2006) in children. Several studies have found associations between NO_2 levels and ED visits and hospitalizations for many respiratory conditions including respiratory infections (Peel et al., 2005). However, a study focused on the relationship between lower respiratory tract infection in children and NO_2 in three European cities (up to 42 ppb) found no association (Sunyer et al., 2004).

There are several possible mechanisms that could lead to increased susceptibility from NO_2 exposure (Chauhan and Johnston, 2003). Exposures to higher than ambient levels of NO_2 and O_3 have, in some human and animal studies, been shown to temporarily reduce the action of the cilia that help capture and expel foreign bodies from the airway. However other studies have not observed this effect. In animal studies, exposure to O_3 has led to damage of the cilia. A number of clinical studies have examined the effects of NO_2 exposure on immune and biochemical responses that could account for more severe symptoms from a respiratory infection among those exposed to NO_2. While some studies have found a significant effect, others have not (EPA, 2008d; Hesterberg et al., 2009).

There is even stronger evidence that exposure to NO_2 can worsen the severity of respiratory infections. Studies of children in England (Chauhan and Johnston, 2003) found that those children exposed to high levels of NO_2 (>7.4 ppb week before) had more severe symptoms and decreased lung function as compared to children with low levels of NO_2 exposure. This study also found that children exposed to higher levels of NO_2 were nearly twice as likely to suffer an asthma exacerbation associated with the infection in the week after the infection had started, as compared to children exposed to lower levels (Linaker et al., 2000). This is considered to be one of the strongest studies in that exposure was carefully measured using detectors pinned to the children's clothing.

There are few studies examining the effects of O_3 or particulate matter on respiratory infection. In one study, visits to physicians in a managed care organization in Atlanta for upper and lower respiratory infection were not related to O_3 levels (Sinclair et al., 2010). While there are a number of studies of particulate matter and respiratory infection and/or pneumonia, two multicity studies in the United States found increases in admissions for respiratory infections or pneumonia (Dominici et al., 2006; Medina-Ramón et al., 2006).

5.3.5.3. HOSPITAL ADMISSIONS FOR RESPIRATORY PROBLEMS

Well over 100 studies have been conducted to assess the relationship of ambient pollutant levels and the number of hospital admissions and/or ED visits for asthma exacerbations, chronic COPD, all respiratory complaints, or other related health problems. There are two types of studies. To look at short-term effects, the numbers of visits are compared to pollutant levels, either maximum levels or 24-hour average concentrations, for the day of the admission or some number of days prior to the admission. The effects of longer-term exposures are measured by using average pollution levels over months or years and subsequent hospital or ED admission rates.

While all these studies share the same general design, there are differences in the population studied, the health endpoints used, the way exposure is measured, and the locations. These differences in study design can lead to variability in the observed results. Overall, such studies are limited in that they can only demonstrate an association between days that have a higher number of admissions and days when pollutant levels are higher; they do not collect additional data from each person about the other factors that can lead to respiratory problems, and as such they cannot control for these factors in the analysis. Nevertheless, these studies show a consistent set of results. In the vast majority of studies, days with higher levels of NO_2, O_3, and/or particulate matter were associated with higher numbers of hospital or ED admissions for all respiratory complaints.

Overall, most of the studies showed that a 20 ppb increase in NO_2 was associated with a 1–25% increase in the number of admissions. Even though high levels of NO_2 tend to occur on the same days as high levels of other pollutants, controlling for the other air pollutants did not significantly change the estimated effect of NO_2 levels. While some of the studies looked specifically at children or people over 65, the results were about the same.

Katsouyanni et al. (2009) recently conducted a study which combined information from several large multicity studies of air pollution and respiratory hospital admissions. Daily increases in O_3 resulted in significant increases in admissions; a change of 40 ppb was associated with a 2–3% increase in admissions. Particulate matter levels were not consistently associated with admissions. While very few studies have looked at the independent effects of the different types of particulate matter, Ostro et al. (2009), using hospital records from six counties in California, found that

many of the specific components of $PM_{2.5}$, including nitrate, were independently related to respiratory hospitalizations among children; the effect of NO_2 was similar to that of $PM_{2.5}$. In one of the only studies of neonates (birth to 27 days after birth), Dales et al. (2006) found that higher O_3 levels were followed by more hospital admissions.

5.3.5.4. EVIDENCE OF THE IMPACTS OF NO_2, O_3, AND PARTICULATE MATTER EXPOSURE ON CARDIOVASCULAR DISEASE

Possible effects of NO_2 on CVD have been investigated using hospital and ED admission data. Many studies found mean 24-hour NO_2 levels or 1-hour maximum levels were related to the number of CVD admissions (Andersen et al., 2007; Jalaludin et al., 2006; Metzger et al., 2004; Tolbert et al., 2007). When only cardiac diseases were considered, almost all studies found statistically significant associations (Chang et al., 2005; Simpson et al., 2005a; Von Klot et al., 2005). Studies examining correlations between NO_2 and stroke were mixed; overall, there was little evidence of an effect.

Several toxicological studies provide evidence that O_3 exposure could be related to cardiovascular problems. While the studies linking O_3 exposure and CVD morbidity did not show a linkage, large multi-country studies (Katsouyanni et al., 2009) and a multicity study in the United States (Zanobetti and Schwartz, 2008) found that during the warm season, short-term increases in O_3 were followed by increases in deaths due to CVD.

Particulate matter is the air pollutant that appears to have the greatest impact on CVD (Wellenius et al., 2012). The US EPA has concluded that both short- and long-term exposures to $PM_{2.5}$ are causally related to cardiovascular effects, and that "a causal relationship is likely to exist" between CVD and larger particles (EPA, 2009b). This conclusion was based on a large number of studies—toxicological, clinical, and epidemiological—that measured impacts of particulate matter exposure on many aspects of cardiovascular health. People with existing CVD and the elderly appear to be at higher risk (Brook et al., 2010).

Over 20 studies examined effects of particulate matter exposure on heart rate variability (HRV; a risk factor for arrhythmias and heart attacks) and most have found exposure to reduce HRV (Brook, 2008). Some studies have found effects of particulate matter, O_3, and NO_2 on arrhythmia, rapid heart rate (tachycardia), ECG abnormalities, increased blood pressure, thrombosis, inflammation markers, and coagulation factors (Brook, 2008). For these outcomes, the results have been inconsistent, with some studies showing an effect of particulate matter exposure (EPA, 2009b). Several studies have shown a relationship to atherosclerosis (Brook and Rajagopalan, 2010; Brook et al., 2010). A number of studies also examined the relationship, usually positive, between long-term particulate matter exposure (1–5 years) and CVD (Baccarelli et al., 2008; Miller et al., 2007; Zanobetti and Schwartz, 2008). For example, Miller et al.

(2007) observed significant effects on stroke, cerebrovascular disease, and all CVD using a prospective study. Likewise, Baccarelli et al. (2008) found significant effects on deep vein thrombosis (DVT), while Zanobetti and Schwartz (2008) reported higher rates of hospitalization for heart attacks among survivors of an initial heart attack in 21 US cities.

Examining the temporal relationship between short-term particulate matter levels and subsequent ED visits or hospital admissions is the most common study design used to look at the risk of particulate matter exposure. Almost every study looking at all CVDs observed a significant association with recent particulate matter levels (Brook, 2007; EPA, 2009b). Several studies established associations with hospital/ED admissions for specific conditions, including heart attacks, ischemic heart disease (lack of blood flow to the heart muscle), and congestive heart failure. Authors of a review of 49 studies concluded that there was evidence that exposure to particulate matter for less than a day can lead to ischemia and heart attacks, especially among the elderly and those who already have heart disease (Burgan et al., 2010). A few studies have tried to determine the form of the concentration–response relationship between particulate matter and CVD. As with many of these studies, Zanobetti and Schwartz (2005) found an almost linear relationship with no threshold, that is, even very low levels of PM_{10} were associated with an increase in the number of hospital admissions for heart attacks.

5.3.5.5. EVIDENCE OF THE IMPACTS OF NO_2, O_3, AND PARTICULATE MATTER EXPOSURE ON CANCER

Exposure to NO_2 is hypothesized to lead to the formation of nitrosamines in the lung. Nitrosamines resulting from ingesting nitrate or nitrite have been shown to be carcinogenic (IARC, 2014). Two studies examining the effect of NO_2 exposure on the risk of developing lung cancer found significant relationships (Nafstad et al., 2003; Nyberg et al., 2000). O_3 and particulate matter are also thought to possibly be involved in the carcinogenic process, possibly through their effects on cells or DNA. One recent study found no association between overall cancer deaths and O_3 but did see an association between $PM_{2.5}$ and lung cancer mortality (Krewski, 2009).

5.3.5.6. EVIDENCE OF THE IMPACTS OF NO_2, O_3, AND PARTICULATE MATTER EXPOSURE ON BIRTH OUTCOMES

Over 20 studies have examined the prevalence of LBW delivery (<2,500 g or 5.5 lbs.), preterm delivery (PTB, <37 weeks) or babies that are small for the number of weeks of gestation (SGA) with ambient air pollutant levels. Almost no studies saw any evidence of an association (Shah and Balkhair, 2011; Stillerman et al., 2008) except two California studies that showed effects on birth weight. NO_x exposure was associated with LBW, PTB, or SGA in a small number of studies; no effect was seen in most of the studies (EPA, 2008d; Shah and Balkhair, 2011). A California-wide study observed a small but statistically significant increase in LBW with increasing NO_2.

Several studies have shown reproductive impacts associated with exposure to $PM_{2.5}$. A study in Southern California, limited to women who lived close to an air monitor, found a significant increase in the risk of an LBW baby and a PTB (Wilhelm and Ritz, 2003). Data from the Child Health Study in Southern California found no impact on birth weight by PM_{10}, but a clear association with O_3 levels, even when accounting for the other air pollutants (Salam et al., 2005). Two California studies found reductions in birth weight with exposure to particulate matter (Parker et al., 2005) and a strong effect of $PM_{2.5}$ exposure on PTB (Ritz et al., 2007).

While there are relatively few studies of air pollutants and stillbirth or infant death, several studies in California found recent increases in PM_{10} levels to be associated with an increased risk of childhood deaths (Barnett et al., 2005; Ritz et al., 2006; Woodruff et al., 2006); other studies, however, have not found significant effects. There are few studies and little evidence of any effect of O_3 or NO_x on infant mortality or stillbirth. There are too few studies of the effects of air pollution on birth defects to draw any conclusions about this health effect.

5.3.5.7. EVIDENCE OF THE IMPACTS OF NO_2, O_3, AND PARTICULATE MATTER EXPOSURE ON MORTALITY

An association between NO_2 exposures and mortality (death) rates was found in five out of six studies and some of the deaths were due to cardiovascular or respiratory causes (Brook, 2007; Burnett et al., 2004; Dominici et al., 2003; Hoek et al., 2002; Samoli et al., 2006; Simpson et al., 2005b; Stieb et al., 2003). At least 15 studies examined the relationship of short-term O_3 levels and all non-accidental deaths, many also looking specifically at deaths due to respiratory and cardiovascular causes (Brook et al., 2010; McClellan et al., 2009). Two studies (Bell et al., 2006 and Katsouyanni et al., 2009) found that excess deaths occurred at O_3 levels below the regulatory limits for the United States. Bell et al. (2006) found that even low levels of O_3 are associated with increased risk of premature mortality and that the risk of mortality is statistically significant when daily average O_3 concentrations are above 80 ppb.

Many studies of particulate matter and mortality, many of which measure very large populations, provide consistent evidence of the effects of short- and long-term exposure to particulate matter (Brook et al., 2010; Dockery et al., 1993; EPA, 2009b). A study of 8011 adults from six US cities (Topeka, KS, Steubenville, OH, St. Louis, MS, Harriman, TN, Watertown, MA, and Portage, WI), which controlled for smoking and other personal factors, found that fine particulate matter was positively associated with death from COPD and lung cancer but not from other causes combined (Dockery et al., 1993). In Brook et al.'s (2010) and EPA's

(2009b) studies, particulate matter was associated with all causes of death (non-accidental), and was related to greater increases in deaths due to ischemic heart disease, COPD, and CVD.

5.3.5.8. SUMMARY OF HEALTH IMPACTS OF NO$_2$, O$_3$, AND PARTICULATE MATTER

There is strong evidence in hundreds of studies since the 1960s that exposure to NO$_2$, O$_3$, and particulate matter leads to respiratory symptoms, reduced lung growth, asthma exacerbations, and respiratory infections in humans (Table 5.3.2). Such exposures may be associated with visits to emergency rooms and hospital admissions for respiratory complaints, asthma, COPD, and cardiac problems. Increases in mortality, in general and specifically for respiratory and cardiovascular causes, occur after high ambient pollutant levels. Children, the elderly, and people with existing COPD or CVD are more likely to be affected.

Many of these epidemiological studies are limited by the way that exposure is measured. Typically when the concentration of one air pollutant is high, so are the concentrations of other air pollutants. They together may contribute to a particular health outcome and only some studies try to account for the presence of other pollutants. Further, many studies use observed or predicted ambient pollutant levels near each case's house, and do not account for temporal and spatial patterns of exposure.

Given the known and suspected relationships of NO$_2$, particulate matter, and O$_3$ with many health problems, the overall public health impact of these exposures is potentially very high. The California Air Resources Board estimates that each year exposure to PM$_{2.5}$ results in 7,300 excess deaths from cardiopulmonary diseases and 5,500 from ischemic heart disease (for exposures greater than 5.8 µg m^{-3}) (CARB, 2011). If O$_3$ was decreased to California's standard, it is estimated that each year 630 deaths, 4,200 hospital admissions for respiratory diseases, 660 ER visits for asthma, and 4.7 million days of missed school among children would be averted (Ostro et al., 2006).

5.3.5.9. RESEARCH NEEDS

As this review has shown, there are many health impacts that result from exposure to NO$_2$, O$_3$, and particulate matter in the air. Certainly more studies are needed to determine the extent to which these air pollutants might be related to other diseases, such as birth defects and stillbirth, lung cancer, and respiratory infection. Better information is needed about the actual levels of exposure among the population, and differences in exposure by location, income, race, and ethnicity. For almost all of the health impacts associated with air pollutants, much better data are needed describing the relationship between the concentration in the air and the number of people affected to help determine public health improvements resulting from changes in air quality

standards and air contaminant levels. More studies are needed to better quantify the proportion of PM$_{10}$ and PM$_{2.5}$ that are principally made up of nitrogen containing compounds, and the regions and conditions under which such particles are formed. Such data would also be useful if combined with studies estimating the health-care costs resulting from health impacts of nitrogen-derived air pollutants.

5.3.6. Economics of Air Quality

Muller and Mendelsohn (2007) estimated air pollution damages of NO$_x$, NH$_3$, PM$_{10}$, PM$_{2.5}$, as a whole from over 10,000 point and aggregated nonpoint sources in the contiguous United States using the Air Pollution Emission Experiments and Policy (APEEP) analysis model. Based on their estimates, damage costs due to mortality and morbidity account for about 71% ($53 billion) and 23% ($17 billion), respectively, of the total costs (Muller and Mendelsohn, 2007). Overall, damage costs due to human health account for 94% of total damages with the remaining 6% due to visibility impairment ($2.7 billion), reduction in agricultural yield ($1.2 billion), reduction in timber production ($80 million), depreciation of man-made materials ($100 million), and diminished forest health (Muller and Mendelsohn, 2007). These different types of damage costs are not available specifically for California.

The APEEP model was also used to estimate county-level damage costs associated with mobile, point, and nonpoint sources based on data from the US EPA's 2002 National Emissions Inventory (Muller and Mendelsohn, 2009). Map 5.3.3 shows that the highest marginal damage costs (in $ ton^{-1} of pollutant) due to NO$_x$ in California occurred in San Joaquin, Sacramento, Stanislaus, Yolo, Solano, Napa, Sonoma, Merced, Fresno, and Marin Counties (Muller and Mendelsohn, 2009). Damages due to NH$_3$ and PM$_{2.5}$ had a notably different geographic distribution with Los Angeles, San Francisco, Contra Costa, Orange, Alameda, Santa Cruz, Santa Clara, Marin, and San Diego having the highest marginal damage costs (Map 5.3.3).

Considerable uncertainty exists in virtually any estimate of the economic damages associated with air pollution. A statistical analysis of the uncertainty in air pollution damages from 565 electric generating units in the United States found that the largest sources of uncertainty were due to high variance in adult mortality dose–response relationships, mortality valuation, and the methods of air quality modeling (Muller et al., 2011). Also, the estimated marginal damage (per ton damage) distributions were found to be positively skewed and are more variable in urban rather than in rural areas. Likewise, the European Nitrogen Assessment (Sutton et al., 2011a) suggested that variation in the estimation of economic damage costs associated with the dose–response function and a lack of comparability among "willingness to pay (WTP) studies" are key sources of uncertainty and thus pose important constraints for economic research.

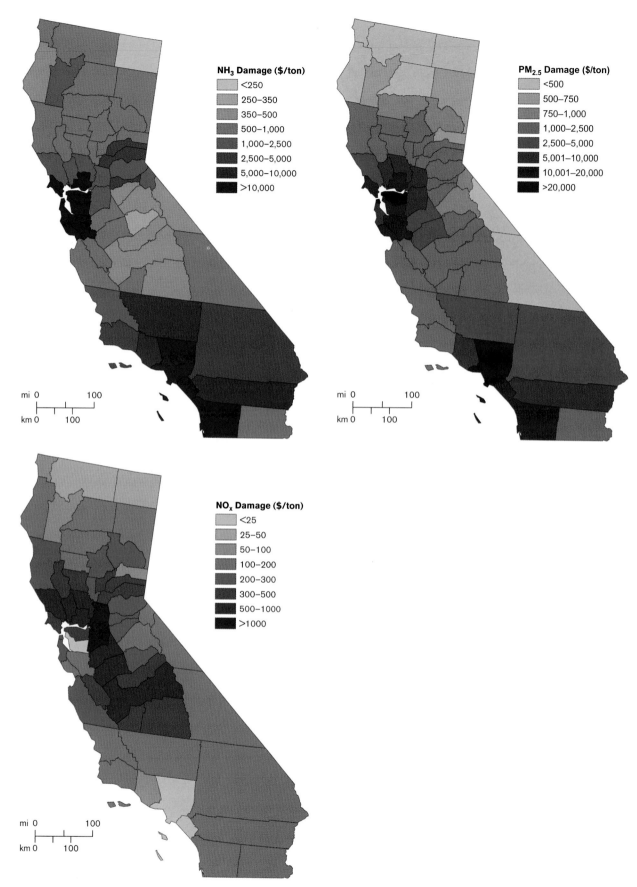

MAP 5.3.3. Marginal benefits of abatement for ground level NH₃, NOₓ, and PM₂.₅ emissions in California. Damage costs are expressed in dollar per ton of each pollutant (adapted from Muller and Mendelsohn, 2009; map by R. Murphy).

TABLE 5.3.4
Health outcomes and economic values related to premature deaths from particulate matter (PM_{10} and $PM_{2.5}$) and minor restricted activity days (MRADs) in the South Coast Air Basin, California, 1989 and 2008

Particulates	1989 study (1988 $)	2008 study (1988 $)[a]	2008 study (2007 $)	1989 study (2007 $)
Premature deaths	1617		3000	
value of statistical life	3.7 million	3.54 million	6.63 million	6.93 million
Total value	5.98 billion		19.88 billion	
Ozone				
MRADs	17.65 million		961,400	
Unit value	34.95[b]	35.08	65.60	65.46
Total value	3.79 million		63.16 million	

a. Adjusted for price level (CPI) and income changes.

b. Commonly used Tolley et al. (1986) value (our value was 21.50).

SOURCE: Adapted from Hall et al. (2010).

5.3.6.1. ECONOMIC COSTS OF AIR POLLUTION ON HUMAN HEALTH

The economic impact of air pollution on the health of Californians has received considerable attention and some of the earliest research on the subject is based in California (Hall et al., 1992).

Economic valuation of the health outcomes resulting from air pollution is typically based on three pieces of information: (1) dose–response relationships; (2) levels of exposure; and (3) demographic information (e.g., population by age and other characteristics). The dose–response relations and estimated exposures are used to calculate the costs associated with hospital visits, treatment, and or mortality for various health outcomes.

Romley et al. (2010) examined private and public insurer spending on hospital admissions and emergency room visits related to illnesses with respiratory, cardiovascular, and asthma causes. They found that poor air quality which fails to meet NAAQS and CAAQS for O_3 and $PM_{2.5}$ has likely contributed to nearly 30,000 hospital admissions and emergency room visits, costing more than $193 million between 2005 and 2007 (Romley et al., 2010). Three-quarters of these incidents are attributable to high $PM_{2.5}$, while the rest were due to high O_3 levels.

Much of the economic research in California focuses on health outcomes related to O_3 and $PM_{2.5}$ exposure in the South Coast basin and SJVAB due to the intensity of the concerns in these locations (Hall et al., 2003). For example, Romley et al. (2010) found that the majority of air-pollution-related hospital visits in California occurred in the SJVAB and SoCAB. Because failing to meet NAAQS had the greatest effect on Hispanic and African Americans communities, with most of the patients poor, a disproportionate share of the cost burden falls on public insurers, such as Medicare and Medi-Cal (Romley et al., 2010). The incidence of O_3 and $PM_{2.5}$ combined is estimated to cause nearly $6 and $19 billion per year (2007 $) of negative health effects in SJVAB and SoCAB, respectively (Table 5.3.4; Hall et al., 2010). These values likely underestimate actual cost because of the inability to catalog or quantify all adverse effects of exposure in economic terms.

Rather than estimating costs, Ostro et al. (2006) examined the economic health benefits from a decrease in the incidence of premature morality, hospital admissions, emergency room visits, lost school days, and minor reduced activity days of attaining the national and the more restrictive 8-hour O_3 levels throughout California. They estimated benefits equaling $2.8 and $4.5 billion per annum (2000 $) for obtaining the NAAQS and CAAQS, respectively.

Children (aged 5–18) are one of the populations most sensitive to high levels of O_3 (Hall et al., 2003). For children aged 5–18 in the SoCAB, exposure to O_3 exhibited a downward trend from 1990–1999 and was associated with a decrease in O_3-related illness school absences (Table 5.3.5). The total economic value (benefit) of differences between 1990–1992 and 1997–1999 in annual 8-hour O_3-related all-illness school absences for children aged 5–18 was $245 million ($75 per capita) in the SoCAB (Hall et al., 2003) (Tables 5.3.5 and 5.3.6). Los Angeles County ($147,689,000) and San Bernardino County ($45,666,000) benefit the most from a reduction in school absences (Table 5.3.6).

Due to limited availability of data on medical treatments associated with school-related absences, the cost estimates of the economic benefit of a reduction in school absences by Hall et al. (2003) only included indirect costs (i.e., value of the caregiver's time). Brandt et al. (2012), on the other hand, considered both the direct and indirect health-care costs of childhood asthma as a result of exposure to traffic-related pollution (TRP) (e.g., NO_2 and O_3) in Riverside and Long Beach. Direct health-care costs include the costs of bronchitis episodes, ED visits, and hospitalization, whereas the indirect costs include value of time loss (e.g., forgone income) by parents/caregivers as a result of the child's

TABLE 5.3.5

Differences in O_3-related all-illness school absences in the SoCAB over time for the 1998 population aged 5–18

Time period	1990–1992 to 1997–1999	1991–1993 to 1997–1999	1992–1994 to 1997–1999	1993–1995 to 1997–1999	1994–96 to 1997–1999	1995–1997 to 1997–1999	1996–1997 to 1997–1999
	Millions of annual absences						
Decrease in all-illness absences	3.19	2.84	2.47	1.70	1.43	0.984	0.480

SOURCE: Data from Hall et al. (2003).

TABLE 5.3.6

Economic value of differences ($) in annual 8-hour O_3-related all-illness school absences from 1990–1992 to 1997–1999 in the 1998 population aged 5–18

School absences	SoCAB	Los Angeles County	Orange County	Riverside County	San Bernardino County
Total all illness	245,048,000	147,689,000	21,584,000	30,109,000	45,666,000
Per capita all illness (2010 $)	75	74	39	91	114

SOURCE: Data from Hall et al. (2003).

TABLE 5.3.7

Cost of health outcomes attributable to pollution exposure for children in Riverside and Long Beach

	Riverside		Long Beach		
	Count	Cost per outcome ($)	Count	Cost per outcome ($)	Total cost per year ($)
Attributable asthma cases	690 (630–750)	4,008	1,600 (1500–1800)	3,819	8,875,920
NO_2: attributable exacerbations of other-cause asthma					
Emergency room visits	40 (5–70)	956	150 (19–280)	944	179,840
Inpatient hospitalizations	8 (6–10)	13,282	27 (22, 32)	13,227	463,385
Clinic office visits	190 (38–340)	158	440 (80–780)	153	97,340
Bronchitis episodes	1500 (440–2300)	975	3,100 (1000–4,400)	918	4,308,300
O_3: attributable exacerbations of other-cause asthma					
Emergency room visits	230 (150–310)	956			219,880
Inpatient hospitalizations	12 (9–15)	13,282			159,384
Clinic office-visits	190 (25–360)	161			30,020
Bronchitis episodes	2,900 (160–3,900)	975			2,827,500
School days absent	2,966 (2,223–4,685)	230	626 (43–1,114)	205	810,510
Total annual cost					17,972,079

NOTE: The 95% confidence intervals are reported in parentheses. All costs are rounded to the nearest US$ 2010 rate.
SOURCE: Adapted from Brandt et al. (2012).

asthma-specific office visits and school absences due to air pollution (Brandt et al., 2012). The annual cost for a case of childhood TRP-related asthma was estimated to be $4,008 in Riverside and $3,819 in Long Beach, with indirect costs due to school absences making up the largest share of cost (34% for Riverside and 32% for Long Beach) (See Table 5.3.7). Total annual costs of TRP-related asthma cases were estimated to be $2,808,300 and $6,120,000 (in 2010) in Riverside and Long Beach, respectively (Brandt et al., 2012). Since these populations constitute only 7% of California's total population, statewide health-care costs due to NO_2 and O_3 will be considerably higher (Brandt et al., 2012).

Evaluations of the economic benefits of controlling air pollution must be made with caution and based on an understanding of concentration-response functions, population growth and mobility (exposure), and economic valuation. In one of the few rigorous comparisons of economic outcomes and health benefits of O_3 and $PM_{2.5}$ reductions in California, looking at the period of 1989–2008, human exposure to O_3 and $PM_{2.5}$ concentrations was found to have declined dramatically (Hall et al., 2010). Furthermore, while the health literature is constantly evolving, the economic values estimated for adverse health outcomes have stayed relatively constant (Hall et al., 2010).

5.3.6.2. ECONOMIC COSTS OF AIR POLLUTION ON CROP PRODUCTION

EFFECTS OF O_3 AND ACID PRECIPITATION ON CROPS

The damaging effects of O_3 on agricultural crops are well established (Benton et al., 2000; Booker et al., 2009; Emberson et al., 2009; Fuhrer, 2003; Sandermann, 1996). Plant injury occurs when O_3 enters the plant leaf cells via the stomata, where it can oxidize and degrade cell membranes, pigments, and proteins. Common visible symptoms include early leaf senescence, leaf chlorosis, and lower root mass fractions (Felzer et al., 2007; Grulke et al., 1998; Thomas, 1951; Wang and Taub, 2010). Leaf tissue damaged by O_3 exposure tends to have lower CO_2 assimilation capabilities, caused by a decrease in Rubisco activity (Booker et al., 2009; Felzer et al., 2007; Grantz and Shrestha, 2006). Exposure to O_3 can also reduce drought tolerance (Feng et al., 2008), increase disease vulnerability (Calvo et al., 2007), and decrease the yield of many grain and vegetable crops (Bender et al., 1999; Feng and Kobayashi, 2009; Hassan et al., 1999; Holmes and Schultheis, 2003).

In California, the high levels of O_3 occurring in regions of high agricultural productivity suggest that California's agricultural economy is particularly vulnerable to O_3 (Muller and Mendelsohn, 2009). Evidence of O_3 impacts on crops has been reported in alfalfa (Mutters and Soret, 1998), almonds (Retzlaff et al., 1990), apples (Retzlaff et al., 1992), beans (Mutters and Soret, 1998), cantaloupe (Mutters and Soret, 1998), citrus (e.g., oranges, lemons; Mutters and Soret, 1998), cotton (Grantz, 2003), grapes (Mutters and

Soret, 1998), lettuce (Heck et al., 1982), onions (Mills et al., 2007), rice (Sawada and Kohno, 2009), stone fruit (e.g., apricots, plums, prunes; Retzlaff et al., 1992), pears (Retzlaff et al., 1992), potatoes (Vorne et al., 2002), tomatoes (Calvo et al., 2007), and wheat (Feng et al., 2008). In general, the regions of California with a combination of severe O_3 pollution and high agricultural production value (e.g., San Joaquin Valley) have received the greatest attention (Booker et al., 2009; Rowe and Chestnut, 1985). However, since toxicity can occur on sensitive species at relatively low O_3 concentrations, damages due to minor yield losses in other regions are thought to be widespread (Grantz and Shrestha, 2005).

In California, the extent of yield reduction directly attributable to O_3 is highly uncertain for most crops, but estimates suggest that O_3-related yield losses in excess of 10% are not uncommon for particularly sensitive crops (Table 5.3.8; Murphy et al., 1999; Mutters and Soret, 1998). Yield reductions vary by crop species tolerance, severity of exposure, and location, with the greatest estimated losses predicted in the San Joaquin basin and SoCAB. Yields of cantaloupe and table grapes were reduced by approximately 33% and 30%, respectively, while fresh market tomatoes were less than 1% (Table 5.3.8; Mutters and Soret, 1998). These differences in crop sensitivity add considerable uncertainty to econometric efforts to model losses in crop yield and revenue.

Early literature reviews by Irving (1983) and Jacobson (1984) concluded that the level of acidity in precipitation and fog is seldom sufficient to cause acute injury to crops and natural vegetation. Studies carried out in California, however, indicate that certain crops may suffer injury from exposure to acidic fog if the pH of water droplets is 2.5 or below, e.g., for pinto beans (Bytnerowicz et al., 1986), lettuce (Granett and Musselman, 1984), alfalfa (Temple et al., 1987), and green peppers (Takemoto et al., 1988). Thus, while very acidic fog and rain pose a possible threat to crop growth, the risks are likely to be relatively small under ambient field conditions.

EFFECTS OF O_3 ON COSTS TO AGRICULTURAL PRODUCERS AND CONSUMERS

The economic impact of increasing concentrations of O_3 has most often been assessed in terms of lost revenue to agricultural producers due to lower yields (Howitt and Goodman, 1989; Howitt et al., 1984; Kim et al., 1998). To estimate economic impacts, biophysical dose–response models calibrated with crop-specific data from empirical experiments are coupled with economic models. Using the economic model developed by Howitt et al. (1984), Rowe and Chestnut (1985) estimated that between $42 and $117 million in economic benefits from improved crop yield in the San Joaquin Valley could be achieved if California's state air quality standards for O_3 were attained. A study by Howitt and Goodman (1989) suggested that various

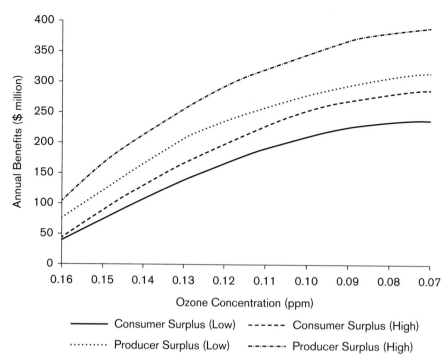

FIGURE 5.3.6. Annual agricultural benefits for consumers and producers for O$_3$ control in the San Joaquin Valley, 1998.

SOURCE: Reprinted from Kim et al. (1998). Copyright *Journal of Agricultural and Resource Economics.* Used by permission.

TABLE 5.3.8

Estimated statewide yield loss due to ground-level O$_3$ in 1993, using 7-hour (27.2 ppb) and 12-hour (25 ppb) mean O$_3$ exposure crop-loss models

Crop	Yield loss (%)
Study	Mutters and Soret (1998)
Cantaloupe	32.8
Grape, table	29.9
Grape, raisin	26.2
Cotton, upland	23.3
Grape, wine	22.8
Bean, dry	17.5
Orange	14.0
Onion	10.6
Alfalfa	9.5
Lemon	8.4
Tomato, processing	6.8
Wheat	6.7
Rice	3.9
Corn, field	1.2
Tomato, fresh-market	0.6

SOURCE: Adapted from Grantz and Shrestha (2005).

policy response scenarios to control O_3 concentrations in California could yield between \$50 and \$333 million per year in economic benefits, with approximately half of those benefits going to agricultural producers.

Since higher food prices are also an outcome of lower crop production, the negative consequences of O_3 exposure on crops are not limited to producers (Howitt et al., 1984) but also affect consumers who bear between 25% and 50% of the costs (Howitt and Goodman, 1989; Kim et al., 1998). Results from Kim et al. (1998) suggest that the benefits of reducing O_3 pollution for crop producers and consumers will likely range from \$50 to \$400 million per year depending on the level of O_3 reduction achieved (figure 5.3.6).

5.4. Climate Regulation

Main Messages

Human activities that increase reactive N have numerous competing effects on the ecosystem and biogeochemical processes that regulate the Earth's climate. Some processes have net warming effects that exacerbate climate change, while other processes have net cooling effects that partially offset the prevailing trend of a warming climate.

Emissions of N_2O have a long-term warming effect on global climate change. As the third most important greenhouse gas (GHG) behind CO_2 and CH_4, N_2O accounts for approximately 8% of total global and 3% of total statewide GHG emissions. The vast majority of N_2O emissions emitted globally and in California come from agricultural sources (N fertilizers, livestock, N_2-fixing crops), while fossil fuel combustion, sewage treatment, and industrial sources are also minor sources.

N deposition and fertilization tends to have an overall cooling effect on climate by enhancing terrestrial C sequestration in plant biomass and soils. Increased C sequestration due to N input has been documented for many forest, grassland, wetland, and agricultural ecosystems in North America (24–177 kg C kg^{-1} N deposited per year), a trend which has also been observed in California.

The formation of O_3 from NO_x has both warming and cooling effects on the Earth's climate. Increased ground-level O_3 has adverse effects on plant photosynthesis and CO_2 uptake, which decrease C sequestration by crops and natural vegetation. While estimates suggest that O_3 decreases plant C sequestration by 14–23% globally, more research is needed to quantify the extent of this impact in California. In contrast, O_3 can also have a small cooling effect on climate by increasing the concentration of hydroxyl radicals (OH), which in turn reduce the lifetime and overall burden of CH_4 in the atmosphere.

Atmospheric aerosols formed from NO_x and NH_3 have a short-term cooling effect on climate by reflecting and scattering solar radiation and stimulating cloud formation and the albedo effect. Since the formation of aerosols from NO_x and NH_3 are generally linked to different pollution sources (e.g., fossil fuel → NO_x, livestock → NH_3), the relative contribution of each pollutant and the chemical composition of resulting aerosols is likely to vary considerably across California's landscape.

Estimates suggest that anthropogenic sources of N have a modest net-cooling effect on the Earth's climate in the near term (20 years), but a net warming effect in the long term (100 years) as the prolonged effects of N_2O dominate the radiative balance. It should also be noted that the overall effects of N on the climate are relatively small compared to CO_2 from fossil fuel combustion (8% globally; 3–4% in California).

5.4.0. Introduction

Nitrogen plays a well-established role in regulating the Earth's climate. Human activities that increase the amount of reactive N that enters terrestrial and aquatic ecosystems can alter many biogeochemical processes that affect the Earth's climate balance (figure 5.4.1). Most notably, increases in reactive N can change the emission and uptake of the three important GHGs: carbon dioxide (CO_2), methane (CH_4), and nitrous oxide (N_2O). Emissions of N oxides (NO_x) and NH_3 also have important impacts on climate, since they are chemical precursors to ozone (O_3) and various atmospheric aerosols (Box 5.4.1). The ecosystem processes and atmospheric feedbacks involving N are complex, and thus the magnitude of their effects on the global climate are often uncertain. In this section, we examine the effects of reactive N on the Earth's climate balance paying particular attention to California's contribution to climate change over time through human activities and ecosystem processes that effect global N and carbon (C) cycles.

5.4.1. Measures of the Radiative Forcing and Global Climate Change

Several important metrics are used to quantify the effects of ecosystem processes that regulate the Earth's climate. The two most commonly used measures are radiative forcing (RF) and global warming potential (GWP). RF is a measure of the influence that a factor (e.g., GHG, atmospheric aerosol) has in changing the balance of energy in the atmospheric system and is expressed in watts per square meter (W m^{-2}) of the Earth's surface. More specifically, the Intergovernmental Panel on Climate Change (IPCC) calculates RF as the change in W m^{-2} relative to preindustrial conditions (i.e., pre-1750) (IPCC, 2007a). The amount of heat trapped by a particular gas depends on its absorption of infrared radiation, the absorption wavelength, and the atmospheric lifetime of the gas species (IPCC, 2007a). By integrating the RF caused by a 1 kg pulse of a given gas over a standard time period (20, 100, 500 years), its absolute GWP (also expressed in W m^{-2}) can be calculated. While

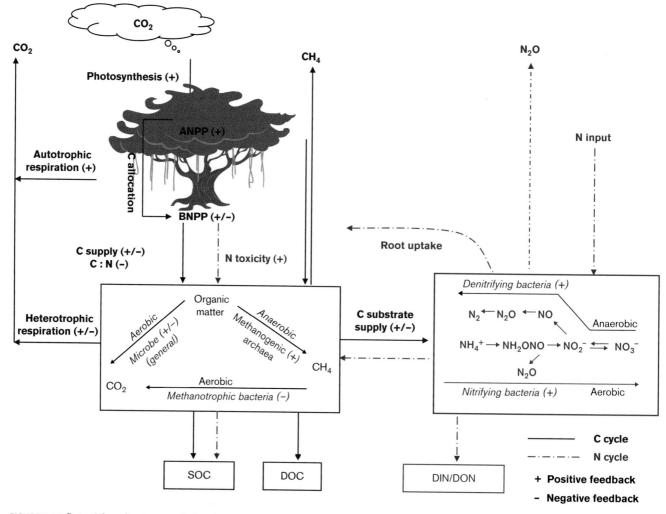

FIGURE 5.4.1. Potential mechanisms regulating the terrestrial ecosystem responses of CO_2, CH_4, and N_2O production and consumption to increased availability of reactive N.

ANPP = aboveground net primary productivity; BNPP = belowground net primary productivity; SOC = soil organic carbon; DOC = dissolved organic carbon; DIN = dissolved inorganic nitrogen; DON = dissolved organic nitrogen.

SOURCE: Reprinted from Liu and Greaver (2009). Copyright 2009 Blackwell Publishing Ltd/CNRS. Used by permission.

the IPCC uses the GWP over a 100-year time period in its policy frameworks, the shorter and longer time frames have also proven useful for detailed modeling studies (IPCC, 2001; Shine et al., 2007). Another convention adopted by the IPCC is to express the GWP of a gas relative to an equivalent mass of CO_2 (IPCC, 2007a). As such, CO_2 is standardized to a GWP value of 1 and other gases are expressed in terms of carbon dioxide equivalents (CO_2e) integrated over a 100-year time period. For example, the GWP of 1 kg of N_2O over 100 years is 298 kg CO_2e (Table 5.4.1; IPCC, 2007a). Due to its computational simplicity, GWP has been widely adopted by scientists and policy-makers working under the IPCC framework. More recent work by Shine et al (2005, 2007) has introduced the global temperature potential (GTP) as an additional measure for consideration by the IPCC. The GTP is calculated as the ratio between the global mean surface temperature change (change in degrees K per

kg^{-1} gas) at a given future time point following a pulse or sustained emission of a gas relative to CO_2 (Shine et al., 2005). Since both GWP and GTP are measures of RF relative to CO_2, they share many advantages as metrics useful for policy-making. The main difference between GWP and GTP is that while the GWP integrates the temperature change over a standard time period (i.e., the contribution of the RF at the beginning and end of the time horizon is exactly the same), the GTP calculates the actual temperature change between the time emitted and a precise future end point (i.e., the RF closer to the end point contributes relatively more) (Shine et al., 2005). While GTP has yet to gain widespread usage by the IPCC, several recent studies by the US Environmental Protection Agency have employed this metric to assess the overall impacts of reactive N species on both short-term and long-term changes in climate (Pinder et al., 2012a, 2012b).

Nitrous oxide (N_2O) contributes to two distinct global environmental issues: climate change and stratospheric ozone depletion. As a greenhouse gas, N_2O is 298 times more potent than carbon dioxide and is the third most abundantly emitted greenhouse gas (after carbon dioxide and methane) (Forster et al., 2007). Overall, N_2O is responsible for approximately 8% of the global anthropogenic greenhouse gas emissions that contribute to climate change (Forster et al., 2007). While N_2O is a relatively long-lived greenhouse gas, in the stratosphere solar radiation eventually degrades the molecule to form the free radical nitric oxide (NO) which has additional implications for the separate issue of ozone depletion. Free radicals such as NO catalyze secondary reactions that convert ozone (O_3) and a single oxygen atom (O) into two molecules of oxygen gas (O_2) (e.g., $O + O_3 \rightarrow 2O_2$). Consequently, increased N_2O emissions also ultimately lead to a buildup of NO in the stratosphere, the depletion of the Earth's ozone layer, and the subsequent loss of protection from the sun's ultraviolet light. At present, N_2O is the most abundantly emitted ozone depleting substance, and will be for the remainder of the twenty-first century if emissions continue at their current pace (Ravishankara et al., 2009).

N_2O is not the only substance to have these dual properties. Many ozone depleting substances, including chlorofluorocarbons (CFCs), methyl bromide, and several other gas species, are also highly potent greenhouse gases. And yet N_2O and CFCs are controlled under two different treaties: CFCs under the 1987 Montreal Protocol and N_2O under the 1997 Kyoto Protocol. The Montreal Protocol is considered a model of global environmental cooperation. Under the Montreal framework, the production and commercial use of ozone depleting substances (with the notable exception of N_2O) have been reduced by approximately 97% globally (Velders et al., 2007). This success in reducing the emissions of other ozone

depleting substances is the main reason why N_2O is now the dominant remaining ozone depleting substance. In contrast, the Kyoto Protocol significantly undershot its first phase targets and the parameters of its second commitment period (with a diminished membership) remain unclear. Kyoto's uneven track record has motivated some to explore how the Montreal Protocol might be expanded to include N_2O and thus maximize the duel ozone and climate benefits of the existing policy framework (Kanter et al., 2013).

In California, the use of N fertilizers and livestock manure in agriculture are the largest sources of anthropogenic N_2O emissions. Other sources of N_2O include fossil fuel combustion, nitric and adipic acid production, biomass burning, and wastewater. Chapters 7 and 8 of this assessment evaluate a range of technical and policy strategies for mitigating N_2O emissions from these economic sectors. Practices to improve fertilizer use efficiency by optimizing N rates, split applications, controlled release fertilizers, nitrification inhibitors, fertigation technologies, and other types of precision agriculture have the potential to reduce N_2O emissions without adverse effects on crop yields (Robertson and Vitousek, 2009). In particular, a recent review of 35 studies found that nitrification inhibitors and controlled release fertilizers on average can reduce N_2O emissions by approximately 38% and 35%, respectively (Akiyama et al., 2010), and are steadily increasing their market share (USDA ERS, 2012). Considerable progress has also been made with the adoption of catalytic conversion for both stationary and mobile combustion sources and in the processes used for nitric acid production, with the added benefit of also often reducing NO_x emissions (Wiesen, 2010). Likewise, any of the agricultural strategies for reducing N_2O are also likely to have positive environmental co-benefits by reducing other forms of nitrogen pollution that diminish California's water and air quality.

5.4.2. Effects of Reactive N on the Global Climate

It is well established that while some human activities involving reactive N result in warming effects that can amplify climate change, others lead to cooling effects that minimize or offset the prevailing global warming trend (figure 5.4.2; Erisman et al., 2011; Pinder et al., 2012a, 2012b). For example, emissions of N_2O have a strong long-term warming effect due to the atmospheric lifetime of the gas and its high GWP (Table 5.4.2; Parry et al., 2007; Smith et al., 2007). In contrast, sequestration of C by natural vegetation caused by increased N deposition typically has a

long-term cooling effect on climate (Table 5.4.2). The effects of atmospheric aerosols also tend to cool the climate, but since they only remain in the atmosphere for a short time period (hours–weeks) their effects are limited in duration (Table 5.4.2; Liu and Greaver, 2009; Shindell et al., 2009).

While there is considerable uncertainty in estimating the magnitude of these countervailing processes, several recent studies have developed methods for quantifying the net effects of reactive N on the global climate (figure 5.4.2). Principal among these is a study by Erisman et al. (2011) which concluded that reactive N has an overall net cooling effect on

TABLE 5.4.1
Atmospheric lifetimes and global warming potential (GWP) values for CO_2, CH_4, and N_2O

Greenhouse gas	Radiative efficiency	Atmospheric lifetime	Global warming potential by integration time period		
			GWP 20 years	GWP 100 years	GWP 500 years
	W m^{-2} ppmv^{-1}	Years	CO_2 equivalents		
CO_2	0.01548[a]	~100 (5–200)[b]	1	1	1
CH_4	0.00037	12	72	25	7.6
N_2O	0.0031	114	289	298	153

a. IPCC (2007a) does not give a radiative efficiency for CO_2. IPCC (2001) lists the radiative efficiency of CO_2 as 0.01548 W m^{-2} ppmv^{-1}, but emphasizes this figure is to be used only for the computation of GWPs.

b. Precise estimation of the atmospheric lifetime of CO_2 is complicated by the multitude of removal mechanisms involved. Accepted values are around 100 years, with a wide error range.

SOURCES: IPCC (2001, 2007a).

RF for the Earth's present climate (–0.24 W m^{-2}), albeit with a wide uncertainty range (–0.5 to +0.2 W m^{-2}) (figure 5.4.3). Using the same methodology, Butterbach-Bahl et al. (2011) also found a net cooling effect of anthropogenic N from European sources on global RF. In contrast, Pinder et al. (2012a) used an alternative method to calculate the change in GTP due to reactive N sources in the United States over time. Consistent with previous work they also found a modest cooling effect of reactive N in the near term (20 years) due mostly to the short-lived effects of O_3 and N-derived aerosols, but indicated that warming will likely occur in the longer term (100 years) when the prolonged effects of N_2O dominate the radiative balance (figure 5.4.4). However, it should also be noted that the net effects of N on climate are very small compared to CO_2 from fossil fuel combustion. Thus, the modest cooling effect of reactive N in the near term is thought to provide only a slight offset to the significant warming trend that is driven mostly by global CO_2 emissions.

Thus far, the results of these global and national studies have not been downscaled or disaggregated for California. However, the convergence of evidence indicates that the climate forcing effects of reactive N in California are likely similar to those observed in the US and Europe. While a full radiative balance for reactive N in California is beyond the scope of this analysis, the following sections summarize the existing data and knowledge on the main N-related processes in California that influence climate change in both the short term and the long term.

5.4.3. Effects of Reactive N on N_2O Emissions

As indicated above, emissions of N_2O have a long-term warming effect on the global climate that is well established in the scientific literature (figure 5.4.2; Forster et al., 2007). The primary biochemical mechanisms that produce N_2O are nitrification and denitrification, which are mediated by aerobic nitrifying bacteria and anaerobic denitrifying bacteria

(figure 5.4.1). Recent estimates suggest that approximately 57–62% of global N_2O emissions come from natural sources (10.5–11 Tg N yr^{-1}), with the remaining 38–43% attributed to anthropogenic sources (6.7–7.8 Tg N yr^{-1}) (Forster et al., 2007; Syakila and Kroeze, 2011). Furthermore, human efforts to fix atmospheric N into usable reactive forms through the Haber–Bosch process and the cultivation of N_2-fixing crops have increased N_2O concentrations in the atmosphere by about 16% relative to preindustrial times (Forster et al., 2007; Park et al., 2012). Emissions of N_2O come from multiple anthropogenic sources including N fertilizers, N_2-fixing crops, livestock urine and manure, sewage and wastewater, biomass burning, and fossil fuel combustion (Parry et al., 2007; Smith et al., 2007). Terrestrial and aquatic ecosystems are also sources of non-anthropogenic N_2O emissions, but significant losses of N from anthropogenic sources to natural and semi-natural ecosystems (e.g., via NO_3^- leaching, NH_3 volatilization, NO_x emissions, N deposition) make it difficult to accurately determine whether a molecule of reactive N originates from natural or human fixation.

Globally N_2O accounts for about 8% of total anthropogenic GHG emissions if all are expressed in CO_2e, making it the third most important GHG behind CO_2 (77%) and CH_4 (14%) (figure 5.4.5; IPCC, 2007a; Smith et al., 2007). In California, a relatively small fraction (3–4%) of the state's total GHG emissions are attributed to N_2O, and these emissions have remained at a relatively stable level between 1990 and 2009 (Table 5.4.3; figure 5.4.2; CARB, 2010b). This stable trend for N_2O has also been observed nationally and has been attributed to widespread adoption of catalytic converters in recent decades (which reduce both NO_x and N_2O) and has offset small increases in N_2O from fertilizer consumption. During the 1990s and 2000s, N_2O emissions on the order of 15–16 MT CO_2e were emitted in California each year, with 68% of N_2O emissions coming from agriculture (Tables 5.4.3 and 5.4.4). The remaining N_2O emissions in California come from sewage treatment (7%) and fossil

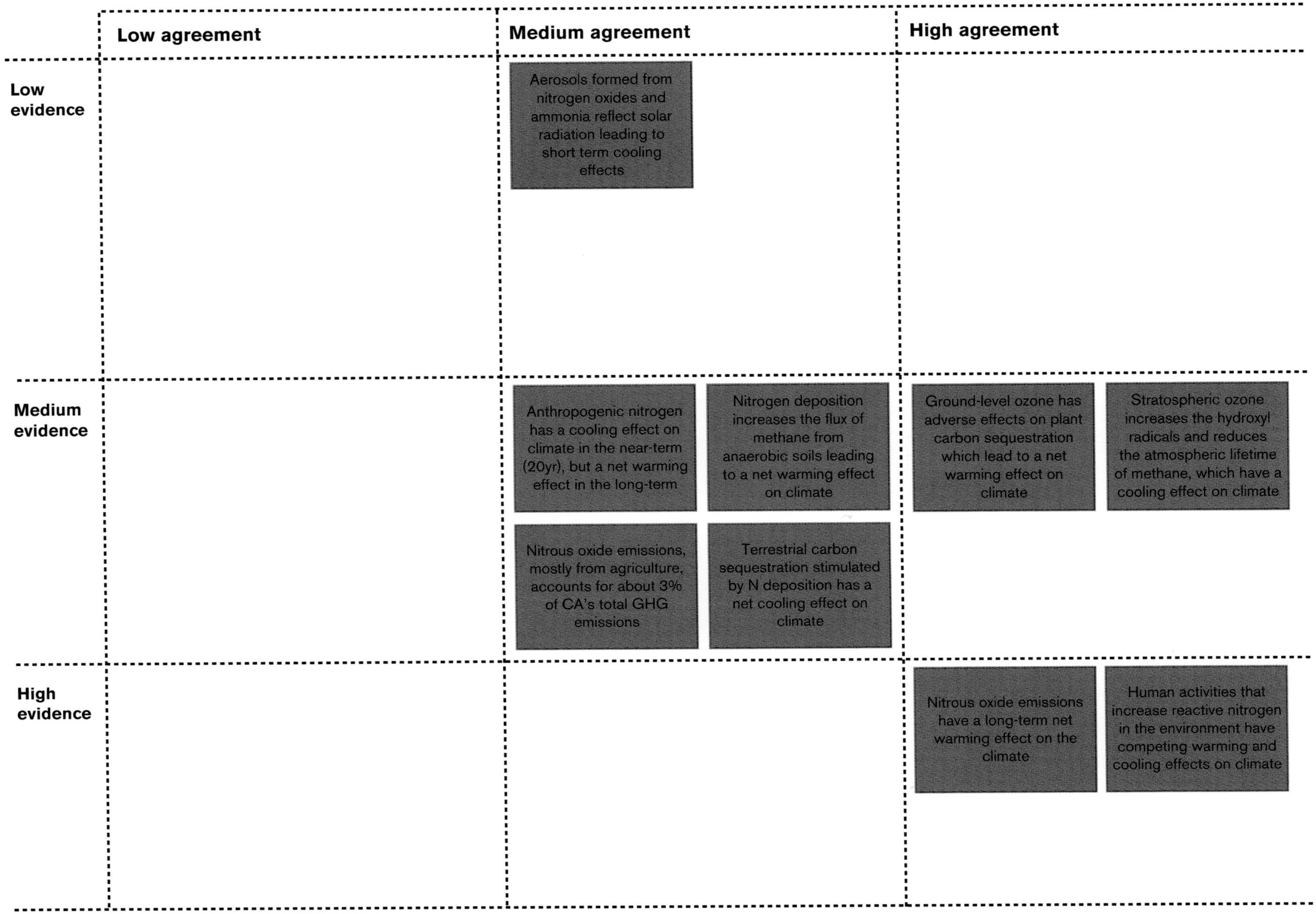

FIGURE 5.4.2. Types of uncertainty in nitrogen's impact on global climate regulation. This figure reflects the amount of evidence and level of agreement for the various nitrogen-related biogeochemical processes that influence climate regulation over time.

TABLE 5.4.2
Processes altered by reactive N that have radiative forcing effects. The size of the short-term (20 years) and long-term (100 years) effect is relative to other processes altered by reactive N

Process altered by reactive N	Radiative forcing effect	Relative size of effect		Description	References
		Short-term	Long-term		
N_2O	Warming	Large	Large	Potent and long-lived greenhouse gas from agriculture, fossil fuel combustion, and sewage	Parry et al. (2007); Smith et al. (2007)
N deposition/fertilizer → CO_2 uptake by plants	Cooling	Large	Large	Generally increases C stored in vegetation and soils of natural and agroecosystems	Liu and Greaver (2009); Sutton et al. (2008); Thomas et al. (2010)
N deposition/fertilizer → CH_4 efflux from soil	Warming	Small	Small	Increases CH_4 emissions and reduces CH_4 oxidation	Liu and Greaver (2009)
NO_x → ground level O_3 → CO_2 uptake by plants	Warming	Large	Large	NO_x forms tropospheric O_3 that damages plant foliage and decreases C storage	Arneth et al. (2010); Felzer et al. (2004); Pan et al. (2009); Sitch et al. (2007)
NO_x → O_3 and CH_4 in atmosphere	Cooling	Large	Small	NO_x effects formation and destruction of O_3 and CH_4 in upper atmosphere	Shindell et al. (2009)
NO_x → aerosols	Cooling	Medium	Small	Aerosols reflect and scatter solar radiation	Shindell et al. (2009)
NH_3 → aerosols	Cooling	Small	Small	Aerosols reflect and scatter solar radiation	Shindell et al. (2009)

SOURCE: Adapted from Pinder et al. (2012b).

fuel combustion in the transportation, energy, and industrial sectors (24%) (Table 5.4.4).

EMISSIONS OF N_2O FROM CALIFORNIA AGRICULTURE

Despite N_2O being a small fraction of the state's overall GHG emissions, between 30% and 40% of the emissions attributed to agriculture in California come from N_2O emissions (CARB, 2010b; Haden et al., 2013). California's statewide estimates for the various agricultural sources of N_2O are based on emissions inventory guidelines developed by the IPCC (2006). The IPCC's Tier 1 methods use default emission factors (EFs) derived from a "bottom-up" assessment of field experiments covering a wide range of global crops, environments, water management regimes, N sources, and nutrient management practices (Bouwman et al., 2002a, 2002b; Stehfest and Bouwman, 2006). These default EFs calculate emissions using a mean value for the

proportion of applied N from synthetic fertilizer, N_2-fixing crops, organic fertilizer, and manure that is directly and indirectly emitted as N_2O (Table 5.4.5). Direct emissions, in this case, refer to those which arise from the soil where the N is applied, whereas indirect emissions are those that occur elsewhere in the environment subsequent to leaching or volatilization losses. Using this approach, the EF for direct N_2O emissions from most agricultural soils is 1% of applied N, with an additional 0.35–0.45% of applied N emitted indirectly following leaching and volatilization (IPCC, 2006). However, it should be noted that considerable natural variation in N_2O flux measurements across many environmental conditions and cropping systems introduces a high degree of uncertainty in the default EFs. For example, the uncertainty in direct N_2O emissions from agricultural soils ranges from 0.003 to 0.03 kg N_2O-N kg^{-1} N applied (Table 5.4.5; see Data Table 21 in Appendix 4.1). Increasingly, region-specific EFs derived from local

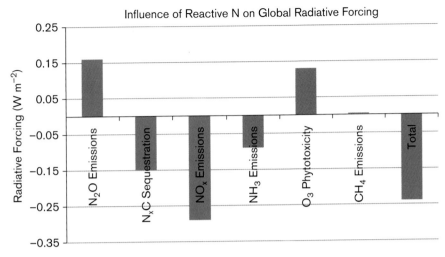

FIGURE 5.4.3. Influence of anthropogenic reactive N on global radiative forcing on the present climate. Radiative forcing values above zero imply a warming effect on the present global climate, while negative values indicate a cooling effect. "N × C Sequestration" includes effects of N deposition on C sequestration and CO_2 efflux in terrestrial and aquatic systems and mineralization in soil. "NO_x emissions" contains effects of O_3 and other aerosols formed from reactions with NO_x. "NH_3 emissions" are particulates and other aerosols formed from NH_3 in the atmosphere. "CH_4 emissions" contains only the effects of N on CH_4 from soils, while the effects of N on the atmospheric lifetime of CH_4 are included in NO_x emissions. Uncertainty for total global radiative forcing ranged from –0.5 to +0.2 W m^{-2}.

DATA FROM: Erisman et al. (2011); Shindell et al. (2009).

agricultural experiments and activity data (i.e. Tier 2 methods) or more sophisticated process-based biogeochemical models (i.e. Tier 3 methods) are being used to further improve the precision and reduce the uncertainty of N_2O estimates (Haden et al., 2013). The region-specific EF used in the California N Assessment to estimate N_2O emissions from 33 California crop categories is a recent example of the former approach (see Chapter 7, figure 7.4).

In contrast to the "bottom-up" emissions inventories that have been used by most national and subnational governments, recent studies by Crutzen et al. (2008), Davidson (2009) and Smith et al. (2012) that employ "top-down" accounting methodologies inclusive of both direct and indirect N_2O emissions suggest that not all of the N_2O emitted over the life cycle of a newly fixed reactive N molecule is accounted for in IPCC's default EFs. It should also be noted that the CO_2, CH_4, and N_2O emitted during the Haber–Bosch process used to manufacture N fertilizers are not included in either the top-down or bottom-up accounting frameworks for agricultural emissions referred to above (Box 5.4.2). In the Crutzen et al. (2008) study, which uses an N budget approach informed by data on the size of global N sources and sinks and the known rate of N_2O accumulation in the atmosphere, approximately 3–5% of newly fixed N is ultimately emitted as N_2O. Building on this approach, Smith et al. (2012) found a good fit between observed atmospheric N_2O concentrations from 1860 to 2000 and estimates of N_2O emissions based on a 4% EF by using estimates of reactive N entering the agricultural cycle that account for both mineralization of soil

organic N following land use change and NO_x deposited from the atmosphere. In the approach used by Davidson (2009), they assume that N molecules in the fertilizer used to produce animal feed are later recycled in manure applied to soil. The results of their analysis indicate that approximately 2% of N in manure and 2.5% of N in fertilizer is eventually converted to N_2O. It is important to note that while these top-down methods indicate twofold to threefold higher N_2O emissions than the IPCC's Tier 1 approach, the contribution of N_2O to California's total GHG emissions (<5–10%) is still relatively minor compared to CO_2. These recent studies also highlight the fact that while the IPCC default EFs provide a computationally simple way to estimate N_2O emissions, the high degree of uncertainty can restrict the precision of national and regional inventory estimates (Smith et al., 2010).

In an effort to improve estimates of N_2O emissions from California agriculture, state agencies (e.g., California Energy Commission, California Air Resources Board) have commissioned a number of recent field studies to measure emissions for California cropping systems with goals of (1) calibrating and validating soil biogeochemical models (e.g., DAYCENT model, Denitrification-Decomposition model) and (2) assessing the impact of alternative agricultural practices (Burger et al., 2005; De Gryze et al., 2009, 2010; Horwath and Burger, 2012). As a case in point, Horwath and Burger (2012) recently published a report of California-based field studies measuring N_2O emissions in tomato, wheat, alfalfa, and rice cropping systems. They found that cumulative N_2O emissions from furrow-irrigated tomatoes

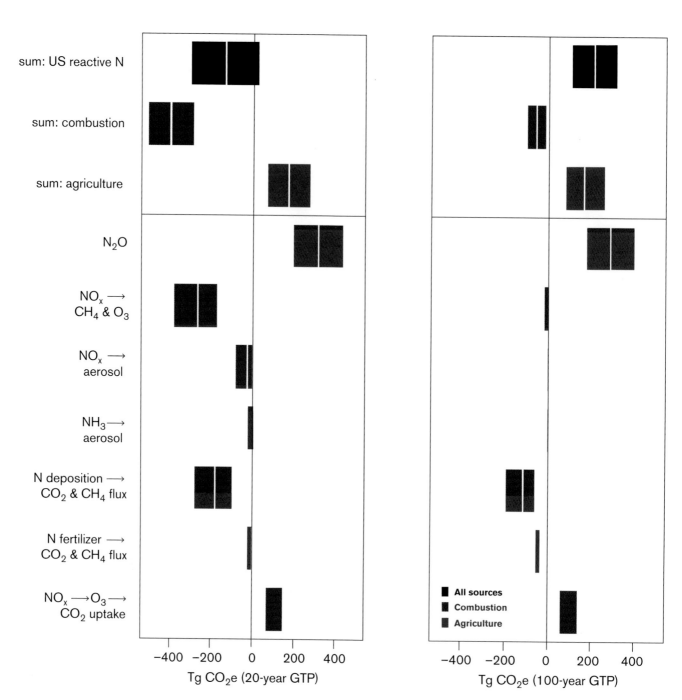

FIGURE 5.4.4. The climate change impacts of US reactive nitrogen emissions from combustion and agriculture, in common units of equivalent Tg of CO_2 (Tg CO_2e) on a 20- and 100-year global temperature potential (GTP) basis. The width of the bar denotes the uncertainty range; the white line is the best estimate; and the color shading shows the relative contribution of combustion and agriculture.

SOURCE: Reprinted from Pinder et al. (2012a). Copyright R. W. Pinder et al.

ranged between 0.67 and 4.69 kg N_2O-N ha^{-1} and had EFs between 0.92% and 2.08% of applied N. In wheat, they obtained EFs ranging from 0.24% to 0.98% of applied N (Horwath and Burger, 2012), results that were consistently lower than the well-established 1% IPCC default EF and the 1.21% mean EF derived from 25 global wheat studies (Linquist et al., 2012). Horwath and Burger (2012) also found that EFs for alfalfa, an N-fixing crop, ranged from 4.5% in a 1-year-old stand to 12.06% in an adjacent 5-year-old stand.

The annual N_2O emissions measured in rice systems ranged from 0.26 to 0.85 kg N_2O-N ha^{-1} and EFs between 0.12% and 0.74%, and were similar to the mean EF estimated in a recent meta-analysis of 17 other rice experiments conducted globally (Horwath and Burger, 2012; Linquist et al., 2012). While a few studies have also begun to examine N_2O emissions in California orchards and vineyards, more experimental data are needed to validate biogeochemical models for the state's diverse perennial and annual crop-

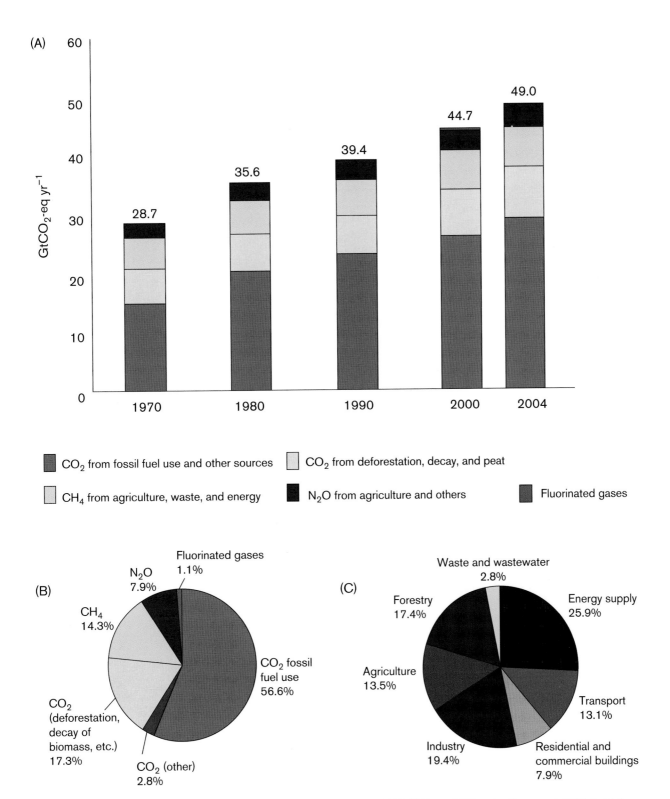

FIGURE 5.4.5. (A) Global annual emissions of anthropogenic GHGs from 1970 to 2004. (B) Share of different anthropogenic GHGs in total emissions in 2004 in terms of CO_2e. (C) Share of different sectors in total anthropogenic GHG emissions in 2004 in terms of CO_2e. Forestry includes deforestation.

SOURCE: Reprinted from IPCC (2007a).

TABLE 5.4.3

California greenhouse gases and percentage of annual total emitted in 1990, 2000, and 2009

	1990		2000		2009	
Greenhouse gas	Emissions (MT CO_2e)	% of annual total	Emissions (MT CO_2e)	% of annual total	Emissions (MT CO_2e)	% of annual total
CO_2	390.0	90	408.9	88.1	393.2	86.1
CH_4	25.1	5.8	28.0	6.0	32.0	7.0
N_2O	16.5	3.8	16.0	3.5	15.2	3.3
SF_6	<1.0	<1.0	1.3	<1.0	1.1	<1.0
Other HFCs*	<1.0	<1.0	10	2.1	15.3	3.3
Total	433.3		463.7		456.8	

*Other HFCs includes all other halogenated fluorocarbon gases.
SOURCE: CARB (2010b).

TABLE 5.4.4

Estimated emissions of N_2O, NO_x, and NH_3 by source and fate in California during 2005

	Statewide emissions					
	N_2O		NO_x		NH_3	
Source and fate of N emissions	Gg N yr^{-1}	%	Gg N yr^{-1}	%	Gg N yr^{-1}	%
Source of N emissions						
Fossil fuel combustion	9	23.7	359	88.4	36	13.4
Soil	24	63.2	24	5.9	67	25
Manure	2	5.3	0	0	141	52.6
Upwind sources	0	0	20	4.9	20	7.5
Wastewater	2	5.3	0	0	0	0
Fire	0	0	3	0.7	3	1.1
Surface water and groundwater	2	5.3	0	0	0	0
Total	38	100	406	100	268	100
Fate of N emissions						
N deposition in California	0	0	135	33.3	67	25.0
Export to atmosphere or beyond California	38	100	270	66.7	201	75.0

NOTE: Percentage of each gas emitted by various sources is also provided. All N_2O emitted was assumed to remain in the atmosphere. NO_x and NH_3 form secondary chemicals, aerosols, and particulates, and a certain fraction of the N in these constituents are deposited in California and the remaining are exported beyond state boundaries. As such, export to the atmosphere or beyond the state boundary was calculated as the difference between total emissions and deposition. This table was developed using input and output data from the California N mass balance developed in Chapter 4 of this report.

TABLE 5.4.5

Default values and uncertainty range for IPCC emission factors used to calculate direct and indirect N_2O emissions from agricultural soils in the California Greenhouse Gas Emissions Inventory

Category	Emission factor description	Default value	Uncertainty range
Direct N_2O emissions	Proportion of N applied to soils via synthetic fertilizer, organic fertilizer, manure, N-fixing crops that is emitted as N_2O	0.01	0.003–0.03
	Proportion of N deposited by livestock on pastures, rangeland, and paddocks that is emitted as N_2O	0.02	0.007–0.06
	N emitted as N_2O per unit area of cultivated organic soils (kg N ha^{-1})	8	2–24
Indirect N_2O emissions	Fraction of synthetic fertilizer N that volatilizes	0.1	0.03–0.3
	Fraction of organic fertilizer and manure N that volatilizes	0.2	0.05–0.5
	Leaching rate: fraction of applied N lost to leaching and runoff	0.3	0.1–0.8
	Proportion of N volatilized and redeposited on soils that is emitted as N_2O	0.01	0.002–0.05
	Proportion of N lost to leaching and runoff that is emitted as N_2O	0.0075	0.0005–0.025

SOURCE: CARB (2009c) and IPCC (2006).

ping systems (Smart et al., 2011a; Steenwerth and Belina, 2010; Suddick et al., 2011).

EMISSIONS OF N_2O FROM SEWAGE TREATMENT, TRANSPORTATION, ENERGY GENERATION, AND INDUSTRY

As with agricultural soils, the N_2O emissions that arise from sewage and wastewater treatment facilities are associated with the breakdown and mineralization of N compounds present in the wastewater and effluent. Again, nitrification and denitrification are the main mechanisms driving N_2O emissions. In general, the N_2O emissions occurring directly within the treatment facilities are relatively small compared to the effluent that is discharged into surface water bodies. Overall, sewage and wastewater treatment accounts for approximately 7% of California's N_2O emissions (Table 5.4.4). In addition to significant amounts of CO_2, the burning of fossil fuels also produces N_2O and accounts for approximately 24% of California's N_2O emissions (Table 5.4.4). In the industrial sector, the production of nitric acid via the oxidation of NH_3 and the application of concentrated nitric acid as an oxidizer for production of various chemicals both result in N_2O emissions. In California, N_2O emissions from nitric acid production are small and have declined from 535,000 MT CO_2e in 1990 to less than 200,000 MT of CO_2e in 2008 (CARB, 2010b). Together, N_2O emissions from wastewater treatment, fossil fuel combustion, and nitric acid production amount to less than 1% of California's total anthropogenic GHG emissions, which is dominated mostly by CO_2 and to a lesser extent by CH_4 (CARB, 2010b).

5.4.4. Effects of Reactive N on Biosphere C Stocks and the Efflux of CO_2 and CH_4

EFFECTS OF N ON TERRESTRIAL C SEQUESTRATION

One of the most prominent cooling effects of reactive N is its stimulation of plant growth in terrestrial ecosystems, which removes CO_2 from the atmosphere and sequesters C in plant biomass (Butterbach-Bahl et al., 2011; Pinder et al., 2012b). It is well-established that the net primary productivity (NPP) of most ecosystems is limited by the availability of N (figure 5.4.1; LeBauer and Treseder, 2008). Consequently, estimates suggest that N deposition in temperate North American forests increases aboveground and belowground C stocks by 24–177 kg C kg^{-1} N deposited per year (de Vries et al., 2009; Liu and Greaver, 2009; Sutton et al., 2008; Thomas et al., 2010). While generally lesser in magnitude, N deposition also increases C sequestration in the vegetation of temperate grasslands and wetlands (LeBauer and Treseder, 2008).

Soils are an even larger sink for carbon, with 2–3 times more C stored in soils than in vegetation globally (White et al., 1999). Most evidence from forest ecosystems suggests that sequestration of C in soil increases with higher rates of N deposition (Fenn et al., 1996). In agroecosystems, the application of N fertilizers and manure can also build soil C stocks (Ladha et al., 2011; Pinder et al., 2012b). The main mechanisms responsible for N-induced C storage in soil are greater inputs of organic matter through leaf litter, crop residues, decreased rates of litter decomposition, and decreased soil respiration (Janssens et al., 2010; Knorr et al., 2005).

Several studies of California forest ecosystems have found that N deposition contributes to increased C storage

BOX 5.4.2.   CLIMATE IMPACTS OF FERTILIZER MANUFACTURE

Use of synthetic nitrogen fertilizer typically accounts for a high percentage of greenhouse gas (GHG) emissions in conventional crop production. In addition to soil emissions resulting from the application of fertilizers on crop fields, the production of the fertilizer itself accounts for a large portion of total emissions. According to a selection of food life cycle assessment studies (Blengini and Busto, 2009; Carlsson-Kanyama, 1998), the proportion of total GHG emissions attributable to fertilizer production can range from less than 5% to 18%, depending on the type of food product being analyzed and the system boundaries of the study. For some horticultural crops, it can account for up to 70% of CO_2 emissions (not counting other GHGs) up to farm gate (Lillywhite et al., 2007). Moreover, fertilizer production has been estimated to account for 1.2% of the world's total energy use and 1.2% of the total GHG emissions (Kongshaug, 1998).

Nitrogenous fertilizer production relies on the Haber–Bosch process, an industrial process that synthesizes ammonia (NH_3) by mixing nitrogen from the air with hydrogen under high-temperature and high-pressure conditions, requiring a large amount of energy—approximately 25–35 GJ t^{-1} NH_3 (IPCC, 2007b). Approximately 80% of the world's NH_3 production uses natural gas as both the hydrocarbon feedstock (as the source of hydrogen) and the energy source, resulting in CO_2 emissions as the dominant component of GHG emissions (Wood and Cowie, 2004). Various studies have estimated emissions in NH_3 production to range from 1,150 to 2,800 g CO_2-equivalent (CO_2-e) per kg NH_3, or 1,402 to 3,415 g CO_2-e kg^{-1} N content (P.E. International, 2009; Wood and Cowie, 2004). Wood and Cowie (2004) report that the transparency of reviewed reports was inadequate to explain these large differences between estimates, but that a large portion of the variation is likely due to variation in plant efficiencies, the use of alternative fossil fuels, and differences in methods for accounting for the steam exports that result from the process (which can be used for electricity generation and therefore can be counted as a credit toward the fertilizer emissions, due to offsetting of other electricity generation). A study commissioned by Natural Resources Canada (NRCan, 2007) confirms that measures taken by Canadian urea production plants to improve overall energy efficiency and reuse CO_2 (see below) result in substantially lower emissions relative to facilities in other parts of the world.

Manufacture of ammonium nitrate, a fertilizer used commonly in California and around the world, results in even higher GHG emissions because, after the initial synthesis of NH_3, it must be oxidized at high temperature and pressure to create nitric acid, a process that while being exothermic (heat-releasing, and thus a net energy generator) also results in N_2O as a by-product. NH_3 is then neutralized with aqueous nitric acid (HNO_3) to produce ammonium nitrate (NH_4NO_3). The total CO_2-e emissions, therefore, essentially consist of both the CO_2 emissions from the initial NH_3 production and the N_2O emissions from the subsequent nitric acid production. Different studies have calculated emissions ranging from 1,000 to 2,381 g CO_2-e kg^{-1} NH_4NO_3, or 2,985 to 7,109 g CO_2-e kg^{-1} N (P.E. International, 2009; Wood and Cowie, 2004).

Urea is synthesized by combining NH_3 and CO_2 under high pressure to form ammonium carbonate, which is then dehydrated under heat to form urea and water. Emissions calculations vary depending on whether CO_2 use in the production process is subtracted from the CO_2 emissions from NH_3 production, and are reported as 420–1,849 g CO_2-e kg^{-1} product, or 913–4,018 g CO_2-e kg^{-1} N (P.E. International, 2009; Wood and Cowie, 2004). However, while urea may have a lower footprint in its production phase, due to the capture and reuse of CO_2, Snyder et al. (2009) point out that some or all of this CO_2 may be emitted again from the soil after application.

For perspective, we can calculate rough estimates for CO_2-equivalent soil emissions resulting from application of these fertilizer products to agricultural fields. The IPCC default emissions factor for direct N_2O emissions (the emissions resulting from direct nitrification and denitrification pathways for fertilizer N) is 1% of applied N (IPCC, 2006, p.11.11). In the case of NH_3, which is 82% N by mass, the emissions would thus be estimated as 8.2 g N_2O-N kg^{-1} ammonia (calculated as 1% of 820 g NH_3-N). This figure translates into 12.89 g N_2O, and 3,816 g CO_2-e kg^{-1} product, or 4,654 g CO_2-e kg^{-1} N applied (using the 100-year global warming potential of 296 for N_2O). One should note that these calculations do not include indirect emissions—those arising from volatilization, leaching, and runoff of fertilizer N, some of which later becomes N_2O in off-site locations. As with direct emissions, these processes can vary tremendously according to soil type, climate, and management practices, but the default factors suggested by IPCC (2006) for use when these specific variables are unknown would raise the total emissions estimates by an additional 33%. At any rate, the GHG emissions produced in manufacture of the fertilizer range from 20% to 150% of emissions produced after field application, or 23% to 60% of total emissions attributable to N fertilizer use, depending on fertilizer product, manufacturing efficiency, and additional variables. Given that many field studies have found that field emissions in specific locations and conditions may in fact be much lower than the IPCC default factors (Snyder et al., 2009), the proportion of life cycle GHG emissions attributable to fertilizer production may be correspondingly higher. Nitrogenous fertilizer production, therefore, contributes substantially to total climate change impacts of synthetic fertilizer use.

in both vegetation and soil organic matter. For example, in Southern California's San Gabriel and San Bernardino Mountains, higher rates of N deposition were associated with increased rates of biomass accumulation in overstory trees, leaf litter on the forest floor, and soil organic matter levels (Fenn et al., 1996). Working in the Sierra Nevada Mountains, Powers and Reynolds (1999) found that water is generally the most limiting factor for forest growth, but also documented tree growth responses and C sequestration with increased N inputs. While the above-mentioned studies linking reactive N to increased C storage in California's vegetation and soils are broadly consistent with national and international studies (Erisman et al., 2011; Pinder et al., 2012a, 2012b), no efforts thus far have been made to quantify the total amount of C sequestered in California due to N deposition and application nor its relative contribution to global climate change.

In some cases, excessive N inputs can lead to "N saturation" where the negative effects of soil acidification, base-cation leaching, and aluminum toxicity are thought to overwhelm the positive effects of N fertilization, ultimately leading to forest decline and loss of stored C (Aber et al., 1998; Bowman et al., 2008; Matson et al., 2002). The adverse effects of excess N are also likely to differ among plant species and ecosystems. Instances of N saturation have been reported in the forests of California's San Bernardino Mountains, though measurable losses of C from forest decline were not observed in this case (Fenn et al., 1996). Furthermore, these authors hypothesize that the high base saturation of arid forest soils in the Western US make decline less likely than in other forest ecosystems (Fenn et al., 2003). While more long-term research is needed to determine if N saturation from continued air pollution will eventually lead to losses of stored C, it is provisionally agreed upon by most that inputs of reactive N generally have a positive effect on C sequestration in California's ecosystems (figure 5.4.2). That said, it is important to note that plant biomass and soils are not permanent sinks for carbon and a significant fraction of the carbon will eventually be returned to the atmosphere following plant death, harvest, decomposition, and wildfire. Future research conducted in California is therefore critically important, not just to understand the N response from different ecosystems and plant species, but also to determine the timescales of carbon storage and loss to the atmosphere.

EFFECTS OF N ON CH_4 EMISSIONS IN TERRESTRIAL AND WETLAND ECOSYSTEMS

In most terrestrial and wetland ecosystems, CH_4 production and oxidation by soil microbes occur concurrently, and the balance of these processes regulates the net flux of CH_4 to the atmosphere (Bowman et al., 2008; Liu and Greaver, 2009). As discussed above, enrichment of N generally increases the growth of vegetation and the amount of organic matter present in soil. The increased availability

and mineralization of organic C subsequently drives greater O_2 consumption and creates a more anaerobic environment in the soil. At the same time, higher levels of NH_4^+ in soil may also reduce the rate of CH_4 oxidation to CO_2 by methanotrophic bacteria. Together these processes provide more C substrate and more suitable redox conditions for methanogenic bacteria, thus stimulating the net production of CH_4 (figure 5.4.1; Liu and Greaver, 2009).

Liu and Greaver (2009) carried out a meta-analysis of existing studies which found that N addition via fertilization and deposition increased CH_4 production by 95% and decreased CH_4 oxidation by 38% when averaged across grassland, wetland, and anaerobic agricultural systems. However, when each ecosystem type was analyzed separately, the effect of N addition on both CH_4 production and CH_4 oxidation was only significant in anaerobic agricultural fields. When combined, these processes result in more total CH_4 entering the atmosphere, albeit only a very small amount more since agricultural soils tend to be weak sinks. Consequently, when the Liu and Greaver estimates were used in recent studies the overall warming effect of N on net CH_4 emissions occurring in anaerobic agricultural fields was found to be virtually negligible at both the US national scale (figure 5.4.4; Pinder et al., 2012b) and the global scale (Erisman et al., 2011). At present, no studies have quantified the overall effect of reactive N on CH_4 emissions in California. However, it is reasonable to assume that the universal biochemical processes that govern CH_4 emissions in soils will yield results in California that are similar to those observed in other wetland, forest, grassland, and agricultural ecosystems (figure 5.4.2).

5.4.5 Effects of Reactive N on Atmospheric Gases and Aerosols

EFFECTS OF N AND O_3 ON ECOSYSTEMS AND THE ATMOSPHERE

As discussed in Chapter 5.3, emissions of NO_x and NH_3 have a multitude of effects on the chemistry of atmospheric gases and the formation of O_3 and aerosols, many of which have important implications for the Earth's climate. Most importantly, NO_x and volatile organic compounds play a role in the formation of tropospheric O_3, which has a warming effect on the climate (Pinder et al., 2012b) through its adverse effects on plant photosynthesis and CO_2 uptake (Arneth et al., 2010; Felzer et al., 2004; Pan et al., 2009; see Chapter 5.3). The damaging effects of tropospheric O_3 on plant growth are estimated to decrease the ability of the world's vegetation to sequester atmospheric CO_2 by as much as 14–23% (figure 5.4.4; Pinder et al., 2012a; Sitch et al., 2007). While NO_x and O_3 concentrations throughout much of California have been declining since the 1970s, O_3 levels during the spring and summer months are still among the highest in the United States (Felzer et al., 2004). Experimental and modeling studies indicate that sensitivity to tropospheric O_3 is generally highest for crops followed by deciduous vegetation, with coniferous vegetation generally more tolerant (Felzer et al., 2004; Grantz and Shrestha, 2006; Shrestha and Grantz, 2005). Recent studies by Felzer et al. (2004, 2005) suggest that in spite of high O_3 levels the overall effect on C storage in Southwestern ecosystems (inclusive of California) is lower than in the Midwestern and Southeastern regions of the United States. This result is due mainly to California's arid summer climate which constrains NPP and thus the total amount of C sequestered by native vegetation. While these studies are inclusive of California, no focused efforts have been made to quantify the impact of O_3 on statewide C sequestration.

In the atmosphere, O_3 also has a direct warming effect on the climate since it is a short-lived GHG that effectively traps heat. At the same time, there is also a small feedback chemical reaction involving interactions between O_3 and CH_4 which occur throughout the atmosphere and have a small cooling effect on the climate (Holmes et al., 2013). This is because increases in O_3 also increase the concentrations of hydroxyl radicals (OH), which in turn reduce the lifetime and overall burden of CH_4 in the atmosphere (Butterbach-Bahl et al., 2011; Holmes et al., 2013). Since O_3 itself has a short atmospheric lifetime, the cooling effect is very small and more important in the short term (20 years) and almost negligible in the long term (100 years) (figure 5.4.4; Pinder et al., 2012b).

Overall, the warming associated with the adverse effects of O_3 on vegetation and the direct trapping of heat by O_3 dominate the radiative balance (Erisman et al., 2011; Shindell et al., 2009). Understanding these processes, as well as the small counteracting effects of O_3 on atmospheric CH_4, is an emerging area of research; thus, considerable uncertainty remains regarding the net effects of O_3 on the Earth's radiative balance and how the effects change over time (Forster et al., 2007). In California, more studies are needed to quantify the total amount of O_3 formed in the atmosphere, evaluate its competing effects, and assess how strategies to reduce O_3 might impact global climate change.

EFFECTS OF N ON ATMOSPHERIC AEROSOLS

While the amount of data and level of agreement regarding the direct climate effects of GHGs (CO_2, CH_4, N_2O) are considered to be high among scientists, there remains considerable uncertainty about the effects of atmospheric aerosols on the global climate due in large part to the complexity of the interactions which occur between aerosolized chemicals and clouds (Anderson et al., 2003; Forster et al., 2007). In this context, aerosols refer to any fine particulate matter or liquid droplet that is suspended in the gaseous environment. The limited evidence that is available on aerosols and particulate matter derived from NO_x and NH_3 emissions are tentatively agreed by most to have a short-term cooling effect on the climate by scattering solar radiation and stimulating cloud formation and the albedo effect (figure 5.4.2; Erisman et al., 2011; Shindell et al., 2009). The main aerosols formed from

chemical reactions with NO_x and NH_3 include: ammonium nitrate (NH_4NO_3), ammonium sulfate (NH_4SO_4), ammonium bisulfate ((NH_4)HSO_4)$_2$), calcium nitrate ($Ca(NO_3)_2$), and sodium nitrate ($NaNO_3$). Each of these aerosols form under different conditions, and holds differing amounts of water which leads to a range of effective sizes, optical properties, and radiative effects (Butterbach-Bahl et al., 2011).

The chemical complexity of atmospheric aerosols and the paucity of empirical data on their effects results in a high level of uncertainty regarding the magnitude and duration of their impact on the Earth's radiative balance (Forster et al., 2007). Globally, aerosols derived from NO_x and NH_3 are together estimated to have an RF of -0.38 W m^{-2}, with the negative value indicative of a modest cooling effect on the present climate (Erisman et al., 2011; Shindell et al., 2009). For the United States, Pinder et al. (2012b) found a similar cooling effect of N-derived aerosols in the near term (20 years), but considered their long-term impact on climate to be negligible relative to CO_2, CH_4, and N_2O emissions and terrestrial C sequestration (figure 5.4.4). Recent studies also suggest that the short-term cooling effect of NO_x-derived aerosols is 2–4 times as large as the effect of NH_3-derived aerosols (Pinder et al., 2012b; Shindell et al., 2009). However, since the formation of aerosols from NO_x and NH_3 are generally linked to different pollution sources (e.g., fossil fuel → NO_x, livestock → NH_3), the relative contribution of each pollutant is likely to vary considerably across California's landscape (see Chapter 5.3). For example, recent studies indicate that in the San Joaquin Valley, where fossil fuel combustion and agricultural activities produce a balanced mix of NO_x and NH_3, NH_4NO_3 tends to be the most abundant chemical species among atmospheric aerosols (Battye et al., 2003). In parts of California where NH_3 emissions are low, NH_4NO_3 is a much more minor component of aerosols. At present, only a small number of studies have tried to quantify California's total contribution of NO_x- and NH_3-derived aerosols to the atmosphere, and no available statewide studies are known to have estimated their net effects on global climate change.

5.4.6. Future Research Needs in California

The recent efforts to quantify the short- and long-term climate impacts of reactive N at the national and global scales have established sound methodologies for understanding the anthropogenic sources and environmental fate of various forms of reactive N, as well as their relative contribution to climate change (Butterbach-Bahl et al., 2011; Erisman et al., 2011; Pinder et al., 2012a, 2012b). Across these regional scales, it is provisionally agreed upon by most that the modest cooling effect of reactive N in the near term is likely to be short-lived as increasing concentrations of N_2O build up in the atmosphere and exacerbate the overall warming trend driven primarily by CO_2 (figure 5.4.2).

While California's contribution to Earth's radiative balance has not been fully assessed in the scientific literature, it is likely that the sources of N, climate forcing processes, and overall trends will be similar to those reported in the large-scale regional assessments discussed above. That said, future studies are still needed to confirm whether or not this is true for California in particular. For those interested in pursuing a more comprehensive statewide analysis, the data presented in the preceding mass-balance on key flows of reactive N (e.g., N_2O, NO_x, and NH_3 emissions; N deposition) may serve as a useful starting point for future studies (see Chapter 4; Table 5.4.4). With an eye towards future research, the following studies would be needed to complete a full assessment of California's contribution to climate change through anthropogenic sources of reactive N.

- Field and modeling studies measuring N_2O emissions from California's diverse annual and perennial cropping systems. These are needed to improve the calibration of soil biogeochemical models and thus reduce the uncertainty of statewide N_2O emissions estimates.
- Ecological studies quantifying the effects of N deposition on both C sequestration and CH_4 emissions in California's natural and agricultural ecosystems.
- Ecological and atmospheric studies quantifying the effects of tropospheric O_3 on C sequestration by native vegetation and crops in California.
- Atmospheric studies evaluating the effects of NO_x emitted in California on the formation of O_3, hydroxyl radicals, and the decay of CH_4.
- Atmospheric studies quantifying the radiative effects of aerosols formed from emissions of NO_x and NH_3 in California.

5.5. Cultural Services

Main Messages

Human-induced changes in the N cycle have numerous positive and negative effects on the cultural services that are provided to society through natural and working landscapes. Key services influenced by reactive N include the aesthetic value, recreational value, cultural heritage values, and spiritual and religious values of certain landscape elements and characteristics.

Shifts between natural, agricultural, and urban land uses, all made possible through N fertilizers and fossil fuel, have significant impacts on the aesthetic appearance of both natural and man-made environments in California. Studies suggest that most people prefer the visual appearance of environments along the following land-use gradient: natural habitat → diversified agricultural → agricultural monoculture → urban → industrial.

Loses of N to aquatic and terrestrial ecosystems through runoff and air pollution have a number of adverse effects on recreational opportunities in California. Recreational opportunities such as fishing, hunting, hiking, and bird watching are diminished because N losses tend to promote ecologically harmful eutrophication and anoxia in surface water bodies, and increases in N deposition on native grassland and forest ecosystems. These changes in N availability generally reduce native biodiversity and subsequent recreational opportunities.

Agritourism, culinary travel, and other rural recreational activities (e.g., vineyards, U-pick farms) are examples of some of the benefits of N fertilizer and fossil fuel use. Recent research indicates that opportunities for agritourism have been expanding in recent years with numerous ancillary benefits for job creation and economic growth in California's rural areas.

Excess N in the environment can have detrimental impacts on native species, biodiversity, and natural and working landscapes, thus diminishing their natural heritage value to society. Many of these elements of our natural environment are prominent subjects of nature study, literature, and other aspects of our cultural heritage.

Many religious traditions consider important species, locations, or geographic features to be "sacred." To the extent that N impacts biodiversity and ecosystem change, the spiritual and religious value that people derive from these species and places may be diminished.

Shared cultural and spiritual values can also be a key source of motivation and inspiration for environmental stewardship. While this potential exists, more work is needed to determine effective ways to couple local cultural and spiritual values with sound science and public policy.

Studies in this field rarely attach monetary (or even quantitative) values to cultural services. Like much of the rest of the world, there is very little quantitative evidence for California on cultural services generally and even less on cultural services specifically linked to N flows. The authors have made an effort to include in the text all those cases where they have found quantitative evidence, which is presented along with appropriate use of controlled vocabulary regarding uncertainty. The authors believe this approach is preferable to omitting these important (yet difficult to quantify and monetize) considerations.

5.5.0. Introduction

The scenic beauty of California's landscape is a vital part of our natural and cultural heritage. Prominent features of California's environment, both natural and man-made, play a central role in the formation of our individual and collective values as a society. Urban, rural, and wilderness settings also provide the backdrop for shared experiences with others, which over many generations have resulted in the distinctive regional culture and subcultures for which California is known the world over. These and other "non-material benefits that people obtain from ecosystems" are defined in the Millennium Ecosystem Assessment as cultural services (Alcamo and Bennett, 2003; MA, 2005b).

While the cultural services offered by ecosystems are often difficult to characterize and quantify, there is broad agreement that they encompass (1) aesthetic value, (2) recreation, (3) cultural heritage, and (4) spiritual or religious values (Daniel et al., 2012b; MA, 2005b). These are the deep but intangible values that John Muir described in *The Mountains of California* when he wrote, "Everybody needs beauty as well as bread, places to play in and pray in, where nature may heal and give strength to body and soul" (Muir, 1894).

At present, very little research has been done specifically examining the effects of N on the cultural services provided by ecosystems. Given that the links between N and cultural ecosystem services are for the most part indirect, this lack of coverage in the scientific literature is understandable. While noting "the importance of cultural services has consistently been recognized," Daniel et al. (2012b) summarizes some of the major challenges in quantitatively valuing cultural and religious services.

That said, a recent review of the ecosystem services altered by increases in reactive N in the United States has drawn needed attention to the dearth of information that exists regarding the cultural services potentially affected by N (Compton et al., 2011). These authors primarily highlight the adverse effects of N pollution on fishing, hiking, and other recreational activities through declines in air quality, water quality, and biodiversity (Table 5.0.1). However, they also suggest that N-related impacts on ecosystem quality and biodiversity may have ramifications for other cultural and spiritual values as well.

In this section, we expand on this nascent effort and assess how the uses of N (and its losses to the environment) affect the cultural services that California's ecosystems provide to society. Since impacts of N on land use and biodiversity constitute two important avenues through which N indirectly affects cultural services, we begin by examining these land-use and biodiversity effects first, followed by an exploration of various types of cultural services, including the aesthetic and recreational value of California's landscapes, cultural heritage, and spiritual and religious values, of which the latter two have thus far received little attention in the scientific literature. To close the chapter, we then consider some of the ways in which cultural and spiritual values, when coupled with sound science, can help society to address the consequences of N pollution by motivating people from diverse belief systems to adopt sustainable practices that are aligned with their shared values of environmental stewardship, social justice, and community.

5.5.1. Effects of Nitrogen on Land Use and Biodiversity

5.5.1.1. LAND USE AND AGROBIODIVERSITY

The use of fossil fuels and N fertilizers has in large part facilitated the expansion of urban and suburban land uses and the intensification of agricultural land uses. Since the end of the Second World War, the availability and low cost of N fertilizers have largely decoupled crop and livestock systems and allowed for less diverse and more intensive crop rotations (Russelle et al., 2007; Sulc and Tracy, 2007). This specialization of crop and animal production systems has notable economic advantages, but has also had important effects on land-use decisions and agrobiodiversity, and has posed challenges in managing fertilizers and manure so as to protect air and water quality (Russelle et al., 2007).

5.5.1.2. AQUATIC BIODIVERSITY

In California's aquatic ecosystems, eutrophication (excess nutrients) and hypoxia (low levels of dissolved oxygen) are two of the most direct consequences of N losses to surface water bodies. The main causes of eutrophication are elevated levels of nitrate (NO_3^-) (and phosphorous [P]) in agricultural runoff, and high ammonium (NH_4^+) loads in effluent from wastewater treatment plants. Eutrophication can lead to population shifts within native plant and animal communities and have ramifications for the entire aquatic food web (Glibert, 2010). For example, diatoms generally prefer NO_3^- over NH_4^+, unlike many algae which preferentially use NH_4^+ (Berg et al., 2001; Brown, 2010; Glibert, 2010). Thus, as NO_3^- has become less available relative to NH_4^+ in the San Francisco Bay-Delta, the structure of phytoplankton communities has shifted from diatoms to other algal species (Glibert, 2010; Jassby, 2008). Diatoms are considered a higher quality food source for higher order aquatic species, and the shift in species composition has been correlated with declines in pelagic fish populations in the San Francisco Estuary (Glibert, 2010; Jassby, 2008).

Fish are also particularly sensitive to high levels of dissolved NH_4^+, which can affect the central nervous system and ultimately lead to death (Randall and Tsui, 2002). Nutrient loading and harmful algal blooms have also contributed to episodic and seasonal occurrences of hypoxia in many of California's major coastal estuaries and waterways (e.g., San Francisco Bay, San Diego Bay, Monterey Bay, Los Angeles Harbor, Alamitos Bay, Anaheim Bay) (CENR, 2010; Table 5.5.1). While oxygen levels have improved in some water bodies over recent decades (e.g., South San Francisco Bay, Los Angeles Harbor, Alamitos Bay), several recent episodes of hypoxia have led to fish kills in the North San Francisco Bay (Bricker et al., 2007; Lehman et al., 2004). Physical and biological processes occurring in the ocean, such as shoaling from oxygen minimum zones of the California Current can also be an important cause of hypoxia off the California Coast (Bograd et al., 2008; Chan et al., 2008). The relative importance of nutrient loading from local anthropogenic sources versus shifts in ocean currents as factors contributing to hypoxia off the California Coast merits further research.

Oxygen depletion also affects biodiversity in inland freshwater bodies. For example, oxygen depletion in the

TABLE 5.5.1

Location, date, hypoxic and eutrophic status, and cause of nitrogen related biological impacts to surface water bodies in California

Location	Decade	Status of water body	Cause	Biological impacts	References
Los Angeles Harbor	1950	*Hypoxic (improved):* seasonal hypoxia observed since 1950s. Recent improvements due to nutrient management in the watershed.	Not reported.	Hypoxic events have caused mass mortality at the sea bottom (benthic zone), requiring multiyear recovery.	Collias (1985); Reish (1955, 2000)
Long Beach Harbor	1960	*Hypoxic (improved):* water quality has improved recently as a result of increased runoff controls in the drainage area.	Not reported.	Not reported.	Collias (1985); Whitledge (1985)
South San Francisco Bay	1960	*Hypoxic (improved):* seasonal hypoxia, observed since the 1960s, has been nearly eliminated with the construction of modernized sewage treatment plants	Sewage discharge.	Not reported.	Nichols et al. (1986); Bricker et al. (2007)
Coyote Creek	1970	*Hypoxia (episodic)*	Partly caused by sewage spills.	Fishermen report absence of fish and pelagic invertebrates, with fish returning when hypoxia ends.	Cloern and Oremland (1983)
San Joaquin River	1970	*Hypoxic (episodic)*	In part by sewage discharge from the Stockton Regional Wastewater Control Facility and agricultural runoff from further upstream.	Low oxygen conditions (<6 mg L^{-1}) interfere with spawning and migration of fish, in particular the Chinook Salmon. In 2003, fish kills of steelhead and salmon reported as the result of a hypoxic event.	Jassby and Van Nieuwenhuyse (2005); Lehman et al. (2004); Bricker et al. (2007)
Tomales Bay/ Bodega Harbor	1980	*Eutrophic:* eutrophication has been a concern in the bay since the 1980s.	Sources include runoff from animal waste (dairies and rangelands), failing septic systems, streambank and road erosion, storm drains, and boating activities.	Poor water quality causes seasonal closure of shellfish beds, high bacterial counts in swimming areas along tributaries to the bay.	Collias (1985)
North San Francisco Bay Estuary	1980s to 2000s	*Hypoxic (episodic):* seasonal hypoxia first observed in 1980s.	Nutrient sources include discharge from sewage treatment plants and urban and agricultural runoff.	Recently, seasonal hypoxia has resulted in fish kills.	Lehman et al. (2004)
Alamitos Bay	1990	*Hypoxic (episodic):* oxygen levels improved from 1990s to 2000s, but an estimated 2 km² still affected by episodic hypoxia since 2000.	High population density, leading to high levels of nutrient runoff.	Not reported.	Rabalais (1998)

(continued)

TABLE 5.5.1 (continued)

Location	Decade	Status of water body	Cause	Biological impacts	References
Elkhorn Slough	1990	*Hypoxic (improved)*	Located within a highly productive agricultural landscape and receives high nutrient inputs from agricultural runoff.	Eutrophication has led to high phytoplankton populations, persistent macroalgal mats, and hypoxia.	Bricker et al. (2007), 1999; Collias (1985); Sanger et al. (2002)
Newport Bay	1990	*Hypoxic (improved):* dissolved oxygen levels improved from 1990s to 2000s; less than 1 km² has been affected by periodic hypoxia since 1990s.	Urban runoff is the primary source of nutrient loads. Nutrient levels expected to decrease in the future due to diversion and treatment of stormwater.	Large macroalgae blooms occur, especially after heavy rainfall.	Collias (1985); Rabalais (1998)
San Diego Bay	1990	*Hypoxic (periodic):* estimated 4.5 km² affected by periodic hypoxia since 1990s.	High population density, leading to high levels of nutrients.		Collias (1985); Rabalais (1998)
Santa Monica Bay	1990	*Hypoxic (unknown):* wastewater treatment plants began improvements in the 1960s, and by 2002 both plants had fully upgraded to secondary treatment.	Receives direct sewage discharges from two wastewater treatment plants.	Shift in the community of benthic organisms living in the sediments to only the most pollution-tolerant species. Improvements in benthic diversity observed since 1995.	EPA (2007)
Tijuana Estuary	1990	*Hypoxic (improved):* estimated 0.1 km² affected by periodic hypoxia in last decade; oxygen levels improving since 1990.	Untreated sewage from Tijuana, Mexico.	Area is an essential breeding, feeding, and nesting ground for over 370 species of migratory and native birds, including six endangered species.	Bricker et al. (2007); Sanger et al. (2002)
Anaheim Bay	2000	*Eutrophic:* moderate eutrophic areas.	Urban runoff and agriculture in the watershed.		Bricker et al. (2007)
Monterey Bay	2000	*Hypoxic (seasonal)*	Natural and anthropogenic factors.	Not reported	Okey (2003)
Central San Francisco/San Pablo/Suisun Bays	2000	*Eutrophic*	Agriculture, urban runoff, and insufficient wastewater treatment in the region.	Moderate eutrophication and algal blooms.	Bricker et al. (2007)

SOURCE: Adapted from CENR (2010).

Merced, Tuolumne, Stanislaus, and San Joaquin Rivers has been found to interfere with the migration of fall run Chinook salmon and in some cases lead to fish kills among popular sport fish (e.g., salmon, steelhead) (Hallock et al., 1970; Jassby and Van Nieuwenhuyse, 2005; Volkmar and Dahlgren, 2006). In the Stockton Deepwater Ship Channel, which is a stretch of the San Joaquin River, Jassby and Van Nieuwenhuyse (2005) found that NH_4^+ loading from a regional wastewater treatment facility was the primary factor controlling year-to-year variability in dissolved oxygen and indicated that NH_4^+ loads have been increasing over the long term. However, at the monthly timescale they found that dissolved oxygen concentrations were also driven by patterns of reservoir release and the overall amount of river discharge, with levels of dissolved NH_4^+ and phytoplankton biomass having a somewhat smaller effect (Jassby and Van Nieuwenhuyse, 2005). Thus, while it is generally accepted that N pollution is often a factor in the recent episodes of hypoxia (and fish kills), more work is needed to determine the relative extent to which point and nonpoint sources contribute to the problem.

5.5.1.3. TERRESTRIAL BIODIVERSITY

It is well established that increased N deposition caused by air pollution is among the most important factors driving long-term changes in plant species diversity across many global and local ecosystems (Bobbink et al., 2010; De Vries et al., 2010; see Chapter 4 for an accounting of N deposition and flows in California). Annual deposition of N varies widely throughout California, with the highest rates (e.g., 25 kg N ha^{-1} yr^{-1}) occurring downwind of major urban centers (Fenn et al., 2010; Map 5.5.1). It is generally accepted that the increased availability of N drives competitive interactions among plant species, alters community composition, and makes conditions unfavorable for many native or rare plant species (Bobbink et al., 1998; Fenn et al., 2003; Map 5.5.2), particularly in naturally N-limited environments. In some habitats, reduced forms of N (e.g., NH_3, NH_4^+) can be toxic to sensitive plant species, particularly when soils are acidic and weakly buffered (Kleijn et al., 2008). The process of nitrification, which converts NH_4^+ to NO_3^- can also lead to long-term soil acidification, leaching of base cations, and increased concentrations of potentially toxic metals (e.g., aluminum), all of which can degrade soil quality and limit plant growth (De Vries et al., 2003). While less widespread, certain N gases and aerosols (e.g., nitric acid [HNO_3] vapor) can have direct toxic effects on plants growing near point sources of air pollution (Fenn et al., 2003; Pearson and Stewart, 1993).

In California's Mediterranean climate, cool moist winters coupled with summer drought tend to generate soils that are rich in base cations. Consequently, increased competition for N from nitrophilous and invasive plant species are more important ecological issues than soil acidification and leaching of bases (Fenn et al., 2003, 2008). For example,

high rates of N deposition (10–15 kg N ha^{-1} yr^{-1}) near San Jose, California, have contributed to the invasion of exotic grasses at the expense of native forb species that are key host plants for the endangered checkerspot butterfly (Weiss, 1999). Similarly, N deposition and invasion by annual grasses have contributed to a major decline of the native coastal sage scrub habitat in the Riverside–Perris Plain (Padgett and Allen, 1999; Padgett et al., 1999). Nagy et al. (1998) suggest that encroachment by invasive annual grasses in the Mohave Desert also diminishes the quality and availability of forage for threatened desert tortoise species (e.g., *Gopherus agassizii*). Studies conducted in California grasslands that use fertilizer to simulate atmospheric deposition provide further experimental evidence that N addition favors colonization by invasive plants and a decline in native N-fixing species (Huenneke et al., 1990; Zavaleta et al., 2003).

The effects of increased N deposition and air pollution on the biodiversity of lichen species in California's chaparral and forest ecosystems are well established in the scientific literature. At a relatively low N deposition rate of 6 kg N ha^{-1} yr^{-1}, Fenn et al. (2008) observed shifts in species composition from those naturally dominated by N-sensitive lichen species (e.g., *Letharia vulpina*) to communities dominated by more N-tolerant lichen species. Similar shifts in lichen communities in response to N deposition have recently been documented in the Sacramento and San Joaquin Valleys, the Coast Ranges, and the Sierra Foothills (Fenn et al., 2010, 2011). These are important ecological findings because large areas of California's chaparral and forest ecosystems are exposed to N deposition rates in the 3–5 kg ha^{-1} yr^{-1} range (Fenn et al., 2008, 2010; Map 5.5.1). Overall, N and O_3 pollutants are estimated to have contributed to the disappearance of up to 50% of the lichen species that occurred in the Los Angeles Air Basin in the early 1900s (Fenn et al., 2003; Fenn et al., 2011; Riddell et al., 2008; Map 5.5.2). How this shift in lichen species affects the larger food web is unclear and will require further study, but many pollution-sensitive macrolichens are known to be important forages for birds, small mammals, and deer (McCune and Geiser, 1997).

It is generally accepted that N deposition in combination with other N-related air pollutants (e.g., ozone [O_3]) is also adversely affecting plant communities in mixed conifer forests (Takemoto et al., 2001). It should be noted that high levels of O_3, which is formed from emissions of nitrogen oxides (NO_x) and volatile organic compounds (VOC), are widely considered to have the most severe impacts on plant growth in natural ecosystems relative to other air pollutants (Fenn et al., 2003; Campbell et al., 2007). Several studies in the San Bernardino Mountains have indicated that a combination of O_3 exposure and N deposition is disrupting the physiology of ponderosa pine (*Pinus ponderosa*) by reducing fine root growth and increasing aboveground wood and foliage production (Fenn et al., 2003, 2008; Takemoto et al., 2001). The authors suggest that this

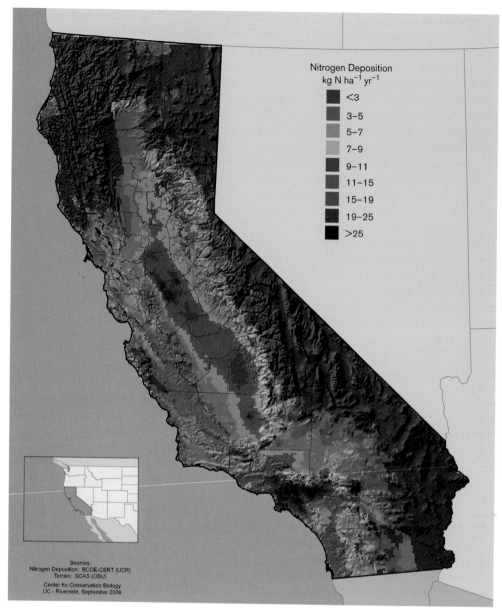

MAP 5.5.1. Map of total annual N deposition in California based on simulations from the US EPA Community Multi-Scale Air Quality (CMAQ) model. Simulated N deposition in forested areas has been adjusted based on the linear relationship with empirical throughfall data (reprinted from Fenn et al., 2010).

physiological change in plant biomass allocation increases the amount of litter and fuel wood on the forest floor, thereby increasing the risk of severe fire damage (Fenn et al., 2003). In southern California's San Bernardino Mountains, the number of understory plant species in mixed conifer forests declined by 20–40% between 1973 and 2003 in two of the most polluted sites, while invasive species became more abundant (Allen et al., 2007). However, multiple confounding factors (e.g., several air pollutants, O_3, and precipitation differences) occurring across the six study sites make it difficult to attribute the impacts specifically to N deposition (Allen et al., 2007). While studies suggest that California's chaparral plant communities are less prone to

changes in species composition and invasive species, N enrichment of soils has been associated with declines in the diversity and productivity of plant-symbiotic arbuscular mycorrhizal fungi near Los Angeles (Egerton-Warburton et al., 2001; Egerton-Warburton and Allen, 2000; Fenn et al., 2011).

While most anthropogenic changes to the N cycle tend to decrease natural biodiversity, there are several instances where increased N availability from anthropogenic sources offers important benefits to biodiversity in California. Perhaps the most noteworthy example is the provision of food and habitat to migratory birds that overwinter in flooded rice fields following harvest (Hill et al., 2006). In this rice

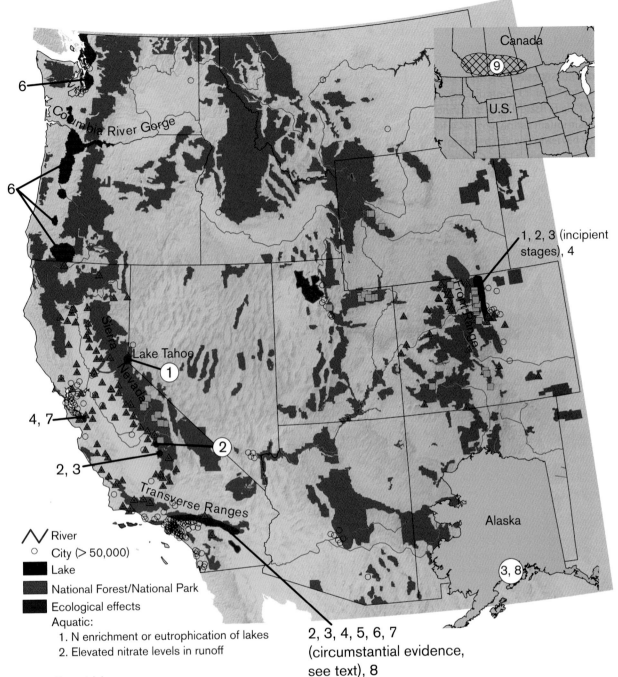

6

Columbia River Gorge

6

1, 2, 3 (incipient
stages), 4

Front Range

Sierra Nevada

Lake Tahoe

① 1

4, 7

② 2

2, 3

Transverse Ranges

Alaska

③ 3, 8

2, 3, 4, 5, 6, 7
(circumstantial evidence,
see text), 8

Canada

⑨ 9

U.S.

Legend:

〰 River

○ City (> 50,000)

■ Lake

■ National Forest/National Park

■ Ecological effects

Aquatic:
1. N enrichment or eutrophication of lakes
2. Elevated nitrate levels in runoff

Terrestrial:
3. N enrichment or N saturation (e.g. soil, vegetation, & water are N enriched; increased fluxes of nitrogenous trace gases)
4. Altered plant communities in response to N enrichment
5. Physiological perturbation of plants; combined effects of ozone and N deposition
6. Impacts on lichen communities
7. Evidence that threatened and endangered species impacted
8. Decreased diversity of mycorrhizal communities
9. Forest expansion into grasslands (preliminary evidence for)

▲ Plot with lichen community affected by air pollution with major N deposition component

■ High-elevation lake with elevated nitrate, reportedly from N deposition

■ Available data indicate elevated N deposition, but ecological effects have not been studied

MAP 5.5.2. Western United States and the primary geographic areas where nitrogen (N) deposition effects have been reported. Areas where effects of air pollution on lichen communities have been reported in California and in Colorado are represented by orange triangles. The areas shown in red in Oregon and Washington (lichen communities affected by N deposition) are kriged data. Only lakes at an elevation greater than 1,000 m and with a nitrate (NO_3^-) concentration of more than 5 μeq L^{-1} (0.07 mg N L^{-1} measured in fall surveys or on an annual volume-weighted basis) are shown in this map. Other high-elevation lakes in the West also had elevated NO_3^- concentrations but were excluded, because N sources other than N deposition may have contributed to the elevated concentrations of NO_3^-. Also see Data Table 23 in Appendix 4.1 (reprinted from Fenn et al., 2003).

cropping system, the residual grain that falls to the ground during rice harvest provides high-quality forage that attracts numerous bird species that migrate along the Pacific Flyway. A study of food abundance and feeding behavior among various bird species concluded most species had slightly higher feeding efficiency in seminatural wetland than flooded rice fields, but that flooded rice fields offered reduced risks of predation (Elphick, 2000). Overall, these authors concluded that flooded rice fields provide functionally equivalent foraging habitat relative to seminatural wetlands and better foraging habitat than non-flooded fields (Elphick, 2000; Elphick and Oring, 1998).

These findings highlight some of the primary effects of anthropogenic N additions on the biodiversity of California's aquatic and terrestrial ecosystems, but the relatively small number of local studies limits our ability to comprehensively assess the geographic and temporal trends that may exist throughout much of the state. Moreover, to our knowledge, there have been no studies in California that have systematically examined the social or economic aspects of biodiversity decline. However, the role of biodiversity in the cultural value derived from California's aquatic and terrestrial resources is further explored in the following sections.

5.5.2. Aesthetic Value

The study of aesthetics is primarily interested in the creation, perception, and appreciation of beauty, particularly in response to art or nature. In the field of landscape aesthetics a key question is how people appreciate urban, rural, and wilderness landscapes (Home et al., 2010; Howley, 2011; Parsons and Daniel, 2002). As global society becomes increasingly urban, research has also begun to examine how shifts in land use affect the aesthetic value of certain landscape qualities (Nohl, 2001). Some of the aesthetic qualities that can be enhanced or diminished through shifts in land use include the variety or diversity of landscape features (e.g., water bodies, landforms, built structures, vegetation types), the naturalness of the landscape, the rural or agrarian characteristics of a locality, the regional identity of a place, and the vista quality (Daniel, 2001; Nohl, 2001). While cultural factors no doubt play an important role in the appreciation of landscapes, it is also well established that people in general have a strong aesthetic preference for landscapes that would have provided ancestral humans with good habitat (Dutton, 2003; Orians and Heerwagen, 1992; Tress et al., 2001). As such, people across cultures tend to prefer the appearance of landscapes with vistas of savannas, open-space, water, green vegetation, wooded areas, and environments that are likely to offer plentiful food and shelter (Dutton, 2003; Howley, 2011; Kaplan and Herbert, 1987).

What influence might human activities that alter the N cycle have on these aesthetic landscape characteristics? The use of fossil fuels and N fertilizers has in large part facilitated the expansion of urban and suburban land uses and the intensification of agricultural land uses. In terms of aesthetics, it is well-established in the social science literature that people's affinity for a landscape tends to be inversely related to the intensity of the land use (Arriaza et al., 2004; Dramstad et al., 2006; Howley, 2011; Lindemann-Matthies et al., 2010). For example, studies indicate that imageries of wilderness and agrarian landscapes are consistently preferred over visual depictions of urban and industrial settings (Howley, 2011). Thus, the judicious use of inorganic and organic N fertilizers may be seen by many as providing an important benefit to society by supporting aesthetically appealing and agriculturally productive "working landscapes."

However, even among agricultural landscapes, psychological studies also show that people typically prefer diverse and lower intensity cropping systems as opposed to monocultures (Lindemann-Matthies et al., 2010). Likewise, aesthetic and ecological concerns have also been raised about the extent to which unique or highly valued wilderness areas are displaced by agricultural and urban land uses. Clearly any land-use choice involves a complex mix of benefits and trade-offs between aesthetic values and other important ecosystem services such as food provisioning or shelter. The forester, farmer, and author Aldo Leopold (1949) put it this way: "The true problem of agriculture, and all other land uses, is to achieve both utility and beauty, and thus permanence. A farmer has the same obligation to help, within reason, to preserve the biotic integrity of his community as he has, within reason, to preserve the culture that rests on it. As a member of the community, he is the ultimate beneficiary of both." In the case of N, this would imply balancing its utility as an input with its disutility as a pollutant that can also diminish the beauty of a landscape by degrading the quality of air, land, and water resources or by altering the biodiversity of native ecosystems.

In Section 5.3 of this chapter we examine the adverse effects of N on air quality and human well-being in California, due mainly to emissions of NO_x and NH_3 and the formation of secondary air pollutants (PM, O_3, smog). Several recent studies have also shown that air pollution and reduced visibility have undesirable effects on the aesthetic value of the places where people reside and enjoy recreational activities (Abt Associates Inc., 2000). The National Park Service categorizes daily visibility measurements into three groups: (1) good visibility are days in the lowest 20th percentile; (2) mid-range visibility days are in the middle of the 40–60th percentile; and (3) poor visibility are days above the 80th percentile (Abt Associates Inc., 2000). Air quality improvements in the San Bernardino Mountains between 1988 and 1998 have reduced the number of poor visibility days in the San Gorgonia Wilderness Area (Abt Associates Inc., 2000). During the same period, no significant improvements in visibility were observed in Yosemite National Park (Abt Associates Inc., 2000). Overall, Abt Associates Inc. (2000) valued the economic impact of air pollution in California on residential and recreational visibility at approximately $61 and $219 million, respectively. This suggests that despite the improvements in California's air quality over recent decades, poor visibility continues to degrade the beauty and aesthetic value of residential and recreational areas. It also suggests that nitrogen's contribution to this degradation in the state's air quality comes with considerable cultural and economic cost.

5.5.3. Recreational Value

In addition to diminishing the aesthetic value of a landscape, losses of N to the environment can also affect recreational activities (e.g., fishing, hiking, hunting, etc.) by changing the biodiversity, species composition, and ecological function of aquatic and terrestrial ecosystems (Compton et al., 2011; Smart et al., 2011b). The N-induced impacts of hypoxia and eutrophication on aquatic biodiversity, as noted in Section 5.5.1.2, especially affect the recreational value of aquatic ecosystems in California when they reduce populations of sport fish, such as salmon and steelhead. The economic value of California's recreational salmon fishery alone has been estimated at $205 million (Business Forecasting Center, 2010).

Recreational swimming is also affected by eutrophic conditions and high bacterial counts commonly found at many of California's coastal and inland swimming areas, particularly following heavy storm events that carry runoff from urban and rural land uses (Collias, 1985; Noble et al., 2003). A survey of water quality at 254 shoreline sites between Santa Barbara (CA) and Ensenada (Mexico) found that 60% of all sites (and >90% near urban areas) failed to meet state water quality standards for three bacterial indicators (enterococci, fecal coliforms, and total coliforms) during storm flow events, while only 6% of these sites were above the threshold during dry weather (Noble et al., 2003). However, a recent study at Mission Bay, California, found that these traditional bacterial indicators were poor predictors of adverse health outcomes (e.g., diarrhea and skin rash) among swimmers exposed to polluted water at beach sites, but identified a novel viral indicator (coliphage) that was significantly associated with increased health risks among male swimmers (Colford et al., 2007). Blue-green algal blooms are also associated with health risks to swimmers and others who come into contact with affected water bodies, with health effects ranging from rashes, skin and eye irritation, allergic reactions, and gastrointestinal upset to liver toxicity and neurotoxicity (CDC, 2012). According to a 2013 report by the California Department of Public Health, recent blue-green algal blooms were reported in rivers and lakes in eight different counties throughout California plus the San Francisco Bay Delta (CDPH, 2013). These studies show the uncertainty associated with attributing detrimental impacts to specific pollutants (be they N-related or otherwise) and highlight the complex effects that infra-

structures for water storage, sewage treatment, and surface runoff have on ecosystem health and downstream recreational activities.

The long-term changes in plant species diversity and ecosystem function caused by increased availability of N in terrestrial ecosystems, as documented in Section 5.5.1.3, can affect the recreational value of these ecosystems as well. Several studies have highlighted the value of recreation in biodiverse landscapes as a means of promoting people's psychological and physical well-being (Daniel et al., 2012b). Fuller et al. (2007) found that psychological well-being, assessed using surveys gauging park visitors' reflection and attraction to certain landscape features, was positively correlated with habitat diversity and species richness. Biodiversity may also enhance recreational experiences by increasing the likelihood of memorable wildlife encounters (Harrod, 2000; Naidoo and Adamowicz, 2005). For example, the level of satisfaction experienced during many recreational activities (e.g., hunting, fishing, bird watching) is often highly dependent on the abundance of particular species of interest as well as the overall species richness of a landscape (Stallman, 2011).

On the other hand, as noted in Section 5.5.1.3, the widespread presence of winter-flooded rice fields in northern California, made possible in part by the availability of cheap N fertilizers, raises the recreational value of landscapes that might otherwise be lost to other uses not supportive of wildlife habitat. The economic value of recreational hunting and bird watching on agricultural lands has not been quantified, but a 2002 survey of 179 rice producers documented that 75% allowed hunting on their land, with annual hunting fees ranging from $1,000 to $3,000 per hunter (Garr, 2002, cited in Eadie et al., 2010).

Inorganic and organic N fertilizers also facilitate the cultivation of a diverse range of perennial and annual crops (>300 crops) that have made California a national and international destination for agritourism and culinary travel. Prime examples are the state's many vineyards and wineries, "U-pick" farms with fruit and berries that visitors can harvest themselves, and livestock operations that offer visitors the opportunity to learn how food is produced by participating in seasonal management activities (e.g., calving, shearing, cheese making) (Rilla et al., 2011). Each year more than 2.4 million people participate in agritourism activities in California and generate approximately $35 million in revenue; this sector is expected to continue to grow (Rilla et al., 2011; USDA, 2009b). Moreover, the agritourism sector is expected to be a significant source of economic growth and jobs for California's rural communities in the coming years (Rilla et al., 2011).

The role of N in shaping the recreational value of the state's natural and agricultural ecosystems is complex. Additional work will be needed to address the gaps in our knowledge of how N pollution impacts recreational usage of California's natural ecosystems and to develop management strategies that enhance recreational experiences.

5.5.4. Cultural Heritage Values

Cultural heritage is defined as "the legacy of biophysical features, physical artifacts, and intangible attributes of a group or society" that helps define the identity of the individual or group and provide experience shared across generations (Daniel et al., 2012b). California's natural and managed landscapes and ecosystems have great value to society as locations where a variety of expressions of cultural heritage take form, including knowledge acquisition and transfer, traditional livelihood practices, and artistic expression. Human interactions with these landscapes and ecosystems over time define and reinforce important cultural constructs and identities. For example, culturally valuable species and places include the iconic bald eagles and condors that soar above the Sierra Nevada, symbolizing freedom and independence, and the magnificent groves of redwoods standing tall along California's North Coast, reminders of endurance and the slow evolution of nature. Cultural values and identities are also associated with managed landscapes, as exemplified by the idyllic vineyards that are synonymous with the Napa and Sonoma Valleys, and the oak savannas and rangelands of the Coast Ranges that support "happy cows." Given that these landscapes provide numerous cultural heritage services to society, the degree to which human-induced changes to the N cycle either support or diminish these ecosystems (and their resulting services) merits closer examination.

Knowledge, generated through scientific as well as other means, is one important aspect of cultural heritage. Natural and working landscapes provide a place where observation, measurement, and critical assessment can help society accumulate knowledge about species, ecosystems, and N cycling that has both practical and scientific values. Roughly 30% of California's native plant species are found nowhere else in the world (Barbour et al., 1993) and these, and other forms of California's biodiversity, can be adversely affected by N pollution and loss of native habitat. These reductions in biodiversity are likely to limit opportunities for nature study, hinder scientific discovery, and in some cases limit the practical application of new knowledge (Compton et al., 2011; Smart et al., 2011a). For instance, the Northern California black walnut (*Juglans hindsii*) is often used as a root stock for commercial English walnut varieties that are broadly cultivated on approximately 280,000 acres in California (Ramos, 1997; USDA NASS, 2011b). Despite its wide use in the walnut industry, *J. hindsii* is classified as a vulnerable species since only a few native stands remain in its native riparian woodland (IUCN, 2013). More broadly, loss of native species due to N pollution (or land-use change enabled by fuel and fertilizer) can diminish the genetic resources available to future generations of scientists, plant and animal breeders, and other industry innovators.

Biodiversity changes can have large implications for the maintenance of unique livelihoods and cultural identities of California's indigenous peoples. Extensive research (e.g., Anderson, 2005) has documented the reliance of California's Indian tribes on hundreds of species of plants and animals for food, cordage, firewood, basketry, and construction. In almost every contemporary California Indian tribe, one can find individuals of all ages who fish and hunt for food and gather native plants for food, medicinal, handicraft, and ceremonial uses. The economies of some northern California tribes are built substantially or entirely around fishing and hunting. Salmon, in particular, often play a significant role in these economies (Anderson, 2005). However, the viability of salmon fishing is threatened by hypoxia, as discussed above.

An upsurge in interest in some traditional practices in recent decades is evidenced by the founding of the California Indian Basket Weavers Association in 1992 (www.ciba.org), which works with land management agencies to promote access to traditional gathering lands and use of management practices that promote populations of native plant species used in basket making. Some of the many species used include native perennial bunchgrasses, such as deer grass, which, as noted above, have been supplanted in many areas by nonnative annual grasses and are more difficult to find. The causes of these species shifts are varied and include prominent factors such as cessation of controlled burning, a practice traditionally used by many California tribes (Anderson, 2005), and therefore the role specifically of N deposition in these shifts is suggested but unproven. However, in so far as any of the N-related biodiversity losses noted above affect species directly used by California's indigenous people, it poses a threat to the lifestyles, economies, and cultures of these groups, as well as to their physical health, which is often predicated on access to wholesome traditional foods (Lynn et al., 2013).

The cultural values imbued in California landscapes are also evident in the abundance of artistic expression that has arisen over the last one and a half centuries of Euro-American settlement of the West Coast. For example, values of nature as an "antidote" to civilization arise in the prose writings of the famous California naturalist and essayist John Muir, who wrote:

> It is a good thing, therefore, to make short excursions now and then to the bottom of the sea among the dulse and coral, or up among the clouds on mountain-tops, or in balloons, or even to creep like worms into dark holes and caverns underground, not only to learn something of what is going on in those out-of-the-way places, but to see better what the sun sees on our return to common everyday beauty. (MUIR, 1894)

Muir's emphasis on the need for immersion in and preservation of California's "pristine" wild places encapsulates the spirit of a growing interest in preserving natural areas, which would later come to be called "wilderness," from overexploitation of economically useful resources. For example, at one point he wrote that "thousands of tired, nerve-shaken, over-civilized people are beginning to find out that going to the mountains is going home; that wildness is a necessity; and that mountain parks and reservations are useful not only as fountains of timber and irrigating rivers, but as fountains of life" (Muir, 1997). This sentiment regarding the value of California's unique landscapes and species played a central role in the early advocacy for the United States national park system led by John Muir, Theodore Roosevelt, and others (Roosevelt, 1985). Eventually, Muir's work culminated in his founding of the Sierra Club in San Francisco in 1892, a group that has been involved in public legislative efforts to preserve wilderness areas and conserve land, air, and water resources ever since. With a current membership of more than 100,000 people, the Sierra Club remains one of the largest and most influential environmental organizations in the United States.

While a thorough survey of the arts and humanities is beyond the scope of this assessment, several other prominent themes that emerged in mid-twentieth-century American nature writing exemplify a growing national movement valuing nature as something to be loved and stewarded instead of feared as dangerous or viewed only as a resource for economic exploitation. For example, in *A Sand County Almanac*, Aldo Leopold extends the definition of human community to include the natural world and the land itself, a definition that entails certain ethical obligations to the land (Leopold, 1949). From his perspective as both forest ecologist and farmer, Leopold (1949) writes:

> All ethics so far evolved rest upon a single premise: that the individual is a member of a community of interdependent parts. His instincts prompt him to compete for his place in that community, but his ethics prompt him also to co-operate; perhaps in order that there may be a place to compete for. The land ethic simply enlarges the boundaries of the community to include soils, waters, plants, and animals, or collectively: the land.

Moreover, Leopold's "land ethic" attempts to balance the practical and intrinsic values of the land. For instance, he writes that "a land ethic of course cannot prevent the alteration, management, and use of these 'resources,' but it does affirm their right to continued existence, and, at least in spots, their continued existence in a natural state" (Leopold, 1949).

If Leopold was the father of the land ethic, Rachel Carson (a marine biologist) was most certainly the mother of the sea ethic. While Carson does not use the term sea ethic, a recent critique of Carson's *Under the Sea-Wind* by Bratton (2004) suggests that there are several parallel concepts that are shared between the two authors. In particular is Leopold and Carson's joint recognition that (1) humans need to understand the complexity of aquatic and terrestrial ecosystems; (2) human activities can disrupt ecosystem process

through overextraction, degradation, and pollution; and (3) human imagination and scientific inquiry can help us more fully value life, nature, and its key ecological processes (Bratton, 2004; Carson, 1941; Leopold, 1949). Toward the end of her career Carson also became concerned with the use of pesticides and their growing impact on terrestrial ecosystems, and on birds in particular. This concern is the basis of her final book *Silent Spring*, where she laments that "over increasingly large areas of the United States, spring now comes unheralded by the return of the birds, and the early mornings are strangely silent where once they were filled with the beauty of bird song" (Carson, 1962). Through their literature, Carson and Leopold link science-based ecological knowledge with the aesthetic beauty of nature, and in so doing became voices of the modern land stewardship movement, whose proponents place a high value on both natural and working landscapes that preserve the integrity of their respective ecosystems. In so far as N pollution threatens the health of these ecosystems, it threatens the material basis of these social values. Conversely, in so far as prudent use of N enhances agricultural landscapes, it can support these stewardship values.

Another key theme in contemporary American literature is people's experience of the human and natural world through one's "sense of place" (Lopez, 1996; Snyder, 1993). This branch of nature writing is dependent on the existence of distinct cultural landscapes and regional subcultures, and its popularity illustrates the continued value of these entities to many people within contemporary society. David Masumoto, a Fresno-based peach farmer of Japanese-American decent, is one contemporary author whose writing reflects a strong sense of place and land stewardship from an agrarian viewpoint. In his book *Epitaph for a Peach*, Masumoto (1995) chronicles his efforts to rescue the Sun Crest peach, "one of the last remaining truly juicy peaches," from commercial obsolescence due to the industry's preference for a uniform and less perishable product rather than overall flavor and quality. But perhaps more importantly, he also illustrates the curiosity, artistry, and traditional wisdom that are part of his agrarian cultural values. One way that he does this is through his description of why he uses legume cover crops and their role in the nitrogen cycle.

> Some farmers question the value of cover crops. How much nitrogen do they produce? Do they consume huge volumes of water? What plants attract which beneficial insects? All valid questions that need research, these issues will take years to determine and may never be clear. But the benefits of my fall planting go beyond making interesting plant mixtures and achieving proper nitrogen levels. Every fall I plant seeds of change for the next year. I am an explorer and adventurer, a wild man in the woods. No one can know the exact benefits of my cover crops; they are a blend of artistry and the wisdom of experience, a creation and reaffirmation of tradition. (MASUMOTO, 1995)

In a similar vein, Wendell Berry, icon of the land steward-ship movement at the national level, argues that the rural exodus and urbanization of America, and our subsequent loss of connection to the land, has diminished our cultural and spiritual identity (Berry, 1977). In his case, he uses the example of waste recycling as the agrarian basis of fertility to critique modern urban culture.

> Ninety-five percent (at least) of our people are also free of any involvement or interest in the maintenance phase of the cycle. As their bodies take in and use the nutrients of the soil, those nutrients are transformed into what we are pleased to regard as "wastes"—and are duly wasted. This waste also has its cause in the old "religious" division between body and soul, by which the body and its products are judged offensive. (BERRY, 1977)

One implication of this class of literature is that some farming practices are especially imbued with cultural val-ues relating to land stewardship of certain types of agricul-tural landscapes—including those occupied by locally ori-ented, regionally diversified family farms such as Masumoto's—that embody the cultural significance of a sense of place. By extension, then, any trends that threaten the viability of these farming practices or these agrarian landscapes can potentially threaten the preservation of these cultural values. However, the relationship between these trends and the use of N is complex. On the one hand, the rapid increase in availability of cheap inorganic N ferti-lizers (see Chapter 3) may have reduced the use of certain traditionally important farming practices. On the other hand, the role of easily available N in supporting the con-tinued existence of widespread agricultural landscapes that are key to cultural identity in California also merits consideration.

5.5.5. Spiritual and Religious Values

Since many religious traditions were established during pre-scientific times, it is perhaps not surprising that direct refer-ences to N and its role in the environment are rare among the world's sacred texts and scriptures. Similar to the cul-tural heritage values discussed above, the complex relation-ship that exists between N, the environment, and people's spiritual values has not been rigorously examined. That said, the broader ecosystem services literature has recently begun to examine some interesting lines of inquiry on spiritual and environmental values that are beginning to fill the gap (Bhattacharya et al., 2005; Daniel et al., 2012b; Taylor, 2004). These studies note that spiritual values regarding our relationship to (and appropriate use of) nature have existed since prehistoric times and remain an important part of the contemporary spiritual values of vir-tually all global cultures. Here we specifically examine spir-itual values in the context of N-related impacts on the environment.

As discussed by Daniel et al. (2012b), one of the primary ways that spiritual values are linked to the environment is through the practice of giving "sacred" status to important species, locations, or geographic features. This practice of granting sacred status to species and places is common to many religious traditions practiced in California, though the degree of reverence attached to these rituals can vary widely among religions, subcultures, and time periods. How something becomes viewed as sacred also varies widely, but common ways include significant references to particular species or sites in creation narratives, oral traditions, scrip-tures, and religious rituals. For example, California's Miwok tribes have a creation narrative, preserved through oral tra-dition, which depicts humans first emerging from feathers planted into the soil by supernatural personages with both animal and human characteristics (e.g., coyote, fox) (Kroe-ber, 1907; Taylor, 2005). Likewise, several indigenous tribes hold the belief that certain animals (e.g., buffalo, salmon, whales) offer themselves up as food for humans and thus are given a place of reverence within the spiritual tradition (Daniel et al., 2012b; Harrod, 2000). Consequently, Native American hunting rituals often include moral requirements for how to treat the bodies of animals after they are killed. There are also spiritual dimensions to the indigenous practices used to tend and harvest important plant species including the careful use of fire to maintain meadows and the gathering of tubers, acorns, and buckeyes (Anderson, 2005; Kroeber, 1907). Some believe that failure to observe these rituals and practices will cause animals to withdraw from humans and thus lead to suffering and starvation (Harrod, 2000). Hence, many rituals are specifically intended to renew both the animals and the landscape in anticipation of future hunting seasons (Harrod, 2000).

Sites where these mythical events or religious rituals take place also have deep spiritual attachments among contemporary indigenous groups and are often considered sacred ground. For example, the Miwok roundhouses erected at sacred sites located on reservations, and in Point Reyes and Yosemite National Parks, are still used for important rituals and dances (Taylor, 2005). Likewise, in some indigenous oral traditions the Great Spirit is said to have lived on Mount Shasta and thus the mountain is revered as the center of creation by the Shasta, Modoc, Ajumawi, Atsuwegi, and Wintu tribes. These beliefs regarding sacred species and sites are also part of a broader world view that does not draw a conceptual distinction between the natural and spiritual worlds, but rather sees the land itself as a sacred being (Brady, 1999). These beliefs and sacred sites remain a vital part of the contemporary worldview held by many Native Americans in California. As a result, the impacts of N-related air and water pollution on the sacred sites (and to lesser extent the species) of tribal communities are now being addressed through the environmental justice components of the US Clean Air Act and the California Environmental Protection Act (EPA, 2013; Gatto and Alejo, 2014; NEJAC, 2002).

While prominent in the Native American worldview, attributing some measure of sacred value to the natural environment is common in many other spiritual traditions as well, be they rooted in traditional religious beliefs or other forms of contemporary spirituality (Fick 2008; Hitzhusen et al., 2013; Taylor 2004, 2009). These values from other cultural and spiritual traditions also merit consideration when assessing the broader impacts of N use and pollution on society. That said, much more research is needed to determine which of these diverse spiritual values are most relevant and how they might be influenced by the complex interactions between N and the natural environment.

5.5.6. Cultural and Spiritual Values as Motivators for Addressing N Issues

In this section, we have focused mainly on how nitrogen can affect various landscapes, natural resources, ecosys-

tems, and species that have important cultural and spiritual value to society. Given how closely entwined cultural and spiritual values are with nature, it is perhaps not surprising that they can often play a central role in shaping the environmental ethics of our communities (Leopold, 1949; Taylor, 2005). Likewise, such values can also be a key source of motivation and inspiration for environmental stewardship (Hitzhusen et al., 2013; Posey, 1999; Taylor, 2004). Indeed, there is growing evidence from the social sciences that people's perception of the scientific facts related to nutrient cycling, air and water pollution, and climate change are strongly mediated by their cultural and spiritual perspectives, which therefore have large implications for their individual and collective responses (Bickerstaff, 2004; Schweizer et al., 2013). Thus, a better understanding of the values that motivate, or possibly deter, environmental stewardship is likely to be a useful complement to the scientific knowledge, technologies, and policies that are assembled to address environmental problems (Hitzhusen et al., 2013).

Table 5.5.2 illustrates the wide range of cultural and spiritual values that have played a key role in motivating society in California (and beyond) to respond to environmental issues related to land stewardship and pollution, and to N in particular. Most of these examples are drawn from the literature discussed in the sections above. Here we suggest that these values in many instances have helped to shape the specific goals of large social movements that bring people together at the local, national, and international scales and raised awareness of pressing social and environmental issues. And while conflicts on priorities and policies can often arise among people who emphasize different values, common ground can often be cultivated by considering our shared values and our collective link to the local landscape and ecology (Snyder, 1993).

Of particular relevance to our discussion of N management and agriculture, sets of shared cultural and religious values led to a joint environmental stewardship "covenant" that was made between Christian watermen and farmers in Maryland's Chesapeake Bay watershed (Emmerich, 2009; Hitzhusen et al., 2013). In this example, the watermen promised that they would abide by various crab harvesting regulations, while the farmers upstream promised to adopt improved nutrient management practices to reduce eutrophication and water pollution. Hitzhusen et al. (2013) argue further that while such anecdotes are instructive on their own, the key to developing transformative models of environmental stewardship is to identify ways to synergistically couple cultural and spiritual values with sound science and effective public policy. One instance of positive synergism between cultural values and science education is a nationwide conservation program known as "Soil Stewardship Sundays" (Hitzhusen et al., 2013). The movement was initiated in the 1920s and 1930s during the Dust Bowl by the National Catholic Rural Life Conference, but it is now supported by a wide range of faith communities as well as scientists at the

TABLE 5.5.2

Examples of cultural values that have played a role in motivating society to respond to environmental issues related to nitrogen, land stewardship, and pollution

Cultural values	Region	Issues or movement	Response	References
Spiritual value	Global	Nature and religion, environmental ethics	Religious leaders from Buddhism, Christianity, Hinduism, Islam, and Judaism gathered in 1986 to issue joint Assisi Declarations on humanity's spiritual relationship to nature.	ARC (1986)
Spiritual value	Global & United States	Environmental ethics climate change	Faith-based organizations from many religions are active participants in global efforts to address the causes and consequences of climate change through religious education, community action projects, interfaith partnerships, and United Nations initiatives.	Faith-350 (2013); Posas (2007)
Cultural heritage Recreational value	Global & United States	Organic, local and slow food movements	California-based writers M. Pollan and E. Schlosser are prominent figures in these contemporary movements, which critique how food and culture intersect on our plate.	Pollan (2008); Schlosser (2001)
Cultural heritage Aesthetic value	United States	Land ethics, conservation, pesticides, pollution	Nature writing of A. Leopold and R. Carson helped raise awareness of environmental issues (pesticides, degradation of aquatic and terrestrial ecosystems), and the resulting movement led to the federal Clean Air and Water Acts of the 1970s.	Carson (1941, 1962); Leopold (1949)
Cultural heritage Spiritual value	United States	Soil conservation	Soil Stewardship Sundays initiated in the 1920s by the National Catholic Rural Life Conference to address declines in rural culture and soil quality. The movement expanded and is currently supported by various faith communities in the United States, and the secular National Association of Conservation Districts.	Hitzhusen et al. (2013); Woods (2009)
Cultural heritage Recreational value Spiritual value	United States & California	Nature conservation Transcendental philosophy	The nature writing and activism of R.W. Emerson, H.D. Thoreau, and J. Muir emphasized the transcendental unity of man and nature, and helped to establish the US National Park system as a place for recreation and reflection.	Emerson (1849); Muir (1894, 1901); Rettie (1996)
Cultural heritage Spiritual value	United States & California	Environmental justice, air and water quality	Policy engagement by Native American communities has helped incorporate their cultural concerns and spiritual values into the policy frameworks of the California Environmental Protection Act and federal Clean Air and Water Acts.	Gatto and Alejo (2014); NEJAC (2002)
Cultural heritage: Aesthetic value Spiritual value	United States & California	Sense of place, agrarian & urban values, bioregionalism	The literature of W. Stegner, G. Snyder, W. Berry, D. Masumoto, and many others established a style that is rooted in (and contributes to) one's "sense of place" and the environmental, agrarian, urban, and spiritual values that exist in regional subcultures.	Berry (1977); Masumoto (1995); Snyder (1993)
Recreational value Aesthetic value	California	Air quality legislation, bird habitat conservation, agricultural practices	Following the Rice Straw Burning Act of 1991, collaborative efforts by California farmers, scientists, and bird conservation organizations helped to reduce straw burning, improve air quality, expand habitat for migratory waterfowl, and support bird watching and hunting.	Elphick (2000); Hill et al. (2006)
Spiritual value	Chesapeake Bay Watershed	Eutrophication, soil nutrient management, overexploited fishery	Shared religious values led to a joint "covenant" between Christian watermen in the Chesapeake Bay and farmers in Pennsylvania. Watermen promised that they would abide by various crab harvesting regulations and upstream farmers promised to adopt nutrient management practices to reduce eutrophication and water pollution.	Emmerich (2009); Hitzhusen et al. (2013)

National Association of Conservation Districts who provide technical training to farmers on soil stewardship and nutrient management (Woods, 2009). Specific initiatives within California similarly combine religious or spiritual practice with land stewardship. For example, the Green Gulch Farm in Marin County, part of the San Francisco Zen Center, conducts Buddhist training and public teachings, while also encouraging volunteers to work on the organic farm and in maintenance of the larger watershed that the farm occupies, as part of their Zen practice. Examples of more secular efforts to instill stewardship values in California include the Center for Land-Based Learning (CLBL), which engages high school students in hands-on farming and conservation projects, with the explicit goal of addressing "the need to instill conservation and stewardship values in high school students", in order to meet the "needs for healthier land and more wildlife habitat" (CLBL, 2015).

In addition to shared values, a shared sense of place and a keen awareness of the local culture and ecology can also form the basis for individual and collective responses to environmental challenges. Notably, empirical evidence from social psychology studies has found that sense of place or "place attachment" can be a significant determinant of sustainable behavior for several N-related issues including water quality (Stedman, 2002; Lubell et al., 2002), climate change (Schweizer et al., 2013), and other aspects of biodiversity and ecosystem management (Cantrill, 1998). In particular, regional watershed partnerships and other cooperative institutions based on the inherent ecological boundaries of natural resources are viewed by some as an effective complement, or alternative, to the role played by federal or state regulatory agencies (Lubell et al., 2002; Snyder, 1993). Cultivating a shared sense of place can also foster a closer connection between local farmers and consumers, spur economic enterprise associated with "civic agriculture" (e.g., farmer's markets, CSAs, etc.), and ultimately encourage wider adoption of sustainable farming practices (DeLind, 2002; Rilla et al., 2011; Thayer, 2003). These lines of research highlight the merits of rigorously examining the role of cultural values in the context of sustainable natural resource management, and indicate that additional work is needed to effectively address the coupled social-ecological challenges associated with N pollution.

Of course the language of science often fails to fully capture and convey the deep cultural values that tie us to the land and motivate our collective decisions regarding our use of natural resources. This deficiency of science underscores our intrinsic need for poetry, stories, and songs that are rooted in the local landscape. Our local beat poet Gary Snyder no doubt had California's diverse cultural and ecological landscape in mind when he wrote: "If the ground can be our common ground, we can begin to talk to each other (human and non-human) once again" (Snyder, 1993). He also believed that "this sort of future culture is available to whoever makes the choice, regardless of background. It need not require that a person drop his or her Buddhist, Voudun, Jewish, or Lutheran beliefs, but simply add to his or her faith or philosophy a sincere nod in the direction of the deep value of the natural world" (Snyder, 1993).

California is gold-tan grasses, silver grey tule fog,
olive-green redwood, blue-grey chaparral,
silver-hue serpentine hills.
Blinding white granite,
blue-black rock sea cliffs,
– blue summer sky, chestnut brown slough water,
steep purple city streets – hot cream towns.
Many colors of the land, many colors of the skin
(GARY SNYDER, 1993)

CHAPTER SIX

Scenarios for the Future of Nitrogen Management in California Agriculture

Authors:
S. BRODT AND T. P. TOMICH

Contributing Authors:
J. BARNUM, C. BISHOP, AND G. HARRIS

What Is This Chapter About?

Scenarios can help stakeholders deal with controversy and complexity, and they are particularly useful in cases where there is a large amount of uncertainty, as is the case in this assessment. This chapter describes the process (overview in Section 6.1) and results (Sections 6.2 and 6.3) of a scenarios development workshop involving a diverse group of stakeholders who were asked to creatively think about the following question: *How will we manage nitrogen (N) in California agriculture over the next 20 years?* Although the starting perspectives were quite diverse, the stakeholders collectively identified two areas of high uncertainty and great influence: future profitability of California agriculture and the future course of agricultural policy, including mechanisms for implementation. This exercise led to the stakeholders developing four plausible futures of how N-relevant technologies and policies might unfold and how these would affect N management and impacts, based on different possible profitability and policy trajectories.

Stakeholder Questions

The California Nitrogen Assessment engaged with industry groups, policy-makers, nonprofit research and advocacy organizations, farmers, farm advisors, scientists, and government agencies. This outreach generated more than 100 N-related questions which were then synthesized into five overarching research areas to guide the assessment (figure 1.3). While this chapter presents possible "scenarios" of the future of N in California, it provides some insights related to the following stakeholder questions[1]:

---

1. These questions are addressed in more detail in other chapters based on historical evidence, trends, and analysis of current conditions in California.

- To what extent would policies designed to reflect the public health and environmental costs of nitrogen pollution affect food prices and farm revenues?
- How can policies account for the trade-offs between costs and benefits of N use?
- How might policy be used more effectively to both monitor and address nonpoint source agricultural pollution?

Main messages

Participants in the scenarios workshops identified the profitability of farming and environmental regulations as two of the most uncertain forces and important drivers affecting N management in California over the next two decades.

Based primarily on variations in these two attributes of profitability and regulation, stakeholders determined four potential futures for N in California agriculture. The four scenarios are the following:

1. *End of agriculture:* Rising cost and declining competitiveness for California farmers, with mandates and regulation running ahead of technological capabilities to address N issues.
2. *Regulatory lemonade:* Good prices and strong competitiveness for California farmers, with strict mandates and regulations to control N tempered by flexible implementation to allow technological capabilities to catch up.
3. *Nitropia:* Farming economics are favorable, and technological innovation spurs controls of N before there is need for regulation.
4. *Complacent agriculture:* Rising costs and declining competitiveness for California farmers, with incentives and regulation lagging behind technological capabilities to address N issues.

The four scenarios show that the environmental and human health impacts of agricultural N use could vary substantially depending on regulatory responses and the competitiveness of California's agriculture industry in the global context. The worst-case scenario, from the perspective of outcomes for agriculture, the environment, and human health, evolves from a combination of low agricultural competitiveness and low regulatory pressure to adopt better management practices and technologies, which leads to poor outcomes for the agricultural sector and mixed outcomes for the environment and human health. The two best-case scenarios in terms of outcomes involve high agricultural profitability, which stimulates investment in better management options, and either strict regulations that are rolled out in a flexible and timely manner or government policies and consumer-driven certification schemes that provide incentives for adoption, resulting in better environmental and human health outcomes.

The four scenarios collectively suggest that multiple pathways could lead to positive environmental and human health outcomes around N. On the one hand, strict regulations can force more monitoring, information management, and technology adoption, as happens in Scenarios 1 and 2, while on the other hand, agricultural profitability, often driven by consumer demand and possibly price premiums for best management practices, can also drive industry investment in development and adoption of better practices, as in Scenario 3.

The scenarios suggest that the manner in which regulations are implemented can be as important as the actual extent of regulations, and that farm profitability can be an enabler of both better N management and an outcome of N management policies. In Scenario 2, regulations are implemented with flexibility and with more advance notice and involvement from agricultural producers, allowing producers to maintain profitability while changing practices. In Scenario 1, rapid imposition of regulations decreases profitability and farmer buy-in, resulting in good environmental outcomes but poor economic outcomes for the farm sector. Differences in scenarios suggest that proactive industry participation may help agriculture to adapt successfully to a highly-regulated environment. Moreover, the scenarios suggest that farm profitability can also be an important driver or at least a critical precursor to innovation in N management, suggesting multiple feedback loops between regulatory policies, farm profitability, and N management.

None of these scenarios by themselves lead to sufficient improvement in groundwater quality to fully address human health concerns by 2030. This shortcoming is primarily due to the fact that N leaches through the soil profile at very slow rates, often taking decades to reach the groundwater. Therefore, even if all agricultural N inputs were 100% ended in 2010, the N that had already been added in prior years would continue to accrue in groundwater in 20 years' time. For this reason, regulation of agricultural N management alone is unlikely to fully address human health concerns in only 20 years, although it could improve the condition of groundwater over a longer time frame.

A historic drought affected water supply across the entire state for several years immediately following the creation of these scenarios in 2010. That extreme weather event—combining low rainfall with historic high average temperatures, has raised awareness of prospects for extreme fluctuations in climate going forward, particularly among the agricultural communities of California's Central Valley. *While these prospects for greater uncertainty about climate and water supply accentuate their importance, recent events do not significantly affect how these scenarios would be formulated.*

6.1. Using Scenarios to Establish a Common Understanding around N in California

This chapter describes a set of four scenarios developed by a diverse group of stakeholders, representing a wide range of perspectives in California agriculture, public health, and environmental advocacy (see Appendix 6.1 for a complete list of participants). These scenarios represent plausible alternative futures for the "story" of N in California, stemming from stakeholder workshops in June and September 2010, in which participants were asked to think creatively about the following question: *How will we manage nitrogen in California agriculture over the next 20 years?*

6.1.1. The Logic behind the Scenarios

Environmental scenarios are "plausible, provocative, and relevant stories of how the future may unfold" (Bennett et al., 2005) based on an internally consistent set of assumptions about how key driving forces will interact. The use of formalized scenarios development and analysis to deal with uncertainties in future trends and events began more than 50 years ago and has increasingly been used for addressing environmental uncertainties since the 1980s and 1990s. A formalized scenarios development process offers multiple benefits to stakeholders facing complex, uncertain, and potentially nonlinear changes in the environment. With uncertainty, it becomes imperative to develop adaptive decision-making that is flexible to unexpected changes, and scenarios can be an effective tool to facilitate that process (Aggarwal, 2010). They play a useful role in raising awareness and educating people about the dynamics of environmental problems, they provide scientists opportunities to explore the inter-connectedness of information from different disciplines (social and biophysical sciences), and they support strategic planning by stakeholders and decision-makers by providing insight into possible future developments (Alcamo and Henrichs, 2008).

The primary objectives for constructing scenarios as part of the California Nitrogen Assessment were to foster creative interaction and to build a common understanding around the dynamics and consequences of nitrogen management among diverse stakeholders who often hold very different views on the subject. The scenario construction process itself can enable negotiation on different views and

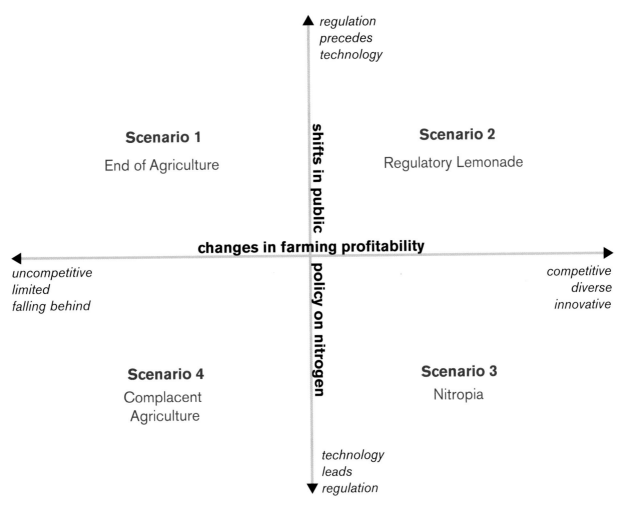

regulation
precedes
technology

shifts in public

Scenario 1

End of Agriculture

Scenario 2

Regulatory Lemonade

changes in farming profitability

policy on nitrogen

*uncompetitive
limited
falling behind*

*competitive
diverse
innovative*

Scenario 4

Complacent
Agriculture

Scenario 3

Nitropia

technology
leads
regulation

FIGURE 6.1. Four scenarios for nitrogen in California agriculture, 2010–2030.

can also build competencies in recognizing potential consequences of different driving forces and stakeholders' own actions and responses (Wiek et al., 2006). Secondarily, the objectives of this exercise were to create scenario storylines that could inform policy thinking on strategies to effectively address nitrogen-related problems.

The California Nitrogen Assessment scenarios are grounded in consideration of the many driving forces that are likely to shape the future use of N in California agriculture. During the June 2010 scenarios workshop, participants developed a list of driving forces that were grouped into seven main categories (see Appendix 6.2 for additional details about process). Many of these driving forces were similar to ones identified by the California Nitrogen Assessment (see Chapters 2 and 3). Participants then selected two driving forces as simultaneously highly uncertain and highly important: (1) changes in farming profitability and economic competitiveness and (2) shifts in the public policy of N management. These two attributes were used by four teams of stakeholder participants to populate four quadrants in which the horizontal axis represents differences in

farming profitability and the vertical axis the range of possible public policy approaches to agriculture in California (figure 6.1). The four scenarios reside within these four quadrants, differing in their driving forces and their subsequent divergent outcomes. Building the scenarios on two critical uncertainties that influence most or all of the others follows a model employed by other scenario exercises, notably the Intergovernmental Panel on Climate Change (IPCC, 2000) and the Millennium Ecosystem Assessment (Bennett et al., 2005) (see also Henrichs et al., 2010; Schwartz, 1996).

6.2. Our Four Scenarios: An Overview

Below is a brief summary of the core ideas in each of the four scenarios developed by the stakeholder groups: (1) end of agriculture; (2) regulatory lemonade; (3) nitropia; and (4) complacent agriculture. The full scenarios, describing plausible futures extending over the next 20 years, can be found in Section 6.3 of this chapter. Each of these scenarios were assessed along common characteristics that affect the nature of future conditions—these include key drivers such

TABLE 6.1
Defining characteristics of nitrogen scenarios

	Scenario 1: end of agriculture	Scenario 2: regulatory lemonade	Scenario 3: nitropia	Scenario 4: complacent agriculture
Economic landscape	High production costs	Higher global food prices; growing demand for diverse foods	Higher global demand for diverse foods; consumers drive outcomes	Dominated by global drive toward cheap food
Regulatory landscape	Increasingly strict regulations	Flexible regulatory implementation schedule	Policies are targeted based on economic feasibility	Policy driven by global focus on maintaining cheap food and high quantities; some incentive-based regulations
Public landscape and consumer behavior	Public increasingly interested in environmental and public health issues	Public helps to pay for cost of compliance	Global demand for high-end CA products increases	Consumers eat more imported food; increased public demand for cheap food
Agricultural sector outcomes	CA's total farm gate revenue declines; small farms struggle to afford cost of compliance; farm consolidation	Precision agriculture grows; farms remain viable	CA farmers maintain diverse base of crops and keep up with shifting market demand	Increase in food imports; less farming in CA as farmers leave state; farm consolidation
Health and environmental outcomes	Improved air and water quality due to fewer farm acres	Emergence of precision agriculture results in a decline in the total amount of nitrogen leaked into environment	More efficient use of nitrogen results in a decline in the total amount of nitrogen leaked into environment	Some late gains in the understanding of nitrogen science and health impacts; less farming leads to less nitrogen leakage

as the economic and regulatory landscapes, public opinion and consumer behavior, and outcomes to the agricultural sector, human health, and the environment (Table 6.1). It should be noted that, in terms of regulatory policies, the scenario-building groups focused more heavily on the outcomes of different regulatory directions and did not delve very deeply into describing the specific types of regulations that might be enacted within each scenario. More details about different possible policy approaches to managing N and an assessment of their potential effectiveness can be found in Chapter 8.

6.2.1. Scenario 1: End of Agriculture

Scenario 1 is a world in which California agriculture becomes significantly less competitive over the next 29 years, as farmers incur higher production costs driven by tighter regulations being implemented faster than farmers are able to adapt to. Due to growing environmental and public health concerns, policy-makers and regulators mandate changes in agricultural practices to reduce N applications in California.

The technology to do so proves costly as technological solutions develop slowly, with few if any clear incentives for technological innovation being offered. As scientific knowledge of health impacts becomes clearer, regulations address water quality and the buildup of greenhouse gases (GHGs) in the atmosphere. As the state's total farm gate revenue declines, farmers are unable to invest in new technology and innovative farming methods. Small farmers struggle to afford the cost of compliance. These developments trigger rounds of consolidation, a decline in crop diversity, rising unemployment for farmworkers, and a rise in the number of larger farms. Many dairies leave California, and move to states with weaker regulatory environments. Total N use declines in California agriculture as farming acres and crops decrease in number, and the state's air and water quality has improved.

6.2.2. Scenario 2: Regulatory Lemonade

Scenario 2 is a world in which California agriculture benefits from higher global food prices and growing demand for the diverse and environmentally "clean" foods grown in the state over the next 20 years. Advances in N science and public awareness drive agricultural policy, but California farmers remain competitive—with ample public investment and incentives for research and development and a favorable regulatory implementation schedule to reduce N applications that takes a long-term approach to address environmental and health impacts. New regulations also make it possible for costs to be shared by consumers, and farmers are able to meet requirements with new investments that stimulate innovation in farming practices. In this sense, they are successful in turning the "lemons" of a strong regulatory environment into the "lemonade" of new innovations that keep them competitive and their products in demand. Issues of water scarcity, population growth, and continued monitoring of N in the environment are balanced so that California agriculture is protected as a resource vital to the state's long-term economy. Nitrogen is used more efficiently and with improved scientific understanding of smart use, leading to a long-term decline in the total amount of N leaked into the environment. A form of precision agriculture expands in California farming. In the short term, excess N remains in the environment, but specific interventions increasingly protect public health and reduce GHG emissions.

6.2.3. Scenario 3: Nitropia

Scenario 3 is a world in which California agriculture benefits from two complementary trends over the next two decades: higher global demand and prices for its production, and the private and public development of new technologies and farming methods that result in sustainable N management. Regulators as well as strong consumer interest and willingness to pay provide farmers with incentives to make adjustments and invest in effective monitoring and management tools, which lead to efficient use of N and cost reductions. California farmers maintain their diverse base of crops and keep up with shifting market demand for sustainably grown and high-end foods. Air and water quality has improved due to the precision management of N fertilizer and advanced N management on dairies. While some N issues persist over the long term, policy and investment establish a clearly positive path to improve the efficiency of N management.

6.2.4. Scenario 4: Complacent Agriculture

Scenario 4 is a world in which California agriculture is unable to offset growing international price competition and high production costs with innovative farming practices. New farming technology, although available, proves expensive to farmers, which reduces the global competitiveness of California agriculture. Lacking incentives to adopt new technology and practices, many farmers leave the state or sell to larger players, resulting in higher levels of consolida-

tion. Dairies also increasingly leave California in search of more industry-friendly regulatory environments. Marketing agreements and private branding agreements emerge as a way for farmers to promote their sustainable practices, and these agreements later become the template for public policy. But policy later in this scenario is guided by three dominant themes: a worldwide focus on low prices and high quantities, policies protective of agriculture, and the consolidation of agriculture which leads to consolidation of political power among a handful of large operations. While these developments prevent the implementation of punitive regulations around N, they also fail to spur creation of adequate positive financial or other incentives for on-farm adoption of practices to address excess N application. With cheap food being the primary societal concern, farm gate revenue is low, suppressing interest and capacity to develop new practices. Water availability and land use issues continue to cause shifts in the state, affecting where and how food is grown, the most visible impacts being a reduction in the diversity of crops grown in the state. With California (and US) farming less competitive, imported food has a significant place on the American dinner plate.

6.3. Alternate Futures for Nitrogen in California Agriculture 2010–2030

Here, each of the four scenarios is described in greater detail, covering three distinct time periods from 2010 to 2030. Each scenario has distinct outcomes regarding the economic health of the agricultural sector and environmental and human health impacts of N (Table 6.2). Outcomes for the agricultural sector include factors such as crop value, crop diversity, and technology development. These economic and technological outcomes affect the flows of N in California, which in turn affect environmental and human health outcomes, including groundwater quality, air quality, and GHG emissions, among others.

6.3.1. Scenario 1: End of Agriculture

Abstract. Costs are rising and competitiveness is declining for California farmers, with mandates and regulation preceding technological capabilities to address N issues.

6.3.1.1. EARLY YEARS: 2010–2017

THE SHAPE OF THINGS TO COME: ENVIRONMENTAL REGULATIONS AND FARM CONSOLIDATION

Scenario 1 is a world where trends that emerge in 2010 gather momentum and eventually lead to a significant restructuring of California agriculture. Those emerging trends include growing competition from abroad, intense pricing pressure, and tighter environmental restrictions to address concerns for excess N in the environment and lack of voluntary adoption of better N management practices. As these trends

TABLE 6.2

Outcomes of the four scenarios affecting changes in nitrogen flows in California agriculture by 2030

	Scenario 1 End of agriculture	Scenario 2 Regulatory lemonade	Scenario 3 Nitropia	Scenario 4 Complacent agriculture
Agricultural competitiveness	↓	↑	↑	↓
Environmental regulations	↑	↑	↓	↓
Agricultural sector outcomes				
Crop value		P	P	
Livestock value		?	P	
Total farm gate revenue		P	P	
Nitrogen use efficiency in agriculture	?	P	P	?
Public investment in agriculture	?	P	P	?
Private investment in agriculture		P	P	
Agricultural technology development and adoption		P	P	
Environmental and human health outcomes				
Reducing N leakage	P	P	P	M
Groundwater quality	P	P	P	M
Groundwater quality impacts on health	?	?	?	?
Surface water quality	P	P	P	M
Air quality	P	P	P	M
Reducing GHG emissions and ozone depletion	P	P	P	M

Key		Changes in N flows in this scenario produce
Positive	P	Beneficial impact on agriculture, the environment, or human health
Mixed or neutral	M	Mixed or neutral impact on agriculture, the environment, or human health
Negative		Negative impact on agriculture, the environment, or human health
Uncertain	?	Uncertainty about whether impact exists or whether it is positive or negative

develop, farmers are forced out of the most N-intensive production systems, such as dairy, and those crops that face the most foreign competition, such as fresh-market tomatoes and other vegetables, and are pushed into consolidating.

During the early years of this scenario, California agriculture is caught up in the larger adjustments occurring in the global economy. These include economic policies to address the recession following the credit crisis of 2008 and 2009. Some of those policies are highly deflationary and, along with higher unemployment, dampen overall demand and put downward price pressure on most services and products, including food. The United States, as the world's largest economy, stands by a more open trading policy and keeps import barriers low, thereby allowing agricultural products to flow freely into the country.

THE DATA IS IN: NITROGEN'S DANGERS ARE QUANTIFIED AND THE PUBLIC PUSHES FOR CHANGE

California plays a leading role in the United States in addressing environmental issues, and, based on consumer activism, pays increased attention to the impacts of excess N in the environment. The effects of N on groundwater and air quality are more heavily monitored, and research based on these data increasingly points to negative health and environmental impacts. Over time, increased activism to protect the environment leads to increasing mandates for agriculture in the state to aggressively reduce N leakage and clean up its effects. One of the new initiatives is a tax on N inputs, levied across the board on farms and ranches. Increased expenditures for monitoring equipment are also mandated, and various other environmental taxes are imposed to fund clean-up projects. Policies tend to favor punitive regulations over innovation incentives, thus driving up operating costs for producers. These increased costs make it harder for California farmers to compete and prosper with the intense competitive pressures they face, and smaller operations increasingly consolidate or are bought out by larger farmers who can bear the costs over the long term. California's agricultural industry begins to shrink as operations move out of state, many to Mexico and Central America, where land is cheaper, labor costs are lower, regulations are fewer, and transportation into the US market is easy.

6.3.1.2. MIDDLE YEARS: 2017–2024

REAPING WHAT THEY'VE SOWN: POLICIES LEAD TO FARM SHUTDOWNS

During these years, there is considerable frustration in the California agriculture industry across all stakeholders. Farmers and dairy operators are upset over rising costs, many of them related to attempts to address environmental issues, and find it difficult to pay for adoption of the monitoring and N management practices required by new regulations. Regulators are unhappy that despite the changes made on farms and dairies, N issues are not showing dramatic improvement, essentially because it is impossible to erase over a century of synthetic N use in such a short time. Activists are unhappy because they are still seeing the public health and environmental impacts continue. Farmworkers are experiencing rising rates of unemployment as farms are closed and sold to developers. Many dairies leave California and move to states with more favorable regulatory environments. Pockets of poverty increase in Central Valley communities formerly dependent on the farm economy. Consumers are also unhappy as imported food is sometimes of lower quality and lacking in freshness. When food safety issues arise, it is often impossible to trace distribution chains and clearly identify problems.

PAST THE POINT OF NO RETURN: LAND IS REPURPOSED AND CROP DIVERSITY SUFFERS

The factors driving the changes leading to the frustration, however, are now firmly in place and in fact are gathering momentum during these years. Foreign investments into US farms outside of California are now producing returns. Federal policy supporting open trade is now locked into international agreements. Prime farmland in California is being developed into solar power installations, suburban subdivisions, and in some cases, protected habitats. Farmworkers have also migrated out of the state. Larger farms have emerged and rely on policy instruments such as incentives for conservation practices to lower some of their taxes and make ends meet. California's crop diversity has dropped significantly as production of some fruits and vegetables disappear from the state.

THE WRITING ON THE WALL: REGULATIONS AIMED AT BIG FARMS ALSO HURT SMALL-SCALE FARMS, AND TECHNOLOGICAL SOLUTIONS DEVELOP TOO SLOWLY TO HELP

California's attempt to address the N issue also runs into some challenges during these years as original cost estimates prove overly optimistic. Technological challenges also emerge as some of the hoped-for innovation proves less effective than forecast, and other precision agriculture solutions arrive too late, due at least in part to a lack of clear incentives for faster and more far-reaching innovation, and lack of sufficient public or private sector investment in research and development. Policy-makers push ahead and find ways to put additional pressure on what are now larger farms, who they believe can handle the increased costs and can seek federal assistance. Activists continue to point to "Big Ag" as the problem and continue to lobby for more regulation, such as design standards for dairy lagoons and manure handling systems, and performance standards that restrict the amount and seasonal timing of fertilizer applications and require time-consuming documentation (for a description and assessment of design and performance standards, refer to Chapter 8). These new regulatory pressures have the unintended consequence of squeezing small farms that cannot afford the cost of compliance. Because of this, many of the small-scale farmers still in business see the writing on the wall, and most of them rush to sell their land.

6.3.1.3. END YEARS: 2024–2030

THE NEW ORDER: ONLY LARGE-SCALE FARMS WITH SELECT CROPS HAVE SURVIVED

In urban areas, some people will be growing fruits and vegetables in home or community gardens. These will make up an insignificant amount of the food most people consume. But even those gardens will be restricted in their use of chemical fertilizers; gardeners will be heavily urged to use compost. On the other hand, both the conventional and organic farms that remain in California will be mostly large scale.

BETTER LIVING, AT A COST: HIGH-TECH AGRICULTURE, REDUCED GREENHOUSE GAS EMISSIONS, AND IMPROVED HUMAN HEALTH

California will also be a national leader in the implementation of environmental technologies, some which have met their promise and some which have not. The environmental technologies that failed will be seen as wasteful experiments and painful lessons learned. However, total use of N in California agriculture will have declined significantly. In most cases, California's production costs for agriculture will be among the highest in the nation, due in large part to the high costs of mandatory monitoring and precision agriculture technologies to reduce N use, and the additional fees and taxes levied on agriculture to help fund N pollution clean-up projects.

In its long-term fight to reduce the impacts of excess N in the environment, California will be able to claim some important victories. A large portion of California agriculture's GHG emissions will have been mitigated, and transportation and energy sources in the state will also contribute less to atmospheric and ground-level N pollution.

WAS IT WORTH IT?

Looking back over the past two decades, California farmers and state officials will wonder whether the big changes they have gone through were all worth the results. Food may be affordable for consumers, but food quality and safety will not have improved much, and in many cases will have declined. On the other hand, with a smaller agricultural base and some of the former cropland going into solar power generation and protected habitat, statewide water use has declined, easing some of the urban versus rural and north versus south conflicts over water allocations. However, the diversity of crops grown in California will be much lower, and many small and medium-sized farms will be lost to consolidation. Fertilizer usage on a global basis

will be much larger—but outside of California. Years of intensive scientific study has resulted in a clear understanding of nitrogen's effects on human health and the environment, but it is unclear if the responses have been proportional to the problem.

6.3.2. Scenario 2: Regulatory Lemonade

Abstract. California farmers benefit from strong prices and competitiveness, while mandates and regulation lead technological capabilities to address N issues.

6.3.2.1. EARLY YEARS: 2010–2017

SETTING THE STANDARD: HEALTHY PEOPLE AND HEALTHY FARMS

Scenario 2 is a world in which California agriculture continues to set the standards for the nation in terms of environmental safety, food quality, and the integration of technology into farming. Just as the state was the leader in setting standards for automobiles to address environmental concerns, it will also be a leader in moving the nation to precision agricultural practices. The state will combine tougher regulatory oversight, advanced technology, and consumer-supported standards to lead to a competitive, more specialized, and high-value agricultural sector. This sector feeds a population that is increasingly concerned about healthy food and is willing to pay for higher costs.

FARMERS RIDE THE WAVE OF REGULATION: PUBLIC INTEREST DRIVES POLICY CHANGES THAT THE AGRICULTURAL INDUSTRY HELPS TO SHAPE

Indications of the emerging future for California agriculture during these years include the drive for more monitoring and measuring of N. Public awareness of the health and environmental effects of excess N leakage is a key driver for policy change. Regulators take a long-term view to address problems connected to N, and believe that more work needs to be done to thoroughly understand the science of nitrogen's effects on human and environmental health. They are careful to build some flexibility and feedback loops into new regulations and set up monitoring and analysis procedures within a long-term strategy.

Farmers get a sense of what might be emerging and increase their level of understanding and sophistication in N management. In many cases, the steps needed can be readily implemented, based on already proven best management practices that until now have been poorly adopted. Farmers who are already specializing in high-value crops are taking the initiative. They take the risks that higher prices and growing markets will bear out in the long term. Larger farm operators see long-term advantages in proprietary processes that allow them to outcompete others, so some private investment is also supporting the evolving new standards. The agricultural industry stays involved in shaping new regulations, and it pays off with an implementation schedule that allows farmers plenty of lead time to make changes. Collaboration and advance notice of new regulations mean there are opportunities to prepare for the changes and compete more effectively.

6.3.2.2. MIDDLE YEARS: 2017–2024

FORGING A NATIONAL AND GLOBAL MODEL: CALIFORNIA DEFINES SUSTAINABLE AGRICULTURE

During these years, it becomes increasingly clear how strong California's influence in agricultural innovation is. The fact that the state serves such a large consumer base, has such a diversity of crops, and has the scientific and educational resources to apply to agricultural innovation and improvement becomes a dominant factor for the nation. California standards are copied by other states that do not have its advantages but want the benefits of its know-how. Global companies also take note of California's innovation and its ability to make the state's products more competitive internationally. The state's practices begin to define what sustainable agriculture is. It is information intensive, science based, and comprehensive.

THE KEYS TO SUCCESS: AN EASY-TO-USE REPORTING SYSTEM, ENGAGEMENT, AND COLLABORATION WITH FARMERS, AND INTEGRATED POLICIES BASED ON SMART SCIENCE

California takes a big step in implementing an easy-to-use online nutrient-use reporting system that contains high levels of integrated information. This is used to enforce new rules as well as reward those who comply. Over time, the system weeds out poor performing or noncomplying farming operations and leads to some consolidation. Outreach to farmers and ranchers increases as funding flows into University of California Cooperative Extension. Cooperative Extension personnel also interact more with the public to increase public awareness of critical agricultural needs and trends in use of better nutrient management practices. Although efforts are made to ease in new requirements with sensitivity and flexibility to address the needs within different crops and regions of the state, changes are still mandatory. Consumers also play an important role. New food labeling rules allow California-produced products that meet or exceed stringent new nutrient reporting and management requirements to bear "eco-California" labels that enhance consumer interest and willingness to pay higher prices. Alternative energy technologies are favored in tax policies so as to reduce energy costs.

A comprehensive view of agricultural activity in California emerges in a way that allows detailed analysis and "smart" public decision-making. An integrated approach to public policy has also taken shape, with regulatory silos consolidating and allowing for elimination of contradictory regulations and more streamlined enforcement and compliance. Measuring and monitoring instruments blanket the state, leading to real-time, in-the-field management information across a wide range of variables. This allows new guidelines and applications to be developed and used wisely. Some private firms invest profitably in the new technology and systems. Air and water quality is also monitored and integrated into local land use and transportation policies. For farmers, a new world of nutrient management is technologically enabled, implemented, and enforced. It is worth noting that this will also require an educated and trained workforce that can develop and operate the technologies and systems described.

PUBLIC BACKING FOR ENVIRONMENTAL STEWARDSHIP: STRONG CONSUMER SUPPORT FOR CALIFORNIA-GROWN PRODUCTS

The net environmental impact of much of this change has yet to show big results at this stage. Still, the true costs of cleaning up the environment are better understood and communicated to the public through increased Cooperative Extension and industry outreach to the public. Clean water infrastructure is under construction in many places and new practices are taking hold. The historical use of N remains an issue and a public health concern. Voters give a groundswell of support to the notion of California as a national leader in environmentally friendly agriculture. New policies level the playing field in California as all producers are required to meet the new nitrogen monitoring and management standards. Consumers respond with a willingness to pay more for California-grown products. Legislators pass bonds to finance the provision of water treatment infrastructure in underserved communities to address nitrate accumulated in drinking water from prior decades of agricultural N applications, and approve funding increases to UC Cooperative Extension.

6.3.2.3. END YEARS: 2024–2030

STAYING AHEAD OF THE CURVE: CALIFORNIA BEGINS DEVELOPING THE NEXT GENERATION IN PRECISION AGRICULTURE AND OTHER TECHNOLOGY INNOVATION

California agriculture is widely recognized as the leader in precision farming practices and in high-quality food products. Industry and public sector investments pay off in strengthening a world-class food industry. Other states and countries are working to catch up with California. The state begins movement into the next generation of technology and systems, which it believes will lead to a long-term reduction of excess N in the environment and reduce public health risks. Next-generation technologies have the potential to reduce the cost of, expand the use of, and improve the effectiveness of nutrient management systems.

REAPING THE REWARDS: GOOD FOOD AT REASONABLE PRICES

The changes in California farming have delivered substantial benefits to the state's consumers. Even though there has been some reduction in numbers of crops and producers who could not keep up with the changes, on balance the food choices remain high and quality and availability are unsurpassed. Jobs in the state's agricultural sector have been sources of steady employment and the state's positive trade balance in agricultural products has supported economic growth. Highly trained agricultural "knowledge workers" are well paid and are in high demand.

STORM CLOUDS ON THE HORIZON: POPULATION GROWTH, LAND-USE TENSIONS, AND COMPETITION FOR WATER CAN'T BE IGNORED

Environmental challenges remain for California, despite its progress. The overall growth in population puts growing pressure on the state's infrastructure. Water is in higher demand, its allocation is contentious, and land prices and land-use tensions increase as the state's population nears 50 million. Meeting the state's energy demand has also put pressure on land use in the case of alternative energy technologies that require large amounts of land. Removing excess N from groundwater remains a challenge, even though new technology and practices are moving in the right direction. Furthermore, while some of the technologies that have improved N management in agriculture have involved increased irrigation efficiency, increasing impacts from climate change place even more pressure on water supplies and on agricultural producers to adapt. Political activism on the N issue remains strong, and policy-makers have a larger body of science to draw upon for decision-making.

6.3.3. Scenario 3: Nitropia

Abstract: Farming economics are favorable, and technological innovation spurs control of N before there is need for regulation.

6.3.3.1. EARLY YEARS: 2010–2017

BUILDING NITROPIA: INNOVATIVE TECHNOLOGY, A THRIVING FARM ECONOMY, AND SMART POLICY

Scenario 3 is a world in which innovative technology, smart agricultural policy, and strong consumer demand for high-quality food and environmentally-sound production practices combine to usher in a new age of food production in California. Key positive trends lead to a more modern, efficient, and higher quality food system for the nation, where N is efficiently used and well monitored.

If agriculture is to be reinvented in the United States, there is no better place for it to begin than California. The state has a combination of all of the key factors: a research

and technological base in its great universities; a diverse crop base from which to learn and experiment; consumers interested in food quality and willing to pay enough to encourage growers to respond; private venture capital constantly searching for new innovation; and highly qualified regulators with the desire and capability to use science-based interventions and incentives to achieve objectives. All of these factors combine into a vibrant and innovative environment where agriculture is moved onto a more sustainable path.

A SHARED GOAL WITH A SHARED COST: CONSUMERS PAY HIGHER PRICES WHILE INCENTIVES AND REGULATIONS TAKE SHAPE

An indicator of things to come occurs in the early years, as both the state government—in response to public concern—and the agricultural sector—in response to consumer demand as well as to internal concerns about long-term business viability—invest in research focused on improving N management and N use efficiency. This investment meets with some early successes as management, monitoring, measurement, and information sharing technologies lead to better farming methods and reduced economic and environmental costs. Consumer demand for high-quality food keeps California's farmers economically competitive. New policy focuses on incentives by offering cost-sharing arrangements for farmers to adopt new technologies for monitoring crop N needs and applying fertilizer and irrigation water. Rather than mandating the use of any one technology that may not work in all cropping systems, these incentives give farmers a range of choices and enable continued diversity among the crops grown in California. Incentives also are focused most strongly in areas where public health impacts are most acute and where technical interventions are likely to be the most successful. Policy-makers also invest in new water system projects where needed, but cost-effectiveness and public health and safety issues are kept in balance. High-cost projects promising high-end results are studied closely and their risks are identified. Many high-end projects are rejected for lower-cost approaches. In addition, early research results show that urban uses of N need to be managed as much as farm-based uses, which in turn opens the eyes of consumers to N management issues and increases support for remediation projects.

6.3.3.2. MIDDLE YEARS: 2017–2024

EVEN BETTER THAN EXPECTED: TECHNOLOGY GREATLY IMPROVES N MANAGEMENT

Innovative energy-efficient technologies, new genetic research, and improved information technologies lead to a revolution in food production and consumption. The new technologies and methods exceed expectations because they are able to combine with existing processes that lead to new efficiencies and capabilities. Farmers are increasingly able to target markets, improve quality and safety, and manage their whole enterprise on a real-time basis. Biotechnology results in crops with better nutritional content and drought and pest resistance, which will allow crops to grow better under adverse conditions and recover applied N more reliably. Information and monitoring systems also allow farmers to use fertilizers more precisely by adjusting the rate and timing so that the exact quantity is applied only when needed according to the development stage of the crop. Changes in equipment also improve placement of fertilizer and expansion of minimum tillage techniques, for a combined effect of lower N applications, less N leakage into water and air, and cost savings to producers. Livestock systems, especially dairies, also benefit from cost-sharing policies that assist producers in adoption of more efficient manure management technologies. Advances in information technology enable consumers to know which crops and producers achieve the highest levels of N efficiency, thereby enabling those producers to be rewarded with customer loyalty and higher profits. These practices begin to define sustainable agriculture for the twenty-first century.

BETTER LIVING THROUGH SCIENCE: ESTABLISHING THE IDEA OF A SUSTAINABLE N BALANCE

A concept of sustainable N balance emerges in California agriculture. This idea becomes practical as information and monitoring systems are designed with a deeper understanding of the N cycle in the environment and nature's ability to recycle N. Policies and plans emerge that over the long term will slowly and eventually reverse the contamination of groundwater. Better understanding of N use emerges from science, and with the right economic incentives, proper changes can be made in agricultural practices. Farmers have so much information on the state of their crops that they are able to manage N much more efficiently with lower costs and improved food quality. Soil management also improves significantly based on research conducted in earlier years. Farmers are increasingly able to manage both soil quality and plant health.

THE WILL OF THE PEOPLE: PUBLIC HEALTH CONCERNS OVER N CONTINUE TO DRIVE CONSUMER CHOICE, AND FARMERS RESPOND TO THE CHANGING TASTES OF CONSUMERS

Progress in managing N, not only in agriculture but also in energy and transportation, proceeds as public health concerns continue to drive policy and consumption patterns. Just as people are driving more hybrid and electric vehicles during this time, they are also opting for more organic, high-quality, and resource-conserving food. Farmers are responsive to those demands because food prices allow them to succeed in meeting the changes.

During these years, momentum gathers from the positive results in new technology and farming methods. These new approaches expand rapidly in the state and throughout the nation. California becomes a world leader in innovative agricultural technology and sustainable practices. The state benefits by having continued high crop diversity, more choice for consumers, and higher food quality. California's economy benefits as agriculture, jobs, and food exports expand. The cycle of research and innovation, venture capital investment, and new business development continues to thrive in the state, with agricultural innovation playing an important role.

6.3.3.3. END YEARS: 2024–2030

TOWARD SUSTAINABLE N USE: COMBINING MONITORING, MANAGEMENT, AND TECHNOLOGY HELPS TO IMPROVE AIR AND WATER QUALITY

California moves to more efficient N management during these years. The combination of increased N use monitoring, more efficient use of fertilizers and organic N sources across the board, improved N management on dairies, reformulated fertilizers, and reduced urban use of fertilizers has begun to have an impact. Just as air quality was greatly improved with technology and changed behaviors, water quality is following suit and quality is no longer degrading. However, a few important targets remain—such as addressing rural septic systems and water treatment systems in small communities that cannot afford to finance advanced treatment on their own.

Over the previous 20 years, California agriculture has been restructured into a more high-technology, high-quality, and market-interconnected sector. Farming proves to be both profitable and innovative. The consumer market is diverse and demand is strong with both national and international sales. Demand for organic food grows so sharply that organics now account for one-third of the market share, but conventional crops also perform well by meeting increasing global demand.

FARMING EVEN FURTHER OUT ON THE CUTTING EDGE

New technology allows a balance to be achieved in keeping food costs low, while making farming more profitable in many ways. Quick-response information systems at every stage, operated by an educated and trained workforce, help direct behavior and activity. Fertilizer is applied more precisely, with application of excess N reduced by at least 50%, saving farmers money and resulting in positive environmental impacts. As a result, the total amount of synthetic N fertilizer sold in California decreases. With better information technology, food waste is reduced at the production, processing, and wholesale stages, resulting in less unharvested N staying in crop fields and less food N being sent to landfills. A sustainable food system has emerged as a balance of smart farming methods, environmental monitoring, and distribution efficiency.

6.3.4. Scenario 4: Complacent Agriculture

Abstract. Costs are rising and competitiveness is declining for California farmers, with incentives and regulation lagging behind technological capabilities to address N issues.

6.3.4.1. EARLY YEARS: 2010–2017

A SWIFTLY TILTING MARKETPLACE: HIGH IN-STATE PRODUCTION COSTS AND SLIM MARGINS KEEP CALIFORNIA FARMERS FROM CHANGING QUICKLY ENOUGH TO COMPETE GLOBALLY

Scenario 4 is a world in which slim economic margins drive how N is used in California agriculture. Few new regulations are written, and those that do emerge are paired with increasingly capable technology to monitor the environment. Despite a relatively lax regulatory landscape, rising production costs keep many California crops from competing on a national and global basis. Farmers must change their crop mixes, leave the state, and/or sell out to larger players. Federal trade and agricultural policies allow increasing imports and competition to keep food costs low.

SLOW RESPONSE TIME: POLICY IS FOCUSED ON HELPING THE INDUSTRY TREAD WATER

During these years, policy-makers shelve talks on incentives that would take aim at N management. Instead, policy is focused on sustaining a farm industry that maintains crop diversity and produces a wide range of products that consumers want. There is hope among agricultural leaders that a science-based approach will allow the state to maintain a thriving farm economy—one which will develop more sustainable methods of farming. This approach to agriculture relies on sound science, data collection, monitoring, and enforcement of existing standards. The scientific understanding of the full N cycle is progressing, but many significant questions remain. The first stages of technological research are primarily focused on monitoring and measuring, but new tools to improve N management are slow to develop and farmers lack incentives to adopt best management practices already identified. The public has an increasing interest in the monitoring of groundwater and surface water quality as it relates to N, and policy-makers again discuss incentives as a possible key to mitigation.

6.3.4.2. MIDDLE YEARS: 2017–2024

A RELUCTANCE TO CHANGE: MOST FARMERS MAINTAIN THE STATUS QUO BECAUSE OF COMPETITION FROM IMPORTS

These are the years when new tools and techniques are demonstrated and adopted by some farmers. Some attempts are made to encourage farmers to implement some of the new approaches. Where farmers see cost, marketing, or other competitive advantages, they quickly make changes accordingly. But other farmers are reluctant to change, due to their concern about higher costs which they are unable to pass on to consumers because of competition from imports. Farmers who are especially sensitive to environmental and public health concerns adopt the new approaches on long-term sustainability grounds and trust that the economics will work out. A limited pool of federal funding and regional pilot projects help support the limited spread of new farming techniques, but most farmers are unwilling and unable to change their practices without effective incentives.

SETTING A PRIVATE STANDARD: SOME FARMERS DEVELOP PRIVATE MARKETING AGREEMENTS TO PROMOTE THEIR SUSTAINABLE PRACTICES

While the pace of progress moves slowly for some, other farmers look for opportunities to innovate and compete. They also seek public acceptance of new technologies such as genetically modified crops that might be more efficient and better for the environment. Farmers who want to stay on the leading edge of farming practices forge ahead without policy-makers, and in the absence of regulations or incentives, these farmers develop private marketing agreements to promote their sustainable practices. These private standards later become the template for public policy.

THE IMPORTANCE OF GLOBAL FORCES: A WORLD FOCUS ON LOW PRICES AND HIGH QUANTITY PUTS CONTINUED PRESSURE ON CALIFORNIA FARMERS

Meanwhile, price and cost competition continues to drive the global food business. While the state features prominently in the world market for select crops, farmers in those crops find it increasingly difficult to compete in the global marketplace. Food distributors increasingly view all food products as a commodity, and strive to keep food prices as low as possible. Even though some consumers are dedicated to more costly organic and specialty foods, the majority of people on the planet are not. Most consumers are unwilling to pay premium prices for food, especially if they are not sure it has health or nutritional benefits worthy of the higher price. Government policy largely supports this consumer paradigm, with a policy focus on maintaining high quantities and low prices.

TESTING THE MARKET: FARMERS GRASP FOR OPPORTUNITIES TO TARGET LIMITED MARKETS

A pattern emerges during these years of targeting new technologies and practices to limited markets where they might be most readily accepted. This extends from biotechnology to alternative fuels and N management practices. Crops benefitting the most from these approaches and those able to pass on the higher costs are selected first for innovations.

Time and testing will tell whether innovation might expand to other areas or find limited applications only in select crops.

FEWER FARMS, LESS N: A SHRINKING AGRICULTURAL SECTOR MEANS LESS POLLUTION

Land-use patterns shift in the state due to population growth and loss of farms and crops that were unable to compete effectively on the world market. California agriculture contracts during this period, leading to lower demand for N. Additionally, dairies begin to leave the state in search of more favorable economic environments in which to operate. In the short-term, however, groundwater quality does not improve significantly because of historical accumulation of N that continues to flow downward. Pressure on regulators to address excess N remains and drives expansion of monitoring of both surface water and groundwater. Air quality issues are also a hot area of activism, and public health impacts are becoming better understood.

6.3.4.3. END YEARS: 2024–2030

BIGGER, FASTER, STRONGER: THE CONSOLIDATION OF AGRICULTURAL POWER RESULTS IN A WEAK REGULATORY LANDSCAPE

During these years, the dominance of the food industry by food retailers increases. The industrialization of food is global, and all crops are essentially commodities outside of small protected local areas where specialized quality and features command a premium price. The diversity of crops grown in the state has greatly declined and larger industrial farms, with long-term contracts and real-time information systems tied to big distributors, govern the way food is grown. Only a few small-scale farms and ranches remain viable in the state, capitalizing on their ability to exploit niche markets.

Having significant economic power, large farmers wield significant political and market power as well, and as a result, regulatory changes are negotiated to fit the needs of dominant players. Regulatory mandates are rare. Instead, incentive-based systems that leave lots of room for choice are the predominant approach. Only major health-related issues can invigorate public discourse and dramatically change the rules that govern agriculture. In this arena, food safety and availability are more powerful considerations than concerns over long-term environmental damage.

EATING OUT: FOOD IMPORTS NOW PLAY A MAJOR ROLE IN CONSUMER DIETS

With the changes that have occurred, N use in California agriculture has significantly declined, driven primarily by the overall decline in farming activity in the state. In-state crop diversity has declined and imported food has increased in market share. Food products from China, Mexico, and South America have significant places on the American dinner plate.

A SILVER LINING TO A CLOUDY ECONOMIC OUTLOOK: SOME GAINS IN AGRICULTURAL TECHNOLOGY AND N SCIENCE, ALTHOUGH PROGRESS HAS BEEN SLOW

Nevertheless, some technological advances have emerged in California's farming sector. Nitrogen is being better managed in the soil and in crops. The N cycle is better understood and its lessons applied in areas where the impacts are the greatest and where they help manage production costs. An incentive-based regulatory regime exists in the state and it is working well in many locations. There is real-time monitoring and a continuous flow of information about N application and management. However, few scientists would argue that what has been achieved is a model of environmentally sustainable agriculture. Public health risks also remain to be addressed completely. More scientific research is also needed to improve and deepen the understanding of the effect N has on human and environmental health. The political will for this additional work is yet to emerge. Complacency is reinforced by low food costs.

6.4. Discussion

6.4.1. Climate Change and Water Availability

One issue notably absent in any detail from these scenarios is the potential future effect of climate change on agriculture. Climate change is already affecting California—with sea levels on the California coast having risen by as much as seven inches over the last century, and the state's snowpack and water supply shrinking under even the most conservative climate change scenario (CARB, 2009b). Although neither the possible future effects of further climate change on possibilities for extreme events (both droughts and floods) nor the plausible impacts on water supply in California received detailed attention by our stakeholder group in these scenario exercises, these topics are covered in Chapter 2 (Section 2.3).

Although competition for water resources was mentioned as a future concern in Scenarios 2 and 4, the details of this competition and the related issues of water scarcity were not described. Legislation already in place (the "20x2020" plan, formally enacted as Senate Bill x7-7 2009) requires that state agencies must implement strategies to achieve a statewide reduction of 20% in per capita urban water use by 2020, and requires agriculture to implement efficient water management practices. The economic impact of this or future legislation on agriculture is unclear.

Additionally, other factors make the full effect of climate change on the state's agricultural system hard to predict (Jackson et al., 2009). Agriculture may experience some benefit from higher levels of CO_2, as well as longer growing seasons and the related decrease in the occurrence of freezing temperatures for sensitive crops. However, higher average temperatures may also increase pest, weed, and invasive pressures on agriculture, disturb winter dormancy in tree and vine crops, and disrupt the timing of crop pollination. Rising temperatures can also increase livestock mortality and/or decrease their productivity (CARB, 2009b).

While the effect of climate change on agriculture is not detailed in these scenarios, the scenarios suggest that agriculture may have some positive effects on climate change mitigation efforts. Most of the scenarios make some mention of a reduction in GHG emissions, but exactly how this happens—beyond the generic development of new technologies that increase N use efficiency and improve overall N management—is unclear. Presumably, such a reduction would allow agricultural GHG emissions to remain below the regulatory radar. Currently, agriculture is an unlikely regulatory target for future GHG emissions (Jackson et al., 2009) because it accounts for only 6% of the state's total emissions (CARB, 2008)—although agriculture contributes more than any other economic sector to GHG emission relative to its contribution to the economy (AIC, 2006). Moreover, agriculture may stand to benefit from climate change mitigation efforts, by sequestering carbon (C) and reducing methane (CH_4) and nitrous oxide (N_2O) emissions (CARB, 2008).

6.4.2. Trigger Point Analysis: What Could Move Our Future from One Scenario to Another?

To get the most benefit from these scenarios as "thought tools," it is useful to consider what specific trigger points or conditions would result in a hypothetical future shift from one scenario to another. Identifying such triggers builds our understanding of the defining features of each scenario, and also helps us to consider what types of real-world trends or events might be most likely to lead to substantially different future conditions.

Several participants expressed the opinion that, from among the four scenarios presented here, Scenario 2, regulatory lemonade, at its starting point, seemed to be the closest to current conditions in California, and therefore could serve as a useful baseline for comparisons. While the details may differ substantially, what Scenario 2 shares with the current situation is a combination of a comparatively strict regulatory environment and an agricultural industry that has by and large succeeded in innovating and adapting to regulations and has maintained its global competitiveness. Therefore, we use Scenario 2 as the starting point in the following analysis, in which we examine the key trigger points that would move conditions from one scenario to another.

6.4.2.1. SCENARIO 2 TO SCENARIO 1: END OF AGRICULTURE

Scenarios 1 and 2 both involve strong regulatory environments, but a key difference between them is that in Scenario 1, regulations are applied broadly without regard for differences between regions or crops, while in Scenario 2 they are implemented more flexibly, and with more advance notice and involvement from agriculture, so producers have more time to prepare and contribute to the search for workable solutions. This difference suggests that the manner in which regulations are implemented can be as important as the actual extent or "strictness" of regulations. Important triggers to transform Scenario 2 into Scenario 1 include a refusal or failure of agricultural industry groups and public agencies to work together in shaping regulations and their implementation schedules. A lack of flexibility among government agencies to be able to delegate some implementation decisions to local authorities could also be important in hindering regulations from being better adapted to different regions and crops. Pressure from the public or environmental and health advocates to apply stringent restrictions on a statewide basis could hinder government flexibility. Opposition of industry groups to all regulations, regardless of their scope, or to voluntary self-policing efforts, would also lead to a situation in which agricultural groups miss an opportunity to commit to a series of earlier, smaller, or easier-to-implement regulations that might obviate the need for harsher or broader regulations later when environmental conditions have been allowed to deteriorate further.

Consumers can also play important trigger roles. In Scenario 2, consumers are eager to purchase California products, because they understand the environmental advantage of doing so, and are willing to help pay the extra costs incurred by regulations on agriculture. In Scenario 1, cheap food imports compete with California products, and consumers apparently lack awareness, information, and/or motivations and incentives to preferentially purchase California products over imported ones. A downturn in the economy that limits consumers' willingness and ability to spend more, and advertisement that focuses on the low cost of food rather than the public health and environmental advantages of "greener" products, could reduce consumer support for farmers' costs to implement new regulations.

6.4.2.2. SCENARIO 2 TO SCENARIO 3: NITROPIA

A crucial focus in Scenario 3, in which farming remains economically strong, is that early efficiency-related technologies become available that significantly lower net costs to producers, allowing food prices to remain relatively cheap as well. These technologies help producers to remain economically viable even when some regulations do get implemented in later years. In fact, the success of N management in this scenario really hinges on the development of revolutionary new technologies that exceed all prior expectations in their capacity to improve the efficiency of N management. One crucial trigger to attain this situation is strong public and private sector investment in agricultural research and development. Additional triggers could

TABLE 6.3

Responses of different constituent groups to scenarios, and relative importance of their actions to shaping each scenario in early versus later stages of the scenario timelines

Constituent groups	Scenario 1: End of Agriculture	Scenario 2: Regulatory Lemonade	Scenario 3: Nitropia	Scenario 4: Complacent Agriculture
Agricultural Producers	Struggle to adapt to inflexible and strict regulations, or go out of business.	Adapt to regulations by adopting improvements over time; invest in some of their own tech improvements.	Invest strongly in tech improvements throughout.	Driven by import competition to increase production efficiencies over time, but improvements are small.
Consumers	Prefer cheaper food imports over CA products.	Become willing to pay for environmental and health protections over time.	Exert strong demand and willingness to pay for environment and health protections throughout.	Prefer cheaper food imports over CA products
Public sector research and extension	Develop and extend monitoring and precision ag technologies.	Lead initial innovation development and extension.	Unclear role.	Constituency needed for public investment is lacking.
Private sector technology developers	Largely absent.	Support later tech improvements.	Support later tech improvements.	Inadequate response.

Key:
◼ Crucial leading role in shaping this scenario from early years.
▨ Important role in maintaining scenario trajectory in later years.
☐ Passive, nonreactive role in shaping scenarios.

include policies that favor establishment of incentive programs, both for the development of efficiency-boosting technologies and practices, as well as for their adoption on California farms. Such incentives could be market-based (eco-labeling and branding) or could involve private and public sector competitions that reward technology developers and the producers who adopt them and can document the highest increases in N utilization efficiency. Another important trend to consider is coupling the development and release of N regulating and monitoring technologies with efficiency-boosting technologies (which may or may not be the same technologies or techniques), so that producers may be able to adopt them as a package and benefit from a boost to their bottom line, while minimizing N pollution. If the implementation of investments and incentives described above were to succeed in spurring development as well as producer adoption of new or existing approaches that significantly increase the efficiency of N management early on, then the highly regulatory approach of Scenario 2 would be unnecessary. If increases in on-farm resource use efficiency alone do not sufficiently compensate producers for costs to implement new approaches, then early and committed consumer buy-in and willingness to pay would also be an essential trigger to attain Scenario 3.

6.4.2.3. SCENARIO 2 TO SCENARIO 4: COMPLACENT AGRICULTURE

In Scenario 4, a complacent California public and its policy-makers do not follow through on emerging environmental concerns. Instead, cheap food prices and competition from imports are defining aspects of Scenario 4. Although farm profitability is not hampered by costly environmental regulations, California farmers still face difficulties competing with the large volume of cheap imported food. The trigger point in this case is marked agricultural expansion in other countries with low costs of production, as well as a consumer preference for these imported products and a lack of willingness to pay for any special "California-grown" characteristics. Another key trigger to switch from Scenario 2 to Scenario 4 would be either a cessation of a policy focus on actively incentivizing adoption of the new technologies and practices that are being developed or implementation of perverse policies that get in the way of incentives for adoption. A shift between scenarios might also hinge on large-scale farm consolidation, which solidifies the political power of a relatively small group of dominant players. Successful alliances between these players and politicians from the powerful and more liberal urban centers of the state would likely be necessary to trigger a shift to lower-intensity agricultural regulation.

As shown by these three analyses, competition from cheap imports but also consumer interests and awareness of distinguishing qualities of California-produced food can be critical trigger points that can affect the nature of future conditions. In addition, the way regulations are implemented—with sensitivity to geographic and crop variability and with adequate time for adaptation—could be just as important as what the regulations specifically require. Finally, the nature of new technology developments can also greatly influence future conditions. If technologies are as effective at increasing farm efficiency as they are in limiting N use, then they may be able to pay for themselves in terms of allowing farmers to adopt them without risking much reduction in overall farm profitability.

Finally, the role of agricultural research and how it is funded merits its own consideration. The fact that the positive aspects of Scenario 2 (regulatory lemonade) and Scenario 3 (nitropia) strongly depend on new technologies becoming available to monitor N status and regulate N management means that such research must be supported by adequate funding, from both public and private sources. Both these scenarios entail strong economic conditions for agriculture, but it is uncertain what the situation in the public sector will be. Currently, agricultural research receives support from private interests, including commodity boards, and the public sector, with the latter's share declining. With a strong farm economy, research funding generated directly from agricultural assessments, or even in-house research by agricultural and food companies, may increase. Public sector funding could also conceivably increase under Scenario 2, which has a consuming public that is highly engaged and interested in agricultural and environmental outcomes. Under Scenario 3, technology development seems to be spurred more from within the agricultural sector, and the role of the public sector in funding research is less clear. In the case of an economic downturn that cuts agricultural profits, would research and technology development continue to be funded in this scenario? Even in Scenario 2, significant strains on public coffers might constrain the otherwise good intentions of the public and policy-makers. Under such situations, the continued success regarding agricultural and environmental outcomes might hinge on new public–private partnerships that could engage new or different sources of funding, such as the food industry and private foundations.

6.5. Responses

The differences between the scenarios illustrate contrasting responses from agricultural producers, consumers, public sector research and extension, and private sector technology developers (Table 6.3). In Scenario 3, nitropia, the positive environmental and human health outcomes that stem from proactive, market-driven adoption of practices and technologies by farmers minimize the need for strict regulations as the scenario unfolds. In this scenario, farm profits obtained in early years can fund continued private sector investment in research and development for further improvements in later years. However, continued success in this scenario hinges on consumers continuing to demand, and pay for, increasing levels of environmental and human health protections associated with the food products they buy. In contrast, in Scenario 2, regulatory lemonade, environmental regulations are very strong from the beginning but are phased in to allow for adaptation. This scenario spurs technological innovation in the agricultural sector, which may initially need to be led by the public sector. Over time, successful development and adoption of innovations allow farms to remain profitable as the scenario progresses, even within a challenging regulatory environment. Success in this scenario hinges on rapid technology development and effective public and private sector extension.

The lack of profits within the agricultural sector in Scenarios 1 and 4 requires more public sector investment to stimulate progress toward environmental and human health goals. In Scenario 1, however, the emphasis on regulation without accompanying increases in farm profitability means that, in the end, large parts of the agricultural sector are lost from the state. In Scenario 4, agriculture limps along, but without regulation, environmental health outcomes also suffer, and farming operations cannot afford to make needed technical improvements. Obtaining better outcomes in each of these scenarios might hinge on better coupling of regulatory policies with opportunities to increase farm profitability over time, for example, by designing environmental policies that provide more financial incentives for farmers to adopt specific practices or achieve specific measurable environmental outcomes.

CHAPTER SEVEN

Responses: Technologies and Practices

Lead Author:
T. S. ROSENSTOCK

Contributing Authors:
HAROLD LEVERENZ, DEANNE MEYER, SONJA BRODT,
AUBREY WHITE, AND CHRISTINA ZAPATA

What Is This Chapter About?

Management practices, and their underlying technologies, together with land-use decisions, have a dramatic influence on the total amount and ultimate fate of nitrogen (N) in the environment. Based on the California nitrogen mass balance (Chapter 4), nine critical areas for intervention in the nitrogen cascade were identified. This chapter reviews these critical control points and evaluates related mitigative strategies and technological options to reduce emissions of nitrogen. This chapter also evaluates the potential for synergies and trade-offs that may occur from adopting these strategies.

Stakeholder Questions

The California Nitrogen Assessment engaged with industry groups, policy-makers, nonprofit organizations, farmers, farm advisors, scientists, and government agencies. This outreach generated more than 100 nitrogen-related questions that were then synthesized into five overarching research areas to guide the assessment (figure 1.3). Stakeholder-generated questions addressed in this chapter include the following:

- From a systems perspective, where are the control points for better management of N?
- Are there trade-offs between reduced N application and other cropping considerations? Will deviating from current N applications affect product quality, increase pest pressure, etc.?
- Are there current management practices that would increase N use efficiency and reduce N pollution?

Main messages

Today, countless technologies and practices are available to optimize reactive nitrogen (N) use and change the way Californians interact with the nitrogen cascade. Knowledge and tools to limit the introduction of new reactive N into the cascade; mitigate the exchange of N among the biosphere, hydrosphere, and atmosphere; and adapt to the increasingly N-rich environment already are widely available for agriculture, transportation, industry, water treatment, and waste processing. With current technology, we estimate that strategic actions could reduce the amount of reactive N in the environment significantly.

Limiting the introduction of new reactive N—through improving agricultural, industrial, and transportation N efficiency—is the most certain way to create win-win outcomes. Increasing efficiency would decrease the amount of N per unit activity (potentially decreasing costs) and decrease emissions. Fortunately, practices are available to increase fertilizer and feed N use efficiency for virtually every agricultural commodity. Our conservative estimate suggests gains in efficiency could result in 36 Gg less fertilizer N use per year and 82 Gg less feed N demand per year without compromising productivity. By comparison to agricultural practices, the efficacy of engineering solutions to increase efficiency is well established.

Because a single source category is generally responsible for the majority (>50%) of each N transfer among environmental systems, priorities to mitigate N emissions are clear. These include manure management (to reduce ammonia [NH_3] to air), soil management (to reduce nitrate [NO_3^-] to groundwater), fertilizer management (to reduce nitrous oxide [N_2O] to air), fuel combustion (to reduce nitrogen oxide [NO_x] to air), and wastewater treatment (to reduce ammonium [NH_4^+] to surface water). Though these activities are the most obvious, a diverse number of additional actions also contribute to these transfers and it will take a systemic perspective to reduce N emissions. Further, because reactive N is intrinsically mobile in the environment, a narrow focus on a specific mitigative action will tend to cause secondary emissions,

TABLE 7.1

Critical control points for reactive nitrogen in California

Control points to limit new N inputs

1. Agricultural N use efficiency

2. Consumer food choices

3. Food waste

4. Energy and transportation sector efficiency

Control points to reduce N transfers between systems

5. Ammonia volatilization from manure

6. Nitrate leaching from croplands

7. Greenhouse gas emissions from fertilizer use

8. Nitrogen oxide emissions from fuel combustion

9. Wastewater management

thereby transferring the burden possibly with more harmful environmental and human health outcomes.

Reactive N is already changing California's air, water, soils, and climate, and dynamics of the N cascade dictate that further degradation will continue to occur for some time. Moving forward, Californians will have to adapt systems and behavior to maintain productivity, minimize exposure, and relieve further pressure on the environment. Adaptation will be especially important as populations grow and concentrations of reactive N in the environment increase. There already is a need to treat drinking water to the regulated level (45 mg L^{-1} as NO_3^- or 10 mg L^{-1} NO_3^--N) in many parts of the state, with this need projected to increase in the future. Ozone, groundwater NO_3^-, and increased deposition may all cause changes in productivity and management.

7.0. Introduction: Critical Control Points of California's Nitrogen Cascade

Californian activities mobilize more than 1 Tg of N each year (Chapter 4). In the environment, it is transformed through physical, biological, and chemical processes enabling it to move back and forth repeatedly among the hydrosphere, biosphere, and atmosphere, where it affects human health and the environment, in both positive and negative ways. That continuous multimedia cycling is referred to as the "Nitrogen Cascade" (Galloway et al., 2003). At certain points in the N cascade, human actions or environmental conditions can modify N transformations or transfers between environmental systems. Because of their strategic importance in regulating the N cascade, these points are collectively referred to as "critical control points" (Table 7.1). Critical control points are activities, not specific technologies. Selection of the appropriate technology to accomplish the activities will be subject to constraints on prices, land, labor, and the N intensity of the activity.

Critical control points of the N cascade have been identified at national (United States), continental (Europe), and global scales (EPA SAB, 2011; Galloway et al., 2008; Oenema et al., 2011b). These assessments indicate key actions targeted at the critical control points could significantly alter the relationship humans have with the N cycle. Estimates suggest that increasing fertilizer N use efficiency, treating wastewater, reducing emissions from fuel combustion, and improving manure management would reduce the amount of reactive N released into the environment by 25–30% (EPA SAB, 2011; Galloway et al., 2008). These studies raise the question: is technology available to achieve similar or even greater control of California's N cascade, without compromising benefits of N in California?

Based on California's N mass balance, we identified nine critical control points to manage its N cascade (figure 7.1). Four of these change the demand for new reactive N and therefore alter multiple emissions pathways simultaneously. Three of these four control points affect the total amount of N required for food production through changes in agricultural N use efficiency, consumer food choices, and amount of food wasted. The fourth control point acts to reduce fossil fuel burning by improving transportation and energy sector efficiency. The remaining five control points target specific transfers of N between environmental systems, including NH_3 volatilization from manure, NO_3^- leaching from croplands, greenhouse gas (GHG) emissions from fertilizer use, NO_x emissions from fuel combustion, and wastewater management. In addition, adaptive responses to the plausible future N-rich environment are considered, including treating for NO_3^- in groundwater used for drinking, and designing N-smart agricultural systems. When possible, we provide a first approximation of the mitigative potential attainable. Additionally, we discuss the potential for synergies and trade-offs that may occur.

Two appendices support Chapter 7. Appendix 7.1 reviews specific agricultural practices and technologies that alter N cycling on farms and ranches. Appendix 7.2 outlines the calculations that support the estimates of decreases in N emissions.

7.1. Limit the Introduction of New Reactive Nitrogen

Food production, fuel combustion, and feed importation represent the three primary sources of new N inputs into California's N cascade (Chapter 4).

7.1.1. Agricultural Nitrogen Use Efficiency

Inefficient agricultural N use increases total N demand, because less of the N applied achieves its intended purpose of producing a harvestable product. Unassimilated N represents a waste of resources used to fix atmospheric N and causes indirect emissions beyond rootzone and field boundaries, with the threat increasing exponentially with excessive use

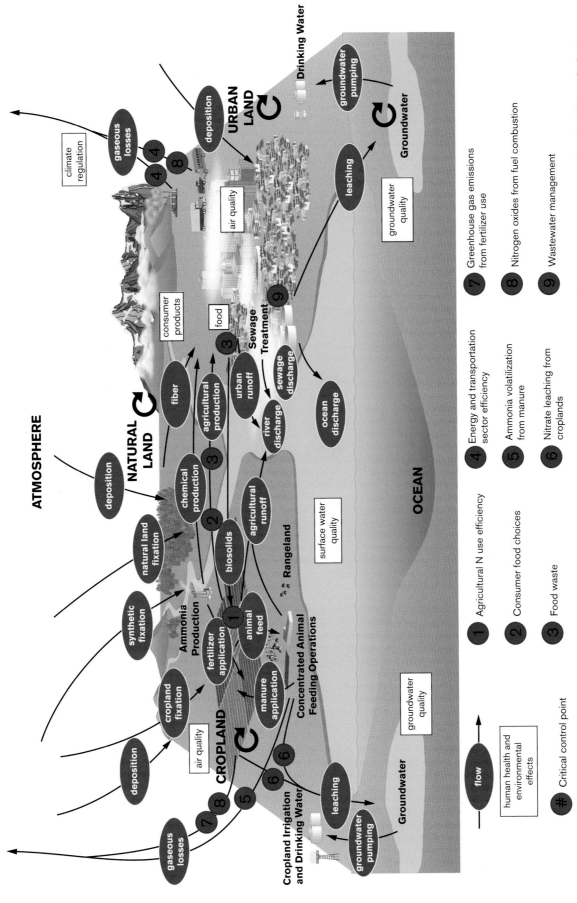

FIGURE 7.1. Critical control points for reactive nitrogen in California. See also Figure 1.1. Symbols courtesy of the Integration and Application Network, University of Maryland Center for Environmental Science.

TABLE 7.2

The mitigative effects of cropland management practices on the fate of N

Cropland management goal	Yield	Direct mitigative effects[a]				Confidence[b]		Applicable system[d]	Barriers[e]
		$\uparrow NH_3$	$\uparrow N_2O$	$NO_3^-\downarrow^c$	$NO_3^-\rightarrow^c$	Evidence	Agreement		
Nutrient management									
Reducing N rate	±	+	+	+	+	***	**	v, tv, e	Δ $_i$?
Switching N source	n	+	±	±	±	*	***	all	Δ
Changing N placement and timing	+	+	±	+	+	***	**	Lim.	$_i$? r
Water management									
Switching irrigation technology	n		±	±	+	***	***	v, tv, sb	$_i$
Increasing soil drainage	+	+	+	−	+	***	***	f	$_i$ Δ ?
Soil management									
Conservation tillage	n	−	±	+	+	**	**	f, v	$_i$ Δ t
Organic amendments and practices	±	−	±	±	±	**	*	all	t $_i$? Δ r
Diversify crop rotations	n	n	±	+	+	*	**	f, v	$_i$
Manage fallow periods	n	−	±	+	+	**	***	f, v	$_i$ $_o$
Edge of field	n	n	−	+	+	***	***	f, tv, sb, e	Δ
Agricultural residue	−	+	−	−	+	**	**	f, r	t Δ
Genetic improvement	+	−	±		+	***	***	Lim.	$_i$?

a. Mitigative effects: + = positive effect; − = negative effect; ± = uncertain; n = no effect.

b. Confidence: relates to the amount of evidence (increasing with more) available to support the relationship between practice and fate of N and the agreement within the scientific literature (* = contrasting results, *** = well established).

c. $NO_3^-\downarrow$= nitrate leaching to groundwater; $NO_3^-\rightarrow$ = nitrate surface runoff.

d. Applicable cropping systems: f = field crops, r = rice, tv = trees and vines, v = vegetables, sb = small fruit and berries, e = nursery, greenhouse, floriculture, Lim. = limited applicability.

e. Barriers to adoption: t = science and technology, $_i$ = cost of implementation, $_o$ = opportunity cost, ? = information, Δ = logistics, L = labor, r = regulations.

SOURCES: Appendix 7.1, CNA farm operator discussions, and expert opinion.

(Broadbent and Rauschkolb, 1977; van Groenigen et al., 2010). Because of inherent and to a certain extent unavoidable inefficiencies, some N leakage will occur, although clearly some individual operators are more efficient than others (Chapter 3; Breschini and Hartz, 2002; Lopus et al. 2010). More efficient use of N would enable producers to cut leakage while maintaining productivity and benefiting both farmers' bottom line and the environment (Hartz et al., 1994a, 1994b). However, fertilizer N costs are but a small portion of total operating costs (<5%) for many crops (Hutmacher et al., 2004; Jackson et al., 2003; Medellín-Azuara et al., 2011).

There are many practices and technologies to manage N in agriculture (Table 7.2). Advanced irrigation systems, crop growth and development models, reduced tillage systems, enhanced efficiency fertilizers, precision feeding, staged feeding, hormones, breeding, and animal husbandry are only a few of the available approaches that have been tested. Today, producers can select from a diverse menu of options to fine-tune N use in their systems (Hristov et al., 2011; Nahm, 2002; Ndegwa et al., 2008). Production decisions, however, are subject to multiple constraints—land, water, economic costs and returns, regulations, technology, etc. Until recently, fertilizer and feed N were relatively cheap production insurance, and little attention was paid to the environmental externalities and social costs resulting from N pollution (Meyer, 2000; VandeHaar and St-Pierre, 2006). At present, control of N pollution is a major driver of production decisions for only a few systems in California (e.g., dairy). For N use efficiency to increase, consideration of N leakage will have to be integrated into operational decision-making.

Organic fertilizers are thought by some to be more environmentally benign than synthetic fertilizers. While their advantages are not always so clear-cut (see Appendix 7.1), a simple mass balance exercise to consider what it would take for California agriculture to "go organic" reveals that current statewide manure production would only provide 45% of the N needed by the current crop mix, assuming a 54% partial nutrient balance (Chapter 3). Making up the difference with leguminous crops and green manures would require from 1.6 to 6.9 million hectares planted to these crops (depending on species and N fixation rates), or 32–141% of the currently irrigated cropland. In short, converting all or a substantial percentage of California agriculture to organic N sources would require significant changes in cropping systems, livestock populations, or both (see Appendix 7.3 for more details on this analysis).

For nearly all cropping systems, N use efficiency is consistently higher in plot and field-scale research trials than the documented statewide average, frequently considerably so (see Table 7.2 in Appendix 7.2). These data suggest that it is possible to increase agronomic N use efficiency, expressed as a partial nutrient balance, significantly.[1] Assuming yields do not change, raising N use efficiency even half[2] as much as this amount could decrease inorganic N fertilizer demand (and application) by 36 Gg N yr^{-1}. As a result, it is reasonable to expect at least proportional reductions in emissions (8 percentage points), much of which would likely be translated into reduced NO_3^- loading to groundwater, given the relatively large fraction of applied N lost directly through this pathway. Because emissions increase exponentially after N application rates exceed crop uptake, this may be a conservative estimate (Sections 7.2, 7.3, Appendix 7B). The fact that recorded statewide average N use efficiencies are almost universally less, across crops, than efficiencies achieved in research trials, suggests that neither technology nor scientific information is the primary impediment to N-efficient California croplands.[3] Future efforts to increase N use efficiency will have to extend beyond the development of new technological innovations to include socio-economic drivers of technology adoption and use (e.g., Jackson et al. 2003).

It appears feed N utilization efficiency in California animal production systems can also be improved, at least incrementally. Because data are sparse, we conservatively assume that the increase could be at least 4 percentage points. Even such a modest increase would have significant consequences for feed N demand and management of manure N. Assuming that product yield and N concentration remain constant, feed demand would decrease to 85% of current levels (equivalent to an 82 Gg N decrease). At the same time, emissions reductions from avoided fertilizer use and biological N fixation in feed production and the manure N burden would be reduced proportionately.

Increasing agricultural N use efficiency has the potential to create win-win outcomes for the producer and the environment. Better N management may add to labor and material costs for producers, however. Some studies suggest that incremental improvement may be achieved with little added investment (Medellín-Azuara et al., 2013; Schaap et al., 2008). And it is likely that the total investment would be less than the potential resource degradation and health costs caused by N overuse (Chapter 5). Therefore, agricultural N use efficiency appears to be the cornerstone of any strategy to slow the N cascade.

7.1.2. Consumer Food Choices

Demand by US and global consumers shapes farmers' decisions on both what crops to produce and how to produce them. Because foodstuffs differ in their N content and in the amount of N required to produce them, consumer preferences for specific commodities can have a large influence on local, statewide, national, and global N cycling.

Animal products are the least technically efficient foods in terms of the amount of N required to produce each unit of final food N (or protein) consumed, due to the basic biological inefficiencies that occur when animals that have consumed plants are in turn consumed by humans. These inefficiencies are due to the fact that the majority of the N used to produce feed crops—estimates indicate that it can be over 90% (Galloway and Cowling, 2002)—is lost to physiological maintenance of the animal, manure, and other avenues, with only a small amount making it all the way to the consumer's plate (see Box 5.1.1. for more detailed estimates on the percent of feed N that is eventually consumed as meat products). For this reason, consumer demand for animal products is one of the most important factors affecting the introduction of new N. Three distinct sets of consumer choices about animal products would yield considerable reductions of new N. First, consumers could limit their choices to those animal products that are physiologically more N efficient (e.g., require less N per unit of final food product produced), such as poultry (Pelletier, 2008). Second, consumers could choose foods from livestock that are raised using lower inputs of new synthetic N, such as livestock finished on unfertilized rangeland rather than in confined facilities requiring fertilized feed crops. The drawbacks of this option might include limitation in available rangeland (likely only an issue in the case of very widespread consumer adoption of this option), higher production costs leading to higher food prices, and potentially

---

1. Calculations based on Tables 3.1 and B7.2 suggest a potential increase in NUE of 16 percentage points, based on an area-weighted average for 33 crops. Potential increases vary significantly among crops, with some being far less.

2. NUE in research trials is always greater than that obtained in field production, sometimes considerably so, because of technical inefficiencies. To account for this, we have suggested that technical potential of increasing NUE are half of the calculated differences. This is likely a conservative estimate but represents a starting point for discussions.

3. Results must be interpreted with caution. Estimating NUE by partial nutrient balance (PNB) is unable to distinguish between soil and fertilizer N in the plant. Indigenous soil N contributes variable quantities of N depending on the fertility of the soil potentially confounding the comparison. Research sites may perform better due to underlying soil fertility. Regardless, in virtually every crop examined, statewide average partial nutrient balances were lower than recent research using feasible production practices, sometimes by quite significant amounts, irrespective of crop type.

higher GHG emissions, especially methane, from range-fed cattle compared to feedlot cattle. When compared with beef cattle raised on highly managed pastures, some studies showed that those finished in feedlots resulted in lower system-wide emissions (Pelletier et al., 2010), while some studies of dairy systems (O'Brien et al., 2012; Rotz et al. 2009) found that the pasture-based systems resulted in lower overall GHG emissions. On the other hand, Arsenault et al. (2009) found no major differences in emissions between pasture-based and confined dairy systems. To date, similar comparisons have not been examined for nonruminant livestock, such as chicken.

The third consumer option is to lower animal protein intake. Average US consumers, and by likely extension Californians, consume more than double the recommended levels of annual protein intake, 63% of which comes from animal products (USDA, 2011b, p.20). Moreover, dietary patterns that include less processed meat and red meat, and more plant foods, are generally accepted in the medical literature as being associated with decreased risk of cancer, cardiovascular disease, and other diseases and mortality risk factors (Kushi et al., 2012), providing a health incentive for this choice.

Lowering animal protein consumption would not likely reduce N loss in California proportionally, however. Often diets low in animal protein contain greater proportions of fruits, vegetables, and nuts, many of which require high N inputs and are grown in California. In contrast, slightly over one-third of the N fed to California livestock comes from feed crops imported from other states (Chapter 4). Thus, decreasing animal protein intake may lead to trade-offs, especially pertinent to the California agricultural landscape. (It should be noted, however, that reliance on imported feed does not really eliminate the N impacts; it only exports them out of California.) Nevertheless, because the quantity and quality (e.g., more proteins, fruits, and vegetables) of food demand increase with population growth and affluence (Dawson and Tiffin, 1998), both of which are projected to increase significantly in the future, the importance of shaping diets towards low resource intensity foods is clear (Hall et al., 2009).

7.1.3. Food Waste

Addressing food losses may also play an important role in reducing the N loading in the cascade. Food losses represent a waste of fixed N since the fertilizer and feed N either are not consumed or are discarded into the environment. Nitrogen released from decomposing organic materials in the field or landfill contributes to air and water pollution and climate change. Reducing losses, therefore, shrinks resource demand and decreases pressure on the environment.

Food losses occur across every stage of the supply chain: from production through consumption. Food losses at retail and consumer stages alone have been estimated to reach approximately 27% in the United States (Kantor et al., 1997). Food losses for individual highly perishable products—such as ones produced in California—can be even higher. Dairy

products and fresh fruits and vegetables accounted for half of retail losses in 1995 (Kantor et al., 1997). Consumer losses for dairy products range from 8% to 51% and for fresh fruits from 8% to 54% (Muth et al., 2011). Though the extent of food losses and waste in California have not been quantified, these findings clearly indicate that when farm, retail, and home wastes are added together, a nontrivial fraction of agricultural products go uneaten.

Not all food loss is suitable for consumption, thus N wastage via this mechanism cannot be reduced to zero. However, clearly there are opportunities to recover food at most stages of the supply chain. Although data are unavailable to estimate exactly how much food goes unharvested, California crop producers often abandon significant fractions of production due to pests, costs, market, or weather constraints. Creation of incentives to harvest less desirable products would increase the quantity of food in the market and potentially have ramifications for N cycling. Recent interest in capturing on-farm food losses has catalyzed charitable "gleaning" crews across the state. Farmers who donate production that would have otherwise gone to waste can receive tax benefits. Gleaning results in greater export of N off-site, reducing the soil pool of N and the environmental N burden. But current levels of such harvest are miniscule relative to the total amount of loss.

Consumers' waste, rather than retail waste, dominates postproduction food waste, comprising 96% by one estimate (Kantor et al., 1997). The consequence is loading of landfills with food waste. In California, food waste accounts for 24% of landfilled materials, despite extensive composting and recycling efforts (Adams et al., 2009; Integrated Waste Management Consulting, 2010). Finding ways to reduce waste would reduce the N load in landfills and recycle food-N to the soil. A diversity of issues contributes to high consumer food waste, including over-preparation, cooking losses, spoiled leftovers, and faulty packaging. Two mechanisms of behavioral change would have a positive effect. First, reducing the amount of food that enters the waste stream could be achieved through education on storage times, improved packaging, and smaller portions. Second, education on composting and disposal would also be beneficial. Engineered behavioral solutions are another option. For example, cafeterias that eliminate the use of trays (reducing the customer's ability to carry more than one plate at a time) have documented reductions in food waste (Hackes et al., 1997).

7.1.4. Energy and Transportation Sector Efficiency

Reactive N released from fuel combustion has far-reaching consequences on air quality, human health, and downwind ecosystems, especially with California's hot and dry climate and in highly N-limited ecosystems (Chapter 5, Section 5.3). California has long recognized the major impact of fossil fuel combustion on air quality and has led the nation in combating emissions, primarily of NO_x. However, NO_x as well as secondary air pollutants derived from N emissions (i.e., ozone and $PM_{2.5}$ and PM_{10}) continue to plague the health of Californians and our natural ecosystems (Chapter 5, Section 5.3). Thus, decreasing emissions further remains a critical goal.

Efforts to minimize these nitrogen emissions can be divided into two major categories—decreasing emissions from fuel combustion and decreasing the overall amount of fuel combusted. Control technologies decrease emissions by transforming nitrogen emissions into nitrogen gas (N_2) or filtering nitrogen-containing particulate matter (PM) out of the exhaust before release into the atmosphere. Major steps have been taken to reduce tailpipe emissions—from 1999 to 2011, particulate matter dropped by 47% in the Los Angeles air basin and by 26% in the San Joaquin Valley (CARB, 2013). The potential for further improvements in these control technologies is limited (Section 7.2.4).

For more drastic change, like that proposed in California's plan to reduce GHG emissions to 1990 levels by 2020 (CARB, 2014a), it is generally agreed by most that decreasing fuel combustion altogether will be key to major reductions in GHG emissions and other N-based pollutants. Alternative fuels and alternative vehicles offer promising pathways to such improvements, and will be required to achieve deep reductions in N emissions without reducing vehicle demand. But such improvements are complicated by upstream emissions from power generation. Research to understand how nitrogen emissions are affected upstream is still cursory, as life cycle assessments of emissions generally focus on CO_2 and N_2O, and often do not include other nitrogen species. The nitrogen-relevant factors of these technologies are assessed below, with particular attention paid to upstream emissions that can be decreased by improved efficiency in electricity generation.

7.1.4.1. FOSSIL FUEL USE SUBSTITUTION IN VEHICLES

Technologies currently on the market or on the horizon include hybrid electric vehicles (HEVs), plug-in hybrid electric vehicles (PHEVs), full electric vehicles (EVs), fuel cell vehicles (FCVs), flex-fuel vehicles (FFVs) (designed to run on gasoline of a blend of up to 85% ethanol), and compressed natural gas (CNG) vehicles. In addition to these alternative designs, the use of ethanol and biodiesel fuel blends is expanding as a GHG reducing measure. The timeline between research and development of new vehicles and 50–75% market penetration may be as long as 50 years, and requires policy development to both push for technology improvement and create the infrastructure to support major changes in the vehicle fleet, including sufficient charging stations for EVs and hydrogen storage for hydrogen FCVs (Ogden and Anderson, 2011).

While CO_2 emissions are relatively simple to estimate (as they are directly related to the carbon content of fuel), nitrous oxide is significantly more difficult to calculate and makes estimating the emissions of alternative fuels and vehicles hard to track. N_2O emissions are dependent on fuel combustion temperature, pressure, and air-to-fuel ratio. Despite decreases in direct emissions from alternative-fuel

vehicles and technologies, additional emissions stem from a variety of upstream processes such as resource extraction, electricity production, fuel transport, and fuel distribution. The time of day vehicles are charged presents a major uncertainty in measuring emissions. If the majority of PHEVs are charged at night, as many studies assume, their emissions will be dependent on the fuel mix used in the marginal electricity generated at the end of the day or at nonpeak times. If marginal electricity is derived from renewable sources, emissions will be lower than if marginal electricity comes from coal-fired power plants or similar sources. Other variations in emissions can stem from driving patterns (such as length of trip) as well as the size of the vehicle itself (Lipman and Delucchi, 2010).

Numerous life cycle assessments have been conducted to assess emissions levels from alternative fuel vehicles and the potential reduction that can come from improved fuel sources. Table 7.3 compares several studies' estimates of the decrease in emissions from different vehicle types compared to the conventional internal combustion engine.

While some life cycle assessments account solely for carbon dioxide emissions, the GREET model, created by the Argonne National Laboratory (https://greet.es.anl.gov/), accounts for N_2O emissions as well as other GHGs, and represents cumulative emissions decreases as carbon dioxide equivalent (CO_2e) amounts. While N_2O is included in the GREET model, individual pollutants are generally not described in well-to-wheel vehicle studies.

The GREET model estimates that, with the existing California energy mix, which is largely produced by natural gas and renewable fuel sources, EVs can reduce life cycle GHG emissions compared to conventional internal combustion engine vehicles by about 60%, while FCVs using H_2 derived from natural gas can reduce life cycle emissions by 50% (Lipman and Delucchi, 2010). However, if that grid mix has a higher dependence on coal-based electricity generation than the California mix, EVs could result in an overall *increase* in GHG emissions. With an entirely renewable fuel source, EVs and FCVs could nearly eliminate GHG emissions (Lipman and Delucchi, 2010).

Electric vehicles also reduce NO_x emissions. The American Council for an Energy-Efficient Economy estimates that an all-EV fleet powered by the average California power mix generates 2.3 lb (1 kg) NO_x over the course of a year (12,000 mi),[4] compared to 16–20 lb (7–9 kg) NO_x emissions from conventional vehicles. Hybrid vehicle NO_x emissions are estimated at 11 lb yr^{-1} (5 kg yr^{-1}), and PHEVs using a California energy mix see a 40% reduction in NO_x emissions from today's hybrid vehicles (those with an average range of 50 mpg). These estimates, however, can be affected by the fuel efficiencies of different vehicles, as well as the time of day vehicles are recharged (unaccounted for in these estimates) (Kliesch and Langer, 2006).

---

4. Estimates do not include emissions upstream from electricity generation, such as mining and material transport.

TABLE 7.3

Estimates of emissions reductions of select alternative fuel vehicles compared to standard vehicles with gasoline internal combustion engines (ICE)

Vehicle type	Pollutant	Grid	% decrease from ICE	Source
HEV	Annual NO_x	CA	41%	Kliesch and Langer (2006)
HEV	Life cycle CO_2e	Avg. US	20–25%	Samaras and Meisterling (2008)
HEV	Life cycle CO_2e	Low carbon US	30–47%	Samaras and Meisterling (2008)
PHEV	Life cycle CO_2e	Avg. US	32%	Samaras and Meisterling (2008)
PHEV	Life cycle CO_2e	Low carbon US	51–63%	Samaras and Meisterling (2008)
PHEV	Annual NO_x	CA	65%	Kliesch and Langer (2006)
EV	Annual NO_x	CA	88%	Kliesch and Langer (2006)
EV	Life cycle CO_2e	CA	60%	Lipman and Delucchi (2010)
FCV	Life cycle CO_2e	CA	50%	Lipman and Delucchi (2010)

NOTE: Comparisons of CO_2e emissions are based on whole vehicle life cycles, including both manufacture of the vehicle and standard mileage for a lifetime of usage. Comparisons of NO_x emissions are based on annual standard mileage assumptions only, not counting upstream emissions.

HEV = hybrid electric vehicles; PHEV = plug-in electric vehicles; EV = full electric vehicles; FCV = fuel-cell vehicles.

Despite these uncertainties, it is generally agreed that the use of renewable energy sources will decrease life cycle emissions, but that using an electricity mix derived largely from coal-fired power plants and other nonrenewable sources has the potential to increase GHG emissions. Long-term modeling using the Lifecycle Emissions Model (LEM) shows the greatest potential GHG reductions from hydroelectric, nuclear, and biomass energy sources (Lipman and Delucchi, 2010). California's 2013 in-state power generation included 60.5% natural gas, 8.9% nuclear power, 10.4% large hydropower, 0.5% coal power, and 19.6% renewable power (CEC, 2015). Statewide use of renewable power (in-state generation and imports from out of state) totaled 18.7% of total electricity use in 2013. Governor Jerry Brown has mandated an increase to 33% renewable power use by 2020, which will bring significant increases in the efficiency of HEVs, PHEVs, and EVs.

7.1.4.2. WELL-TO-WHEELS ANALYSIS OF BIOFUELS

Biofuels are frequently discussed as a renewable fuel source and a potentially GHG-neutral alternative to fossil fuels (Chum et al., 2011). Substituting biofuels for gasoline potentially can reduce GHG emissions if one only focuses on the potential of feedstocks to replace fossil fuels and sequester carbon during the plant growth phase (Searchinger et al., 2008). However, when examining soil N_2O emissions induced by fertilizer use, all the upstream emissions associated with inputs, as well as other indirect effects of biofuel production, it is generally accepted that life cycle GHG emissions for common biofuels, especially corn ethanol, can be higher than those for fossil fuels, especially when considering global land-use changes (NRC, 2011; Searchinger et al., 2008). For example, Searchinger et al. (2008) found that the diversion of existing cropland into biofuel production trig-

gers rising crop prices which in turn induce farmers around the world to convert forest and grasslands (i.e., systems that are already providing carbon storage and sequestration potential) into cropland. Similarly, assuming a conversion factor of 3–5% from synthetic fertilizer N to nitrous oxide (N_2O) from crop production systems, it is agreed but unproven that the next-generation unfertilized cellulosic crops, such as perennial grasses and woody plants, are likely to provide substantial positive net benefits in reducing GHG emissions from fuel use (Adler et al., 2007; NRC, 2011).

It is suggested but unproven that some of these same alternative biofuel crops in California could help manage additional environmental problems and contribute to overall agricultural sustainability. For example, switchgrass, one of the perennial cellulosic crops, is very salt tolerant and therefore useful in areas such as the western San Joaquin Valley, where high salinity impedes production of other crops (Kaffka, 2009). In addition, safflower and sugarbeets, two alternative biofuel crops, can play a useful role in rotation with more valuable crops (e.g., tomatoes or cotton) as they can better utilize water and N stored at greater soil depths (Kaffka, 2009).

7.1.4.3. FUEL COMBUSTION IN STATIONARY SOURCES

Like mobile sources, stationary modes of fossil fuel combustion will benefit from increased use of renewable energy. That energy can come from new electricity sources—including wind power, solar, hydro and fuel cell. Improvements in power plant design that incorporate cogeneration or a gas-fired combined cycle can also increase overall efficiency. Both systems are designed to use excess heat as steam power. Reductions in NO_x from these designs will depend on what technology is being replaced (Bradley and Jones, 2002).

7.1.4.4. REDUCTION IN TRAVEL DEMAND

AB32 mandates that GHG emissions levels in California decrease to 1990 levels by 2020. Additionally, California set a goal to drop emissions by 80% of current levels by 2050—a goal often referred to as 80in50 (Yang et al., 2009). Yang et al. (2009) model different strategies by which emissions could be reduced so drastically. Their scenarios, which model reductions only for in-state emissions (travel that originates and terminates within California), show that no single strategy for emission reductions can meet the 80in50 requirements, but that there are multiple strategies that can succeed together. In all three strategies examined, Yang et al. found that light-duty vehicle technologies will need to bring the majority of changes, using a combined strategy of fuel efficiency of vehicles and carbon intensity of fuel generation. Biofuel- and EV-heavy scenarios bring the most significant change to GHG emissions. However, as stated above, heavy reliance on biofuels may have trade-offs in N emissions.

A key element in one of Yang et al.'s scenarios is a decrease in travel demand. A reduction in travel demand is one alternative to reduce GHG emissions without changing fuel, mode, or vehicle technology. The scenario suggests that a decrease in travel demand should account for nearly one quarter of emission decreases (based on Yang et al.'s reference scenario). Achieving such dramatic decreases will require changes in the built environment that allow people to travel more easily without the use of passenger vehicles—including building more densely, increasing access to public transportation and potentially adding costs to driving (higher taxation on gasoline and parking costs). Bringing significant change from these measures will not be easy. Heres-Del-Valle and Niemeier (2011) suggest that decreasing vehicle miles traveled (VMT) by as little as 4% may require residential density increases of up to 29%, or increases in gasoline prices by 27%. Other studies show that public responsiveness to increases in gasoline prices is limited, and has reduced over time (Hughes et al., 2008; Small and Van Dender, 2007). In addition, without improved public transportation infrastructures, higher gasoline prices may disproportionately affect lower-income households who lack access to public transportation or must commute long distances to work. To adequately address emissions from fossil fuel combustion, however, will require a suite of changes not only to the technologies we use to combust fuel, but also in the lifestyles that depend heavily on fossil fuel combustion for transportation.

7.2. Mitigate the Movement of Reactive Nitrogen among Environmental Systems

Critical control points (Table 7.1) exist in other parts of the N cascade, beyond the introduction of new reactive N. Once N has already been "fixed," by natural or industrial means, or released via fuel combustion, it is still possible to mitigate its impact. Generally, each of the major N transfer pathways is dominated by a single activity. For example, animal manure management and fuel combustion are the primary sources of NH_3 volatilization and NO_x emissions, respectively. The overwhelming importance of certain activities for specific N species suggests clear research, outreach, or policy priorities to target these concerns.

7.2.1. Ammonia Volatilization from Manure

Manure N that results from dairy, beef, egg, and meat bird production contributes a large portion of NH_3 emissions to California's atmosphere and impacts air quality and the health of downwind ecosystems (Chapters 4 and 5). Therefore, becoming more N sustainable in California requires reducing NH_3 volatilization from manure. Fortunately, many tactics already exist to reduce NH_3 emissions from animal manures, including frequent manure collection, anaerobic storage, composting, precision feeding, and use of nitrification inhibitors (Ndegwa et al., 2008; Xin et al., 2011). Unfortunately, relative changes in emissions rates from either the common manure management systems (Chapter 3) or "alternative practice"' are not well understood for the climatic and production conditions characteristic of California animal production systems (Table 7.4) (San Joaquin Valley Dairy Manure Technology Feasibility Assessment Panel, 2005b). For example, on the one hand, more frequent flushing of freestalls transfers reactive urea N to the lagoon where depth and pH restrain volatilization. On the other hand, manure deposited in freestalls is collected with recycled wastewater, spreading urea and NH_4^+ thinly over the concrete/soiled surface and creating conditions conducive to NH_3 emissions (expansive boundary layer, wind, increased total ammoniacal N). Levels of uncertainty about emissions from open lot dairies or poultry facilities are similar. Effects on NH_3 emissions of hot, arid conditions in dairy corrals of the Tulare Lake Basin, or of changes in layer housing structures are not well understood. One study from Canada, which has similar poultry production systems as California, has shown that layers housed in larger cages, where birds had more space, had a similar nitrogen utilization efficiency (35%) as layers housed in conventional cages (36%) (Neijat et al., 2011), but it is unclear how specifically NH_3 emissions would be affected by the change in housing.[5] So while there are many possible actions operators might take to control NH_3 (Rotz, 2004), the extent of their applicability to California production systems is suggested but unproven.

In spite of the uncertainty in emission rates and the variation among operations, evidence suggests there are opportunities to reduce NH_3 emissions from manure management in California. Dairy production creates 79% of

---

5. As of January 1, 2015, the California Shell Egg Food Safety regulation (3 CCR 1350) requires egg producers to provide a new minimum amount of floor space per egg-laying hen (see CDFA, 2013).

TABLE 7.4
Anticipated effects of dairy manure management technologies

Animal management goal	Yield	↑NH₃	↑N₂O or NOₓ	NO₃⁻↓ᶜ	NO₃⁻→ᶜ	Evidence	Agreement	Potential systemᵈ	Barriers to adoptionᵉ
		*Mitigative effects*ᵃ				*Confidence*ᵇ			
Feed management									
Precision feeding	+	+	+	+	+	**	***	d, b, p	Δ $ᵢ ?
Supplements and hormones	+	+	+	+	+	**	***	d, b, p	r
Manure storage and treatment									
Frequent manure collection		+	±	+	+	*	***	d, b, p	$ᵢ
Solid–liquid separation		+	+			***	***	d	$ᵢ Δ ?
Composting manure solids			>			**	*	d, b, p	$ᵢ Δ L
Biological additives for wastewater		±	±						$ᵢ, t
Anaerobic digestion of wastewater		+	±			**	***	d	$ᵢ, r
Storage cover for wastewater ponds		+				*	***	d	$ᵢ
Land application of manure									
Measured applications and flow meters		±	±	+	+	**	***	d	$ᵢ
Split applications		±	±	+	+	**	**	d	$ᵢ Δ
Incorporation below surface		+	+	−	+	***	***	d, b, p	?
Species improvement									
Genetic improvement	+					***	***	p	$ᵢ t?

a. Mitigative effects: + = positive effect; − = negative effect; > = minimal impact; ± = uncertain; n = no effect.

b. Confidence: Depends on the amount of evidence available to support the relationship between practice and fate of N and the amount of agreement within the scientific literature (* = little evidence or contrasting results, *** = strong evidence and high level of agreement).

c. NO₃⁻↓ = nitrate leaching to groundwater; NO₃⁻→ = nitrate surface runoff.

d. Potential systems: d = confined dairy; b = beef feedlot; p = poultry; c = grazing cattle.

e. Barriers to adoption: t = science and technology; $ᵢ = cost of implementation; $ₒ = opportunity cost; ? = information; Δ = logistics; L = labor; r = regulations.

SOURCE: San Joaquin Valley Dairy Manure Technology Feasibility Assessment Panel (2005a). See Appendix 7.1 for detailed discussion of practices.

statewide manure N and hence dominates NH₃ production. The University of California Division of Agriculture and Natural Resources Committee of Experts reported estimates of NH₃ losses on a typical dairy in the Central Valley, including NH₃ volatilized from the production unit and during land application. While these estimates contain some uncertainty, the reported range of volatilization is approximately 25–50% of excreted manure N, a 100% difference between the least and greatest producers (Committee of Experts on Dairy Manure Management, 2005b). The wide distribution indicates there is substantial room for improvement, especially for the operators with the highest emissions rates. Assuming that extreme rates are not very common (e.g., emissions are normally distributed) and there is a differential in potential improvement because of the wide distribution, we suggest that NH₃ volatilization from manure can be reduced by approximately 4 percentage points on average and in total 10–15 Gg N yr⁻¹ given current manure deposition rates (Appendix 7B).

Reducing NH₃ emissions from animal production units requires a whole-farm approach (Castillo, 2009). Manure management involves a series of complex unit processes that link together to collect, process, treat, and store manure, with volatilization taking place throughout (EPA, 2004b). When volatilization decreases at any stage, N is conserved and transferred to the next process increasing the total N pool and the potential for emissions in subsequent stages of treatment and disposal. While N conservation is a laudable goal, it must be recognized this ultimately increases the N utilization burden on animal production systems and potentially requires more land or capital for distribution. There is a need to better develop and build the evidence base for N conservation throughout entire manure management trains, as opposed to only for individual practices, and to identify the best leverage points to reduce losses. Such a reduction would require a significant effort by dairies to distribute and recycle the additionally conserved N. On a positive note, the N conserved would largely be in the urea or NH_4^+ form, which has higher fertilizer value because it is relatively plant available in comparison to organic N.

One primary constraint to the mitigation of NH_3 emissions from manure management is the cost of control technologies for the producer. Often the changes required increase the producer's cost of production, be it additional labor, more machine operating time, or monitoring and record keeping. The ability for producers to absorb additional costs of NH_3 management is questionable given the thin profit margins characteristic of recent milk markets, evidenced by the decline in numbers of dairies in the state.

7.2.2. Nitrate Leaching from Croplands

It has been well established by a range of studies that historical and contemporary cropping practices are responsible for widespread groundwater pollution in California's agricultural valleys (see Chapter 4; Burow et al., 2007, 2008b; Fogg et al., 1998; Harter et al., 2012; Miller and Smith, 1976; Rosenstock et al., 2014). Because of the long time lag between cause and effect—commonly 5–50 years or more in California—reducing N loading to groundwater from croplands will not decrease groundwater NO_3^- concentrations in the short term (Dubrovsky et al., 2010; Harter et al., 2012). Regardless, reducing NO_3^- leaching losses from croplands is an important strategy to minimize future groundwater degradation and protect drinking water resources in the long term.

It is important to recognize, however, that leaching is an essential part of irrigated crop production in arid and semi-arid climates. Without it, plant-toxic salts tend to accumulate within the rootzone and decrease production (Hanson et al., 2008, 2009). For this reason, continued productivity of many California cropping systems depends on transporting salts below the rootzone, which typically occurs with irrigation or precipitation. In such environments, trade-offs need to be made between management of the soil salt balance for continued viability of farming operations,

on the one hand, and the environmental impacts of NO_3^- leaching, on the other hand.

Although NO_3^- leaching and some groundwater contamination from California crop production is practically inevitable, growers have many options for relieving pressure on the resource (Appendix 7A). A recent review identified over 50 management measures that could help (Dzurella et al., 2012). The fundamental basis of managing leaching is that losses are correlated with N and water inputs (Addiscott, 1996; Letey et al., 1979; figure 7.2). Practices that closely monitor and manage soil water and N status over active cropping and fallow periods are effective at reducing losses (Feigin et al., 1982a, 1982b; Hartz et al., 2000; Jackson et al., 1994; Poudel et al., 2002). Consequently, when N use and irrigation efficiency increase, losses decrease. High N and irrigation efficiency result in a small soil mineral N pool and longer residence times of N in the rootzone. The latter has the dual benefit of increasing the potential for uptake as well as increasing the potential for denitrification because of the high degree of biological activity in this region. Often reducing leaching requires additional labor and capital resources, and possibly the adoption of new or advanced technologies (Addiscott, 1996). However, optimizing the management of existing practices, such as shortening furrows or optimizing drip irrigation technology, can also be an effective strategy (Breschini and Hartz, 2002; Hanson et al., 1997; Jackson et al., 2003, 1994).

Virtually all modern cropping systems in California pose a NO_3^- leaching risk, but certain systems disproportionately affect groundwater. Differences in leaching potential are related to the soil physical properties, irrigation method, crop cultivated, and soil management practices (Pratt, 1984). Though actual leaching rates are location-specific due to the aforementioned factors, certain combinations of technologies, sites, and crop species present greater risk. Researchers at the University of California, Riverside, led an initiative to create a system to identify NO_3^- leaching risk for irrigated crop production in the Western United States. The outcome, called the Nitrate Hazard Index, scores the threat of a cropping system based on soil, crop, and irrigation system characteristics (Wu et al., 2005). Knowledge about the vulnerability of the system can be used to guide management decisions, such as planting deep-rooted crops, or removing a field from production altogether. Indeed, using such tools might help mitigate leaching. But it must be remembered, the Nitrate Hazard Index is simply a planning tool; management ultimately determines the leaching rates (Hanson, 1995; Pang et al., 1997). Arresting cultivation of highly susceptible sites and managing crop–soil–technology combinations that minimize leaching hazard would further reduce NO_3^- leaching.

Our estimates suggest improved fertilizer, water, and soil management could avert at least 7 Gg N leaching losses each year. Reductions represent the minimum expectation when increasing N use efficiency by 8 percentage points (Section 7.1.1). It is entirely plausible that leaching losses would be

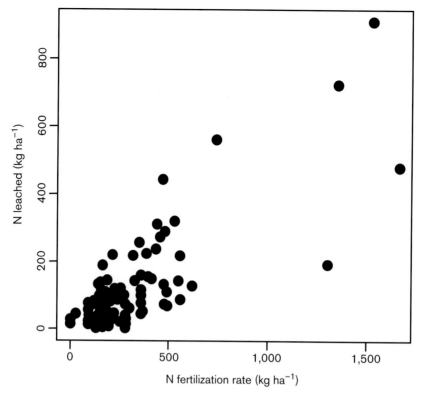

FIGURE 7.2. Relationship between mass nitrogen leaching (kg ha⁻¹) and nitrogen application rates (kg ha⁻¹). Data compiled by the California Nitrogen Assessment. Outliers of high leaching and N application rates are omitted from the graph.

reduced to an even greater extent with improved practice. Surplus soil mineral N is highly susceptible to leaching loss, with potential leaching losses rising exponentially after plant uptake is exceeded. Therefore, reducing the size of the pool by increasing N use efficiency is more likely to have an exponential instead of proportional effect.

But would reducing NO_3^- leaching have negative consequences for farm profits? Practices that reduce leaching often entail more intensive management, adding to production costs. Efforts to estimate costs are complicated by the number of operations that must be included and the uncertainty and variability in actual leaching rates for a given field. However, it appears leaching losses could be incrementally reduced without significantly affecting farm profits (Knapp and Schwabe, 2008; Medellín-Azuara et al., 2013). Adoption of transformative practices in irrigation, manure, and chemical fertilizer management is hindered by numerous barriers on and off the farm, including farm logistical limitations to changing irrigation practices, insufficient development or local adaptation and demonstration of required technologies, insufficient grower education, and land tenure issues. Costs and benefits to individual farmers, however, need to be appraised simultaneously with the costs borne by society at large due to groundwater contamination (e.g., costs of treatment or buying drinking water) and the benefits accruing from cheaper foodstuffs.

7.2.3. Greenhouse Gas Emissions from Fertilizer Use

Use of nitrogenous fertilizers is the primary cause of recent increases in atmospheric concentrations of N_2O globally (Crutzen et al., 2008; Davidson, 2009; Ravishankara et al., 2009; Wuebbles, 2009). In California, inorganic fertilizer use accounts for about 80% of the total N_2O emissions according to California's most recent GHG inventory (CARB, 2014b). When integrated over a 100-year time frame and converted to CO_2e, N_2O emissions amount to approximately 2% of California's total climate forcing emissions.[6]

Unfortunately, due to the complexity of mechanisms driving N_2O evolution in soils, there are no agronomic "silver bullets" that universally, or even consistently, reduce N_2O emissions, other than ceasing all N applications (Appendix 7B). Soil physical and chemical properties (including texture, pH, oxygen and carbon availability, and water holding capacity); management practices (including tillage, irrigation, and fertilizer source and rate, etc.); weather (including temperature and precipitation); and biological activity each affect the magnitude of fluxes and total emissions (Mosier et al., 1998; Stehfest and Bouwman, 2006). Complex interactions among these factors cause large variance in direct emission rates from the field, with

6. This figure ignores the substantial CO_2-equivalent emissions that accrue during out-of-state manufacture of the fertilizer, which increase the total GHG impact of fertilizer use by 20–150% (see Box 5.4.2).

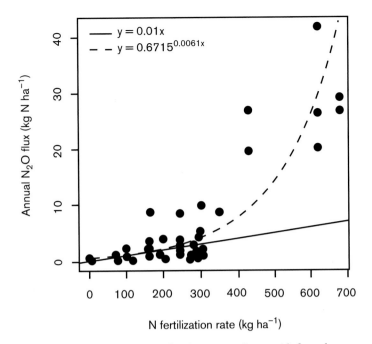

FIGURE 7.3. Impact of nitrogen application rate on nitrous oxide fluxes from California agricultural soils. Data compiled by the California Nitrogen Assessment and Rosenstock et al. (2013). Calculations account for approximately 76% of annual fertilizer sales. Rice is not included due to the negligible amount of N_2O produced under flooded soil conditions.

the Intergovernmental Panel on Climate Change estimating an uncertainty range of 0.003–0.03 kg N_2O-N kg^{-1} of N applied (IPCC, 2006). The considerable spatiotemporal variability, within and among fields and farms—even when seemingly similar production conditions are present—complicates emissions predictions and control. A recent study measuring N_2O emissions from processing tomato systems in Yolo County illustrates the issues well. Kallenbach et al. (2010) compare emissions from treatments using subsurface drip and furrow irrigation with and without leguminous cover crops grown during the winter, between cash crops. Nitrous oxide emissions were greater when leguminous cover crops were planted compared to barren fields in the furrow-irrigated plots, as might have been expected because they are an additional source of N. However, subsurface irrigation negated the effect of the green manure and emitted less N_2O in comparison to the other treatments. Similar interactions have been found in studies of tillage (Mosier et al., 1998; Six et al., 2004; Venterea et al., 2011), as well as fertilizer placement, and other fertility management practices—e.g., the 4Rs[7] (Snyder et al., 2009). With highly site-specific responses, the limited number of field measurements in California, and concerns about measurement protocols and interpretation (see Appendix 4.1), conclusions about the ability of individual or bundles of practices to reduce N_2O production and the consequential magni-

tude of any reduction for specific locations are largely speculative.

Somewhat more certain is that N_2O emissions correlate with N application rates. Therefore, practices that allow growers to reduce N use will generally reduce emissions. The magnitude of the reductions depends on the nature of the relationship between N_2O and N fertilizer rate, with both linear and exponential functional forms being observed, which is controlled by the site-specific conditions identified previously (figure 7.3; Eagle et al., 2010; McSwiney and Robertson, 2005). Expectations about the impact of marginal reductions of N use are then subject to assumptions of the relationship. If linear, then incremental change will have a proportional effect regardless of the magnitude of reduction. But if exponential, then decreases in N use can be expected to dramatically reduce emissions—assuming producers fertilize at rates greater than crop uptake. In this assessment, we assume a linear response function when estimating potential emission reductions. The assumption is reasonable when estimating emissions over scales as significant as California because field-to-field variation averages out once aggregated (figure 7.3). Utilizing the median rate of emissions from California-specific studies (1.4% of N applied) and the increase in N use efficiency discussed above (Section 7.1.1), we might expect to reduce emissions by 0.53 Gg N yr^{-1}.

Field-level emission responses are best described by exponential response functions; thus, relatively small reductions in N application may dramatically decrease emissions. That suggests that growers could participate in

---

7. The 4Rs typify the current N fertilizer management paradigm. Judicious fertilizer applications are those that use the **r**ight source, **r**ight amount, at the **r**ight time, in the **r**ight place (see Chapter 8).

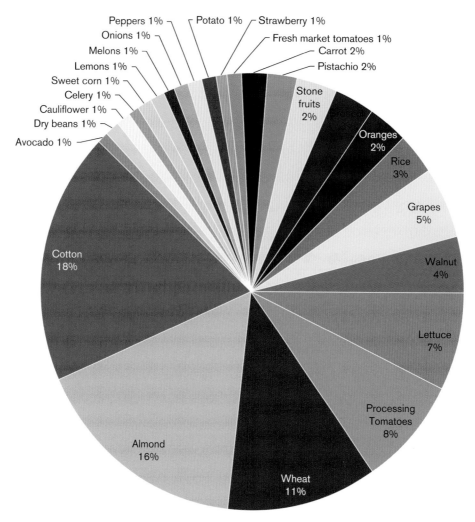

FIGURE 7.4. Relative contribution of N$_2$O emissions for 33 crops in California, based on California-specific emissions factor (1.4% of N applied), fertilizer use data developed by the California Nitrogen Assessment, and USDA Census of Agriculture 2007. The emission factor used for rice is 0.3% of total N applied (IPCC, 2006). Melons include watermelon and cantaloupe; peppers include bell and chili; stone fruits include peaches, nectarines, dried plums, and fresh plums; grapes include wine, table, and raisin.

DATA FROM: Rosenstock et al. (2013).

carbon finance schemes such as the Climate Action Reserve's N fertilizer reduction protocol (e.g., Climate Action Reserve, 2012). In general, development of low N$_2$O production systems is only beginning in California, even though some of the seminal research on N$_2$O evolution from cropland soils occurred in California (Ryden et al., 1981). Recent research has aimed to set a baseline of emission rates for a range of systems. More comparative research is needed. With the diversity of cropping systems, uncertainty of the impacts of specific practices, and differential importance to state production, a targeted approach could set priorities for future research. Based simply on estimates of inorganic N fertilizer use, future research to develop low-emission systems should initially focus on almond, cotton, lettuce, tomato, and wheat (Rosenstock et al., 2013). Indeed, special attention may be paid to almond, cotton, and

lettuce as estimates suggest they are responsible for the largest amount of emissions for their respective crop type: perennials, field crops, and vegetables, respectively (figure 7.4). Lessons learned from these crops can then be transferrable to other production systems with similar characteristics.

It is important to note that the discussion here so far has concentrated on direct emissions alone. Indirect emissions, those that occur after N is transported beyond the field boundaries due to initial volatilization, deposition, or leaching/runoff, represent another source of N$_2$O to the atmosphere, though the expected magnitude of the flux is smaller. For example, the IPCC default emissions factor for N$_2$O-N for N leached is 0.0075 with an uncertainty range 0.005–0.025 (IPCC, 2006), only about 7.5% of expected direct field emissions.

7.2.4. Nitrogen Oxide Emissions from Fuel Combustion

NO_x released into the atmosphere in California from fossil fuel combustion is a major source of N (359 Gg N yr^{-1}) (Chapter 4). The major mobile contributors of NO_x include heavy-duty diesel vehicles, light-duty vehicles, and ships and commercial boats. Stationary sources of NO_x include manufacturing/industrial sources and residential fuel combustion. According to CARB (2006), it is feasible to reduce NO_x emission by more than 60.3 Gg in the South Coast, San Joaquin, and Sacramento Air Basins.

7.2.4.1. MOBILE SOURCES OF NITROGEN EMISSIONS: LIGHT-DUTY VEHICLES

Little nitrogen exists in fuels for light-duty vehicles; rather, N is derived from the N_2 in the air that serves to combust fuel. Emissions from light-duty vehicles are the result of incomplete combustion (releasing PM) and high combustion temperatures (releasing NO_x). The primary way to reduce emissions from this source, without reducing vehicle activity or fuel switching, has historically been to reduce tailpipe emissions. Since the 1960s, a series of technologies have become available that either increase control of the air—fuel ratio and temperature during combustion—or modify gas prior to release, which have had the impact of reducing emissions per vehicle mile traveled. Today, fuel injectors are used in all light-duty vehicles to control the air: fuel ratio in vehicles, which helps to prevent incomplete combustion (Pulkrabek, 2003). Exhaust gas recirculation systems recirculate 5–15% of exhaust back to engine intake, lowering combustion temperatures and decreasing NO_x emissions (Pulkrabek, 2003). Exhaust Gas Recirculation was first introduced in 1973 and is commonplace in passenger vehicles today. Three-way catalytic convertors were added to vehicles beginning in the late 1970s to help lower combustion temperatures and decrease NO_x emissions and have become the standard form of NO_x emission decreases. Catalytic convertors serve to speed the fuel combustion chemical reaction and, in best-case scenarios, can convert 95% of NO_x into inert N_2. Catalytic convertors are the most effective technology to reduce NO_x emissions from light-duty vehicles, but the technology is not without its trade-offs. Catalytic convertors are generally designed to decrease NO_x emissions, but may have a secondary impact on increasing N_2O and NH_3 production (Kean et al., 2009; Lipman and Delucchi, 2010).

While internal combustion engines do not normally reach the high temperatures required to produce N_2O, catalytic convertors, used to lower NO_x emissions, can create N_2O emissions as a by-product. Cold engine starts produce pulses of N_2O that decrease as engines warm up, and aging catalytic convertors emit more N_2O than younger ones. As hybrid vehicles gain market penetration, increasing N_2O emissions are a concern. As hybrid engines cycle on and off when vehicles start and stop, catalytic convertors can cool off enough to produce N_2O emissions multiple times throughout a vehicle's trip. To date, catalytic convertors are not produced to address both N_2O and other NO_x emissions, and the technology's potential requires significant research and development. Potential amendments include electrically heated catalytic convertors, though the heating may result in a small net energy loss for vehicles (Lipman and Delucchi, 2010; Ogden and Anderson, 2011).

The case is similar for NH_3 emissions from light-duty vehicles. Three-way catalytic convertors employ ammonia in the form of urea to help speed reactions and reduce NO_x to a steady state (N_2). Catalytic convertors can over-reduce NO_x beyond N_2, resulting in NH_3 emissions as part of vehicle exhaust. Because three-way catalytic convertors were not introduced until 1981, older vehicles without them produce almost no ammonia. Newer vehicles with efficient catalytic convertors also produce lower emissions, making the problem most abundant in middle-aged vehicles with aging catalytic convertors (Kean et al., 2009). Other materials can substitute for urea to reduce NO_x, and urea injections into catalytic convertors can be measured more precisely (Johnson, 2009), but there is likely a trade-off between lowering ammonia emissions and lowering NO_x emissions using the existing three-way catalytic convertor technology.

7.2.4.2. MOBILE SOURCES OF NITROGEN EMISSIONS: HEAVY-DUTY VEHICLES, OCEAN-GOING VESSELS, AND OFF-ROAD VEHICLES

In the past, emissions controls used for light-duty vehicles could not apply to heavy-duty diesel trucks. Diesel trucks have historically had poor fuel injection control, resulting in poor control of PM emissions. But there are promising advances in control technologies to reduce emissions from diesel trucks. Often, the turnover to newer engine models can effectively lower emissions (Dallmann et al., 2011). Vehicle turnover is slow, but California has mandated upgrades to many heavy-duty vehicles, which, over time, will result in significant emissions reductions in the goods movement industry. Low-sulfur fuel is now mandated for diesel trucks in California, and trucks are being equipped with better fuel injection systems, exhaust gas recirculation to lower combustion temperatures (reducing NO_x emissions), and diesel particulate filters used to trap PM and burn it off intermittently (EPA, 2008e; Pulkrabek, 2003). Diesel particulate filters are required in all new vehicles manufactured, and are a required addition to older engines under CARB's Truck and Bus Regulation (CARB, 2015b). The regulation also includes a scheduled phase-out of engines manufactured prior to 2010: by the end of 2023, all trucks are expected to meet 2010 engine emission standards and to be equipped with a diesel particulate filter. These technology improvements are anticipated to reduce PM emissions from goods movement by 86% by 2020, and NO_x emissions by up to 68% (CARB, 2006). Selective Catalytic Reduction (SCR) is being phased into heavy-duty vehicles (a technology commonly used in stationary sources to reduce NO_x). While heavy-duty trucks do not currently emit

TABLE 7.5
NO$_x$ removal efficiencies (%) for selected primary and secondary technologies

NO$_x$ reduction technology	Fuel		
	Coal	Oil	Gas
Combustion control			
Low-excess air	10–30	10–30	10–30
Staged combustion	20–50	20–50	20–50
Flue gas recirculation		20–50	20–50
Water/steam injection		10–50	
Low-NO$_x$ burners	30–40	30–40	30–40
Post-combustion treatment			
Selective catalytic reduction	60–90	60–90	60–90
Selective non-catalytic reduction		30–70	30–70

SOURCES: EPA (1999b) and World Bank (1998).

a significant amount of NH$_3$ (Kean et al., 2009), the increased use of SCR, which uses urea as a NO$_x$ reducing agent, could contribute to increases in NH$_3$ emissions (Kean et al., 2009).

Ocean-going vessels (OGVs) contribute 10% of California's NO$_x$ emissions (CARB, 2015b), and a negligible amount of N$_2$O. In 2010, the US EPA and the International Maritime Organization officially designated waters within 200 nautical miles of North American coasts, including California, as an Emission Control Area (ECA). Between 2012 and 2016, OGVs operating within the North America ECA are required to reduce their emissions of NO$_x$, sulfur oxides (SO$_x$), and PM$_{2.5}$ through a graduated transition to increasingly lower-sulfur fuels (EPA, 2010). In addition, establishing electrical power for ships to use while docked will decrease emissions further. Ships can also generate their own electrical power through solar panels, fuel cells, or with natural gas engines equipped with SCR technology to control NO$_x$ (CARB, 2006). However, the introduction of catalytic convertors on OGVs will likely add the trade-off of increased N$_2$O and possible NH$_3$ emissions.

Off-road diesel vehicles such as tractors and construction equipment are subject to the same technological needs as heavy-duty trucks in order to improve emissions. Low-cost improvements like adding a diesel oxidation catalyst can cut PM in half, but do not affect NO$_x$ emissions. Adding SCR technology to diesel engines, which can dramatically reduce NO$_x$ emissions, can be cost prohibitive, ranging from \$12,000 to \$20,000 (EPA, 2008e). The EPA also emphasizes shortening of idling times, and replacement of aging fleets as key ways to decrease emissions.

7.2.4.3. STATIONARY SOURCES OF NO$_x$ AND N$_2$O

Stationary sources of fuel combustion, including energy generating power plants and manufacturing, comprise about 13% of California's NO$_x$ inventory (CARB, 2014a; see Chapter 3), and 80% of emissions were derived from only 187 facilities in 2007, so the path to lower NO$_x$ emissions is relatively achievable, though retrofits can be cost prohibitive. NO$_x$ emissions are dependent on a number of factors at industrial facilities including flame temperature, residence time at high temperature, quantity of excess air available for combustion, and nitrogen content of the fuel (Bradley and Jones, 2002). There are a number of combustion and post-combustion technologies in place to control NO$_x$ emissions from stationary sources (Table 7.5). Reducing peak temperatures, gas residence time near the flame, or oxygen concentrations by low excess air; staged combustion; over-fired air; and flue gas recirculation in the zone of combustion are already commonplace measures that achieve substantial reductions in NO$_x$ emissions. SCR and Selective Non-catalytic reduction (SNCR) are both frequently used to reduce NO$_x$ to N$_2$ and water using ammonia as a reducing agent, presenting similar trade-offs as mobile sources. SNCR can reduce NO$_x$ emissions by 60%, while SCR can reduce NO$_x$ emissions by as much as 90% (Bradley and Jones 2002; Table 7.5).

Emissions of N$_2$O from most industrial sources are extremely low (CARB, n.d.). N$_2$O from stationary sources generally originates either as a byproduct of incomplete fuel combustion or as a byproduct of adipic acid (used primarily to make plastics) and nitric acid production (used for fertilizer, plastics, and explosives). N$_2$O originating from adipic acid can largely be reduced by N$_2$O destruction (incineration), while nitric acid-based N$_2$O requires catalytic reduction. Nitric acid facilities generally use the same SCR to control NO$_x$ and N$_2$O emissions, but the system is designed primarily to control NO$_x$ and is therefore significantly less effective at controlling N$_2$O (Johnson, 2009). A third control system, Non-Selective Catalytic Reduction

(NSCR) is very effective at controlling both NO_x and N_2O, but is used by few nitric acid plants because of high-energy costs (CCTP, 2006). The US Climate Change Technology Program emphasizes the need to improve SNCR technologies and encourage research that focuses on simultaneous reduction of N_2O and NO_x.[8]

7.2.5. Wastewater Management

Until recently, wastewater was discharged without specific treatment for N to the detriment of California's drinking water, wildlife, climate, and ecosystems (Boehm and Paytan, 2010; Glibert, 2010; Jassby and Van Nieuwenhuyse, 2005; Seitzinger et al., 2006). Today, about 50% receives treatment to decrease its N load prior to release into soils, freshwater, or coastal regions (Chapter 3). However, traditional notions of wastewater N treatment—removal and discharge—ignore ancillary environmental consequences and the nutritive value of this resource. Wastewater N management could be transformed to expand N removal where appropriate and stimulate recycling when possible.

Technologies capable of reducing the N load from 40% to 99% of untreated levels are well established for wastewater treatment plants (WWTPs) and on-site wastewater treatment systems (OWTS) (Henze, 1991, 2008; Kang et al., 2008). Currently, N treatments largely consist of creating conditions to support microbial nitrification (oxidation of NH_4^+ to NO_3^-) and denitrification (reduction of NO_3^- to N_2 gas). Its effectiveness and relative cost make this the most attractive option (Ahn, 2006). However, drawbacks include that N_2O can be emitted during both nitrification and denitrification (Townsend-Small et al., 2011a) at rates from 0.5% to 14.6% of the N in wastewater at WWTPs (Kampschreur et al., 2009). Similar concerns likely affect OWTS using nitrification–denitrification to an even greater extent since their operators have little or no control over critical environment conditions regulating waste digestion (e.g., chemical composition, pH, flow, organic carbon).

We estimate that improved wastewater management could greatly decrease N in effluent from WWTPs and OWTS. A conservative increase in N treatment at WWTPs (10% of influent) would reduce N discharged into the environment by 15.6 Gg N yr^{-1}. And depending on the extent of OWTS retrofits and operations, an additional 1.3–10.9 Gg N yr^{-1} could be removed.

Source separation of human waste is an emerging strategy to handle N-rich waters stemming from toilets. Most of the constituent mass of N in wastewater is in urine (approximately 70–80% of the total) (Metcalf & Eddy, Inc., 2003). With urine separation technology, N can be recycled back to the landscape more easily, saving energy and recycling

nutrients to the soil. Source separation technology, in which urine is removed from the waste stream and reused as a fertilizer, can be expected to reduce N loading to wastewater treatment systems by about 50%.

High costs significantly constrain advanced treatment applications for large-scale facilities and homeowners alike. A synthesis of costs shows that capital costs and operations and maintenance costs attributed to N removal can range from $1.08 to $8.51 kg^{-1} N removed and $1.08 to $2.00 kg^{-1} N, respectively (Kang et al., 2008). The large range reflects differences in the extent of the retrofit or expansion necessary, the specific technology applied, and the amount of wastewater processed. Economies of scale reduce per unit costs for many of the WWTPs reviewed. Based on a median rate, we estimate that it would cost roughly $214 million in capital expenditures to implement N reduction technologies across untreated wastewater throughout WWTPs in California, plus an additional $69 million annually for operation and maintenance. Relative costs for retrofitting or replacing septic systems are also high. Retrofitting an existing system can be $10,000–20,000 each (Viers et al., 2012). Another option is to treat effluent emerging from septic tanks via biological nitrification and denitrification treatment. Wood chip bioreactors have been shown to reduce influent nitrate by 74–91% (Leverenz et al., 2010), with costs ranging from $10,000 to $20,000 to retrofit existing septic systems.

It is impractical at this time to contend that all California wastewater be treated for N given that more than 50% is discharged into the Pacific Ocean. However, the economics of treatment for WWTPs and homeowners needs to be counterbalanced by acknowledgment of the significant indirect impacts on ecosystems. A thorough assessment of the sensitivity and vulnerability of receiving ecosystems would help to set priorities for future N reductions.

7.3. Adapt to a Nitrogen-Rich Environment

Going forward, Californians will face pressure to adapt their behavior to the new state of air, water, and soil resources to reduce exposure risks, maintain productivity, and relieve pressure on the environment. Drinking water will require treatment for the foreseeable future in some areas because it will take decades before groundwater shows the impact of changes in farm management practices. Agriculture will need to adapt to N already present in irrigation water and to higher ozone levels.

7.3.1. Treatment and Alternative Sources of Drinking Water

Poor water quality disproportionately affects the most vulnerable citizens among us and constitutes an environmental justice concern (Chapter 5). Significant uncertainties still persist about the extent of the concerns and the best

---

8. Other options for addressing N_2O emissions are available through CARB's Clearinghouse of Non-CO_2 greenhouse gas emissions control technologies. http://www.arb.ca.gov/cc/non-co$_2$-clearinghouse/non-co$_2$-clearinghouse.htm#Nitrous_Oxide.

solutions (Honeycutt et al., 2012). Yet the known dynamics of the problem (large N load migrating through soil profile, shallow wells, unequal cost of treatment burden, few resources available to adapt) suggest that the threat is significant and will only worsen and spread to many additional communities (Harter and Lund, 2012).

Though reducing NO_3^- leaching loss will be instrumental for meeting future drinking water needs, the concentration of NO_3^- in drinking water already exceeds safe levels—the legal maximum contaminant level (MCL, 10 mg L^{-1} NO_3^--N)—in many regions and remedial actions are needed to minimize exposure. Options to treat drinking water supplies for NO_3^- that are proven effective include both removal and reduction technologies, but they are highly site-specific. Seidel et al. (2011)[9] thoroughly review the major options including ion exchange, reverse osmosis, electrodialysis, and biological and chemical denitrification. Because each has clear advantages and disadvantages, selecting the "best" option cannot be done a priori. Characteristics of the water system and water quality must be taken into account. Decisions about cost, waste disposal, information demands, size of the facility, and future needs of the community need to be considered, at a minimum. Planning for future needs and local conditions is particularly important because of inherent limitations of treatment systems and the demands they place on the community and/or operators. For example, small water systems often lack technical, managerial, and financial capacity to mitigate NO_3^- issues and the available funding may cover initial capital cost but not operations and maintenance. Moreover, the use of some technologies such as anion exchange—one of the most common in NO_3^- treatment—requires salt and results in a brine which needs to be disposed of, which can be a significant cost especially for inland communities. In many cases, avoiding the challenges of treatment by developing new water resources instead may be more feasible. However, the long-term sustainability of nontreatment options needs to be considered as with the migration of NO_3^- into groundwater increasing with time, some alternatives such as blending or drilling new wells may be feasible now but may not be in the future. While planning for the future, interim solutions including point of use treatment may well be needed to deliver safe drinking water.

Because treating for NO_3^- in drinking water can be quite costly (both in initial capital costs and in operations and maintenance costs) and technically challenging, options for simply avoiding the high NO_3^- water altogether, or adjusting to it in other ways, are often explored first. Commonly used options in California are well inactivation, blending high NO_3^- water with water from other wells in which concentrations are lower, consolidation with nearby water systems, and development of alternative sources. New wells are often drilled deeper than older wells, in order to reach older groundwater containing less NO_3^-. This strategy, besides being more expensive, also often creates other challenges. For example, deeper water more often contains high levels of arsenic, which may need to be treated for in order to make the water safe for drinking.

In summary, an array of management options are available to provide clean drinking water for Californians. Costs, however, can be high. An assessment of the costs for supplying drinking water to populations serviced by high NO_3^- wells in the Tulare Lake Basin and Salinas Valley indicates $12–17 million per year are needed to provide water for only 220,000 people (Honeycutt et al., 2012). More densely populated areas would have a lower per capita cost because of economies of scale, yet low NO_3^- water will not come cheap. When considering all the options for adapting to NO_3^--rich groundwater, care must be taken to evaluate the relative advantages and disadvantages among them, considering appropriate initial and ongoing capital, labor, information demands, timescales, and development scenarios—and not simply relative costs.

7.3.2. Adaptation of Agricultural Systems

Farmers already adapt to N in California's environment. The most obvious example is when growers modify fertility programs to account for NO_3^- levels in irrigation water, allowing it to supplement or completely replace purchased fertilizer N inputs (e.g., Hutmacher et al. 2004). Less attention is paid to airborne N pollutants, despite the prospects for significant economic consequences. Exposure to elevated ambient concentrations of ground-level ozone (O_3) reduces yields, sometimes by nearly 20%, costing producers millions of dollars in lost revenue each year (Grantz, 2003; Kim et al., 1998; Mutters and Soret, 1998). However, few producers select crops or varieties based on O_3 tolerance. If concentrations of N compounds continue to increase in the environment, adapting to these new levels will become a matter of necessity to maintain the productivity of agricultural production systems.

In addition to environmental changes, N-related regulatory changes will also require agriculture to sharpen its adaptive capacity. Concerns of N in the environment are gaining attention in ongoing state and federal policy discourse. The US Department of Agriculture (USDA), State Water Resources Control Board (SWRCB), EPA, CARB, and local counterparts (e.g., Regional Water Control Boards) have recently examined N use in agriculture. On top of the relatively long-standing air and water quality rules that include NO_x emissions and surface water and groundwater maximum contaminant loads, scoping and implementation of new N budgeting requirements and new monitoring and reporting regulations are underway in many parts of the state (e.g., several new orders on waste discharge in the Central Valley under the Irrigated Lands Regulatory Program and the General Order on Dairy Waste Discharge). Since many of the regulations are still being developed, there is considerable uncertainty about

---

9. Readers are directed to Seidel et al. (2011) and Jensen et al. (2014) for detailed analyses of NO_3^- treatment options for drinking water, including applicability, efficacy, costs, trade-offs, case studies, and many examples from California water systems.

how they will affect farmers (for more about uncertainty in regulations, see Chapters 6 and 8).

At the state level, the diversity of California's product mix allows for a certain degree of adaptability. There is a wide range of knowledge and experience within the agricultural sector overall, due to its diverse array of production systems. Therefore, opportunities may exist to move quickly to adapt to changes in N by modifying production practices and moving between crops. That ability relies on information that will need to be organized, generated, and distributed.

Currently, little is known about the thresholds that will trigger a need for large changes. A few bioeconomic models address California agriculture's response to N-rich environments and changing policies. They tentatively suggest that incremental change, such as shifting crop species to adapt to O_3 or changing soil management practices to reduce NO_3^- leaching modestly, is plausible without significant economic loss (Kim et al., 1998; Knapp and Schwabe, 2008).

7.4. Synergies and Trade-Offs among Nitrogen Species

The strategies and practices identified to control the N cascade can have far-reaching effects, for target N species, nontarget N species, and environmental systems. Some actions will cause synergistic responses, reducing multiple N emissions simultaneously while improving the state of additional environmental and health concerns. Often, however, they will induce trade-offs, where reduction of one N concern inflames another. Secondary impacts arise from the ubiquity of N in living things, its presence in day-to-day human activities, and its interaction with the carbon and hydrologic cycles. It is important to identify and use metrics that can capture these trade-offs and cascading impacts (for more on metrics and monitoring strategies, see Appendices 7.5 and 7.6). Understanding the potential positive and negative unintended consequences is essential to evaluating the relevance of any particular N response activity (for more on pollution trading and life cycle accounting, see Appendix 7.4).

So-called "pollution swapping" essentially reallocates the environmental and human health burden from one ecosystem service or economic sector to another, with occasionally more harmful consequences than the original pollution. Each mitigative action that focuses narrowly on a single activity and pollutant poses such threats (see Section 7.2). Some possible trade-offs and synergies are described below,[10] though given the nature of the N cascade others are plausible.

MINIMIZATION OF AMMONIA VOLATILIZATION FROM MANURE: NH_3 (−), NO_3^- (+), N_2O (+)

Avoiding NH_3 volatilization by improving manure management benefits downwind ecosystems and will help decrease

---

10. Signs refer to direction of flow: + = increasing; − = decreasing.

PM formation in the atmosphere. But by reducing NH_3, the likelihood of NO_3^- leaching and N_2O emissions will increase (Velthof et al., 2009), because the manure retains a greater N load than it would have had otherwise. Assuming the additional N is conveyed throughout the manure management train (e.g., collection, processing, and storage facilities), croplands must absorb the additional load. Increased N load requires a larger application area or increases the risk of overapplication, if additional land is not available for distribution. Even when manure N is applied judiciously, the increased N load itself will likely lead to higher fluxes of NO_3^- leaching to groundwater and gaseous N_2O emissions because of the greater loading to the soil. Indeed, a fraction of the original NH_3 emitted would have deposited downwind and been lost via these pathways anyway. However, the relative quantity of losses via leaching and denitrification would be less than expected from the increased N loads applied to crop fields directly; deposition of airborne NH_3 represents only approximately 20% of applied N and only 1% of that amount is lost as N_2O versus 2% from the original load of manure (assuming IPCC 2006 default emissions factors). Therefore, California decision-makers are left weighing the impacts of NH_3 on natural ecosystems (including the potential for fire, invasive species, and biological diversity) and air quality (including $PM_{2.5}$ production) in the case where no additional effort is made to decrease volatilization versus increased climate change impacts, ozone depletion, and groundwater degradation in the case where volatilization is actively minimized.

REDUCTION OF NITRATE LEACHING FROM CROPLANDS: NO_3^- (−), N_2O (+)

Reducing leaching from croplands, without decreasing N application, requires NO_3^- to be better timed with crop demand or remain in the rootzone longer. Greater residence times—through decreased percolation or extending the release of the N—provide additional opportunities for plant roots to seek out and assimilate the NO_3^-, converting it eventually into organic molecules. It also provides a chance for microbes to denitrify the NO_3^- to N_2, especially in clayrich soils. The efficacy of denitrifying bacteria to completely transform NO_3^- to N_2 depends on soil conditions (water content, organic carbon availability, pH, and temperature), and biological activity and organic carbon content are typically highest within the rootzone. In the absence of the appropriate conditions, denitrifying bacteria produce intermediary products of NO and N_2O, instead of the inert and desirable N_2. Wetting and drying cycles consistent with optimal N and water management tend to promote environmental conditions conducive for N_2O evolution. Soil heterogeneity only compounds this problem, making it more difficult to maintain denitrifying conditions and producing hotspots and hot moments of N_2O volatilization. California crop producers (and those that regulate them) must decide between practices that preserve groundwater, a local concern, at the expense of climate change, a global concern.

EMISSIONS REDUCTIONS FROM FUEL COMBUSTION: NO_x (−), NH_3 (+)

Combustion technologies already effectively limit NO_x emissions from transportation and industry. As discussed, additional gains are plausible, especially at unregulated sources or by improving conversion efficiency of technologies. Certain technologies that use post-combustion catalysts to transform NO_x to N_2, however, have the potential to produce NH_3 instead of N_2. This is common in industrial applications, where "ammonia slip" results from aging catalysts or too little reaction time. Potentially more troublesome because of the relative ubiquity of the source activity is the increased production of NH_3 from vehicle engines using three-way catalytic converters. Under today's driving environment (congestion, low speeds), conditions promote less reduction to N_2 and, consequently, NH_3 becomes a larger fraction of tailpipe emissions. In short, efforts to control NO_x contribute to the increase in NH_3 in the atmosphere.

TRANSFORMATION OF WASTEWATER MANAGEMENT: NH_4^+ (−), NO_3^- (−), N_2O (+)

Nitrogen removal from wastewater almost exclusively relies on microbial nitrification and denitrification at this time. Fortuitously, the process tends to result in lower concentrations of NH_4^+ and NO_3^- in wastewater effluent with reduced N loading to the soils, rivers, and ocean environments, assuming discharge patterns remain unchanged. However, a larger amount of the N is released to the atmosphere as N_2O. According to one study of WWTPs in Southern California, emissions of N_2O at WWTPs utilizing advanced technology to remove N can be three times as high as emissions at facilities that do not use advanced N removal technology (Townsend-Small et al., 2011a). A fraction of the emissions occur during nitrification. But most result from incomplete denitrification, as wetting and drying cycles of N- and carbon-rich materials present ideal circumstances for microbial activity. Even under tightly controlled conditions, it is challenging to completely eliminate N_2O. Treatment of wastewater at WWTPs in California serves to protect sensitive aquatic ecosystems and groundwater resources. While essential to avoid degradation, it is important to recognize that this protection is achieved at the expense of negative impacts on climate and the ozone layer.

7.5. Conclusion: The Need for Increased Adoption

In summary, this assessment finds that many management practices and technologies to address N issues are already available. However, continued environmental degradation despite the existence of effective control technologies leads us to conclude that the challenge is only in part technical. Chapter 8 describes policy options to promote the adoption that is needed to create positive changes in California's N landscape.

CHAPTER EIGHT

Responses: Policies and Institutions

Lead Authors:
K. BAERENKLAU AND T. P. TOMICH

Contributing Authors:
S. DAROUB, A. DREVNO, V. R. HADEN, C. KLING, T. LANG,
C.-Y. LIN, C. MITTERHOFER, D. PARKER, D. PRESS, T. ROSENSTOCK,
K. SCHWABE, Z. TZANKOVA, AND J. WANG

What Is This Chapter About?

Because nitrogen emissions from agricultural sources are geographically dispersed, cannot be easily observed, and are difficult to precisely control, this problem presents unique challenges for effective policy design. A suite of integrated practice and policy solutions may be needed to achieve both adequate source control and mitigation of the existing N stock within reasonable time frames. This chapter provides an overview of available policy instruments for non-point source pollution control and examines specific outcomes when these mechanisms have been implemented to control nitrogen pollution in practice. Policy characteristics are then organized into a coherent methodology for assessing candidate policies for controlling nitrogen emissions from agricultural sources in California.

Stakeholder Questions

The California Nitrogen Assessment engaged with industry groups, policy-makers, farmers, farm advisors, scientists, government agencies, and nonprofit organizations (including environmental, community, and agricultural organizations, and commodity groups). This outreach generated more than 100 N-related questions, which were then synthesized into five overarching research areas to guide the assessment (see figure 1.3). Stakeholder-generated questions addressed in this chapter include the following:

- How might policy be used more effectively to both monitor and address non-point source agricultural pollution?
- What are the hurdles to having a coordinated and cohesive nitrogen policy across regulatory jurisdictions?

- What would the impacts of policy regulating nitrogen use be on farm profits, food prices, and rural economic activity?

Main Messages

California's long-term success in achieving environmental goals through regulation of the main sources of nitrogen pollution from combustion (tailpipes and smokestacks) is largely irrelevant to challenges of addressing numerous, spatially dispersed, highly variable, context-specific ("non-point") sources of nitrogen pollution typical of agriculture.

Any successful strategy to reduce nitrogen emissions from agriculture must take a comprehensive approach to the most important forms of nitrogen leakage into the environment, particularly ammonia and nitrate, but also including nitrous oxide. Effort to control any one alone, while neglecting the others, is very likely to be counterproductive—"solving" one problem can worsen others.

There have been apparent improvements in the ability of producers to implement the 4Rs of nutrient stewardship in crop production: right amount, right time, right place, and right form. Overall, however, although technologies and practices that can reduce nitrogen pollution from agriculture certainly do exist, they typically are costly (in money and management) for farmers and ranchers; thus, voluntary adoption tends to be low.

It is well established that voluntary participation in best management practice (BMP) programs typically cannot achieve significant reductions in nitrogen pollution from agriculture.

Dairy waste is a significant source of nitrogen pollution in California, both to water and to air. It is critical to develop and implement cost-effective polices to effectively reduce nitrogen pollution from dairy operations. The California Dairy Quality Assurance Program plays an important role in helping dairies comply with existing regulations. While not a panacea by any means, this is an example of how a voluntary, largely information-based educational program can play a supporting role to other environmental regulations.

Even if policies somehow could perfectly control nitrate leakages from farms and dairies starting immediately, California will be living with the consequences of past nitrate leakages to groundwater for decades to come. Thus, for communities where drinking water supplies are unsafe because of high nitrate concentrations, point-of-use treatment or some other approach will be needed in the short run in order to assure safe drinking water for all California communities.

There is very limited information on the magnitudes of economic benefits that would be achieved through reductions in nitrogen emissions. For this reason it is currently not possible to estimate the economically efficient level of nitrogen emissions—the level that balances marginal benefits and costs—nor the relative efficiency of policy instruments. However, it is possible to compare policy instruments in terms of cost to achieve desired emission levels.

Over the longer term, five types of policy instruments appear to be most promising: emission standards, emission charges, tradable emission permits, abatement subsidies, and auction-based abatement contracts. However, theory provides little guidance on which of these instruments would be most effective under specific circumstances. The general lack of evidence, rigorous experimentation, comparative study, or integrated assessment of the impact of alternative policy instruments for controlling nitrogen pollution from agriculture is a major barrier to development of sound policy.

Given the monitoring challenges presented by non-point source nitrogen pollution, and the importance of having adequate data to enforce pollution control policies, efforts should be made to develop the technologies and tools needed to acquire the necessary data and to appropriately model the movement of nitrogen in the environment. Doing so facilitates the transition of nitrogen from a non-point source problem to a more manageable point source problem. In addition, existing data from the diversity of monitoring sites and programs already operated by state and federal agencies need to be made more accessible and integrated with each other. Comprehensive integration, transparent protocols, and evaluation of uncertainty are key characteristics of an effective statewide platform.

This assessment concludes that integrated policy solutions are needed to take advantage of existing technology and to develop new technologies and practices necessary to transition California to a sustainable nitrogen future. While a necessary step, design and implementation of an integrative strategy for nitrogen policy holds many challenges, including the need to fill key information gaps, address existing administrative rigidities, and identify conflicting policies.

8.1. Framing of the California Nitrogen Policy Problem

This chapter addresses a number of practical issues in framing policy problems related to nitrogen in California, focusing on policy options for the two most prominent nitrogen problems in the agricultural sector: groundwater nitrate and atmospheric ammonia emissions. First, and foremost, in order to identify workable options that will have the intended effects, it is necessary to be clear about specific economic, social, and environmental objectives and possible relationships between them. By design, the previous chapters in this assessment address these questions regarding overall objectives.

8.1.1. Overview of Nitrogen Issues

The California Nitrogen Assessment began with a basic question: Is there a nitrogen problem in California? If so, is the problem about agricultural profitability or production costs, climate change forcing, air pollution, surface water pollution, groundwater pollution, public health threats, all of these, some of these, or none of these? In the course of the consultations described in Chapter 1, California stakeholders raised additional, more specific policy-relevant questions listed in the "Stakeholder Questions" section at the beginning of this chapter.

8.1.1.1. NITROGEN DRIVERS, FLOWS, AND IMPACTS

As shown in Chapters 2–4, two broad categories of human activity dominate California's nitrogen cycle: fossil fuel combustion and agricultural production. Despite increases in fuel combustion in California since 1980, emissions have declined steadily. Over the last 50 years, global demand for food has been a fundamental driver of expansion of agricultural production in California. These effects of demand drivers on California agriculture are likely to continue. Over the same period, long-term reduction of both transport costs and international trade barriers increased access to international markets. Long-term decline in synthetic nitrogen fertilizer prices resulted in a large increase in nitrogen use from the 1950s through the 1970s. Thereafter, nitrogen prices were relatively stable relative to the prices of crops until 2000. The future course of these drivers of production and marketing costs are uncertain, particularly regarding energy prices and trade policy. Synthetic nitrogen fertilizer sales in California have risen dramatically since World War II and increased by at least 40% since 1970. Although consumption of synthetic nitrogen fertilizer has leveled off in the past 20 years, the mass balance calculations in Chapter 4 indicate that synthetic fertilizer was still the largest inflow of nitrogen statewide in the last decade, with manure production in second place—a finding corroborated by Harter and Lund (2012). California's livestock herd has continued to grow, despite some slowdown during the recent recession. Manure monitoring efforts are underway, but the fate of manure is largely unknown at policy-relevant scales. Nevertheless, the assessments of drivers in Chapters 2 and 3 and the mass balance calculations in Chapter 4 suggest that nitrogen leakages from agricultural activities—both cropping and livestock—are unlikely to decrease and in fact it is likely that they will continue to grow absent major technological and/or policy changes.

So while nitrogen is indispensable in California agriculture, much of the nitrogen applied as fertilizer or manure is lost to the environment, resulting in a variety of negative effects on ecosystems and potential risks to human health. Groundwater contamination is the greatest environmental concern directly related to agriculture. Fossil fuel combustion, fertilizer use, and livestock all contribute to air quality concerns. Chapter 5 reviews the evidence linking groundwater contamination and air pollution resulting from nitrogen leakages to human health risks, which are believed to be significant (albeit neither well-established nor generally accepted in all cases). Costs of treating nitrate-contaminated drinking water pose significant financial burdens, especially for low-income households and the water systems that serve disadvantaged communities (Chapter 5).

8.1.1.2. TECHNOLOGICAL OPTIONS

Countless technologies and practices are available today that could improve nitrogen use efficiency and reduce nitrogen leakages in agriculture, industry, transportation, water treatment, and waste management. And, as documented in Chapter 7, priorities for mitigating nitrogen emissions include manure management to reduce ammonia emissions to the atmosphere, soil nutrient management to reduce nitrate contamination of groundwater, fertilizer management and fuel combustion efficiency to reduce nitrous oxide leakage to the atmosphere, and enhanced wastewater treatment to reduce ammonium emissions to surface water.[1] In agriculture, there has not been widespread adoption of available technological options because these measures involve increased costs, greater manage-

ment effort or effort to access new information, or some combination of one or more of these, while the incentives for adoption have been relatively weak.

8.1.1.3. RATIONALE FOR FOCUSING ON AGRICULTURE

While the evidence in Chapters 2–5 indicates existing policy interventions have been increasingly effective in reducing air pollution, especially from fossil fuel combustion, the same cannot be said about reversing groundwater contamination from agriculture. Indeed, because of the long lag times between initial leakage and eventual groundwater contamination, none of the scenarios presented in Chapter 6—not even "the end of agriculture"—reverses the deterioration of groundwater quality in the immediate future. On the other hand, over the longer term, some combination of technological and policy change in agriculture will be necessary to achieve a better overall balance of production, environment, and human health objectives. Thus, while point-of-use treatment will be needed urgently in some communities to address immediate problems of drinking water contamination, this assessment focuses on agricultural source control policy because of the broader benefits to be gained from it (albeit over a much longer time frame).

8.1.2. Policy-Relevant Characteristics of Agricultural Nitrogen Emissions

Nitrogen emissions from agricultural sources present unique challenges for effective policy design. Foremost, agricultural nitrogen emissions are "non-point" emissions: they enter the environment from sources that are geographically dispersed, diffuse, and seemingly random—in other words, emissions emanate from many locations, cannot be easily observed, and are difficult to precisely control. The broad geographical distribution of sources means that the relationships between production, emissions, and abatement are likely to exhibit significant heterogeneity due to differences in factors such as climate, soil type, hydrology, farming practices, local policies and economic conditions, and other site-specific characteristics. Uniform statewide policies thus will operate across many local jurisdictions and will have the potential to concentrate abatement effort (and associated costs) due to spatial heterogeneity. The difficulties associated with observability imply that policies requiring knowledge of source emission levels must be based on estimates rather than direct measurement, and this also implies that even agricultural producers may not have good information about the magnitudes of their own nitrogen emissions.

In addition, the emitted compounds themselves present specific challenges for policy design. Because the main nitrogen species of concern, nitrate and ammonia, are not conservative pollutants (i.e., they tend to react in the environment to form other compounds), policies should account for potential transformations of nitrogen emissions and associated shifting of pollution damages. Such transformations

---

1. It is well established that surface water in California is relatively low in nitrate, with the majority of surface water bodies containing average concentrations well below the EPA drinking water MCL. Thus in this chapter we focus on priorities for mitigating nitrogen emissions to groundwater. See Chapters 4 and 5 for more on surface water.

may be undertaken deliberately by sources in response to policies or may occur naturally in the environment. Furthermore, because it is the stock of nitrogen in the environment (e.g., nitrate concentrations in groundwater) rather than the flow of nitrogen into the environment (e.g., nitrate leaching rates from agricultural fields) that is the proximate cause of pollution problems, policies should account for the uncertainties and time lags that are inherent in the relevant environmental fate and transport mechanisms. Importantly this means that relationships between nitrogen sources and their specific impacts may be both spatially and temporally distributed and not well understood. It also suggests that a suite of policies may be needed to achieve both adequate source control and mitigation of the existing stock within reasonable time frames (see Chapter 5 on spatial and temporal trends and impacts).

Finally, the spatial distribution and incidence of damages caused by the stock of nitrogen pollution have implications for policy design. Nitrate and ammonia are not global pollutants—rather the appropriate spatial scale is local to regional. Nitrates tend to contaminate aquifers and ammonia tends to exacerbate air quality problems relatively close to their sources. Policies therefore should target local conditions rather than aggregate statewide measures of emissions or impacts. This approach not only promotes economic efficiency (Baumol and Oates, 1988), but also helps ensure that environmental improvements are attained where they are most needed.

8.2. Overview of Available Policy Instruments for Non-point Source Pollution Control

Pollution control may be achieved through a variety of policy mechanisms. Each mechanism has its own advantages and disadvantages, some of which may depend greatly on the specific context in which a policy is implemented. However, some generalization is possible and even beneficial for understanding the basic properties of broad categories of mechanisms. A taxonomy often used for this purpose includes three such policy categories: (i) education, (ii) standards, and (iii) economic incentives. In this section we describe each of these classes with attention to both textbook treatments of their properties and published research and assessments. In the following section we examine specific outcomes when various policy mechanisms have been implemented to control nitrogen pollution in practice. Readers who are familiar with textbook presentations of pollution control policy instruments and their economic properties, especially in the context of non-point source pollution control, may find it beneficial to spend relatively less time on this section and relatively more on the subsequent section.

8.2.1. Education

Policies that fall into the education category are information based and are predicated on the idea that when stake-

holders have better information about a pollution problem, they will be inclined to modify their behavior in ways that help to mitigate the problem. The type of information provided can vary, and may include labeling or certification of products, training and outreach for producers, reporting inventories or registries, and awards or recognition. In the training and outreach category, we include both centralized education programs, typically delivered by third parties such as cooperative extension, agricultural commissioners, farm bureaus, or private consultants, and community-based learning efforts in which producers share information with each other about promising production practices (possibly with facilitation by third parties).

Education offers some advantages for addressing non-point source agricultural pollution problems. Information-based policies are flexible: they leave decisions in the hands of stakeholders who tend to have the best information about their own decision environments. Information-based policies tend to be relatively easy to implement and receive relatively little opposition from stakeholders compared to other types of policies. Importantly, education tends to be relatively low-cost for both regulators and producers (Ribaudo et al., 1999). Certain types of labeling programs potentially can incur significant costs due to the need to monitor and certify products. However, if this activity can be sourced to an independent third party that specializes in such activities, such as USDA Accredited Certifying Agents for organic standards, then costs to the regulator can be substantially reduced. Training and outreach programs benefit from the preexisting infrastructure of county extension services, Natural Resource Conservation Service field offices, and land grant universities (Daberkow et al., 2008). These institutions can deliver new content to stakeholders without incurring the potentially large fixed costs that characterize the establishment of such infrastructure. They also can facilitate community-based learning efforts. All types of education-based policies also benefit from the falling costs of generating, storing, processing, and disseminating information (Tietenberg and Sterner, 1998). However, such policies must nonetheless compete with a host of other diverse information flows for the attention of the stakeholders they target, some of whom may hold attitudes or ideologies that are unreceptive to new information or may be averse to the risk and uncertainty that tend to be associated with new information. Awards or other types of recognition can be very inexpensive policies. But if the main benefit of the award accrues to only a small number of recipients out of a large number of stakeholders (e.g., a financial award to the farm with the most sustainable production practices), then its overall effect on behavior can be mitigated by the low probability that any single stakeholder wins the award and the associated low expected benefit of pursuing the award. However, awards like this also can produce ancillary benefits through information dissemination (e.g., an evaluation of the sustainable production practices utilized by the award winner) that may lead to behavioral changes by stakeholders who are not necessarily motivated by the chance to receive recognition.

The potential impact of education-based policies depends heavily on the specifics of each situation. Perhaps the main challenge of using education-based policies to bring about large changes in pollution loading is that such policies tend to be purely voluntary. Even in cases where training is mandatory, stakeholders who are targeted by the education efforts must subsequently choose to act on the new information. This choice can be costly, and stakeholders typically have little, if any, incentive to act; an education-based policy will not alone influence a change in behavior if there is an economic cost to the stakeholder.

Education-based policies tend to be seen as potentially useful only in "win-win" situations—that is, when pollution can be reduced and stakeholder welfare increased—when such outcomes have not already been achieved. One example is certification of a product for which (1) consumers exhibit significant willingness to pay for more sustainable production practices, (2) suppliers can produce sustainably without significant extra cost, and (3) the certification process is cost-effective and reliable. Under such conditions, which may not occur frequently in practice, providing the previously missing information to consumers will likely make consumers and producers better off while also reducing negative environmental impacts. Another example is a production practice that both reduces pollution and increases profits, perhaps by more efficiently utilizing inputs. But because producers already face private incentives—a "green" reputation and increased profit—to use such practices, the role of education-based policies in promoting adoption of established "win-win" technologies tends to be limited (Daberkow et al., 2008). However, an important exception is when new technologies become available. Enhanced education and outreach efforts can hasten the rate of adoption of such technologies, provided they are not perceived by producers to be significantly risk-increasing and they provide similar benefits in practice as they do under the more carefully controlled settings that tend to characterize research and development efforts.

There are many examples of environmental labeling and certification programs (e.g., Energy Star, Green Seal, USDA Organic, Dolphin Safe, Forest Stewardship Council, ISO 14000) as well as inventories and registries (e.g., US Toxics Release Inventory and similar registries in Australia, Canada, Mexico, and the EU). Training and outreach to producers has had a major role in efforts to control agricultural non-point source pollution (Daberkow et al., 2008), even though outcomes have fallen short. Training and outreach efforts include activities like demonstration projects, direct technical assistance, newsletters and seminars, and community-based learning. Successful examples of education-based policies include those promoting the adoption of conservation tillage (Gould et al., 1989), soil and tissue nitrogen testing (Ribaudo et al., 2011; Wu and Babcock, 1998), and farm-level information systems (Knox et al.,

1995). However, there are also examples of education failing to produce significant differences between treatment and control groups for water quality protection practices in Wisconsin (Nowak et al., 1997) and nitrate reduction strategies in California (Franco et al., 1994). Ribaudo et al. (2011) also find evidence that soil and tissue nitrogen testing has much less influence on nitrogen application rates when both commercial fertilizer and manure are applied to crops. Daberkow et al. (2008) conclude that "education by itself cannot be considered a strong tool for water quality protection" (p.904). Daberkow et al. (2008) cite three conditions needed for effective education efforts: (1) a "win-win" scenario, (2) producers with strong altruistic/stewardship motives, and (3) high private costs of water quality degradation (Ribaudo et al., 1999). Unfortunately, the convergence of these factors is not common in practice. Daberkow et al. (2008) recommend that education is probably best used as a component of other pollution control policies, such as a mechanism to help producers meet a pollution standard cost-effectively or to effectively utilize new technologies. For example, Bosch et al. (1995) find evidence that additional education makes producers more likely to utilize soil nitrogen testing information. The IPCC (2007c, Chap. 13) reaches a similar conclusion in the context of climate change, stating that "there is only limited evidence that the provision of information can achieve emissions reductions, but it can improve the effectiveness of other policies (*high agreement, medium evidence*)."

8.2.2. Standards

Standard setting, often referred to as "command-and-control" regulation, is the traditional and still most commonly used form of environmental regulation. Policies that set standards obligate producers to meet certain requirements or face consequences for failing to do so. A standard may specify how certain inputs, technologies, or management practices may be used in a production process (input standard); or that emissions do not exceed a specified limit (emission standard); or that the ambient concentration of a pollutant in an environmental medium does not exceed a specified limit (ambient standard). Input standards are also referred to as "design" standards, while the other types of standards are often referred to as "performance" standards. Consequences for failing to meet a standard can include fines, legal liability for pollution damages and remediation costs, forfeiture of performance bonds, loss of licenses or other privileges (i.e., cross-compliance provisions), and even criminal conviction and imprisonment (Sterner, 2003, p.115).

The fact that standards remain the most common type of environmental regulation suggests that they do offer some advantages. The main advantage may be their simplicity and intuitive appeal: when faced with too much (little) of a bad (good) thing, we are inclined to set a limit and enforce it with a penalty. This is a familiar concept that is easily understood by both those who generate and those who are affected by pollution. It also provides for a seemingly decisive response to an issue that elicits strong public concern, which appeals to policy-makers who operate on relatively short political time frames. Another advantage of design standards specifically is that they are relatively inexpensive to monitor (EPA SAB, 2011, p.59). This can be an important consideration when it is difficult or costly to accurately monitor emissions or ambient concentrations and/or predict the environmental effects of varying emission levels. Such was the case in the United States when the Clean Water Act and Clean Air Act were passed, which helps to explain the reliance of those acts on design standards (EPA SAB, 2011, p.59).

The main drawback of design standards is lack of flexibility in both the short and the long run. In the short run, when producers are required to utilize or avoid certain production practices in order to achieve a desired environmental outcome, they are rendered unable to fully utilize the private information they have about their specific operations—information that could lead to a different set of production choices that would achieve the same environmental outcome at lower cost. In other words, design standards tend to be cost-ineffective for an individual producer (Sterner, 2003, p.76). Furthermore, design standards typically are allocatively inefficient across a group of producers when there is substantial heterogeneity within the group and yet design standards are applied uniformly, as is often done to keep administrative costs low (Sterner, 2003, p.77). In the long run, the rigid nature of design standards may undermine incentives for technological innovation because potential innovators must consider the possibility that their costly research and development efforts will produce technologies that are not incorporated into revised design standards and thus cannot be utilized.

Performance standards provide greater flexibility than design standards because they leave production choices in the hands of producers. Thus, producers are able to take full advantage of their private information and select the most cost-effective set of production practices that meets the performance standard. However, emissions monitoring has presented a considerable challenge for using performance-based standards to mitigate agricultural non-point source pollution (Ribaudo et al., 1999; Shortle et al., 2012). The diffuse nature of non-point source pollution means that monitoring effort cannot be focused on a limited number of pollutant loading points to implement an emission standard. Accurate monitoring of non-point source emissions thus tends to be prohibitively expensive, but advances in environmental monitoring technologies can help reduce this cost. Examples of such technologies include the use of Landsat satellite data to estimate field-level evapotranspiration (Idaho DWR, 2013) and the use of remotely sensed normalized difference vegetation index (NDVI) to estimate field-level nitrogen application rates (Shaver et al., 2011). Such technologies have not yet been applied to agricultural

non-point source pollution control and so, in practice, the traditional monitoring problem remains. However, we discuss the potential for and implications of improved monitoring in Section 8.5.

An alternative to an emissions-based performance standard is to implement an ambient-based performance standard by monitoring pollutant concentrations at a limited number of points in the environment. For example, instead of attempting to monitor runoff and leaching throughout a watershed, surface water and groundwater quality would be monitored only at the base of the watershed. While this approach can reduce monitoring costs, other types of administrative costs can be significant because knowledge of the pollutant transport mechanism would be needed to assess penalties on individual producers if/when the ambient standard is violated. Pollution transport mechanisms for agriculture tend to be characterized by spatial heterogeneity, incomplete observability, time lags, and a great deal of uncertainty; therefore, they are difficult to specify accurately. However, progress continues to be made in this area. For example, Srivastava et al. (2001) integrate a pollutant loading model with a geographic information system (GIS) to assess non-point source pollution potential at the watershed level. Alternatively, Millock et al. (2002) give producers the choice of incurring the cost to monitor and verify their own emissions in exchange for a preferable set of regulations that are not ambient based. Thus, each producer can reveal at least some information to the regulator and effectively convert a non-point source problem into a point source problem. Kurkalova et al. (2004) find evidence that costly source monitoring can produce significant benefits for the case of carbon sequestration by agricultural producers in Iowa.

A potential compromise between design standards (lower information cost for the regulator, higher implementation cost for producers) and performance standards (higher information cost for the regulator, lower implementation cost for producers) is a standard based on estimated emissions (Ribaudo et al., 1999). In this case, rather than attempting to monitor actual pollution emissions, the regulator utilizes a simulation model that approximates the true pollution generation process to an acceptable degree. The model incorporates input and technology choices by producers and generates an estimated level of emissions to which a standard is applied. A producer then uses the model to determine the most cost-effective set of production practices that meet the standard, and the regulator monitors the producer's choices. Penalties can be assessed in two ways. First, the design-based approach is to assess a penalty if the producer does not implement the agreed-upon production practices. Second, the performance-based approach is to assess a penalty if the estimated emission or ambient standard is violated regardless of the producer's choices. Standards based on estimated emissions thus have potentially low information costs because they rely on an approximation of the true pollution mechanism, as well as potentially low implementation costs (and greater incentives for technological innovation) because they allow producers to make production choices based on their private information. However, as Ribaudo et al. (1999) emphasize, the latter is true only to the extent that the pollution model is able to incorporate private information. Thus, the main trade-off associated with standards based on estimated emissions is between incurring the cost of improving the model and incurring the cost of implementing a suboptimal set of production practices. Monitoring costs tend to be similar to those for design standards, though perhaps somewhat higher due to the associated site-specific modeling efforts that are needed to implement the policy. Daberkow et al. (2008) also note that there may be legal problems with using estimated rather than measured emissions if an acceptable level of model accuracy cannot be achieved.

There are many examples of standards-based environmental policies. Examples include the National Ambient Air Quality Standards, motor vehicle emission standards, technology-based emission standards associated with NPDES permitting and state air quality implementation plans, maximum contaminant levels for drinking water, bans on pesticides and other toxic substances, workplace safety standards, and hazardous waste treatment, storage, and disposal standards. However, until recently there have been relatively few examples of standards being used to control agricultural non-point source pollution (Daberkow et al., 2008). An exception is a 1986 Nebraska law requiring BMPs that are tied to local groundwater nitrate concentrations. Bishop (1994) found that areas with the highest nitrate concentrations experienced moderate improvements in groundwater quality after this policy was implemented. Another example is concentrated animal feeding operations (CAFOs) for which many states, including California, have adopted design standards for animal waste storage lagoons. Furthermore CAFOs, although non-point in nature, are regulated as point sources and thus must adhere to NPDES permitting requirements including a nutrient management plan (NMP) that mandates certain production practices (i.e., a technology-based emission standard).[2] In theory, NMPs should be effective in reducing nutrient pollution because they directly limit land application of nutrients in accordance with crop nutrient uptake rates. However, studies have estimated that strict NMP compliance by representative CAFOs would produce potentially significant losses in net revenue, depending on farm location and characteristics (Aillery et

al., 2005; Baerenklau et al., 2008; Huang et al., 2005; Ribaudo and Agapoff, 2005; Ribaudo et al., 2003; Wang and Baerenklau, 2014). Importantly, the relatively larger losses—upwards of 27% of net revenues—estimated by Baerenklau et al. (2008) and Wang and Baerenklau (2014) apply to large California dairies that must limit nitrogen application rates to 1.4 times the agronomic rate of crop nitrogen removal (CRWQCB, 2007). Thus, without adequate enforcement, NMP compliance could be low because of these anticipated loses. Furthermore, anecdotal evidence in California suggests that enforcement may be lacking due to both the complexity of each NMP and limited enforcement resources. Therefore, NMPs, as currently designed and implemented, may fall short of their pollution reduction potential in practice. Several studies have found that using site-specific information to target nutrient restrictions can reduce economic losses; but when information costs are considered, such targeting may not be worthwhile (Carpentier et al., 1998; Moxey and White, 1994).

Another important example of standards applied to non-point source pollution is known as "critical N-loads" in the European Union and Total Maximum Daily Loads (TMDLs) in the United States. A TMDL itself serves as an ambient standard: it limits the daily estimated pollutant load to a receiving water body. Furthermore, implementation of a TMDL may involve specific design or performance standards being placed on individual pollution sources. Although the Clean Water Act of 1972 created the legal framework for TMDLs, states have only more recently begun to implement and enforce TMDLs and therefore more time is needed to monitor and assess their environmental and economic effectiveness.

Ribaudo et al. (1999, p.60) conclude that "performance standards based on runoff or ambient quality are not feasible policies for controlling non-point source pollution, given current monitoring technologies." However, this does not apply to model-based performance standards that are linked to estimated emissions or ambient quality. Such performance standards and design standards are both feasible regulatory instruments for agricultural non-point source pollution control, although each has seen limited use in this context. Sterner (2003, p.77) presents some general conditions under which a design standard may be a preferred policy option:

1. Technical and environmental information is complex.
2. Producers do not hold crucial private information.
3. Producer choices are price inelastic and investments are costly to reverse.
4. Standardizing a technology produces significant ancillary benefits.
5. There exists a clearly superior technology.
6. Monitoring technologies are relatively inexpensive.

The IPCC (2007c, Chap. 13) echoes this reasoning, concluding that standards "generally provide some certainty of

<hr />

2. Point and non-point sources can be distinguished either in legal terms (i.e., applicable regulations) or in physical and economic terms (i.e., whether they emit diffuse pollution that is costly to monitor). Large CAFOs are legally designated under the Clean Water Act as point sources. This has proved effective for reducing emissions to surface waters in part due to the relative ease of observing discharges from manure handling and storage facilities (Doug Patteson, SWRCB Region 5, pers. comm., March 12, 2015). However, nitrogen emissions to groundwater and the atmosphere are more difficult to monitor and remain persistent problems.

emissions levels, but their environmental effectiveness depends on their stringency. They may be preferable when information or other barriers prevent firms and consumers from responding to price signals (*high agreement, much evidence*)." In addition, design standards tend to be more preferable when there is limited scope for input substitution. In such cases, standards applied to a few key inputs and practices (thus requiring less monitoring) can achieve close to the same outcome as a complete set of standards applied to all practices that would produce the theoretically optimal input set.

8.2.3. Economic Incentives

There is a large body of literature on the use of economic incentives for pollution control. Fundamentally, an economic incentive creates scarcity where none previously existed. This typically involves setting a new price or establishing a new market with a limited supply. Most generally, economic incentives include charges, subsidies, or tradable permits for inputs/technologies/practices, emissions, or ambient quality. As with standards, a similar distinction is made between design incentives that focus on inputs/technologies/practices and performance incentives that focus on emissions or ambient quality. Charge and subsidy instruments set prices directly, whereas tradable permits fix the supply of a desirable good and allow the price to be set through market transactions. In each case, the price sends a signal to producers and consumers that certain resources are scarcer than previously thought and provides a financial incentive for using less (or providing more) of the resource. A specific example is an emission charge: pricing emissions signals to the producer that the waste disposal services of the environment are not unlimited. Use of these scarce services incurs an opportunity cost that is represented by the price. Now that the service is priced, the producer will tend to use less of it than if it were free. A similar result is achieved when a cap is placed on total emissions and producers must compete for the right to emit. Competition in the market for emissions rights produces a price for emissions that sends the same signal as a charge.

Economic incentives offer several advantages. One of the main advantages of charges or tradable permits for emissions or ambient quality is allocative efficiency. When diverse producers face the same price for emitting a unit of pollution, each will adjust its production such that the marginal cost of abating pollution is equal to the price of emitting. This means that there will be relatively less high-cost abatement and relatively more low-cost abatement, which facilitates cost-effective attainment of the aggregate pollution level. Another important benefit of economic incentives is that they provide incentives for innovation in the long run. Because producers naturally try to minimize production costs, pricing environmental services tends to encourage them to seek out cheaper ways to effectively abate pollution. Some economic incentives (e.g., charges

and auctioned tradable permits) also provide a revenue stream that can be used to administer environmental policies, improve environmental quality, and potentially reduce the distortionary effects of other revenue generating mechanisms such as income taxes. Subsidies can be appealing when producers have property rights to the polluted medium and/or the polluting industry produces other uncompensated external benefits.

Economic incentives also exhibit certain drawbacks. The fact that diverse producers will select different abatement levels is advantageous from an allocative efficiency standpoint, but it also potentially produces local or regional differences in ambient pollution levels that may be undesirable. Some economic incentives (e.g., charges and auctioned tradable permits) also entail higher costs for producers who must pay for abatement and also for emissions. Although this may not create problems related to economic efficiency, it can be a political issue that affects the policy-making process. On the other hand, a subsidy tends to be more politically feasible because it ostensibly spreads the cost over many people (the public) rather than a small group of producers (Sterner, 2003, p.104); however, the ultimate allocation of costs depends on the specific characteristics of the regulation and how changes in production costs are passed on to consumers. Regardless, either type of price-based instrument may fail to achieve desired levels of environmental quality when—as is common in practice—producers have private information about abatement costs and the regulator has substantial uncertainty about those costs. In such cases, the regulator lacks the information needed to establish an appropriate price level: if costs are higher (lower) than expected, then a given charge (subsidy) level will produce too little (much) environmental quality. Furthermore, a subsidy creates an additional incentive for entry into the industry which can offset pollution reductions at the individual source level. As with standards, monitoring and enforcement also are issues for emission-based economic incentives since the regulator must verify emission levels in order to collect payments. Ambient quality-based economic incentives claim to circumvent this problem, but exhibit other attributes that make them problematic in practice for agricultural non-point source problems (e.g., the need for large side payments for ambient charge schemes; potentially thin markets for ambient permit schemes when spatial heterogeneity matters). As with standards, a model-based approach to developing incentives linked to estimated emissions or ambient quality may be a good alternative provided an acceptable model of the pollution process is available.

There are many examples of incentive-based mechanisms that either have been applied to non-point source agricultural pollution problems in practice or have been studied in the empirical literature. Examples include charges on irrigation water, fertilizer, or runoff; charges on estimated runoff, leaching, or groundwater discharge; point–nonpoint trading schemes; and subsidies for agricultural BMPs. There is relatively little empirical information on performance incentives due to the nature of non-point source pollution and the inherent emissions monitoring problem. However, exceptions include a cropland charge that is tied to phosphorus runoff levels and that has been used successfully to help mitigate phosphorus loading in the Florida Everglades (Light, 2010; Ribaudo et al., 1999) and a point–nonpoint phosphorus trading program in eastern Ontario in which all of the regulated point sources have chosen to purchase non-point offsets rather than pay for costly upgrades (Wainger and Shortle, 2013). There is substantially more information about BMP subsidies (historically the most common policy approach), model-based incentives, and water quality trading.

Examples of BMP subsidy programs include the USDA Environmental Quality Incentives Program (EQIP) and the USDA Conservation Reserve Program (CRP) at the federal level, as well as many state non-point source pollution control programs. Subsidies typically take the form of direct payments, cost-sharing agreements, or cross-compliance provisions that require BMP use as an eligibility condition for other agricultural subsidies (OECD, 2007). Cross-compliance provisions are seen by some as a promising approach to nutrient management in an era of reduced public funding (Claassen, 2004). However, and despite their widespread use, BMP subsidy programs have not achieved significant reductions in agricultural non-point source pollution. Shortle et al. (2012, p.1316) conclude that "a 'pay-the-polluter' approach to getting farmers to adopt BMPs has not succeeded in improving water quality in many impaired watersheds." A 2009 report by the State-EPA Nutrient Innovations Task Group echoes this sentiment, finding that "current efforts to control nutrients have been hard-fought but collectively inadequate at both a statewide and national scale" (EPA, 2009c, p.1). Part of the problem has been that producers are keenly aware of the long-run economic consequences of changing production practices when subsidy programs may be short-lived, combined with the fact that "win-win" scenarios—production changes that reduce pollution without reducing profitability—are uncommon (Daberkow et al., 2008). Shortle et al. (2012) recommend moving away from such voluntary subsidy programs and towards the "polluter-pays-principle" with an emphasis on performance outcomes rather than BMPs.

Daberkow et al. (2008) surveyed a decade of work (1991–2001) on model-based economic incentives for nitrogen pollution control and concluded that both input and emission charges were problematic due to the costs imposed on producers. Those studies suggest that such instruments may be able to bring about moderate reductions in nitrogen pollution, but more significant reductions (e.g., to drinking water standards) likely will entail substantial economic losses. However, more recent work that accounts for the spatial variability of irrigation systems (Baerenklau et al., 2008; Knapp and Schwabe, 2008) and also crop choice and the potential to recycle shallow groundwater for irrigation (Wang and Baerenklau, 2014) suggests that the economic losses associated with charges on estimated emissions may be substantially less than previously thought. With regard to input and output charges, Sterner (2003, p.100) concludes that "if the abatement possibilities are severely limited and monitoring is difficult, then [charges] on inputs or outputs may be good second-best instruments."

Water quality trading is a relatively new policy innovation that has received significant interest from stakeholders and has experienced relatively fast growth in practice. Water quality trading essentially allows sources with high abatement costs to compensate sources with low abatement costs for reducing pollution on their behalf, thus promoting cost-effective pollution abatement. A 2004 survey by Breetz et al. found that 40 trading initiatives had been created in the United States since 1990. Most of these include provisions for trading with agricultural non-point sources; however, the authors find relatively low participation rates by such sources. Several challenges to establishing effective water quality trading programs have been documented, including the following: thin markets; difficulty measuring reductions in non-point source emissions; insufficient compliance by non-point sources with the terms of trade when trading is driven by point source regulations; additional risk borne by point sources when they pay non-point sources to abate but cannot also transfer the legal obligation for abatement; aversion by non-point sources to being labeled as a polluter for participating in such programs; and the requirement that non-point sources satisfy costly baseline conditions (i.e., implementation of prerequisite BMPs) before they can enter the market (Ribaudo et al., 1999, 2011). One area where nutrient trading could overcome some of these barriers is the Chesapeake Bay. This is a large watershed with a regulatory structure in place (TMDL) and the potential for both in-state and out-of-state trading that could overcome the thin markets barrier. An analysis of nutrient trading in the Chesapeake Bay found significant cost savings potential (Van Houtven et al., 2012). An analysis by the EPA SAB (2011) suggests that auction-based contracting, in which BMP subsidies are effectively targeted to desired areas, may be a more appropriate market mechanism than tradable permits for controlling agricultural non-point source pollution. Rabotyagov et al. (2012) demonstrated a model-based approach to such contracting for nutrient management. And Selman et al. (2008) estimated that a BMP auction for phosphorus reduction on Pennsylvania's Conestoga River that did not attempt to target contracts was seven times more cost-effective than the standard BMP subsidy approach utilized by EQIP.

8.2.4. Summary

Performance-based standards and incentives generally are considered to be infeasible for non-point source pollution control problems due to the inherent nature of these problems and the associated monitoring and information costs

(Ribaudo et al., 1999). Design- and model-based policy instruments are feasible options, with education preferably used in a supporting role. A central trade-off to consider is the cost of acquiring information that could be used to incorporate site-specific heterogeneity into the policy design versus the higher abatement costs typically associated with simpler policy instruments that are applied more uniformly across all sources. Both incentives and performance standards create greater flexibility for producers than do design standards and tend to facilitate lower abatement costs and more abatement innovation provided they are costly and enforced. However, the greater flexibility provided by price-based incentives also is associated with greater uncertainty over environmental outcomes. Based on experience to date, it is generally accepted that heavy reliance on voluntary participation in BMP subsidy programs cannot achieve significant reductions in agricultural non-point source pollution. Rather a new approach that relies on a combination of more effective policy instruments appears needed.

8.3. Framing of Assessment Criteria for Policy Instruments

Section 8.2 provides a general overview of available policy instruments for non-point source pollution control, and in the process discusses multiple characteristics to consider when evaluating policies. These include cost-effectiveness, allocative efficiency, administrative cost, complexity, acceptability, environmental effectiveness, feasibility, and uncertainty, among others. The present section considers policy characteristics deliberatively, organizing them into a coherent methodology for organizing evidence about (Section 8.4) and assessing candidate policies for (Section 8.5) controlling nitrogen emissions from agricultural sources in California.

8.3.1. Policy Assessment Criteria

The widely used text by Hanley et al. (1997, p.91) lists four main evaluative criteria: effectiveness, efficiency, equity, and flexibility. The authors also provide (p.95) a list of practical considerations that includes information availability, administrative capability, institutional structure, and political feasibility. The IPCC (2007c, chap. 13.2.2.2) also utilizes a relatively short list of evaluative criteria for climate change policies that is similar to those presented by Hanley et al.: environmental effectiveness, cost-effectiveness, distributional considerations, and institutional feasibility. Sterner's (2003, p.214) list of general policy criteria includes some of these common themes (i.e., distributional considerations, political feasibility, economic efficiency) but also includes some additional considerations: general equilibrium effects, number of dischargers, overall complexity, information asymmetry (i.e., monitoring issues), and risk/uncertainty. The European Nitrogen Assessment (Oenema

et al., 2011a, pp.75–77) adds two more criteria to this list: technological feasibility and enforcement/compliance. Ribaudo et al. (2011, p.47) include geographic coverage for non-point source policies, while Canada et al. (2012, p.19) also consider revenue generation.

We distilled these suggested considerations into a set of six key criteria for evaluating policies for controlling nitrogen emissions from agricultural sources in California. The criteria are as follows:

1. Environmental effectiveness
2. Technological feasibility
3. Cost-effectiveness
4. Distributional effects
5. Institutional compatibility
6. Adaptability

A policy is "environmentally effective" if it has an acceptable likelihood of achieving the desired environmental goal in practice. Here we impose the additional requirement that the policy is not expected to substantially exacerbate the other nitrogen issues identified in this assessment. Therefore, a policy that achieves desired nitrate reductions largely by increasing emissions of ammonia is not environmentally effective; one that does so by increasing emissions of inert nitrogen gas is environmentally effective. It is worth emphasizing that if the policy goal is unrealistic or somehow incompatible with the environmental system, then no candidate policies may be environmentally effective. For non-point source pollution, such a goal might be an ambient standard that neglects the uncertainty or time lags inherent in the pollution transport mechanism.

A policy is "technologically feasible" if it relies on currently available technologies and practices that are suitable for use by the regulated industry and by the regulator. This includes both abatement and monitoring technologies. Clearly there is a need for judgment to define what is "feasible" or "available" or "suitable" which can be related to the willingness to incur costs for environmental improvements. Here we use a relatively generous definition of technological feasibility and note this may increase our assessments of policy-related costs. This definition also is consistent with a primary emphasis on environmental effectiveness when evaluating policies. If currently available technologies are unproven and thus inherently risky, then a policy is still technologically feasible, but the associated risks and learning costs should be considered as additional costs. If currently available technologies can only achieve the policy goal when production is significantly reduced, then this also should be considered as an additional cost. The extent to which a policy drives (and accommodates) technological innovation—a potentially important mechanism for improving cost-effectiveness in the long-run—is more appropriately considered as an element of adaptability.

A policy is "cost-effective" if the total economic cost of implementation is expected to be less than that for other policy alternatives. Policy implementation can have ripple effects throughout an economy and thus a full accounting of the associated costs can be elusive. In practice, judgments must be made about the scope of an economic policy analysis. This includes defining both the set of relevant stakeholders (those to whom costs accrue) and the set of costs that will be evaluated. For our purposes, the set of stakeholders is defined as California's agricultural producers and regulatory agencies. Clearly other residents and businesses in California may incur policy-induced costs, but these will be secondary effects due to responses by producers (e.g., changes in prices for agricultural inputs and outputs). Evaluating welfare effects in secondary markets can be challenging. When those markets operate efficiently, secondary effects often can be ignored (Boardman et al., 1996, p.87), but otherwise secondary effects may be significant. Here we do not attempt to evaluate such effects due to a general lack of good information to do so. Instead, we focus on costs associated with both market (budgetary costs) and non-market goods and that are thought to have potentially significant effects on producers and regulators. Thus, this criterion considers the following: the time cost of producer education and information acquisition; expenditures on abatement technologies; production losses; risk premiums; and education, administration, monitoring, and enforcement costs incurred by regulators. It also implicitly considers the allocation of abatement effort across producers and the extent to which this allocation is sensitive to differences in abatement costs because this can have important implications for cost-effectiveness.

Monitoring and enforcement of policy compliance is complicated by political and institutional considerations. Institutional issues are considered later in this chapter. One key purpose of this assessment is to elevate political debate on solutions. To avoid removing promising options from consideration prematurely (before debate can occur) and because there is little basis for assessing ex ante the politics of specific options, this analysis deliberately abstracts from the use of political capital to shape policy choice, instrument design, and implementation processes. While it is well recognized that this political power to influence policy is concentrated within specific interest groups and that it is an important aspect of the policy process, assessment of these political realities is highly context specific and there is little relevant scientific literature to be assessed for the specific case of nitrogen in California agriculture.

Characterizing the "distributional effects" of a policy typically involves assessing how the significant policy impacts—both costs and benefits—are allocated across specific groups of stakeholders. We also incorporate an evaluation of that allocation in terms of prevailing notions of equity and social justice—in other words, the extent to which we think a policy will be perceived by stakeholders as "fair," particularly in terms of its impacts on disadvantaged groups. While distributional effects and their perceptions also can have a substantial impact on the political viability of a policy, such political calculus is outside the scope of this criterion.

A policy is "institutionally compatible" if its implementation does not conflict with the larger institutional framework in which it exists. This criterion thus considers legislative authority, legal and jurisdictional constraints, and existing policies and administrative structures, as well as historical/cultural expectations. Compatibility should be assessed at federal, state, and local levels. A policy can be institutionally compatible if it creates, for example, the necessary administrative structure when none currently exists; however, this likely entails increased administrative cost that would have implications for cost-effectiveness. A policy that is at odds with cultural expectations might be made institutionally compatible through education and outreach efforts, but again with similar cost implications.

Finally, a policy is "adaptable" if it is flexible enough to accommodate changing economic, political, and environmental conditions across both space and time, while also being resilient enough to maintain its essential characteristics (i.e., environmental effectiveness, cost-effectiveness, etc.). For example, an emission charge that is tied to an inflation index is more adaptable than one that is not; a technology standard that includes provisions for site-specific characteristics is more adaptable than one that does not. Adaptable policies also accommodate, and preferably incentivize, technological innovation. To the extent possible, policy-makers should consider where significant, "game-changing" innovations may occur, and design policies that will both promote and adapt to these changes if/when they occur.

Economic efficiency is notably absent from this list. This is not because it is unimportant: a thorough accounting of both costs and benefits should inform any policy debate. For many pollutants, including nitrogen, benefits assessments can be challenging because they involve valuing changes in environmental quality and the myriad ways those changes may affect individuals, both now and in the future. Often these benefits are realized outside of the market economy, meaning there are no immediately available price signals to use for developing demand functions and assessing welfare effects. Examples of such nonmarket benefits (see Chapter 5) include recreational use of natural resources, the existence value of biotic species, health outcomes, and the desire to leave a cleaner environment for future generations. Furthermore, these benefits often are realized in the future, meaning their actual magnitudes are uncertain and may even accrue to individuals who are not yet alive to express their values. For all of these reasons, formal benefits assessments can be onerous.

Despite these challenges, benefits are considered in various ways during the environmental goal setting process, since it is these goals that largely determine the benefits of a policy. Outcomes from this process can include things like maximum contaminant levels for drinking water and ambient air quality standards, both of which apply to compounds of nitrogen. Due to the difficulties associated

TABLE 8.1
Policy assessment matrix

Selected policy instruments	Evaluation criteria					
	Environmental effectiveness	Technological feasibility	Cost effectiveness	Distributional effects	Institutional compatibility	Adaptability
Education						
Labeling or certification programs						
Training and outreach						
Reporting inventories or registers						
Awards or recognition						
Standards						
Input standards						
Emission standards						
Ambient standards						
Economic incentives						
Input charges						
Ambient charges						
Tradable input permits						
Tradable emission permits						
Tradable ambient permits						
Auction-based abatement contracts						

with undertaking a formal evaluation of benefits, the uncertainties associated with what we currently know about the benefits of alternative levels of nitrogen pollution, and consistent with textbook evaluations of alternative policy instruments, we take the outcomes of the goal setting process as given and focus on evaluating policy approaches for achieving these predetermined goals. Thus, "environmental effectiveness" appears as a criterion, effectively serving as a surrogate for benefits. "Distributional effects" also is a criterion, in which the relative allocation of benefits across groups is considered. However, the aggregate level of benefits, and by extension economic efficiency, is not considered because inadequacy in current data and methods make it impossible to include in this assessment.

8.3.2. Policy Assessment Matrix

Table 8.1 shows the evaluation criteria in table format along with the policy instruments we have selected for evaluation. Labeling and certification programs apply to products and/or producers; thus, the information associated with such programs is targeted at consumers. This contrasts with training and outreach programs, and some reporting inventories or registers, for which information is targeted at producers. Examples of the latter include soil testing results, N application rates and methods, or a nitrogen emission inventory. Participation in education programs can be voluntary or mandatory.

We consider standards applied to inputs, emissions, or ambient quality. Input standards implies a broad definition of "inputs" that includes land, production technologies and practices, variable inputs, and abatement technologies and practices. Thus, zoning and local ordinances would be input standards. We do not consider mandatory training programs to be input standards unless the policy also mandates use of certain production practices; otherwise, use of those practices is unenforceable and thus the program is purely educational. For the case of non-point source pollution, emission standards would involve estimated quantities, whereas ambient standards can be based on actual observations. We do not consider liability rules as a separate policy instrument because we view it instead as a mechanism to enforce ambient standards. Such a mechanism has institutional precedent (i.e., Cleanup and Abatement Orders in Porter-Cologne) but would present significant technological challenges (i.e., adequate environmental modeling).

We consider seven different categories of economic incentives that cover price (charge) and quantity instruments applied to inputs, emissions, and ambient quality; and one auction-based mechanism. As with standards, a broad definition of "inputs" is used here. We consider both positive charges (taxes) and negative charges (subsidies). Thus, input charges also include subsidies for pollution reducing inputs; emission charges also include abatement subsidies; and ambient charges also include subsidies when ambient quality exceeds a specified level. Each of the quantity instruments places an aggregate limit on an activity in the pollution process (i.e., purchase of polluting inputs, generation of emissions, or delivery of pollution to receptor points) and allows dischargers to buy and sell rights to these activities in permit markets. As with standards, emission charges and tradable permits and the auction-based mechanism would involve estimated quantities, whereas ambient charges and tradable permits can be based on actual observations.

We do not consider any hybrid instruments (e.g., an ambient standard that is implemented through enforceable input standards) or combinations of instruments (e.g., input standards with training and outreach) in detail. However, we revisit these possibilities in a subsequent section.

8.4. Experience with Nitrogen Policy Instruments in Practice

The overview provided in Section 8.2 briefly mentions some specific examples of policies that have been used to control nitrogen pollution in practice. The current section revisits some of these examples along with additional case studies that can inform the policy discussion. The presentation is deliberately brief, but the reader can find supporting details in an online appendix.

8.4.1. Case Study Overview

We consider a total of 12 case studies: 5 California programs, 5 nutrient-impaired waterbodies in other states, an overview of European nitrogen policies, and a previously published review of state-level nutrient programs (EPA, 2009c). The last of these differs from the others and includes both program assessments and recommendations for the future. Table 8.2 lists the case studies and their coverage of policy instruments considered in this assessment. There is a heavy emphasis on three policy instruments: training and outreach, input standards, and ambient standards. A fourth instrument, input charges, also appears often. Overall, this is consistent with the standard approach to regulating non-point source agricultural pollution. Section 8.2 argues that this approach has not achieved desired improvements in agricultural nitrogen pollution. The case studies offer insights and lessons learned, as well as information about other less commonly used policy instruments.

8.4.2. Education

Most of the case study evidence for education-based policies applies to training and outreach programs, with each case study including such a component. For two of them that have demonstrated significant nutrient reductions—the Neuse River Basin and the Florida Everglades—education is believed to have played an important role in achieving the program results. More generally, the common theme across case studies is that training and outreach are potentially val-

uable components of a broader regulatory strategy. Reporting inventories are utilized in three case studies (the State EPA review notwithstanding), but two of these instances apply to greenhouse gas emissions (e.g., N_2O) rather than the pollutants that have emerged as the primary concerns in this assessment (nitrate and ammonia). The third instance (the Agricultural Waiver Program) addresses nitrates but was adopted as policy, in 2012. Therefore, the effectiveness of such inventories as a practical regulatory policy cannot be inferred from these case studies. However, the State-EPA Nutrient Innovations Task Group identifies a "nutrient releases inventory" as a potentially useful approach, as well as "green labeling" for proper nutrient management.

8.4.3. Standards

The ubiquitous use of standards is apparent in Table 8.2. Consistent with the diffuse nature of non-point source pollution, input and ambient standards are universal in this set of case studies, while emission standards are much less common. Nitrogen-related emission standards are limited to CAFOs that are classified as point sources under the Clean Water Act, and combustion sources in Europe and the US Standards for nutrient control typically are implemented through a familiar BMP framework: in areas where ambient standards are not met, education and incentives (usually input subsidies) have been offered to producers to encourage the adoption of approved BMPs (input standards). Some of the more salient lessons that can be drawn from this approach are listed below.

- *Technologies that reduce nitrogen pollution exist, but they are costly to implement and produce relatively small private benefits; thus, voluntary adoption tends to be low.* Multiple state-level non-point source programs—including California's—have demonstrated BMP effectiveness in reducing non-point source pollution in specific cases, but BMP implementation and thus pollution reduction have not been widespread. Experience suggests that producers often perceive the economic costs to outweigh the benefits of participating in such programs.
- *Voluntary programs have not been successful enough, but they have helped to inform questions about BMP effectiveness under different conditions.* One of the benefits of subsidized BMP installations in California and elsewhere is a better understanding of how BMPs perform under real operating conditions. This is potentially useful for future policies that may rely on changes in management practices to achieve pollution reductions.
- *A compulsory yet flexible BMP program with ongoing monitoring, research, and education components has proved to be highly effective environmentally in Florida.* Two relatively unique features of Florida's Everglades Regulatory Program are that participation is

TABLE 8.2
Case study coverage matrix

Selected policy instruments	California Non-point Source Program	California Agricultural Water Quality Grants Program	California Central Coast Agricultural Waiver Program	California Dairy Regulations	California Air Regulations	Neuse River Basin	Gulf of Mexico	Chesapeake Bay	Florida Everglades	Conestoga River Watershed	European Experience	State-EPA Nutrient Innovations Task Group
									Case studies			
Education												
Labeling or certification programs												X
Training and outreach	X	X	X	X	X	X				X		X
Reporting inventories or registers			X		X						X	X
Awards or recognition											X	X
Standards												
Input standards	X	X	X	X	X	X	X		X	X	X	X
Emission standards				X	X						X	X
Ambient standards	X	X	X	X	X	X	X	X	X	X	X	X
Economic incentives												
Input charges	X	X					X		X		X	X
Emission charges											X	X
Ambient charges												
Tradable input permits								X				
Tradable emission permits											X	X
Tradable ambient permits												
Auction-based abatement contracts										X		

compulsory rather than voluntary and that monitoring was relatively good due to the existence of a network of drainage canals. And although participation is compulsory it is also flexible: producers must select and implement a minimally sufficient combination of BMPs from an approved menu.

- *Strong collaboration and communication across all parties helps foster success.* North Carolina's Neuse River basin is another BMP success story with similarities to Florida's Everglades. Here again, participation (by both point and non-point sources) was mandatory yet flexible. In this case, flexibility was achieved by affording producers the option of working collectively to achieve an aggregate nitrogen reduction target (similar to a tradable permit instrument).
- *A coordinated mix comprised largely of mandatory standards has produced measurable improvements in Europe.* Nitrogen management in Europe is largely governed by multiple EU policy "directives" aimed at reducing nitrogen emissions to water and air. These directives tend to rely heavily on mandatory standards as well as cross-compliance provisions. Implementation is the responsibility of the member states. Monitoring data show significant reductions in multiple nitrogen loads from 1990 to 2010, but also variability across regions.
- *Regulations have not substantially improved nitrogen-related air pollution from California's agricultural sources.* Policies regulating nitrogen air emissions in California include an agricultural burning policy, NO_x emission limits, and regulations on the disposal of animal carcasses. These policies have had no detectable effect on the number of exceedances of the NO_x standard in agricultural regions.

8.4.4. Economic Incentives

The set of case studies includes a relatively small number of instances where innovative economic incentives have been used to achieve policy goals. However, three important approaches are represented: emission charges, tradable permits, and auction-based contracts. Emission charges were implemented in the EU by the Netherlands through the Mineral Accounting System (MINAS). MINAS levied a tax on estimated excess nitrogen and phosphorus flows through agricultural systems. According to Mayzelle and Harter (2011), this approach was popular for its simplicity and had strong support from the Dutch government. Furthermore, Westhoek et al. (2004) estimate that it reduced the nitrogen surplus on Dutch dairy farms by approximately 50 kg ha^{-1} with very low cost to farmers. However, the EU determined that the approach did not satisfy the EU Nitrate Directive requirements, so it was replaced with nutrient application (input) standards. This appears to be a case in which con-

flicts between federal and state policies undermined an otherwise successful policy.

Tradable permits have been implemented in the Netherlands (prior to MINAS) and currently in the Chesapeake Bay. Although the Dutch trading system achieved measurable load reductions, implementation was burdened by the anticipated change to the MINAS emission tax and associated uncertainty (Wossink, 2003). Trading in the Chesapeake Bay has been very limited, to date. Auction-based contracting was utilized successfully in Pennsylvania's Conestoga River Watershed, which experienced excessive phosphorus loads largely from agricultural producers. Two auctions, conducted in 2006, allowed producers to submit bids for installing and maintaining one or more BMPs on their properties. Bidders worked with Lancaster County Conservation District technicians to estimate with computer models their expected phosphorus reductions based on site-specific characteristics. These estimated reductions were used with the bid prices to determine a cost per pound of phosphorus abatement for each bid. Bids were then ranked by cost-effectiveness from lowest to highest cost per pound, and contracts were awarded in order of cost-effectiveness until the auction budget was exhausted. The auctions mitigated an estimated 92,000 lb of phosphorus. This load reduction would have cost more than seven times as much to achieve using standard EQIP subsidies (Selman et al., 2008). Overall, the auctions were a success, despite the novelty of using this policy instrument. Additional use of this approach would benefit from robust outreach, education, and technical assistance.

8.4.5. Broader Lessons

Several case studies provide broader lessons that are applicable across multiple classes of policy instruments. We list the more salient topics here and direct the reader to the online appendix for additional details.

- *Granting the authority to regulate a pollutant does not mean that the pollutant will be regulated: authority is necessary but not sufficient.* California had the authority to regulate non-point sources of nitrogen pollution for decades, but allocated relatively little attention to the problem until 2004. Similarly, state implementation of federal TMDL legislation has been slow. Because regulatory resources are limited, specific prioritization of issues is needed to achieve progress.
- *Cross-jurisdictional conflicts can seriously impair program effectiveness.* California's Agricultural Water Quality Grants Program requires disclosure of BMP locations and monitoring points, but this conflicts with privacy provisions of the farm bill and has limited participation. Also, California's General Obligation Bond Law requires projects be capital improvements with a useful life of at least 15 years; however, most BMPs have a much shorter useful life

and thus do not qualify for such funding. Furthermore, there are no requirements that matching funds obtained from the federal government through EQIP be used to install desired BMPs.

- *Grant programs are dependent on state financial situations.* The California "bond freeze" of 2008 impaired the ability of grantees and subcontractors in the Agricultural Water Quality Grants Program to complete the work or receive payment for work completed, resulting in a number of stopped or delayed projects.
- *Most programs lack adequate data collection, reporting and evaluation components—particularly of environmental outcomes—but persistent, widespread nutrient problems demonstrate that past programs generally have not achieved the desired results.* The State-EPA Task Group identified this unfortunate information gap in their survey of several programs nationwide.
- *It has been difficult to document environmental progress, particularly over short time horizons, due to time lags and uncertainties in the pollution delivery mechanism.* This is particularly true for larger watersheds with long distances between sources and receptors, and for groundwater.
- *A one-size-fits-all approach at the federal level can undermine otherwise successful local approaches.* This was the case in the Netherlands when the successful MINAS program was deemed insufficient under the EU Nitrate Directive.
- *Flexibility is crucial for cost-effectiveness.* Programs that account for local conditions, that allow producers to make more of their own choices, and/or that allow for coordination and cooperation among sources (e.g., the Conestoga Watershed and the Neuse River Basin) tend to be more cost-effective.
- *Policies should be designed with the complexity of the larger socio-economic-environmental system in mind.* A narrow focus on nitrogen emissions, or a particular type of nitrogen emission, can create additional problems, environmental and otherwise. Coordination of nitrogen policies with other environmental, social, and economic policies is preferable.

8.5. Assessment of Policies for Nitrogen Regulation in California

8.5.1. Policy Assessment Rationale

This assessment has identified numerous nitrogen-related pollution problems in California. Here we consider potential policy approaches for mitigating agricultural contributions to two high-priority nitrogen issues: nitrate emissions to groundwater and ammonia emissions to the atmosphere. For each issue, we rate different policies in each of the six evaluation criteria using a three-point scale: "good" (or

"small" for distributional effects) implies an advantageous or beneficial attribute of a policy, "moderate" implies a generally neutral attribute, and "poor" implies a disadvantage or drawback. We then elaborate on these ratings in the main text. The ratings represent qualitative judgments informed by the available evidence presented in this assessment; therefore, they are best interpreted in relative terms, by comparing policies against each other. Where ratings depend on specific policy attributes (e.g., voluntary or mandatory, uniform or nonuniform) that imply trade-offs across criteria (e.g., environmental effectiveness vs. cost-effectiveness), we select attributes that emphasize environmental effectiveness provided they remain reasonably technologically feasible and institutionally compatible. Elsewhere, where such trade-offs do not exist, we assume policies would be well designed and forego evaluating inferior policies.

As mentioned previously, policy assessment requires first establishing a specific policy goal before proceeding to apply the evaluative criteria. Such specific goals currently are not available for the problems we consider here. While it may seem appropriate to set specific goals of achieving the relevant drinking water and ambient air quality standards, the time frame for doing so and the anticipated contributions from nonagricultural sources remain ambiguous. Therefore, we assess candidate policies that have the potential to successfully reduce emissions from agricultural nitrogen sources to levels compatible with long-run attainment of current environmental standards. We do not define the "long-run" specifically, but we do note that it is shorter for air quality and longer for groundwater quality—the latter may be on the order of many decades. We thus separate the long-run problem of effectively stewarding resources from the short-run problem of remediation.

Two additional comments on the use of these criteria are worth emphasizing. First, the criteria are most useful when applied to a specific policy mechanism under a specific set of conditions. For the case at hand, there is both uncertainty about the details of any future nitrogen policies that might be adopted and heterogeneity in the conditions under which those policies would be applied. Therefore, we necessarily must make some simplifying assumptions and generalizations when evaluating candidate policies. We highlight these where appropriate. Second, the criteria can be used both to evaluate policies and to assess uncertainty about policy characteristics. As in preceding chapters and sections, we do both of these here by providing specific assessments of central tendencies within each criterion and later making more general observations about the level of evidence and consensus.

8.5.2. Groundwater Nitrate Policy Assessment

8.5.2.1. EDUCATION

There is general agreement that education alone is insufficient for mitigating nitrogen pollution problems, including groundwater nitrate. We score each of these instruments relatively low on environmental effectiveness, but also acknowledge that a reporting inventory or register may have somewhat greater effectiveness, as evidenced by the success of the Toxics Release Inventory, provided it is not undermined by moral hazard (Table 8.3). Labeling or certification programs can be somewhat more effective but are dependent on consumer willingness to pay for public goods. For the case of groundwater nitrates, the impact of which is mainly a localized health effect, there is no direct evidence on willingness to pay by the broader public. More generally, compared to products like dolphin-safe tuna, bird-friendly coffee, and sustainably harvested timber, "low nitrogen" farming is arguably less charismatic and suffers from lower levels of public awareness, both of which tend to reduce willingness to pay. Although none of these instruments can be recommended as a cornerstone for mitigating groundwater nitrate, each should be considered in a complementary role to other regulations. The literature on effectiveness of public awareness campaigns (diet, exercise, smoking, drugs and alcohol, texting/cell use while driving) may provide relevant insights, but is beyond the scope of this chapter.

Education programs are generally highly technologically feasible; however, we rate labeling/certification programs and reporting inventories/registries lower due to the associated emissions monitoring problems. Such programs would have to be based on estimated emissions (or perhaps inputs for labeling/certification programs, which is why they rate slightly better), and thus the feasibility of adequately modeling groundwater nitrate emissions from agricultural operations would need to be addressed. Similarly, the cost-effectiveness of education programs tends to be good, with the same caveat for labeling/certification programs. The labor costs associated with labeling/certification and training/outreach tend to increase the costs of these policies.

Because education programs largely depend on the individual choices made by producers, and sometimes consumers (e.g., use of information in a reporting inventory), they tend to generate an uneven distribution of costs and benefits. Because costs tend to be incurred voluntarily, it is the uneven distribution of pollution reduction benefits (a public good) that is of concern. However, because the environmental effectiveness of these programs is relatively poor, we do not expect significant aggregate differences in benefits. Rather such programs are likely to produce small environmental improvements for the vast majority of the population, and substantial environmental improvements for a small minority (e.g., those who happen to rely on drinking water that is affected by producers who respond to the education programs), at little cost to producers, which is why we rate them as having small distributional effects.

Institutional compatibility is generally good for education programs since such programs already exist (e.g., cooperative extension). However, the creation of a reporting inventory/register may require new legislation (EPA, 2009c).

The adaptability of each instrument is generally good, largely because they involve a limited amount of voluntary participation and are based on information transfer; however, the incentives for technological innovation tend to be small.

8.5.2.2. STANDARDS

We consider standards for groundwater nitrate management that are compulsory, enforced, and account for spatial heterogeneity; the last of these implies that input and emission standards would be nonuniform. All types of standards with these properties are potentially effective for groundwater nitrate management. We rate ambient standards somewhat lower because enforcement may be significantly problematic because the regulator knows very little about the contributions of each source.

Input standards that require the use of "best available" technologies should be technologically feasible. Emission standards for groundwater nitrates would be based on estimated emissions, which requires an adequate emissions model; this creates added technical challenges. Ambient standards additionally require an adequate environmental model which further diminishes technological feasibility. It is noteworthy that a TMDL is a type of ambient standard that is being applied to surface water nutrient problems in practice; however, implementation typically relies on enforcement of source-specific input or emission standards rather than on penalties for violating the ambient standard; in this sense, TMDLs are hybrid instruments. The same can be said of the National Ambient Air Quality Standards.

The main drawback of using standards is poor cost-effectiveness. However, for non-point sources, this conclusion is based on experience with input standards (e.g., NMPs); emission standards have not been used and experience with ambient standards is limited to hybrid applications. Applying emission or pure ambient standards to the non-point source groundwater nitrate problem would reduce abatement costs, possibly significantly. However, administrative costs could increase significantly. Overall, we expect the abatement costs savings, which accrue in perpetuity, to outweigh the administrative costs, some of which would be non-recurring fixed costs and others of which would benefit from economies of scale in implementation; however, this is suggested but unproven because of lack of relevant evidence on costs.

Uniform standards would create widespread environmental improvements, implying relatively small distributional effects on the benefit side. However, some producers may incur significant abatement costs and some may find compliance to be substantially costlier than others. Small producers also may be disproportionally impacted by fixed costs. Although some of these costs would be passed on to consumers through higher prices, it is difficult to judge these secondary effects. All of this suggests larger distributional effects on the cost side compared to the benefits

TABLE 8.3

Groundwater nitrate policy assessment matrix

Selected policy instruments	Environmental effectiveness	Technological feasibility	Cost effectiveness	Distributional effects	Institutional compatibility	Adaptability
Evaluation criteria						
Education						
Labeling or certification programs	Poor	Good/moderate	Good/moderate	Small	Good	Good
Training and outreach	Poor	Good	Good/moderate	Small	Good	Good
Reporting inventories or registers	Poor/moderate	Moderate	Good	Small	Good/moderate	Good
Awards or recognition	Poor	Good	Good	Small	Good	Good
Standards						
Input standards	Good	Good	Poor	Moderate	Good/moderate	Poor
Emission standards	Good	Good/moderate	Moderate	Moderate	Moderate	Moderate
Ambient standards	Good/moderate	Moderate	Moderate	Moderate	Poor/moderate	Good
Economic incentives						
Input charges	Good	Good	Poor/moderate	Moderate	Good	Poor
Emission charges	Good	Good/moderate	Moderate	Moderate	Moderate	Moderate
Ambient charges	Good/moderate	Moderate	Moderate	Moderate	Poor/moderate	Good/moderate
Tradable input permits	Moderate	Good	Poor/moderate	Moderate	Good	Poor
Tradable emission permits	Good	Good/moderate	Poor/moderate	Moderate	Moderate	Moderate
Tradable ambient permits	Good/moderate	Moderate	Poor/moderate	Moderate	Poor/moderate	Good
Auction-based abatement contracts	Good	Good/moderate	Moderate	Moderate	Good	Moderate

side. Overall, we rate these effects as being larger than for education-based programs, and thus having moderate magnitude.

Although there is precedent for agricultural input standards (e.g., BMPs, NMPs), Porter-Cologne does not permit the state to prescribe abatement technologies, which reduces the institutional compatibility of this approach. However, Canada et al. (2012) suggest that this obstacle could be overcome fairly easily with well-designed input regulations, so we rate the institutional compatibility of input standards relatively high. We rate emission standards slightly lower because they would have to be based on estimated emissions, which remains a relatively novel concept. Although models are used in other related contexts (e.g., to develop NMPs, establish TMDLs, and set ratios for point–nonpoint surface water trading programs), none of these cases relies on model output to levy a penalty for noncompliance. While this could prove problematic from an institutional perspective, Baerenklau and Wang (2015) argue that a recent legal decision supporting the Chesapeake Bay TMDL bodes well for standards based on estimated emissions. Ambient standards are likely to be more problematic since penalties are not directly related to any source-specific information. Regardless, the authority for enforcing emission and ambient standards is inherent in the Porter-Cologne Act.

Compared to institutional compatibility, the relative ranking is reversed for adaptability. Input standards tend to be the most rigid and inflexible. They are furthest removed from ambient groundwater nitrate concentrations so they are least responsive to changes in factors that subsequently combine with the regulated inputs (e.g., other production and abatement technologies, economic and environmental conditions) to produce the damage-causing concentrations. They also provide the smallest incentive for innovation in abatement technologies. Continually revising standards to keep them appropriate and relevant is a costly process. Emission standards perform better in this regard by accommodating changes in farm-level decision-making without undermining their effectiveness, but they do not accommodate changes beyond the farm scale, such as local environmental conditions or industry size. Ambient standards are the most adaptable because they regulate the environmental medium of concern (groundwater quality) directly; thus, they accommodate changes in broader environmental and economic conditions as well.

8.5.2.3. ECONOMIC INCENTIVES: CHARGES

We focus on positive (nonuniform) charges first before commenting on how the assessment would differ for subsidies. The environmental effectiveness of nonuniform charges for the case of groundwater nitrate is similar to that for standards. Notably water, rather than fertilizer, appears to be the preferable target for an input charge, in terms of both environmental effectiveness and cost to the producer

(Helfand and House, 1995; Knapp and Schwabe, 2008). The ratings for (estimated) emission and ambient charges remain the same as for standards, but in practice their effectiveness may be slightly lower due to the additional uncertainty about abatement cost curves.

Technological feasibility of charges is the same as for standards. Cost-effectiveness should be similar to, but in practice slightly better than, that for nonuniform standards due to improved allocative efficiency. An efficient reallocation of abatement effort potentially can provide substantial improvements in cost-effectiveness, but due to the spatially heterogeneous nature of the nitrate problem, some of this improvement would need to be sacrificed in order to achieve pollution reduction goals statewide. Furthermore, costs will be higher for producers if charge revenues are not invested back into the industry; here we assume such investment would occur since this should reduce distributional effects, and we rate the distributional effects the same as for the analogous standards. Institutional compatibility for charges is rated similar to that for standards, and for similar reasons; but we note that input charges do not conflict with Porter-Cologne as do input standards (in fact, Porter-Cologne implies that dischargers should incur the costs associated with contaminated drinking water), thus improving their compatibility.

Evaluating the adaptability of charges raises some competing issues. On the one hand, price-based instruments generally provide more flexibility for producers than do standards. On the other hand, such flexibility can directly undermine the effectiveness of charges in a spatially heterogeneous environment. Also, being artificially set prices, charges may need to be deliberately revised to retain their initial effectiveness when other economic variables change (e.g., inflation, technological innovation). Such revisions can be costly for regulators to promulgate and for producers to respond to; thus, we do not expect that they would be undertaken as often as they should be. In light of all this, we rate the adaptability of charges about the same as for their analogous standards.

Our assessment of negative charges (subsidies) for pollution-reducing inputs, abatement, or improvements to ambient quality is very similar to that for positive charges. Institutional compatibility would be better due to common past and current experience with agricultural subsidy mechanisms. However, the consequence of this approach, as with other similar subsidy policies, is that it artificially increases profitability in the regulated industry, which can increase the size of the industry. This can exacerbate pollution problems rather than mitigate them, even if individual loadings are reduced, unless steps are taken to diminish the benefits received by new entrants due to the subsidy.

8.5.2.4. ECONOMIC INCENTIVES: TRADABLE PERMITS

The logical alternative to a price-based economic incentive (charges) is a quantity-based economic incentive (tradable permits).[3] For the case of groundwater nitrate, a key challenge facing any tradable permit instrument is again related to spatial heterogeneity and the local nature of nitrate pollution. Specifically, local pollution problems require local permit markets. Local markets imply local prices that mirror the nonuniform charges considered in the preceding section, with the additional advantage that local pollution levels are limited by the total number of allocated permits rather than being dependent upon uncertain abatement cost curves. But local markets also tend to be thin, which undermines the desirable economic properties of tradable permits. For example, the quantity of trades may be small due to the limited number of potential trading partners, or a small number of firms may develop excessive influence in the market. Either case limits the gains from trade in the market. While the problem of thin markets may not have a solution, regulatory supervision can diminish the potential for market power. However, such involvement tends to increase administrative costs. For all these reasons, we rate tradable permits similarly to nonuniform charges, but we expect that they would be somewhat more costly and also somewhat more adaptable because they are quantity based rather than price based.

8.5.2.5. ECONOMIC INCENTIVES: AUCTIONS

Auction-based abatement contracts are implemented with a reverse auction format in which the producers are the sellers and the regulator is the buyer. The producers submit bids (an abatement plan and corresponding compensation) to the regulator who selects the combination of bids that achieves the environmental goal at least cost. Such auctions exhibit properties that are similar to those for abatement subsidies (negative emission charges), which is why we rate the environmental effectiveness, technological feasibility, and distributional effects the same as for abatement subsidies. However, there are also some important differences. First, as with any contract bidding process, there is competition among the sellers to submit bids that are appealing to the buyer. This competition tends to push compensation bids down and thus reduces the surplus of payments received by the producers and the cost incurred by the regulator relative to a standard subsidy mechanism. Second, the regulator is able to deliberately select bids to coordinate abatement efforts across producers and to minimize efforts that are duplicative or even countervailing. So, for example, rather than subsidize many small-scale groundwater capture wells with a uniform subsidy, the regulator might choose to fund a smaller number of large-scale but more cost-effective wells in critical areas of the watershed. This also tends to reduce costs relative to a subsidy mechanism. However, the trade-off is that additional administrative costs are needed to run such a program, including development of a nutrient fate and transport model for the affected areas. For these reasons, we expect that the cost-effectiveness of auctions would be slightly better than for abatement subsidies, with the cost advantage increasing in the long run. The institutional compatibility for auction-based abatement contracts is good because both abatement subsidies (which are typically targeted at management practices) and contract auctions are well-established mechanisms that producers should be familiar with. We rate adaptability as moderate because it is possible that producers will favor the relative certitude of longer-term contracts that would be difficult to modify in the short run.

8.5.2.6. SUMMARY

In order to bring about significant reductions in emissions of nitrate to groundwater, policies must rely on more than just educational efforts. Of the remaining policy instruments, four receive ratings of "moderate" or better for each criterion. These are emission standards, emission charges, abatement subsidies, and auction-based abatement contracts. The first two instruments receive identical ratings, while the latter two receive slightly better ratings than these for institutional compatibility, with the caveat that subsidies potentially increase industry size if steps are not taken to prevent this. Abatement subsidies and contracts are similar policies, which explains their similar ratings, but contracting involves an additional and significant effort to coordinate abatement across sources, whereas the standard subsidy approach does not.

This result may seem at odds with the discussion in Section 8.2 that recommends moving towards the "polluter-pays-principle" based on the lackluster performance of BMP subsidy programs. However, there are noteworthy differences between the subsidy mechanisms we consider and the BMP subsidies that have been implemented in practice. First, the mechanisms we consider are performance based, rather than input based, which is consistent with the recommendations in Section 8.2. Second, the mechanisms we consider would have subsidy levels high enough to induce adequate levels of participation, robust enforcement of abatement obligations, and contracts that guarantee payment of subsidies over time horizons long enough to justify initial capital investments. If there is insufficient public willingness to pay for such mechanisms, or an inability to achieve these conditions for any reason, then instruments that place the payment burden on producers (i.e., emission standards or charges) would be the leading candidate mechanisms.

8.5.3. Ammonia Policy Assessment

The challenges associated with mitigating atmospheric ammonia emissions from agricultural operations are similar

---

3. In a highly cited article, Weitzman (1974) develops a general framework in which quantity-based regulation is preferable to price-based regulation when the slope of the marginal damage curve is steep relative to the slope of an uncertain marginal abatement cost curve. Because we know very little about the marginal damage curve for nitrogen pollution, Weitzman's oft-cited result offers little guidance in this case.

to those associated with nitrate emissions, but with some noteworthy differences (Table 8.4). Ammonia emissions are non-point in nature and thus exhibit the same inherent problems of observability and uncertainty that also characterize nitrate emissions. Ammonia emissions also exhibit spatial heterogeneity, but the scale is substantially larger, whereas only a small number of farms (perhaps just one) may be responsible for nitrate contamination of a groundwater well, most air quality issues are regional, and some (i.e., greenhouse gases) are global. Therefore, policies do not have to be tailored to as many localities. Furthermore, it is not the ammonia emissions themselves that cause local air quality problems, but the interaction of those emissions with sulfur or nitrogen oxides (derived primarily from combustion processes) that creates airborne particulate matter. Therefore, ammonia regulations ideally should be responsive to whether or not a region is nitrogen oxide limited. Finally, ammonia derives primarily from animal manure, and management strategies for reducing ammonia emissions typically involve conserving ammonia in manure and/or converting it to nitrate for subsequent field application. Examples include biofiltration, covering animal housing and manure storage facilities, separating solid and liquid waste streams, and applying chemical additives to stored manure; dietary manipulation (i.e., reducing crude protein in feed) is an exception. Although evidence suggests these practices may be relatively cost-effective (Iowa State University, 2004; Ndegwa et al., 2008), in order for ammonia control policies to be environmentally effective, the additional nitrate emissions from conserving the ammonia must be mitigated as well. Therefore, ammonia control costs would be at least as high as those for nitrate. They also will vary across producers depending on the difficulty of modifying preexisting housing and manure storage facilities to allow implementation of control strategies.

In light of these similarities and differences, we rate education-based policies for ammonia emissions essentially the same as for nitrate emissions. The noteworthy differences between these two pollution problems do not change the fact that education-based policies rely on voluntary actions and thus have generally poor environmental effectiveness. The other evaluative criteria are similarly dependent on characteristics of the education-based policies, not the empirical problem, and thus remain unchanged as well. However, we rate the institutional compatibility of reporting inventories and registries better than for nitrates because ammonia emissions reporting already is required under the Emergency Planning and Community Right-to-Know Act (EPCRA).

For standards we also find broad similarities between nitrate and ammonia policies. However, we rate the institutional compatibility of input and emission standards for ammonia slightly better than for nitrates due to the similarities such regulations would have with existing State Implementation Plans for National Ambient Air Quality Standards; but we leave the evaluation for (pure) ambient standards unchanged due to the additional complications that would arise from using the ambient concentration of a final pollutant (particulate matter) to reduce emissions of a precursor pollutant (ammonia). Despite the preceding logic about higher source control costs, we also note that regional (rather than local) spatial heterogeneity should produce administrative cost savings, and so we leave the cost-effectiveness ratings the same as for nitrates.

For the economic incentives, we also perceive there to be broad similarities compared to nitrate policies. However, we rate the cost-effectiveness of tradable permits slightly better due to the regional scale of the ammonia problem which should improve market efficiency and reduce administrative costs. Given all of these similarities and again using the "moderate or better" selection criteria as for nitrate policies, five instruments appear to be preferred—emission standards, emission charges, tradable emission permits, abatement subsidies, and auction-based abatement contracts—again with the same caveats as for the case of nitrates.

8.5.4. Additional Considerations

The preceding assessment organizes candidate policies and their attributes into a simplified framework for purposes of comparison and evaluation. While useful, there are important additional considerations that should enter any discussion of nitrogen policy. We consider six such issues here: levels of evidence and agreement regarding policy assessment outcomes, emerging abatement technologies, improved monitoring/modeling of non-point problems, point-of-use treatment, hybrid policy instruments, and the potential for integrated nitrogen policy.

8.5.4.1. EVIDENCE AND AGREEMENT

Throughout this assessment, reserved words are used to characterize the levels of evidence and agreement associated with important quantitative and qualitative statements. The policy assessment summary tables contain many such qualitative statements. Generally there is limited empirical evidence to support these statements, particularly for nontraditional policy approaches, as can be seen in the case study matrix in Table 8.2. There are two reasons for this. First, there has not been extensive experimentation with alternative policies for controlling nitrogen pollution. Second, as mentioned in Section 8.3, for policies that have been implemented there have been few formal impact assessments. These factors generally undermine the strength and certitude of any policy implications that may be drawn from experience. Therefore, the policy assessments must rely more on economic theory and intuition than they do on empirical evidence.

Partly because of this, as well as the inherent scale and heterogeneity exhibited by the pollution problems of concern, it is also difficult to gauge the level of agreement for the assessments. While there may be high agreement

TABLE 8.4
Ammonia policy assessment matrix

Selected policy instruments	Evaluation criteria					
	Environmental effectiveness	Technological feasibility	Cost effectiveness	Distributional effects	Institutional compatibility	Adaptability
Education						
Labeling or certification programs	Poor	Good/moderate	Good/moderate	Small	Good	Good
Training and outreach	Poor	Good	Good/moderate	Small	Good	Good
Reporting inventories or registers	Poor/moderate	Moderate	Good	Small	Good	Good
Awards or recognition	Poor	Good	Good	Small	Good	Good
Standards						
Input standards	Good	Good	Poor	Moderate	Good	Poor
Emission standards	Good	Good/moderate	Poor/moderate	Moderate	Good/moderate	Moderate
Ambient standards	Good/moderate	Moderate	Poor/moderate	Moderate	Poor/moderate	Good
Economic incentives						
Input charges	Moderate	Good	Poor/moderate	Moderate	Good	Poor
Emission charges	Good	Good/moderate	Moderate	Moderate	Moderate	Moderate
Ambient charges	Good/moderate	Moderate	Moderate	Moderate	Poor/moderate	Good/moderate
Tradable input permits	Moderate	Good	Moderate	Moderate	Good	Poor
Tradable emission permits	Good	Good/moderate	Moderate	Moderate	Moderate	Moderate
Tradable ambient permits	Good/moderate	Moderate	Moderate	Moderate	Poor/moderate	Good
Auction-based abatement contracts	Good	Good/moderate	Moderate	Moderate	Good	Moderate

among economists regarding the theoretical cost-effectiveness of different policy mechanisms, there may be low agreement among producers regarding abatement costs specifically because of the general lack of evidence and their different operating conditions. Therefore, for purposes of characterizing the evidence and agreement associated with the policy assessments, "tentatively agreed by most" seems appropriate generally.

8.5.4.2. EMERGING ABATEMENT TECHNOLOGIES

The fundamental approach to mitigating nitrate leaching has not changed significantly in the past 20 years. Generally what works is more precise management of water and nitrogen inputs. This includes improved irrigation system uniformity, full accounting of nitrogen sources and sinks, reductions in applied water and N, and proper timing of water and nitrogen applications. Such practices have been called the 4Rs of nutrient stewardship: right amount, right time, right place, and right form. Randall et al. (2008) provide a good overview of management practices commonly used to implement the 4Rs. Some of these strategies were used to successfully reduce P loads in the Imperial Valley in 2004 (SWRCB, 2010). A full accounting of nitrogen sources and sinks also can bring about changes to cropping patterns that can effectively mitigate nitrate leaching. Some cropping changes, such as fallowing, may create relatively large costs for producers; others, such as the creation of nitrate buffer zones, may not (Mayzelle et al., 2015). Similarly, a full accounting can lead to the adoption of improved manure management techniques that reduce volatilization of ammonia from the waste stream and conserve nitrogen on-site for potential use in crop production.

Although the fundamental approach to managing nitrates has not changed significantly, there have been some recent improvements in the ability of producers to implement the 4Rs. For example, nitrification inhibitors, controlled release fertilizers, and precision farming (variable rate) techniques are now commercially available. Recycling of shallow groundwater ("pump and fertilize") also looks effective for both large dairies (Wang and Baerenklau, 2014) and crop operations (Dzurella et al., 2012), but so far this practice has had limited field testing. There is some evidence that constructed wetlands may be effective for smaller animal feeding operations (100–200 head) but not for large-scale operations (1000+ head) that characterize California's dairy industry (Wang, 2012); land costs may be high for large-scale operations, as well. There has been substantial interest and effort in designing treatment technologies for animal manure that would function similarly to municipal wastewater treatment plants (i.e., ultimately disposing of waste nitrogen as nitrogen gas), but so far none has emerged as an economically viable option in practice. Other technologies, such as membrane filtration of aqueous ammonia in waste lagoons (Samani Majd and Mukhtar, 2013) and vermiculture-based technology developed in Chile currently remain experimental.

While currently there does not appear to be an obvious technological solution on the horizon, the future is uncertain and therefore policies that incentivize innovation efforts and are flexible enough to accommodate beneficial new technologies are preferable.

8.5.4.3. IMPROVED MONITORING AND MODELING OF NON-POINT SOURCE PROBLEMS

A comprehensive and integrated monitoring network is required to understand and better focus efforts to address the multimedia impacts of reactive N in the environment. California already has multiple monitoring sites and programs operated by state and federal agencies, including the California Air Resources Board, the State Water Quality Control Board, and the US Environmental Protection Agency. However, the incoherence and inaccessibility of data prevent improved assessment of pollution problems. Moreover, a statewide effort is needed to integrate the diverse air, water, climate, and source activity datasets. Comprehensive integration, transparent protocols, and evaluation of uncertainty in the data are key characteristics to consider in creating a statewide monitoring platform (for more information, see online Supplemental Information).

Despite the existing data, non-point source pollution is still characterized by a relative lack of information: regulators are unable to adequately monitor emissions by individual polluters and also cannot accurately infer from observable ambient pollution levels the contributions by individual emission sources. This contrasts with point sources for which monitoring is relatively easy and inexpensive and thus there is relatively little uncertainty about individual loadings. However, the distinction between point and non-point sources is artificial because monitoring cost is a continuous rather than binary variable. Therefore in reality pollution problems exist along a continuum with some clearly classified as point source problems (easy to monitor) and others clearly classified as non-point problems (very difficult), but many also exist in the middle. As emissions monitoring and modeling technologies improve (i.e., as their accuracies increase and/or their costs fall), more pollution problems can be shifted within the monitoring cost continuum and effectively converted to, and managed as, point source problems.

Examples of recent advances in monitoring technologies include the use of satellite data to estimate evapotranspiration (Idaho DWR, 2013), the use of remotely sensed vegetation indices to estimate field-level N application rates (Shaver et al., 2011), and the use of embedded sensor networks to monitor agricultural water quality (Zia et al., 2013). Although such technologies have not yet been applied to agricultural non-point source pollution control in practice, they are potentially very useful to the extent they can provide more accurate information to regulators about individual source loadings and thus improve the performance of policy mechanisms that are based on estimated emissions.

While this is good news, some non-point source problems—including agricultural nitrate and ammonia emissions—present additional challenges. In both cases, it is the ambient concentration of pollution, not the emissions, that is of concern. And again in both cases, the mechanisms that govern the conversion of emissions to ambient concentrations are neither completely observable nor understood. For nitrates, there is also a significant time lag between emissions entering the environment and arriving at a point where they cause damages. For a regulator who is primarily concerned about improving ambient quality by controlling emissions, all of this means that a good model of the environmental fate and transport mechanisms is needed. A poor model will contribute to a misallocation of abatement effort across sources and thus increased costs and/or reduced environmental effectiveness.

Inadequate modeling, or the perception of it, also can increase policy costs by engendering conflicts between regulators and polluters. A recent example is the case brought by the American Farm Bureau Federation and Pennsylvania Farm Bureau against the USEPA (Copeland, 2012, p.14). In this case, the plaintiffs argued that the Chesapeake Bay TMDL was "arbitrary and capricious on the basis that EPA used models to support TMDL allocations beyond their predictive capabilities" (United States District Court for the Middle District of Pennsylvania, 2013, p.90).

However in 2013, the US District Court ruled in favor of the USEPA, finding that its use of scientific models and data was reasonable, and deferring to the agency's expert judgment in relying on those models to promulgate rules (United States District Court for the Middle District of Pennsylvania, 2013, p.97). The court also emphasized the "heavy burden" of establishing that a model is arbitrary and capricious since it requires establishing "no rational relationship to the realities [it purports] to represent" (United States District Court for the Middle District of Pennsylvania, 2013, p.97). While this opinion bodes well for future reliance on models to regulate non-point source pollution, it also demonstrates another benefit of implementing defensible modeling techniques.

8.5.4.4. POINT-OF-USE TREATMENT

The timescales for the ammonia and nitrate pollution problems differ significantly. If ammonia emissions were to cease today due to a policy intervention, there would be measurable ambient improvements within days or weeks. But this is not the case for nitrate emissions. Due to the very long time lags that characterize transport of nitrate belowground, past emissions of nitrate often will not cause damage at a receptor point for years or decades. In other words, regardless of any nitrate source control policies that might be instituted today, California will be living with the consequences of past nitrate leakages for a long time.

For this reason, along with the source control policy challenges discussed above, it is worth considering point-of-use treatment as a potential pollution control strategy. That is, rather than controlling sources of nitrate *emissions*, nitrate *pollution* could be controlled at receptor points before causing damages. Such an approach could be implemented, for example, as a type of ambient standard where the standard is applied to *produced* rather than in situ groundwater. Although large-scale remediation of California's groundwater basins would cost billions of dollars over several decades (Harter and Lund, 2012), wellhead treatment, blending of available sources, and importing new supplies are potentially cost-effective damage prevention methods (Harter and Lund, 2012). Furthermore, by effectively shifting remediation activities into the future, point-of-use treatment costs are further reduced by the long-term effects of discounting.

However, this approach presents challenges that are similar to, and in some ways more onerous than, those associated with in situ ambient standards. Although long transit times between sources and receptors are beneficial for reducing the present value of abatement costs, long time lags also imply longer distances, greater uncertainties, and thus greater difficulty in adequately modeling the fate and transport mechanisms. This has implications for the cost-effectiveness and technological feasibility of any policy that attempts to regulate current emissions based on estimates of future nitrate concentrations at distant wellheads. An alternative approach would be to charge current producers for current point-of-use treatment costs, effectively passing forward the costs of previous emissions in the same way that some social programs (e.g., Social Security) are structured. While the economic efficiency of such a mechanism is not particularly good (because it fails to internalize the external cost of emissions from each source), it would be a reasonably practical means to fund point-of-use treatment efforts. However, it may not have any effect on current emissions of nitrate to groundwater, particularly if a source's charge is unrelated to its emissions. This means that elevated groundwater nitrate concentrations will persist, potentially causing additional unexpected problems in the future.

For these reasons, a dual approach to the problem has substantial merit. In addition to setting new source reduction policies to create a more sustainable future, a separate additional effort would be made to address acute groundwater nitrate contamination problems that otherwise will persist for many years. This would seem particularly important in areas where drinking water supplies are threatened by nitrate concentrations exceeding the MCL. In such areas, point-of-use treatment will be needed until the ambient effects of new source control policies manifest in the future, at which time treatment efforts can begin to ramp down (see Jensen et al., 2014, for an extensive review of treatment options).

8.5.4.5. HYBRID POLICY INSTRUMENTS

While simple solutions may be adequate for simple problems, more complex problems—such as nitrogen pollution—

typically require more creative and nuanced solutions. For environmental quality problems generally, this often means crafting policies that do not fit neatly into any single category discussed in the preceding evaluations. Hybrid policies that include attributes of multiple policy categories are potentially more effective in practice than a pure policy instrument. Two examples are TMDLs and the National Ambient Air Quality Standards. Both of these policies appear on the surface to be ambient standards, but each relies primarily on source-specific input and emission standards to achieve the desired ambient goals.

While assessing all such hybrid combinations that might be used for nitrogen mitigation is not feasible here, the preceding evaluations can be used to inform discussion of policies that cut across traditional categories. For example, consider a TMDL-type approach for groundwater nitrates. The loading limit might be applied to each domestic well, and a transport model would be used to calculate the allowable loadings throughout the well capture zone. Input and/or (estimated) emission standards would then be established for nitrogen sources to achieve the desired load. Such a policy mechanism largely would exhibit the characteristics of input and emission standards in Table 8.3, but also should exhibit relatively better adaptability presuming the source requirements are driven by observed concentrations at the wells and are thus easier to modify if/when those concentrations exceed the loading limit.

8.5.4.6. POTENTIAL FOR INTEGRATED NITROGEN POLICY

Chapter 7 describes 11 strategic actions that California may take to help solve the nitrogen challenges it faces today. By and large, the actions describe *individual* targets for specific nitrogen sources and impacts. Similarly the present chapter focuses on two critical agricultural nitrogen issues and assesses candidate policy responses for each issue *separately*. While convenient and perhaps even necessary for understanding nitrogen issues and potential responses, such compartmentalization oversimplifies both the issues and the necessary responses. Instead, given what we know about the way nitrogen behaves, efforts to deal with excess nitrogen should be organized in a way that reflects the cross-media nature of the problem. Because of their mobility and multiplicative effects, planning for multiple forms of nitrogen in multiple media simultaneously underlies any successful strategy for management. Efforts to control individual nitrogen species alone neglect the synergies and trade-offs associated with the nitrogen cascade. Poorly designed strategies that fail to account for underlying dynamics of reactive nitrogen in the environment will likely have negative unintended effects. A prime example is the potential trade-off between the two emissions streams highlighted in this chapter (groundwater NO_3 and ammonia), either of which is likely to increase in response to more stringent policies placed on the other (Aillery et al., 2005; Baerenklau et al., 2008). To address this issue, Yeo and Lin (2014) designed a tradable permit system that allows the exchange of nitrogen permits between air and water emissions. With such a permit system, there will be cost savings from trading between air and water since farmers would be able to choose practices that reduce nitrogen emissions to air and water jointly at least cost. There will also be environmental benefits from allowing sources to trade between air and water emissions permits, as a system that accounts for damages to air and water will internalize potential spillovers that would arise if air and water emissions were regulated separately and independently.

Management practices and technologies are already available for virtually every source activity (Chapter 7). Yet concentrations of reactive nitrogen in the environment are increasing and concerns for ecosystems and human health are becoming more severe (Chapter 5). Continued degradation can in part be attributed to increased source activity (Chapter 3) and/or lack of effective regulatory policies (this chapter). In many cases, enhanced nitrogen management is constrained by a lack of information or capital investment. That is, the obstacles to utilization are not technical in nature. Therefore, this assessment concludes that technical solutions to the nitrogen challenge exist for California now and thus integrated practice and policy solutions are needed to transition California to a sustainable nitrogen future.

Design and implementation of an integrative strategy is not without challenges. A multitude of factors constrain such an approach, mostly as a result of actions requiring the crossing of multiple boundaries. Divisions are not only physical but also geographic. An integrated approach would require bridging long-standing separation between ideas and institutions, for example, breaking down regulatory "silos" and identifying conflicting policies. It may require Californians to come to an agreement on the lesser of two pollutants—a judgment which may differ by region. Research would need to change too. It would have to shift its perspective and view whole farming, transportation, and city systems, enabling scientists to look across nitrogen sources and their impacts on society for the greatest potential for nitrogen reductions. Additional boundaries also need to be crossed, including those between science, practice, and policy (e.g., air and water quality policies with tools proven not only to reduce emissions but also to improve the ability to monitor). We also need to consider boundaries between spatiotemporal scales (e.g., from field plots to landscapes and from hours, in the case of ammonia emissions, to centuries, in the case of percolation of NO_3^- to groundwater) and boundaries existing along supply chains (e.g., from pre-farm to fork to human and solid waste disposal). Finally, in order to effectively bridge the aforementioned boundaries, various stakeholder groups, including farmers and low-income communities, need to be able to engage constructively with a range of institutions, including regulatory agencies and research institutions. Currently, data that capture the complexities of nitrogen challenges for both source and impacts and could be used to inform the discussion are only narrowly available and not readily connected (see Boxes 7.4 and 7.5). Reform, expansion, and integration of monitoring systems will be fundamental to providing farmers, scientists, policy-makers, and citizens the information they need for evidence-based decision-making. In short, and consistent with recent recommendations for nitrogen policy in the EU (Bull et al., 2011) and the US (EPA SAB, 2011), we suggest a comprehensive transformation of how nitrogen is thought of, monitored, and managed in California; development of technical solutions by themselves (without appropriate supporting policies and institutional mechanisms) will be wholly inadequate.

8.6. Conclusion and Relevance of California's Nitrogen Policy for the Rest of the World

This chapter has surveyed a variety of environmental policy instruments, both in theory and in practice, and has provided an assessment of several instruments for purposes of controlling agricultural non-point source emissions of nitrate and ammonia in California. It should be clear that there are no simple solutions to these problems. Rather, each candidate solution entails trade-offs as well as increased costs for at least a subset of stakeholders. However, some approaches appear more promising than others, and action is needed given the size and scope of the potential damages.

Business as usual is not an appealing option. Business as usual means continuing to add more than 500 Gg of nitrate and ammonia each year to California's already stressed groundwater and air resources. Free disposal of agricultural nitrogen waste may have been acceptable at some time in the past, but the massive scale of California's modern agriculture means that free disposal eventually leads to widespread degradation and the associated impacts on ecosystems and human health that have been documented in Chapter 5. Absent any regulatory action to mitigate flows of nitrogen into the environment, we will continue to experience these impacts in more locations and at elevated levels, and will pass on to future generations a problem that is more difficult, costly, and urgent than the one we currently face.

A reasonable path forward would mirror the dual policy approach recommended in the preceding section. First, in locations where ambient concentrations pose immediate threats to ecosystems and human health, cost-effective treatment and remediation activities should be undertaken. Second, the process of promulgating long-term source-control policy for nitrogen emissions in California should begin. As is apparent from the empirical cases in Section 8.3 and the policy assessment in Section 8.5, overall there is relatively limited evidence on the effectiveness of specific policy approaches for purposes of mitigating nitrogen emissions. Similarly, Harter and Lund (2012) conclude that "inconsistency and inaccessibility of data prevent effective

and continuous assessment" of nitrate in groundwater. Both of these observations suggest that a valuable first step would be a needs assessment to determine which information gaps must be filled before appropriate policy decisions can be made (Rosenstock et al., 2013). This could include a vulnerability assessment to determine priority areas, similar to the Nitrate Vulnerable Zones that were established under the EU Nitrate Directive. A second step would be to determine how to fill the identified gaps. Some will require the relatively simple task of integrating disparate sources of information that already exist locally; others will require new research to reduce some of the key areas of uncertainty documented in this assessment.

Whatever policy approaches are chosen, there will be a need for regular monitoring of ambient conditions, and review and evaluation to assess and improve policy outcomes. While the state may undertake some assessment activities itself, given the scale and scope of the policy problem and the anticipated widespread effects of the policy response, assessments can be expected to be undertaken independently by researchers at universities, think tanks, and other concerned organizations—similar to the large body of research that has emerged in response to climate change policies. Regardless, the state can facilitate all of these evaluations through careful documentation and maintenance of policy-relevant data, as well as efforts to make the data available and the methods transparent.

The potential benefits of an integrated nitrogen strategy are difficult to measure, but potentially large. This applies to California agriculture and also to other places intensifying production in irrigated systems. Given the ubiquity of nitrogen in the economy and environment, the magnitudes of the stocks and flows, and the potential damages to ecosystems and human health, the distortions that have been created by relatively cheap disposal of nitrogen byproducts are manifold. For example, due to cheap disposal, fuel and fertilizer appear to be artificially inexpensive, which reduces the costs of transportation and food production; this increases the demand for land which consumes excessive amounts of natural habitat; it also increases the supply of agricultural commodities and leads to both water and air pollution and their attendant effects on ecosystems and human health. Therefore, policies that effectively make it more costly to dispose of nitrogen by-products will have ripple effects throughout this chain, potentially affecting everything from agricultural markets to land-use patterns to health outcomes. Similarly, policies that affect other elements in the chain—such as agricultural commodity support programs, for example—will influence the efficacy of nitrogen policies.

Despite distinctive features and crucial details of the nitrogen challenges in California, there is reason for some optimism that insights from future efforts to attain a better balance of benefits and costs of nitrogen flows in the state also can provide useful insights into intensive (and intensifying) agricultural systems in other parts of the world. By the same token, California also can benefit from judicious interpretation of lessons gained elsewhere, often at considerable cost. Europe provides a case in point. Despite major differences between California and the European policy setting across the board—spanning physical, agricultural, environmental, social, cultural, institutional, and political dimensions, to name a few—features of the European case (described in greater detail in the Chapter 8 Appendix A8.2.11; Oenema et al., 2011a and Van Grinsven et al., 2012) suggest some important common lessons and policy parallels. For example, lessons already learned from the European experience indicate that a coordinated mix of mandatory regulatory instruments, including good agricultural practices and nitrogen rate limits, can produce measurable environmental improvements in water and air quality. However, though there is evidence that the integrated European policies have reduced nitrogen pollution, effects are not uniform and addressing nitrate contamination of groundwater seems particularly recalcitrant. These variable outcomes in Europe also appear to reflect political, economic, and environmental differences across regions, providing further support to the point that "one-size-fits all" strategies are unlikely to be effective or efficient. Thus, policy frameworks should embrace the benefits of locally different approaches; otherwise, effective local policies may be undermined. European experience also reinforces the point that outcomes depend not just on policies and practices, but also on trajectories of economic expansion or contraction of polluting sectors. It is highly desirable then that policies are designed to be effective under a variety of uncertain future conditions. More generally, the complexity of the nitrogen cycle presents a formidable challenge, particularly for reducing nitrogen from agricultural systems. This provides additional justification for coordinated policies and cross-compliance requirements.

Is nitrogen the next carbon? As noted in Chapter 1 and examined in various aspects throughout this assessment, nitrogen exhibits qualities similar to carbon. Carbon also is ubiquitous in the economy and environment, is characterized by large stocks and flows, and, while indispensable to life on Earth, also poses a significant threat in terms of potential damages from unintended leakages and emissions into the environment. Hence there has been, and continues to be, much interest in pricing carbon emissions. Doing so not only could mitigate carbon-derived impacts on climate change, but it will also have similar ripple effects as the economy sheds existing carbon-related distortions and readjusts to a new normal in which disposal of carbon is no longer as cheap as it used to be. Pricing carbon emissions thus provides a more integrated, holistic approach than a large number of more narrowly focused policies that would require substantial effort to coordinate. And yet, at the time of this writing, our continuing inability to agree on the need for concerted global action to mitigate climate change by reducing carbon emissions—much less on practical steps needed to design and implement a strategy to avert its risks—raises questions about when (or indeed whether) a global strategy for carbon is politically feasible. As a second-best approach, perhaps some of the nitrogen-oriented practices and policies assessed in Chapters 7 and 8 can contribute workable examples of regional efforts to govern these "common pool resources" (Ostrom, 1990, extended in Dietz et al., 2003 and Stern, 2011) as an inspiration for a more decentralized approach to carbon emissions too. So, in this sense, perhaps nitrogen not only is the next big global concern, but may also hold some practical lessons for breakthroughs in addressing our carbon-based challenges.

REFERENCES

Aber, J.D., McDowell, W., Nadelhoffer, K., Magill, A., Berntson, G., Kamakea, M., McNulty, S.G., Currie, W., Rustad, L., and Fernandez, I. 1998. Nitrogen saturation in northern forest ecosystems. *BioScience* 39: 378–386.

Abshahi, A., Hills, F.J., and Broadbent, F.E. 1984. Nitrogen utilization by wheat from residual sugarbeet fertilizer and soil incorporated sugarbeet tops. *Agronomy Journal* 76:954–958.

Abt Associates Inc. 2000. *Out of Sight: The Science and Economics of Visibility Impairment*. Report prepared for Clean Air Task Force, Boston, MA.

Adamkiewicz, G., Ebelt, S., Syring, M., Slater, J., Speizer, F.E., Schwartz, J., Suh, H., and Gold, D.R. 2004. Association between air pollution exposure and exhaled nitric oxide in an elderly population. *Thorax* 59:204–209. doi:10.1136/thorax.2003.006445.

Adams, L.S., Kuehl, S., and Leary, M. 2009. *California 2008 Statewide Waste Characterization Study*. California Integrated Waste Management Board, Sacramento, 172pp.

Adar, S.D., Adamkiewicz, G., Gold, D.R., Schwartz, J., Coull, B.A., and Suh, H. 2007. Ambient and microenvironmental particles and exhaled nitric oxide before and after a group bus trip. *Environmental Health Perspectives* 115:507–512. doi:10.1289/ehp.9386.

Addiscott, T.M. 1996. Fertilizers and nitrate leaching. *Issues in Environmental Science and Technology* 5: 1–26.

Adler, P.R., Del Grosso, S.J., and Parton, W.J. 2007. Life-cycle assessment of net greenhouse-gas flux for bioenergy cropping systems. *Ecological Applications* 17: 675–691.

Adviento-Borbe, M.A., Pittelkow, C.M., Anders, M., van Kessel, C., Hill, J.E., McClung, A.M., Six, J., and Linquist, B.A. 2013. Optimal fertilizer nitrogen rates and yield-scaled global warming potential in drill seeded rice. *Journal of Environment Quality* 42:1623–1634. doi:10.2134/jeq2013.05.0167.

Aggarwal, R. 2010. *Final report of ASU Workshop Course SOS 594: Future Scenarios for Agriculture and Water in Central Arizona*. School of Sustainability, Arizona State University.

Ahearn, D., Sheibley, R., Dahlgren, R., and Keller, K. 2004. Temporal dynamics of stream water chemistry in the last free-flowing river draining the western Sierra Nevada. *California. Journal of Hydrology* 295:47–63. doi:10.1016/j.jhydrol.2004.02.016.

Ahn, Y.-H. 2006. Sustainable nitrogen elimination biotechnologies: a review. *Process Biochemistry* 41:1709–1721. doi:10.1016/j.procbio.2006.03.033.

AIC (Agricultural Issues Center). 2006. Agriculture's role in the economy. Chapter 5 in *The Measure of California Agriculture*. UC Agricultural Issues Center, Davis, CA.

AIC. 2009. *The Measure of California Agriculture*. UC Agricultural Issues Center, Davis, CA.

AIC. 2011. *Project on Climate Change and Agriculture in Yolo County (Preliminary)*. UC Agricultural Issues Center, Davis, CA.

AIC. 2012a. *Estimating California's Agricultural Exports*, University of California Agricultural Issues Center. Accessed June 6, 2015. http://aic.ucdavis.edu/pub/exports.html.

AIC. 2012b. *The Measure of California Agriculture: Highlights*. UC Agricultural Issues Center, Davis, CA.

Aillery, M., Gollehon, N., Johansson, R., Kaplan, J., Key, N., and Ribaudo, M. 2005. *Managing Manure to Improve Air and Water Quality*, Economic Research Service Economic Research Report 9. United States Department of Agriculture.

Akinbami, L.J., Lynch, C.D., Parker, J.D., and Woodruff, T.J. 2010. The association between childhood asthma prevalence and monitored air pollutants in metropolitan areas, United States, 2001–2004. *Environmental Research* 110: 294–301. doi:10.1016/j.envres.2010.01.001.

Akiyama, H., Yan, X., and Yagi, K. 2010. Evaluation of effectiveness of enhanced-efficiency fertilizers as mitigation options for N_2O and NO emissions from agricultural soils: meta-analysis. *Global Change Biology* 16:1837–1846.

Alcamo, J. and Bennett, E.M. 2003. *Ecosystems and Human Well-being: A Framework for Assessment*. Island Press, Washington, DC.

Alcamo, J. and Henrichs, T. 2008. Towards guidelines for environmental scenario analysis, in Alcamo, J. (Ed.):

Environmental Futures: The Practice of Environmental Scenario Analysis. Elsevier, Amsterdam, pp.13–35.

Alderman, H. 1986. *The Effect of Food Price and Income Changes on the Acquisition of Food by Low-Income Households*. International Food Policy Research Institute, Washington, DC.

Alexander, R.B. and Smith, R.A. 1990. *County-Level Estimates of Nitrogen and Phosphorus Fertilizer Use in the United States, 1945 to 1985*, Open-File Report, USGS Numbered Series No. 90-130. USGS, Reston, VA.

Alexis, N.E., Lay, J.C., Hazucha, M., Harris, B., Hernandez, M.L., Bromberg, P.A., Kehrl, H. et al. 2010. Low-level ozone exposure induces airways inflammation and modifies cell surface phenotypes in healthy humans. *Inhalation Toxicology* 22:593–600. doi:10.3109/08958371003596587.

Allen, E.B., Temple, P.J., Bytnerowicz, A., Arbaugh, M.J., Sirulnik, A.G., and Rao, L.E. 2007. Patterns of understory diversity in mixed coniferous forests of southern California impacted by air pollution. *Scientific World Journal* 7 (Suppl 1):247–263. doi:10.1100/tsw.2007.72.

Alston, J.M., Andersen, M.A., James, J.S., and Pardey, P.G. 2010. *Persistence Pays: U.S. Agricultural Productivity Growth and the Benefits from Public R&D Spending*. Springer, New York.

Altshuler, S.L., Arcado, T.D., and Lawson, D.R. 1995. Weekday v. weekend ambient ozone concentrations: discussion and hypotheses with focus on Northern California. *Journal of the Air and Waste Management Association* 45: 967–972.

Andersen, Z.J., Wahlin, P., Raaschou-Nielsen, O., Scheike, T., and Loft, S. 2007. Ambient particle source apportionment and daily hospital admissions among children and elderly in Copenhagen. *Journal of Exposure Science and Environmental Epidemiology* 17:625–636. doi:10.1038/sj.jes.7500546.

Anderson, K. 1987. On why agriculture declines with economic growth. *Agricultural Economics* 1:195–207. doi:10.1016/0169-5150(87)90001-6.

Anderson, K. 2005. *Tending the Wild: Native American Knowledge and the Management of California's Natural Resources*. University of California Press, Oakland.

Anderson, T.L., Charlson, R.J., Schwartz, S.E., Knutti, R., Boucher, O., Rodhe, H., and Heintzenberg, J. 2003. Climate forcing by aerosols: a hazy picture. *Science* 300:1103–1104.

Anjana, S.U. and Iqbal, M. 2007. Nitrate accumulation in plants, factors affecting the process, and human health implications: a review. *Agronomy for Sustainable Development* 27:45–57. doi:10.1051/agro:2006021.

Anning, D.W., Paul, A.P., McKinney, T.S., Huntington, J.M., Bexfield, L.M., and Thiros, S.A. 2012. *Predicted Nitrate and Arsenic Concentrations in Basin-Fill Aquifers of the Southwestern United States (No. 2012-5065)*, Scientific Investigations Report. USGS, Reston, VA.

Antikainen, R., Lemola, R., Nousiainen, J.I., Sokka, L., Esala, M., Huhtanen, P., and Rekolainen, S. 2005. Stocks and flows of nitrogen and phosphorus in the Finnish food production and consumption system. *Agriculture, Ecosystems & Environment* 107:287–305. doi:10.1016/j.agee.2004.10.025.

API (American Petroleum Institute). 2013. *State Gasoline Tax Reports: Motor Fuel Taxes*. Accessed June 6, 2015. http://www.api.org/oil-and-natural-gas-overview/industry-economics/fuel-taxes.

Arbuckle, T.E., Sherman, G.J., Corey, P.N., Walters, D., and Lo, B. 1988. Water nitrates and CNS birth defects: a population-based case-control study. *Archives of Environmental Health* 43:162–167. doi:10.1080/00039896.1988.9935846.

ARC (Alliance of Religions and Conservation). 1986. *The Assisi Declarations*. ARC, Kelston Park, Bath, UK.

Arjal, R.D., Prato, J.D., and Peterson, M.L. 1978. Response of corn to fertilizer, plant population, and planting date. *California Agriculture* 32(3): 14–15.

Arneth, A., Harrison, S.P., Zaehle, S., Tsigaridis, K., Menon, S., Bartlein, P.J., Feichter, J. et al. 2010. Terrestrial biogeochemical feedbacks in the climate system. *Nature Geoscience* 3:525–532. doi:10.1038/ngeo905.

Arriaza, M., Cañas-Ortega, J.F., Cañas-Madueño, J.A., and Ruiz-Aviles, P. 2004. Assessing the visual quality of rural landscapes. *Landscape and Urban Planning* 69:115–125. doi:10.1016/j.landurbplan.2003.10.029.

Arsenault, N., Tyedmers, P., and Fredeen, A. 2009. Comparing the environmental impacts of pasture-based and confinement-based dairy systems in Nova Scotia (Canada) using life cycle assessment. *International Journal of Agricultural Sustainability* 7:19–41. doi:10.3763/ijas.2009.0356.

ASAE (American Society of Agricultural Engineers). 2005. *Manure Production and Characteristics*, Standard D384.2. ASAE, St. Joseph, MI.

Asano, T., Burton, F., Leverenz, H., Tsuchihashi, R., and Tchobanoglous, G. 2007. *Water Reuse: Issues, Technologies, and Applications*, 1st ed. McGraw-Hill Education, New York.

Aschebrook-Kilfoy, B., Heltshe, S.L., Nuckols, J.R., Sabra, M.M., Shuldiner, A.R., Mitchell, B.D., Airola, M., Holford, T.R., Zhang, Y., and Ward, M.H. 2012. Modeled nitrate levels in well water supplies and prevalence of abnormal thyroid conditions among the Old Order Amish in Pennsylvania. *Environmental Health* 11:6. doi:10.1186/1476-069X-11-6.

Aschengrau, A., Zierler, S., and Cohen, A. 1989. Quality of community drinking water and the occurrence of spontaneous abortion. *Archives of Environmental Health* 44:283–290. doi:10.1080/00039896.1989.9935895.

Aschengrau, A., Zierler, S., and Cohen, A. 1993. Quality of community drinking water and the occurrence of late adverse pregnancy outcomes. *Archives of Environmental Health* 48:105–113. doi:10.1080/00039896.1993.9938403.

Ash, N., Blanco, H., Brown, C., Garcia, K., Tomich, T., Vira, B., Zurek, M. et al. 2010. *Ecosystems and Human Well-Being: A Manual for Assessment Practitioners*. Island Press, Washington, DC.

Avery, A.A. 1999. Infantile methemoglobinemia: reexamining the role of drinking water nitrates. *Environmental Health Perspectives* 107:583–586.

Avery, A.A. and L'hirondel, J.-L. 2003. Nitrate and methemoglobinemia. *Environmental Health Perspectives* 111:A142–A144.

Baccarelli, A., Martinelli, I., Zanobetti, A., Grillo, P., Hou, L.F., Bertazzi, P.A., Mannucci, P.M., and Schwartz, J. 2008. Exposure to particulate air pollution and risk of deep vein thrombosis. *Archives of Internal Medicine* 168:920–927. doi:10.1001/archinte.168.9.920.

Baerenklau, K.A., Nergis, N., and Schwabe, K.A. 2008. Effects of nutrient restrictions on confined animal facilities: insights from a structural-dynamic model. *Canadian Journal of Agricultural Economics* 56:219–241.

Baerenklau, K. A. and Wang, J. 2015. Model-based regulation of nonpoint source emissions, in Dinar, A. and Schwabe, K. A. (Eds): *The Handbook of Water Economics*. Edward Elgar, Cheltenham, UK, pp.313–327.

Baghott, K. G. and Puri, Y. P. 1979. Response of durum and bread wheats to nitrogen fertilizer. *California Agriculture* 33:21–22.

Baghurst, P. A., McMichael, A. J., Slavotinek, A. H., Baghurst, K. I., Boyle, P., and Walker, A. M. 1991. A case-control study of diet and cancer of the pancreas. *American Journal of Epidemiology* 134:167–179.

Baisre, J. A. 2006. Assessment of nitrogen flows into the Cuban landscape, in Martinelli, L. A. and Howarth, R. W. (Eds): *Nitrogen Cycling in the Americas: Natural and Anthropogenic Influences and Controls*. Springer Netherlands, Heidelberg, pp.91–108.

Baker, L. A., Hope, D., Xu, Y., Edmonds, J., and Lauver, L. 2001. Nitrogen balance for the Central Arizona–Phoenix (CAP) ecosystem. *Ecosystems* 4:582–602. doi:10.1007/s10021-001-0031-2.

Balazs, C. L. and Ray, I. 2009. Just water? Environmental justice and drinking water quality in California's Central Valley. Paper presented at *The 137st APHA Annual Meeting and Exposition*, November 7–11, Philadelphia, PA.

Ban-Weiss, G. A., Mclaughlin, J. P., Harley, R. A., Lunden, M. M., Kirchstetter, T. W., Kean, A. J., Strawa, A. W., Stevenson, E. D., and Kendall, G. R. 2008. Long-term changes in emissions of nitrogen oxides and particulate matter from on-road gasoline and diesel vehicles. *Atmospheric Environment* 42:220–232.

Barbone, F., Austin, H., and Partridge, E. E. 1993. Diet and endometrial cancer: a case-control study. *American Journal of Epidemiology* 137:393–403.

Barbour, M. G., Pavlik, B., Drysdale, F., and Lindstrom, S. 1993. *California's Changing Landscapes: Diversity and Conservation of California Vegetation*. California Native Plant Society, Sacramento.

Barnett, A. G., Williams, G. M., Schwartz, J., Neller, A., Best, T., Petroeschevsky, A., and Simpson, R. 2005. Air pollution and child respiratory health: a case-crossover study in Australia and New Zealand. *American Journal of Respiratory and Critical Care Medicine* 171:1272–1278.

Barnett, T. P., Pierce, D. W., Hidalgo, H. G., Bonfils, C., Santer, B. D., Das, T., Bala, G. et al. 2008. Human-induced changes in the hydrology of the Western United States. *Science* 319:1080–1083. doi:10.1126/science.1152538.

Barreiro, T. J. and Perillo, I. 2004. An approach to interpreting spirometry. *American Family Physician* 69:1107–1114.

Barton, L. and Schipper, L. A. 2001. Regulation of nitrous oxide emissions from soils irrigated with dairy farm effluent. *Journal of Environment Quality* 30:1881. doi:10.2134/jeq2001.1881.

Battye, W., Aneja, V. P., and Roelle, P. A. 2003. Evaluation and improvement of ammonia emissions inventories. *Atmospheric Environment* 37:3873–3883. doi:10.1016/S1352-2310(03)00343-1.

Baum, M. M., Kiyomiya, E. S., Kumar, S., Lappas, A. M., Kapinus, V. A., and Lord, H. C. 2001. Multicomponent remote sensing of vehicle exhaust by dispersive absorption spectroscopy. 2. Direct on-road ammonia measurements. *Environmental Science & Technology* 35:3735–3741. doi:10.1021/es002046y.

Baumol, W. J. and Oates, W. E. 1988. *The Theory of Environmental Policy*, 2nd ed. Cambridge University Press, Cambridge.

BAWSCA (Bay Area Water Supply and Conservation Agency). 2012. *Hetch Hetchy Water System*. Available at: http://bawsca.org/water-supply/hetch-hetchy-water-system/.

Bayer-Oglesby, L., Grize, L., Gassner, M., Takken-Sahli, K., Sennhauser, F. H., Neu, U., Schindler, C., and Braun-Fahrländer, C. 2005. Decline of ambient air pollution levels and improved respiratory health in Swiss children. *Environmental Health Perspectives* 113:1632–1637. doi:10.1289/ehp.8159.

Beaulac, M. N. and Reckhow, K. H. 1982. An examination of land use–nutrient export relationships. *JAWRA Journal of the American Water Resources Association* 18:1013–1024. doi:10.1111/j.1752-1688.1982.tb00109.x.

Belanger, K., Gent, J. F., Triche, E. W., Bracken, M. B., and Leaderer, B. P. 2006. Association of indoor nitrogen dioxide exposure with respiratory symptoms in children with asthma. *American Journal of Respiratory and Critical Care Medicine* 173:297–303. doi:10.1164/rccm.200408-1123OC.

Belitz, K., Dubrovsky, N. M., Burow, K. R., Jurgens, B., and Johnson, T. 2003. *Framework for a Ground-Water Quality Monitoring and Assessment Program for California*, Water Resources Investigation Report 03-4166. US Geological Survey, Sacramento, CA.

Bell, M. L., Peng, R. D., and Dominici, F. 2006. The exposure-response curve for ozone and risk of mortality and the adequacy of current ozone regulations. *Environmental Health Perspectives* 114:532–536. doi:10.1289/ehp.8816.

Benbrook, C., Zhao, X., Yáñez, J., Davies, N., and Andrews, P. K. 2008. *New Evidence Confirms the Nutritional Superiority of Plant-Based Organic Foods*, Critical Issue Report March 2008. The Organic Center, Foster, RI.

Bender, J., Hertstein, U., and Black, C. R. 1999. Growth and yield responses of spring wheat to increasing carbon dioxide, ozone and physiological stresses: a statistical analysis of "ESPACE-wheat" results. *European Journal of Agronomy* 10:185–195.

Bendixen, W. E., Hanson, B. R., Hartz, T. K., and Larson, K. D. 1998. *Evaluation of Controlled Release Fertilizers and Fertigation in Strawberries and Vegetables*, FREP Contract # 95-0418. Fertilizer Research and Education Program (FREP), Sacramento, CA.

Benjamin, M. T. 2000. Estimating ammonia emissions in California. *Workshop on Fine Particle Emission Inventories*, September 28, Des Plaines, IL.

Bennett, E., Carpenter, S., Pingali, P., and Zurek, M. 2005. Summary: comparing alternate futures of ecosystem services and human well-being, in Carpenter, S., Pingali, P., Bennett, E., and Zurek, M. (Eds): *Ecosystems and Human Well-Being: Scenarios*. Island Press, Washington, DC, pp.2–17.

Benton, J., Fuhrer, J., Gimeno, B. S., Skärby, L., Palmer-Brown, D., Ball, G., Roadknight, C., and Mills, G. 2000. An international cooperative programme indicates the widespread occurrence of ozone injury on crops. *Agriculture, Ecosystems & Environment* 78:19–30. doi:10.1016/S0167-8809(99)00107-3.

Berg, G. M., Glibert, P. M., Jorgensen, N. O. G., Balode, M., and Purina, I. 2001. Variability in inorganic and organic nitrogen uptake associated with riverine nutrient input in the Gulf of Riga, Baltic Sea. *Estuaries* 24:176–186.

Bernhardt, E.S., Band, L.E., Walsh, C.J., and Berke, P.E. 2008. Understanding, managing, and minimizing urban impacts on surface water nitrogen loading. *Annals of the New York Academy of Sciences* 1134:61–96. doi:10.1196/annals.1439.014.

Berretta, M., Lleshi, A., Fisichella, R., Berretta, S., Basile, F., Li Volti, G., Bolognese, A. et al. 2012. The role of nutrition in the development of esophageal cancer: what do we know? *Frontiers in Bioscience (Elite Edition)* 4:351–357.

Berry, W. 1985. *Enriching the Earth, in: Collected Poems: 1957–1982*. North Point Press, San Francisco, CA.

Berry, W. 1977. *The Unsettling of America*. Sierra Club Books, San Francisco, CA.

Bhattacharya, D.K., Brondizio, E.S., Spierenburg, M., Ghosh, A., Traverse, M., de Castro, F., Morsello, C., and Siqueira, A.D. 2005. Cultural services, in Chopra, K., Leemans, R., Kumar, P., and Simons, H. (Eds): *Ecosystems and Human Well-Being: Policy Responses, Vol. 3: Findings of the Responses Working Group of the Millennium Ecosystem Assessment*. Island Press, Washington, DC, pp.401–422.

Bickerstaff, K. 2004. Risk perception research: socio-cultural perspectives on the public experience of air pollution. *Environment International* 30:827–840. doi:10.1016/j.envint.2003.12.001.

Birch, M.B.L., Gramig, B.M., Moomaw, W.R., Doering, O.C., and Reeling, C.J. 2011. Why metrics matter: evaluating policy choices for reactive nitrogen in the Chesapeake Bay Watershed. *Environmental Science & Technology* 45:168–174.

Bird, J.A., Horwath, W.R., Eagle, A.J., and van Kessel, C. 2001. Immobilization of fertilizer nitrogen in rice. *Soil Science Society of America Journal* 65:1143. doi:10.2136/sssaj2001.6541143x.

Bishop, G.A., Peddle, A.M., Stedman, D.H., and Zhan, T. 2010. On-road emission measurements of reactive nitrogen compounds from three California cities. *Environmental Science & Technology* 44:3616–3620. doi:10.1021/es903722p.

Bishop, R. 1994. A local agency's approach to solving the difficult problem of nitrate in the groundwater. *Journal of Soil and Water Conservation* 49:82–84.

Bleken, M.A. and Bakken, L.R. 1997. The nitrogen cost of food production: Norwegian society. *Ambio* 26:134–142.

Blengini, G.A. and Busto, M. 2009. The life cycle of rice: LCA of alternative agri-food chain management systems in Vercelli (Italy). *Journal of Environmental Management* 90:1512–1522.

Boardman, A.E., Greenberg, D., Vining, A., and Weimer, D. 1996. *Cost-Benefit Analysis: Concepts and Practice*. Prentice-Hall, Upper Saddle River, NJ.

Bobbink, R., Hicks, K., Galloway, J.N., Spranger, T., Alkemade, R., Ashmore, M., Bustamante, M. et al. 2010. Global assessment of nitrogen deposition effects on terrestrial plant diversity: a synthesis. *Ecological Applications* 20:30–59.

Bobbink, R., Hornung, M., and Roelofs, J.G.M. 1998. The effects of air-borne nitrogen pollutants on species diversity in natural and semi-natural European vegetation. *Journal of Ecology* 86:717–738. doi:10.1046/j.1365-2745.1998.8650717.x.

BOE (California State Board of Equalization). 2011. *Annual Report 2010–2011*. Accessed June 7, 2015. http://www.boe.ca.gov/annual/2010-11/appendix.html#stat.

Boehm, A.B., Ashbolt, N.J., Colford, J.M., Dunbar, J.E., Fleming, L.E., Gold, M.A., Hansel, J.A. et al. 2009. A sea change ahead for recreational water quality criteria. *Journal of Water and Health* 7:9–20.

Boehm, A.B. and Paytan, A. 2010. *Understanding Submarine Groundwater Discharge and Its Influence on Coastal Water Quality along the California Coast*. California Sea Grant College Program, UC San Diego.

Boeing, H., Frentzel-Beyme, R., Berger, M., Berndt, V., Göres, W., Körner, M., Lohmeier, R. et al. 1991. Case-control study on stomach cancer in Germany. *International Journal of Cancer* 47:858–864. doi:10.1002/ijc.2910470612.

Boezen, H.M., van der Zee, S.C., Postma, D.S., Vonk, J.M., Gerritsen, J., Hoek, G., Brunekreef, B., Rijcken, B., and Schouten, J.P. 1999. Effects of ambient air pollution on upper and lower respiratory symptoms and peak expiratory flow in children. *Lancet* 353:874–878. doi:10.1016/s0140-6736(98)06311-9.

Bograd, S.J., Castro, C.G., Lorenzo, E.D., Palacios, D.M., Bailey, H., Gilly, W., and Chavez, F.P. 2008. Oxygen declines and the shoaling of the hypoxic boundary in the California current. *Geophysical Research Letters* 35:L12811. doi:10.1029/2008GL034185.

Böhlke, J.K. and Denver, J.M. 1995. Combined use of groundwater dating, chemical, and isotopic analyses to resolve the history and fate of nitrate contamination in two agricultural watersheds, Atlantic Coastal Plain, Maryland. *Water Resources Research* 31:2319–2339. doi:10.1029/95WR01584.

Booker, F., Muntifering, R., McGrath, M., Burkey, K., Decoteau, D., Fiscus, E., Manning, W., Krupa, S., Chappelka, A., and Grantz, D. 2009. The ozone component of global change: potential effects on agricultural and horticultural plant yield, product quality and interactions with invasive species. *Journal of Integrative Plant Biology* 51:337–351.

Bormann, F.H., Likens, G.E., and Melillo, J.M. 1977. Nitrogen budget for an aggrading northern hardwood forest ecosystem. *Science* 196:981–983. doi:10.1126/science.196.4293.981.

Borsa, A.A., Agnew, D.C., and Cayan, D.R. 2014. Ongoing drought-induced uplift in the western United States. *Science* 345:1587–1590. doi:10.1126/science.1260279.

Borzsonyi, M., Pinter, A., Surjan, A., and Csik, M. 1976. The transplacental effect of carcinogenic N nitroso compounds formed in vivo from Carbendazym in Swiss mice. *Magyar Onkologia* 20:163–171. (In Hungarian)

Bosch, D., Cook, Z., and Fuglie, K. 1995. Voluntary versus mandatory agricultural policies to protect water quality: adoption of nitrogen testing in Nebraska. *Review of Agricultural Economics* 17:13–24.

Bosch, H.M., Rosenfield, A.B., Huston, R., Shipman, H.R., and Woodward, F.L. 1950. Methemoglobinemia and minnesota well supplies. *Journal (American Water Works Association)* 42:161–170.

Bourn, D. and Prescott, J. 2002. A comparison of the nutritional value, sensory qualities, and food safety of organically and conventionally produced foods. *Critical Reviews in Food Science and Nutrition* 42:1–34.

Bouwman, A.F., Boumans, L.J.M., and Batjes, N.H. 2002a. Emissions of N_2O and NO from fertilized fields: summary of available measurement data. *Global Biogeochemical Cycles* 16:6-1–6-13.

Bouwman, A.F., Boumans, L.J.M., and Batjes, N.H. 2002b. Modeling global annual N_2O and NO emissions from fertilized fields. *Global Biogeochemical Cycles* 16:28-1–28-9.

Bowman, W. D., Cleveland, C. C., Halada, ʄπ., Hreško, J., and Baron, J. S. 2008. Negative impact of nitrogen deposition on soil buffering capacity. *Nature Geoscience* 1:767–770.

Boyer, C., Brorsen, W., Solie, J., and Raun, W. 2010. Profitability of conventional vs. variable rate nitrogen application in wheat production. Paper presented at *The Southern Agricultural Economics Association Annual Meeting*, February 7–9, Orlando, FL.

Boyer, E. W., Goodale, C. L., Jaworski, N. A., and Howarth, R. W. 2002. Anthropogenic nitrogen sources and relationships to riverine nitrogen export in the northeastern U.S.A., in Boyer, E. W. and Howarth, R. W. (Eds): *The Nitrogen Cycle at Regional to Global Scales*. Springer Netherlands, Heidelberg, pp.137–169.

Boyle, D. B., King, A., Kourakos, G., Lockhart, K., Mayzelle, M. M., Fogg, G. E., and Harter, T. 2012. *Groundwater Nitrate Occurrence. Technical Report 4 in Addressing Nitrate in California's Drinking Water with a Focus on Tulare Lake Basin and Salinas Valley Groundwater*. Report for the State Water Resources Control Board Report to the Legislature. Center for Watershed Sciences, University of California, Davis.

Bradley, M. J. and Jones, B. M. 2002. Reducing global NO_x emissions: developing advanced energy and transportation technologies. *AMBIO: A Journal of the Human Environment* 31:141–149. doi:10.1579/0044-7447-31.2.141.

Bradshaw, T. K. and Muller, B. 1998. Impacts of rapid urban growth on farmland conversion: application of new regional land use policy models and geographical information systems. *Rural sociology* 63:1–25.

Brady, J. 1999. "Land is itself a sacred, living being": native American sacred site protection on federal public lands amidst the shadows of Bear Lodge. *American Indian Law Review* 24:153–186. doi:10.2307/20070625.

Brambilla, G. 1985. Genotoxic effects of drug/nitrite interaction products: evidence for the need of risk assessment. *Pharmacological Research Communications* 17:307–321. doi:10.1016/0031-6989(85)90011-6.

Brandt, S. J., Perez, L., Lurmann, F., and McConnell, R. 2012. Costs of childhood asthma due to traffic-related pollution in two California communities. *European Respiratory Journal* 40:363–370.

Bratton, S. 2004. Thinking like a mackerel: Rachel Carson's Under the Sea-Wind as a source for a trans-ecotonal sea ethic. *Ethics & the Environment* 9:1–22. doi:10.1353/een.2004.0002.

Brauer, M., Gehring, U., Brunekreef, B., de Jongste, J. C., Gerritsen, J., Rovers, M., Wichmann, H. E., Wijga, A., and Heinrich, J. 2006. Traffic-related air pollution and otitis media. *Environmental Health Perspectives* 114:1414–1418. doi:10.1289/ehp.9089.

Brauer, M., Hoek, G., Smit, H. A., de Jongste, J. C., Gerritsen, J., Postma, D. S., Kerkhof, M., and Brunekreef, B. 2007. Air pollution and development of asthma, allergy and infections in a birth cohort. *European Respiratory Journal* 29:879–888. doi:10.1183/09031936.00083406.

Brauer, M., Hoek, G., Vliet, P. V., Meliefste, K., Fischer, P. H., Wijga, A., Koopman, L. P. et al. 2002. Air pollution from traffic and the development of respiratory infections and asthmatic and allergic symptoms in children. *American Journal of Respiratory and Critical Care Medicine* 166:1092–1098. doi:10.1164/rccm.200108-007OC.

Brender, J. D., Olive, J. M., Felkner, M., Suarez, L., Marckwardt, W., and Hendricks, K. A. 2004. Dietary nitrites and nitrates, nitrosatable drugs, and neural tube defects. *Epidemiology* 15:330–336.

Breschini, S. J. and Hartz, T. K. 2002. Presidedress soil nitrate testing reduces nitrogen fertilizer use and nitrate leaching hazard in lettuce production. *HortScience* 37:1061–1064.

Brewer, R. L., Gordan, R. J., and Shephard, L. S. 1983. Chemistry of mist and fog from the Los Angeles urban area. *Atmospheric Environment* 17:2267–2270.

Bricker, S. B., Clement, C. G., Pirhalla, D. E., Orlando, S. P., and Farrow, D. R. G. 1999. *National Estuarine Eutrophication Assessment. Effects of Nutrient Enrichment in the Nation's Estuaries*. NOAA—NOS Special Projects Office, Silver Spring, MD.

Bricker, S. B., Longstaff, B., Dennison, W., Jones, A., Boicourt, K., Wicks, C., and Woerner, J. 2007. *Effects of Nutrient Enrichment in the Nation's Estuaries: A Decade of Change*, National Estuarine Eutrophication Assessment Update (No. 26), NOAA Coastal Ocean Program Decision Analysis Series, NOS Special Projects Office, Silver Spring, MD.

Broadbent, F. E. and Carlton, A. B. 1978. Field trials with isotopically labeled nitrogen fertilizer. *Nitrogen in the Environment* 1:1–41.

Broadbent, F. E. and Rauschkolb, R. S. 1977. Nitrogen fertilization and water pollution. *California Agriculture* 31:24–25.

Broadbent, F. E., Tyler, K. B., and May, D. M. 1980. Tomatoes make efficient use of applied nitrogen. *California Agriculture* 34:24–25.

Brody, J. G., Aschengrau, A., McKelvey, W., Swartz, C. H., Kennedy, T., and Rudel, R. A. 2006. Breast cancer risk and drinking water contaminated by wastewater: a case control study. *Environmental Health* 5:28. doi:10.1186/1476-069X-5-28.

Brook, R. D. 2007. Is air pollution a cause of cardiovascular disease? Updated review and controversies. *Reviews on Environmental Health* 22:115–137.

Brook, R. D. 2008. Cardiovascular effects of air pollution. *Clinical Science* 115:175–187. doi:10.1042/cs20070444.

Brook, R. D. and Rajagopalan, S. 2010. Particulate matter air pollution and atherosclerosis. *Current Atherosclerosis Reports* 12:291–300. doi:10.1007/s11883-010-0122-7.

Brook, R. D., Rajagopalan, S., Pope, C. A., Brook, J. R., Bhatnagar, A., Diez-Roux, A. V., Holguin, F. et al. 2010. Particulate matter air pollution and cardiovascular disease: an update to the scientific statement from the American heart association. *Circulation* 121:2331–2378. doi:10.1161/CIR.0b013e3181dbece1.

Browne, B. A. and Guldan, N. M. 2005. Understanding long-term baseflow water quality trends using a synoptic survey of the ground water–surface water interface, Central Wisconsin. *Journal of Environment Quality* 34:825. doi:10.2134/jeq2004.0134.

Brown, T. 2010. Phytoplankton community composition: the rise of the flagellates. *IEP Newsletter* 22:20.

Bruning-Fann, C. S. and Kaneene, J. B. 1993. The effects of nitrate, nitrite and N-nitroso compounds on human health: a review. *Veterinary and Human Toxicology* 35:521–538.

Brunke, H., Howitt, R. E., and Sumner, D. A. 2004. *Future Food Production and Consumption in California under Alternative*

Scenarios, *California Water Plan Update 2005, Vol. 4: Reference Guide*. UC Agricultural Issues Center, Davis, CA.

Brunke, H. and Sumner, D.A. 2002. *Assessing the role of NAFTA in California Agriculture: A Review of Trends and Economic Relationships*, UC Agricultural Issues Center, Davis, CA.

Bryan, N.S., Alexander, D.D., Coughlin, J.R., Milkowski, A.L., and Boffetta, P. 2012. Ingested nitrate and nitrite and stomach cancer risk: an updated review. *Food and Chemical Toxicology* 50:3646–3665. doi:10.1016/j.fct.2012.07.062.

Bukowski, J., Somers, G., and Bryanton, J. 2001. Agricultural contamination of groundwater as a possible risk factor for growth restriction or prematurity. *Journal of Occupational and Environmental Medicine* 43:377–383.

Bull, K., Hoft, R., and Sutton, M.A. 2011. Coordinating European nitrogen policies between international conventions and intergovernmental organizations, in Sutton, M.A., Howard, C.M., Erisman, J.W., Billen, G., Bleeker, A., Grennfelt, P., Van Grinsven, H.J.M., and Grizzetti, B. (Eds): *The European Nitrogen Assessment: Sources, Effects and Policy Perspectives*, Cambridge University Press, Cambridge, pp.570–584.

Burgan, O., Smargiassi, A., Perron, S., and Kosatsky, T. 2010. Cardiovascular effects of sub-daily levels of ambient fine particles: a systematic review. *Environmental Health: A Global Access Science Source* 9:26. doi:10.1186/1476-069x-9-26.

Burger, M., Jackson, L.E., Lundquist, E.J., Louie, D.T., Miller, R.L., Rolston, D.E., and Scow, K.M. 2005. Microbial responses and nitrous oxide emissions during wetting and drying of organically and conventionally managed soil under tomatoes. *Biology and Fertility of Soils* 42:109–118. doi:10.1007/s00374-005-0007-z.

Burnett, R.T., Stieb, D., Brook, J.R., Cakmak, S., Dales, R., Raizenne, M., Vincent, R., and Dann, T. 2004. Associations between short-term changes in nitrogen dioxide and mortality in Canadian cities. *Archives of Environmental Health* 59:228–236.

Burns, D.A., Lynch, J.A., Cosby, B.J., Fenn, M.E., and Baron, J.S. 2011. *National Acid Precipitation Assessment Program Report to Congress 2011: An Integrated Assessment*. National Science and Technology Council, Washington, DC.

Burow, K.R., Dubrovsky, N.M., and Shelton, J.L. 2007. Temporal trends in concentrations of DBCP and nitrate in groundwater in the eastern San Joaquin Valley, California, USA. *Hydrogeology Journal* 15:991–1007. doi:10.1007/s10040-006-0148-7.

Burow, K.R., Jurgens, B.C., Belitz, K., and Dubrovsky, N.M. 2012. Assessment of regional change in nitrate concentrations in groundwater in the Central Valley, California, USA, 1950s–2000s. *Environmental Earth Sciences* 69:2609–2621. doi:10.1007/s12665-012-2082-4.

Burow, K.R., Jurgens, B., Kauffman, L.J., Phillips, S.P., Dalgish, B.A., and Shelton, J.L. 2008a. *Simulations of Ground-Water Flow and Particle Pathline Analysis in the Zone of Contribution of a Public-Supply Well in Modesto, Eastern San Joaquin Valley, California*, Scientific Investigations Report No. 2008-5035. US Geological Survey, Sacramento, CA.

Burow, K.R., Shelton, J.L., and Dubrovsky, N.M. 2008b. Regional nitrate and pesticide trends in ground water in the eastern San Joaquin Valley, California. *Journal of Environmental Quality* 37:S249–263. doi:10.2134/jeq2007.0061.

Business Forecasting Center. 2010. *Employment Impacts of California Salmon Fishery Closures in 2008 and 2009*. Business Forecasting Center, Stockton, CA.

Butler, A.R. and Feelisch, M. 2008. Therapeutic uses of inorganic nitrite and nitrate from the past to the future. *Circulation* 117: 2151–2159. doi:10.1161/CIRCULATIONAHA.107.753814.

Butterbach-Bahl, K., Nemitz, E., and Zaehle, S. 2011. Nitrogen as a threat to the European greenhouse balance. Chapter 19 in *The European Nitrogen Assessment: Sources, Effects and Policy Perspectives*. Cambridge University Press, Cambridge.

Bytnerowicz, A., Temple, P.J., and Taylor, O.C. 1986. Effects of simulated acid fog on leaf acidification and injury development of pinto beans. *Canadian Journal of Botany* 64:918–922.

CA DOF (California Department of Finance). 2005. *E-6 County Population Estimates and Components of Change—July 1, 1990–2000*. Accessed June 6, 2015. http://www.dof.ca.gov/research/demographic/reports/estimates/e-6/1990-2000/.

CA DOF. 2011. *E-2. California County Population Estimates and Components of Change by Year—July 1, 2000–2010*. Accessed June 6, 2015. http://www.dof.ca.gov/research/demographic/reports/estimates/e-2/2000-10/.

CA DOF. 2013. *Historical County and City Estimates: Reports and Research Papers*. Accessed June 6, 2015. http://www.dof.ca.gov/research/demographic/reports/view.php#objjCollapsiblePanelEstimatesAnchor.

CA DOF. 2014. *E-6. Population Estimates and Components of Change by County—July 1, 2010–2014*. Accessed June 6, 2015. http://www.dof.ca.gov/research/demographic/reports/estimates/e-6/view.php.

CA DWR (California Department of Water Resources). 2003. *California's Groundwater: Bulletin 118 – Update 2003*. CA DWR, Sacramento, CA

CA DWR. 2005. *California Water Plan Update 2005: Selected Water Prices in California*. CA DWR, Sacramento, CA.

CA DWR. 2008. *Managing An Uncertain Future: Climate Change Adaptation Strategies for California's Water*. CA DWR, Sacramento, CA.

CA DWR. 2014a. *California Water Plan Update 2013, Volume 1: The Strategic Plan*. CA DWR, Sacramento, CA.

CA DWR. 2014b. *California Water Plan Update 2013 Investing in Innovation & Infrastructure, Vol. 1: Highlights*. CA DWR, Sacramento, CA.

CA DWR. 2015. *Groundwater*, California Department of Water Resources. Accessed July 26, 2015. http://www.water.ca.gov/groundwater/.

CA EPA Water Resources Control Board. 2015. *Web Data Downloads and Reports from California Water Resources Control Board*. Available at: http://www.waterboards.ca.gov/water_issues/programs/gama/report_depot.shtml.

California Attorney General's Office. 2009. *Climate Change, the California Environmental Quality Act, and General Plan Updates: Straightforward Answers to Some Frequently Asked Questions*. California Attorney General's Office, Sacramento, CA.

California Climate Change Center. 2012. *The Third Climate Change Assessment. California Climate Change Portal*. Accessed June 6, 2015. http://www.climatechange.ca.gov/climate_action_team/reports/third_assessment/index.html.

California Natural Resources Agency. 2009. *2009 California Climate Adaptation Strategy: A Report to the Governor of the State of California in Response to Executive Order S-13-2008.* California Natural Resources Agency, Sacramento.

California Rural Policy Task Force. 2003. *California Agriculture: Feeding the Future.* Governor's Office of Planning and Research, Sacramento, CA.

CalTrans (California Department of Transportation). 2009. *2008 California Motor Vehicle Stock, Travel and Fuel Forecast.* California Department of Transportation, Sacramento.

Calvo, E., Sanz, M.J., and Martin, C. 2007. Ozone sensitivity differences in five tomato cultivars: visible injury and effects on biomass and fruits. *Water Air and Soil Pollution* 186:167–181.

Campbell, S.J., Wanek, R., and Coulston, J.W. 2007. *Ozone Injury in West Coast Forests: 6 Years of Monitoring.* USDA Forest Service, Washington, DC.

Canada, H.E., Harter, T., Honeycutt, K., Jessoe, K., Jenkins, M.W., and Lund, J.R. 2012. *Regulatory and Funding Options for Nitrate Groundwater Contamination: Technical Report 8 in Addressing Nitrate in California's Drinking Water with a Focus on Tulare Lake Basin and Salinas Valley Groundwater.* Report for the State Water Resources Control Board Report to the Legislature, Center for Watershed Sciences, University of California, Davis.

Canfield, D.E., Glazer, A.N., and Falkowski, P.G. 2010. The evolution and future of Earth's nitrogen cycle. *Science* 330:192–196.

Cantrill, J.G. 1998. The environmental self and a sense of place: communication foundations for regional ecosystem management. *Journal of Applied Communication Research* 26:301–318. doi:10.1080/00909889809365509.

CAPCOA (California Air Pollution Control Officers Association). 2012. *California's Progress Towards Clean Air.* CAPCOA, Sacramento, CA.

Caraco, N.F. and Cole, J.J. 2001. Human influence on nitrogen export: a comparison of mesic and xeric catchments. *Marine & Freshwater Research* 52:119–125.

CARB (California Air Resources Board). 1999. *The 1999 California Almanac Of Emissions And Air Quality,* CARB, Sacramento, CA.

CARB. 2004. *Report to the Legislature: Gas-Fired Power Plant NOx Emission Controls and Related Environmental Impacts.* California Air Resources Board, Stationary Source Division, Sacramento.

CARB. 2006. *Emission Reduction Plan for Ports and Goods Movement in California.* CARB, Sacramento, CA.

CARB. 2008. *Climate Change Scoping Plan: A Framework for Change.* CARB, Sacramento, CA.

CARB. 2009a. *CEPAM: 2009 Almanac—Population and Vehicle Trends Tool.* Accessed June 7, 2015. http://www.arb.ca .gov/app/emsinv/trends/ems_trends.php.

CARB. 2009b. *California Climate Adaptation Strategy.* A Report to the Governor of the State of California in Response to Executive Order S-13-2008. CARB, Sacramento, CA

CARB. 2009c. *California's 1990–2004 Greenhouse Gas Emissions Inventory and 1990 Emissions Level: Technical Support Document.* CARB, Sacramento, CA.

CARB. 2010a. *Almanac of Emissions and Air Quality: 2009 Edition.* CARB, Sacramento, CA.

CARB. 2010b. *California greenhouse gas inventory for 2000–2008.* CARB, Sacramento, CA.

CARB. 2011. *Area Designations Maps / State and National.* California Environmental Protection Agency. Available at: http://www.arb.ca.gov/desig/adm/adm.htm.

CARB. 2012. *California Ambient Air Quality Standards (CAAQS).* Available at: http://www.arb.ca.gov/research/aaqs/caaqs /caaqs.htm.

CARB. 2013. *Almanac of Emissions and Air Quality: 2013 Edition.* Accessed July 29, 2015. http://www.arb.ca.gov/aqd /almanac/almanac13/toc13.htm.

CARB. 2014a. *Assembly Bill 32: California Global Warming Solutions Act.* Assembly Bill 32 Overview. Accessed June 6, 2015. http://www.arb.ca.gov/cc/ab32/ab32.htm.

CARB. 2014b. *California Greenhouse Gas Emission Inventory Program.* Accessed March 21, 2013. http://www.arb.ca.gov /cc/inventory/inventory.htm.

CARB. 2015a. *Mobile Source Emissions Inventory (MSEI).* Available at: http://www.arb.ca.gov/msei/msei.htm.

CARB. 2015b. *Truck and Bus Regulation 2014.* California Environmental Protection Agency. Available at: http://www .arb.ca.gov/regact/2014/truckbus14/truckbus14.htm.

CARB. n.d. *Emissions Inventory Data (CEPAM).* Accessed June 12, 2012. http://www.arb.ca.gov/ei/emissiondata.htm.

Carey, A.E., Lyons, W.B., Bonzongo, J.-C., and Lehrter, J.C. 2001. Nitrogen budget in the Upper Mississippi River watershed. *Environmental & Engineering Geoscience* 7: 251–265. doi:10.2113/gseegeosci.7.3.251.

Carlsson-Kanyama, A. 1998. Climate change and dietary choices: how can emissions of greenhouse gases from food consumption by reduced? *Food Policy* 23: 277–293.

Carman, H.F. 1979. The demand for nitrogen, phosphorous and potash fertilizer nutrients in the Western United States. *Western Journal of Agricultural Economics* 4:23–31.

Carpentier, C.L., Bosch, D.J., and Batie, S. 1998. Using spatial information to reduce costs of controlling agricultural nonpoint source pollution. *Agricultural and Resource Economics Review* 27:72–84.

Carriker, G.L. 1995. Factor-input demand subject to economic and environmental risk: nitrogen fertilizer in Kansas dryland corn production. *Review of Agricultural Economics* 17:77–89. doi:10.2307/1349657.

Carson, R. 1941. *Under the Sea-Wind: A Naturalist's Picture of Ocean Life.* Simon and Schuster, New York.

Carson, R. 1962. *Silent Spring.* Houghton Mifflin Harcourt, Boston, MA.

CA SWRCB (California State Water Resources Control Board). 1997. *Wastewater User Charge Survey Report F.Y. 1996–1997.* CA SWRCB, Sacramento.

CA SWRCB. 2008. *Wastewater User Charge Survey Report F.Y. 2007–2008.* CA SWRCB, Sacramento.

CA SWRCB. 2011. *Support for the Sanitary Sewer Overflow Program.* California Integrated Water Quality System Project. Accessed June 7, 2015. http://www.waterboards.ca.gov /ciwqs/chc_sso.shtml.

CA SWRCB. 2015. *GeoTracker.* Available at: http://geotracker .waterboards.ca.gov/.

CASA (California Association of Sanitary Agencies). 2009. *California Biosolids Trends for 2009.* CASA, Sacramento, CA.

Cassel, T., Ashbaugh, L., Flocchini, R., and Meyer, D. 2005. Ammonia emission factors for open-lot dairies: direct measurements and estimation by nitrogen intake. *Journal of the Air & Waste Management Association* 55:826–833. doi:10. 1080/10473289.2005.10464660.

Cassman, K.G., Dobermann, A., and Walters, D.T. 2002. Agroecosystems, nitrogen-use efficiency, and nitrogen management. *AMBIO: A Journal of the Human Environment* 31:132–140.

Castillo, A.R. 2009. Whole-farm nutrient balances are an important tool for California dairy farms. *California Agriculture* 63:149–151. doi:10.3733/ca.v063n03p149.

Castro, M.S., Driscoll, C.T., Jordan, T.E., Reay, W.G., and Boynton, W.R. 2003. Sources of nitrogen to estuaries in the United States. *Estuaries* 26:803–814. doi:10.1007/BF02711991.

Cavagnaro, T., Jackson, L., and Scow, K. 2006. *Climate Change: Challenges and solutions for California agricultural landscapes.* California Climate Change Center White Paper.

Cayan, D.R., Das, T., Pierce, D.W., Barnett, T.P., Tyree, M., and Gershunov, A. 2010. Future dryness in the southwest US and the hydrology of the early 21st century drought. *Proceedings of the National Academy of Sciences of the United States of America* 107:21271–21276. doi:10.1073/pnas.0912391107.

CCTP (Climate Change Technology Program). 2006. *Strategic Plan.* Department of Energy, Washington, DC.

CDC (Centers for Disease Control and Prevention). 2012. *Recent Water-related Response Activities.* Accessed July 10, 2015. http://www.cdc.gov/nceh/hsb/hab/.

CDC (Center for Disease Control and Prevention). 2008. *Behavioral Risk Factor Surveillance System Survey Data.* US Department of Health and Human Services, Atlanta, GA.

CDC and US Department of Health and Human Services. 2009. *Behavioral Risk factor Surveillance System Survey Data.* National Center for Chronic Disease Prevention and Health Promotion, Behavioral Surveillance Branch, Atlanta, GA.

CDC LRP (California Department of Conservation Division of Land Resource Protection). 2007. *Important Farmland Data Availability.* Available at: http://redirect.conservation.ca.gov/DLRP/fmmp/product_page.asp.

CDFA (California Department of Food and Agriculture). 2009. *Fertilizer Tonnage Report by Year.* CDFA, Sacramento, CA.

CDFA. 2010. *California Agricultural Highlights.* CDFA, Sacramento, CA.

CDFA. 2012. *California Agricultural Statistics Review 2012–2013.* CDFA, Sacramento, CA.

CDFA. 2013. *California Dairy Statistics 2012.* California Department of Food and Agriculture, Dairy Marketing Branch, Sacramento.

CDPH (California Department of Public Health). 2013. *Blue-Green Algae (Cyanobacteria) Blooms.* Accessed July 10, 2015. https://www.cdph.ca.gov/HealthInfo/environhealth/water/Pages/Bluegreenalgae.aspx.

CEC (California Energy Commission). 2015. *Electric Generation Capacity & Energy. California Energy Almanac.* Accessed July 26, 2015. http://energyalmanac.ca.gov/electricity/electric_generation_capacity.html.

Cedergren, M.I., Selbing, A.J., Löfman, O., and Källen, B.A.J. 2002. Chlorination byproducts and nitrate in drinking water and risk for congenital cardiac defects. *Environmental Research* 89:124–130. doi:10.1006/enrs.2001.4362.

CENR (Committee on Environment and Natural Resources). 2010. *Scientific Assessment of Hypoxia in U.S. Coastal Waters. Interagency Working Group on Harmful Algal Blooms, Hypoxia, and Human Health of the Joint Subcommittee on Ocean Science and Technology.* CENR, Washington, DC.

Chan, F., Barth, J.A., Lubchenco, J., Kirincich, A., Weeks, H., Peterson, W.T., and Menge, B.A. 2008. Emergence of anoxia in the California current large marine ecosystem. *Science* 15:920–920. doi:10.1126/science.1149016.

Chang, A.C., Adriano, D.C., and Pratt, P.F. 1973. Waste accumulation on a selected dairy corral and its effect on the nitrate and salt of the underlying soil strata. *Journal of Environment Quality* 2:233. doi:10.2134/jeq1973.00472425000200020013x.

Chang, C.C., Tsai, S.S., Ho, S.C., and Yang, C.Y. 2005. Air pollution and hospital admissions for cardiovascular disease in Taipei, Taiwan. *Environmental Research* 98:114–119. doi:10.1016/j.envres.2004.07.005.

Charbonneau, R. and Kondolf, G.M. 1993. Land use change in California, USA: nonpoint source water quality impacts. *Environmental Management* 17:453–460. doi:10.1007/BF02394661.

Chase, L.E. 2011. Maintaining milk yield while lowering dietary protein content. *WCDS Advances in Dairy Technology* 23:153–164.

Chauhan, A.J. and Johnston, S.L. 2003. Air pollution and infection in respiratory illness. *British Medical Bulletin* 68:95–112. doi:10.1093/bmb/ldg022.

Chen, Y., Yang, Q., Krewski, D., Shi, Y., Burnett, R.T., and McGrail, K. 2004. Influence of relatively low level of particulate air pollution on hospitalization for COPD in elderly people. *Inhalation Toxicology* 16:21–25. doi:10.1080/08958370490258129.

Chiu, H.F., Tsai, S.S., Wu, T.N., and Yang, C.Y. 2010. Colon cancer and content of nitrates and magnesium in drinking water. *Magnesium Research* 23:81–89. doi:10.1684/mrh.2010.0203.

Chow, J.C., Watson, J.G., Ashbaugh, L.L., and Magliano, K.L. 2003. Similarities and differences in PM_{10} chemical source profiles for geological dust from the San Joaquin Valley, California. *Atmospheric Environment* 37:1317–1340.

Christensen, L., Bianchi, M., Peacock, W., and Hirschfelt, D. 1994. Effect of nitrogen fertilizer timing and rate on inorganic nitrogen status, fruit composition, and yield of grapevines. *American Journal of Enology and Viticulture* 45:377–387.

Chum, H., Faaij, J., Moreira, J., Berndes, G., Dhamija, P., Dong, H., and Gabrielle, A. 2011. *Bioenergy: IPCC Special Report on Renewable Energy Sources and Climate Change Mitigation.* Cambridge University Press, Cambridge, UK/New York.

Chung, J.-C., Chou, S.-S., and Hwang, D.-F. 2004. Changes in nitrate and nitrite content of four vegetables during storage at refrigerated and ambient temperatures. *Food Additives & Contaminants* 21:317–322. doi:10.1080/02652030410001668763.

CIWMB (California Integrated Waste Management Board). 2005. *2004 Statewide Waste Characterization Study.* California Department of Resources Recycling and Recovery, Sacramento.

Claassen, R. 2004. Have conservation compliance incentives reduced soil erosion? *Amber Waves*. United States Department of Agriculture, Economic Research Service.

Clarisse, L., Clerbaux, C., Dentener, F., Hurtmans, D., and Coheur, P. F. 2009. Global ammonia distribution derived from infrared satellite observations. *Nature Geoscience* 2:479–483. doi:10.1038/ngeo551.

Clarisse, L., Shephard, M. W., Dentener, F., Hurtmans, D., Cady-Pereira, K., Karagulian, F., Van Damme, M., Clerbaux, C., and Coheur, P. F. 2010. Satellite monitoring of ammonia: a case study of the San Joaquin Valley. *Journal of Geophysical Research: Atmospheres* 115:D13302. doi:10.1029/2009JD013291.

CLBL (Center for Land-Based Learning). 2015. *Center for Land-Based Learning*. Available at: http://landbasedlearning.org/.

Climate Action Reserve. 2011. *Rice Cultivation Project Protocol. Climate Action Reserve*. Accessed June 7, 2015. http://www.climateactionreserve.org/how/protocols/rice-cultivation/.

Climate Action Reserve. 2012. *Nitrogen Management Project Protocol. Climate Action Reserve*. Accessed June 7, 2015. http://www.climateactionreserve.org/how/protocols/nitrogen-management/.

Cloern, J. E. and Oremland, R. S. 1983. Chemistry and microbiology of a sewage spill in south San Francisco Bay. *Estuaries* 6:399–406.

Cockcroft, D. W. and Davis, B. E. 2006. Mechanisms of airway hyperresponsiveness. *Journal of Allergy and Clinical Immunology* 118:551–559. doi:10.1016/j.jaci.2006.07.012.

Colford, J. M., Wade, T. J., Schiff, K. C., Wright, C. C., Griffith, J. F., Sandhu, S. K., Burns, S. et al. 2007. Water quality indicators and the risk of illness at beaches with nonpoint sources of fecal contamination. *Epidemiology* 18:27–35. doi:10.1097/01.ede.0000249425.32990.b9.

Collias, E. E. 1985. *Nationwide Review of Oxygen Depletion and Eutrophication in Estuarine and Coastal Waters: Pacific Region*, Report to US Department of Commerce, NOAA, National Ocean Service. Rockville, MD.

Colorado State University. 2012. *Agriculture and Land Use National Greenhouse Gas Inventory Software*. Accessed June 7, 2015. http://www.nrel.colostate.edu/projects/ALUsoftware/index.html.

Committee of Experts on Dairy Manure Management. 2005a. *Managing Dairy Manure in the Central Valley of California*. University of California, Agriculture and Natural Resources, Oakland.

Committee of Experts on Dairy Manure Management. 2005b. *Managing Dairy Manure in the Central Valley of California: Distribution*. University of California, Agriculture and Natural Resources, Oakland.

Compton, J. E., Harrison, J. A., Dennis, R. L., Greaver, T. L., Hill, B. H., Jordan, S. J., Walker, H., and Campbell, H. V. 2011. Ecosystem services altered by human changes in the nitrogen cycle: a new perspective for US decision making. *Ecology Letters* 14:804–815. doi:10.1111/j.1461-0248.2011.01631.x.

Copeland, C. 2012. *Clean Water Act and Pollutant Total Maximum Daily Loads (TMDLs)*, CRS Report for Congress No. R42752. Congressional Research Service, Washington, DC.

Corbett, J. J., Fischbeck, P. S., and Pandis, S. N. 1999. Global nitrogen and sulfur inventories for oceangoing ships.

Journal of Geophysical Research 104:3457–3470. doi:10.1029/1998JD100040.

Correia, M., Barroso, #afA., Barroso, M. F., Soares, D., Oliveira, M. B. P. P., and Delerue-Matos, C. 2010. Contribution of different vegetable types to exogenous nitrate and nitrite exposure. *Food Chemistry* 120:960–966. doi:10.1016/j.foodchem.2009.11.030.

Coss, A., Cantor, K. P., Reif, J. S., Lynch, C. F., and Ward, M. H. 2004. Pancreatic cancer and drinking water and dietary sources of nitrate and nitrite. *American Journal of Epidemiology* 159:693–701.

Cox, P., Delao, A., Komorniczak, A., and Weller, R. 2009. California almanac of emissions and air quality. California Air Resources Board, Sacramento, CA.

Craun, G. F., Greathouse, D. G., and Gunderson, D. H. 1981. Methaemoglobin levels in young children consuming high nitrate well water in the United States. *International Journal of Epidemiology* 10:309–317.

Crews, T. E. and Peoples, M. B. 2005. Can the synchrony of nitrogen supply and crop demand be improved in legume and fertilizer-based agroecosystems? A review. *Nutrient Cycling in Agroecosystems* 72:101–120. doi:10.1007/s10705-004-6480-1.

Croen, L. A., Todoroff, K., and Shaw, G. M. 2001. Maternal exposure to nitrate from drinking water and diet and risk for neural tube defects. *American Journal of Epidemiology* 153:325–331. doi:10.1093/aje/153.4.325.

Cross, A. J., Freedman, N. D., Ren, J., Ward, M. H., Hollenbeck, A. R., Schatzkin, A., Sinha, R., and Abnet, C. C. 2011. Meat consumption and risk of esophageal and gastric cancer in a large prospective study. *American Journal of Gastroenterology* 106:432–442. doi:10.1038/ajg.2010.415

Crutzen, P. J., Mosier, A. R., Smith, K. A., and Winiwarter, W. 2008. N_2O release from agro-biofuel production negates global warming reduction by replacing fossil fuels. *Atmospheric Chemistry and Physics* 8:389–395.

Cruzate, G. and Casas, R. 2012. Extraccion y balance de nutrientes en los suelos agricolas de la Argentina. (In Spanish)

CRWQCB (California Regional Water Quality Control Board). 2007. *Waste Discharge Requirements General Order for Existing Milking Cow Dairies*, Technical Report Order No. R5-2007-0035. California Regional Water Quality Control Board, Central Valley Region.

CWTRC (California Wastewater Training and Research Center) and EPA (US Environmental Protection Agency). 2003. *Status Report: Onsite Wastewater Treatment Systems in California*. California State University at Chico.

Daane, K. M., Johnson, R. S., Michailides, T. J., Crisosto, C. H., Dlott, J. W., Ramirez, H. T., Yokota, G. Y., and Morgan, D. P. 1995. Excess nitrogen raises nectarine susceptibility to disease and insects. *California Agriculture* 49:13–18. doi:10.3733/ca.v049n04p13.

Daberkow, S., Ribaudo, M., Doering, O., Schepers, J. S., and Raun, W. R. 2008. Economic implications of public policies to change agricultural nitrogen use and management. Chapter 22 in *Nitrogen in Agricultural Systems*. American Society of Agronomy, Madison, WI, pp.883–910.

Dahl, M., Bauer, A. K., Arredouani, M., Soininen, R., Tryggvason, K., Kleeberger, S. R., and Kobzik, L. 2007. Protection against inhaled oxidants through scavenging of oxidized lipids by macrophage receptors MARCO and SR-AI/II. *Journal of Clinical Investigation* 117:757–764. doi:10.1172/jci29968.

Dalal, R.C. and Allen, D.E. 2008. Turner Review No. 18. Greenhouse gas fluxes from natural ecosystems. *Australian Journal of Botany* 56:369–407.

Dales, R., Doiron, M.S., and Cakmak, S. 2006. Gaseous air pollutants and hospitalization for respiratory disease in the neonatal period. *Environmental Health Perspectives* 114:1751–1754. doi:10.1289/ehp.9044.

Dales, R., Wheeler, A., Mahmud, M., Frescura, A.M., Smith-Doiron, M., Nethery, E., and Liu, L. 2008. The influence of living near roadways on spirometry and exhaled nitric oxide in elementary schoolchildren. *Environmental Health Perspectives* 116:1423–1427. doi:10.1289/ehp.10943.

Dallmann, T.R., Harley, R.A., and Kirchstetter, T.W. 2011. Effects of diesel particle filter retrofits and accelerated fleet turnover on drayage truck emissions at the Port of Oakland. *Environmental Science & Technology* 45:10773–10779. doi:10.1021/es202609q.

Dangour, A.D., Dodhi, S.K., Hayter, A., Allen, E., Lockhart, K., and Uauy, R. 2009. Nutritional quality of organic foods: a systematic review. *The American Journal of Clinical Nutrition* 90:680–685.

Daniel, C.R., Cross, A.J., Graubard, B.I., Park, Y., Ward, M.H., Rothman, N., Hollenbeck, A.R., Chow, W.H., and Sinha, R. 2012a. Large prospective investigation of meat intake, related mutagens, and risk of renal cell carcinoma. *American Journal of Clinical Nutrition* 95:155–162. doi:10.3945/ajcn.111.019364

Daniel, T.C. 2001. Whither scenic beauty? Visual landscape quality assessment in the 21st century. *Landscape and Urban Planning* 54:267–281. doi:10.1016/S0169-2046(01)00141-4.

Daniel, T.C., Muhar, A., Arnberger, A., Aznar, O., Boyd, J.W., Chan, K.M.A., and Costanza, R. 2012b. Contributions of cultural services to the ecosystem services agenda. *Proceedings of the National Academy of Sciences of the United States of America* 109:8812–8819.:doi:10.1073/pnas.1114773109.

Dauchet, L., Amouyel, P., Hercberg, S., and Dallongeville, J. 2006. Fruit and vegetable consumption and risk of coronary heart disease: a meta-analysis of cohort studies. *Journal of Nutrition* 136:2588–2593.

David, M.B. and Gentry, L.E. 2000. Anthropogenic inputs of nitrogen and phosphorus and riverine export for Illinois, USA. *Journal of Environment Quality* 29:494–508. doi:10.2134/jeq2000.00472425002900020018x

Davidson, E.A. 2009. The contribution of manure and fertilizer nitrogen to atmospheric nitrous oxide since 1860. *Nature Geoscience* 2:659–662. doi:10.1038/ngeo608.

Davidson, E.A., David, M.B., Galloway, J.N., Goodale, C.L., Haeuber, R., Harrison, J.A., Howarth, R.W. et al. 2012. Excess nitrogen in the US environment: trends, risks, and solutions. *Issues in Ecology* 15.

Davidson, E.A. and Kingerlee, W. 1997. A global inventory of nitric oxide emissions from soils. *Nutrient Cycling in Agroecosystems* 48:37–50. doi:10.1023/A:1009738715891.

Davis, D.R. 2009. Declining fruit and vegetable nutrient composition: what is the evidence? *HortScience* 44:15–19.

Davis, F.R. 2013. Worlds built on avian excrement. *Science* 340:1525–1526. doi:10.1126/science.1239339.

Davis, S.J. and Caldeira, K. 2010. Consumption-based accounting of CO_2 emissions. *Proceedings of the National Academy of Sciences of the United States of America* 107:5687–5692. doi:10.1073/pnas.0906974107.

Dawson, P.J. and Tiffin, R. 1998. Estimating the demand for calories in India. *American Journal of Agricultural Economics* 80:474–481. doi:10.2307/1244550.

De Gryze, S., Albarracin, M.V., Catalá-Luque, R., Howitt, R.E., and Six, J. 2009. Modeling shows that alternative soil management can decrease greenhouse gases. *California Agriculture* 63:84–90. doi:10.3733/ca.v063n02p84.

De Gryze, S., Wolf, A., Kaffka, S.R., Mitchell, J., Rolston, D.E., Temple, S.R., Lee, J., and Six, J. 2010. Simulating greenhouse gas budgets of four California cropping systems under conventional and alternative management. *Ecological Applications* 20:1805–1819.

De Roos, A.J., Ward, M.H., Lynch, C.F., and Cantor, K.P. 2003. Nitrate in public water supplies and the risk of colon and rectum cancers. *Epidemiology* 14:640–649. doi:10.1097/01.ede.0000091605.01334.d3.

De Vries, W., Reinds, G.J., and Vel, E. 2003. Intensive monitoring of forest ecosystems in Europe: 2: atmospheric deposition and its impacts on soil solution chemistry. *Forest Ecology and Management* 174:97–115. doi:10.1016/S0378-1127(02)00030-0.

De Vries, W., Solberg, S., Dobbertin, M., Sterba, H., Laubhann, D., Van Oijen, M., and Evans, C. 2009. The impact of nitrogen deposition on carbon sequestration by European forests and heathlands. *Forest Ecology and Management* 258:1814–1823.

De Vries, W., Wamelink, G.W.W., Van Dobben, H., Kros, J., Reinds, G.J., Mol-Dijkstra, J.P., Smart, S.M. et al. 2010. Use of dynamic soil-vegetation models to assess impacts of nitrogen deposition on plant species composition: an overview. *Ecological Applications* 20:60–79.

DeBano, L.F., Rice, R.M., and Eugene, C.C. 1979. *Soil Heating in Chaparral Fires: Effects on Soil Properties, Plant Nutrients, Erosion, and Runoff*, Research Paper No. SW-RP-145. USDA, Forest Service, Pacific Southwest Forest and Range Experiment Station, Berkeley, CA.

DeClerck, F. and Singer, M.J. 2003. Looking back 60 years, California soils maintain overall chemical quality. *California Agriculture* 57:38–41. doi:10.3733/ca.v057n02p38.

Deen, A., Estrada, T., Everts, C., Farina, S., Gibler, J., Guzman, M., Levine, J. et al. 2005. *Thirsty for Justice: A People's Blueprint for California Water*. Environmental Justice Coalition for Water, Sacramento, CA.

Delfino, R.J., Staimer, N., Gillen, D., Tjoa, T., Sioutas, C., Fung, K., George, S.C., and Kleinman, M.T. 2006. Personal and ambient air pollution is associated with increased exhaled nitric oxide in children with asthma. *Environmental Health Perspectives* 114:1736–1743. doi:10.1289/ehp.9141.

DeLind, L.B. 2002. Place, work, and civic agriculture: common fields for cultivation. *Agriculture and Human Values* 19: 217–224. doi:10.1023/A: 1019994728252.

Dellavalle, C.T., Daniel, C.R., Aschebrook-Kilfoy, B., Hollenbeck, A.R., Cross, A.J., Sinha, R., and Ward, M.H. 2013. Dietary intake of nitrate and nitrite and risk of renal cell carcinoma in the NIH-AARP diet and health study. *British Journal of Cancer* 108:205–212. doi:10.1038/bjc.2012.522.

Dellavalle, C.T., Xiao, Q., Yang, G., Shu, X.O., Aschebrook-Kil-foy, B., Zheng, W., Lan Li, H. et al.. 2014. Dietary nitrate and nitrite intake and risk of colorectal cancer in the Shanghai women's health study. *International Journal of Cancer* 134:2917–2926. doi:10.1002/ijc.28612.

Delwiche, C.C. 1970. The nitrogen cycle. *Scientific American* 223:136–146. doi:10.1038/scientificamerican0970-136.

Denbaly, M. and Vroomen, H. 1993. Dynamic fertilizer nutrient demands for corn: a cointegrated and error-correcting system. *American Journal of Agricultural Economics* 75:203–209. doi:10.2307/1242968.

Díaz, F.J., Chow, A.T., O'Geen, A.T., Dahlgren, R.A., and Wong, P.K. 2008. Restored wetlands as a source of disinfection byproduct precursors. *Environmental Science & Technology* 42:5992–5997. doi:10.1021/es800781n.

Dietz, T., Ostrom, E., and Stern, P. 2003. The struggle to govern the commons. *Science* 302:1907–1912.

Dijkstra, J., Oenema, O., and Bannink, A. 2011. Dietary strategies to reducing N excretion from cattle: implications for methane emissions. *Current Opinion in Environmental Sustainability, Carbon and Nitrogen Cycles* 3:414–422. doi:10.1016/j.cosust.2011.07.008.

Dillon, J., Edinger-Marshall, S., and Letey, J. 1999. Farmers adopt new irrigation and fertilizer techniques: changes could help growers maintain yields, protect water quality. *California Agriculture* 53:24–31. doi:10.3733/ca.v053n01p24.

Doane, T.A., Horwath, W.R., Mitchell, J.P., Jackson, J., Miyao, G., and Brittan, K. 2009. Nitrogen supply from fertilizer and legume cover crop in the transition to no-tillage for irrigated row crops. *Nutrient Cycling in Agroecosystems* 85:253–262. doi:10.1007/s10705-009-9264-9.

Dobermann, A. 2007. Nutrient use efficiency: measurement and management. Paper presented at *The IFA International Workshop on Fertilizer Best Management Practices*, March 7–9, Brussels, Belgium.

Dockery, D.W., Pope, C.A., Xu, X., Spengler, J.D., Ware, J.H., Fay, M.E., Ferris, B.G., and Speizer, F.E. 1993. An association between air pollution and mortality in six U.S. cities. *New England Journal of Medicine* 329:1753–1759. doi:10.1056/NEJM199312093292401.

DOL BLS (US Department of Labor, Bureau of Labor Statistics). 2015a. *Producer Price Index-Industry Data (customized data search)*, US Department of Labor. Accessed June 7, 2015. http://data.bls.gov/cgi-bin/dsrv?pc.

DOL BLS. 2015b. *Producer Price Index Commodity Data: Not Seasonally Adjusted (customized data search)*, US Department of Labor. Accessed June 7, 2015. http://data.bls.gov/cgi-bin/dsrv?wp.

DOL WHD (US Department of Labor, Wage and Hour Division). 2015. *Changes in basic minimum wages in non-farm employment under state law: selected years 1968 to 2013*. Accessed June 7, 2015. http://www.dol.gov/whd/state/stateMinWageHis.htm.

Domene, L.A.F. and Ayres, R.U. 2001. Nitrogen's role in industrial systems. *Journal of Industrial Ecology* 5:77–103. doi:10.1162/108819801753358517.

Dominici, F., McDermott, A., Zeger, S.L., and Samet, J.M. 2003. Airborne particulate matter and mortality: timescale effects in four US cities. *American Journal of Epidemiology* 157:1055–1065. doi:10.1093/aje/kwg087.

Dominici, F., Peng, R.D., Bell, M.L., Pham, L., McDermott, A., Zeger, S.L., and Samet, J.M. 2006. Fine particulate air pollution and hospital admission for cardiovascular and respiratory diseases. *Journal of the American Medical Association* 295:1127–1134. doi:10.1001/jama.295.10.1127.

Donham, K.J., Cumro, D., Reynolds, S.J., and Merchant, J.A. 2000. Dose-response relationships between occupational aerosol exposures and cross-shift declines of lung function in poultry workers: recommendations for exposure limits. *Journal of Occupational Environmental Medicine* 42:260–269.

Donham, K.J., Thorne, P.S., Breuer, G.M., Powers, W., Marquez, S., and Reynolds, S.J. 2002. Exposure limits related to air quality and risk assessment. Chapter 8 in *Iowa Concentrated Animal Feeding Operations Air Quality Study*. ISU/UI Study Group, University of Iowa College of Public Health, Iowa City.

Dorsch, M.M., Scragg, R.K., McMichael, A.J., Baghurst, P.A., and Dyer, K.F. 1984. Congenital malformations and maternal drinking water supply in rural South Australia: a case-control study. *American Journal of Epidemiology* 119:473–486.

Dramstad, W.E., Tveit, M.S., Fjellstad, W.J., and Fry, G.L.A. 2006. Relationships between visual landscape preferences and map-based indicators of landscape structure. *Landscape and Urban Planning* 78:465–474. doi:10.1016/j.landurbplan.2005.12.006.

Drinkwater, L.E., Letourneau, D.K., Workneh, F., van Bruggen, A.H.C., and Shennan, C. 1995. Fundamental differences between conventional and organic tomato agroecosystems in California. *Ecological Applications* 5:1098–1112. doi:10.2307/2269357.

Drinkwater, L.E. and Snapp, S.S. 2007. Nutrients in agroecosystems: rethinking the management paradigm, in Sparks, D.L. (Ed.): *Advances in Agronomy*. Academic Press, San Diego, CA, pp.163–186.

Drinkwater, L.E., Wagoner, P., and Sarrantonio, M. 1998. Legume-based cropping systems have reduced carbon and nitrogen losses. *Nature* 396:262–265. doi:10.1038/24376.

Dubrovsky, N.M., Burow, K.R., Clark, G.M., Gronberg, J.M., Hamilton, P.A., Hitt, K.J., Mueller, D.K. et al. 2010. *The Quality of Our Nation's Waters: Nutrients in the Nation's Streams and Groundwater, 1992–2004*. US Geological Survey Circular 1350.

Dubrow, R., Darefsky, A.S., Park, Y., Mayne, S.T., Moore, S.C., Kilfoy, B., Cross, A.J. et al. 2010. Dietary components related to N-nitroso compound formation: a prospective study of adult glioma. *Cancer Epidemiology, Biomarkers & Prevention* 19:1709–1722. doi:10.1158/1055-9965.EPI-10-0225.

Durant, J.L., Ash, C.A., Wood, E.C., Herndon, S.C., Jayne, J.T., Knighton, W.B., Canagaratna, M.R. et al. 2010. Short-term variation in near-highway air pollutant gradients on a winter morning. *Atmospheric Chemistry and Physics* 10:5599–5626.

Dutton, D. 2003. Aesthetics and evolutionary psychology, in Levinson, J. (Ed.): *The Oxford Handbook for Aesthetics*. Oxford University Press, New York.

Dzurella, K.N., Medellin-Azuara, J., Jensen, V.B., King, A.M., De La Mora, N., Fryjoff-Hung, A., Rosenstock, T.S. et al. 2012. *Nitrogen Source Reduction to Protect Groundwater Quality. Technical Report 3 in: Addressing Nitrate in California's Drinking Water with a Focus on Tulare Lake Basin and Salinas Valley Groundwater*. Report for the State Water Resources

Control Board Report to the Legislature. Center for Watershed Sciences, University of California, Davis.

Eadie, J. M., Elphick, C. S., Reinecke, K. J., and Miller, M. R. 2010. *Wildlife Values of North American Ricelands*. The Rice Foundation, Stuttgart, AR.

Eagle, A. J., Bird, J. A., Horwath, W. R., Linquist, B. A., Brouder, S. M., Hill, J. E., and Van Kessel, C. 2001. Rice yield and nitrogen utilization efficiency under alternative straw management practices. *Agronomy Journal* 92: 1096–1103.

Eagle, J. E., Henry, L. R., Olander, L. P., Haugen-Kozyra, K., Millar, N., and Robertson, G. P. 2010. *Greenhouse gas mitigation potential of agricultural land management in the United States: A Synthesis of the Literature*, Technical Working Group on Agricultural Greenhouse Gases (T-AGG) Report. Nicholas Institute for Environmental Policy Solutions, Duke University, Durham, NC.

Egerton-Warburton L. M. and Allen E. B. 2000. Shifts in arbuscular mycorrhizal communities along an anthropogenic nitrogen deposition gradient. *Ecological Applications* 10:484–496. doi:10.1890/1051-0761(2000)010[0484:SIAMCA]2.0.CO;2.

Egerton-Warburton, L. M., Graham, R. C., Allen, E. B., and Allen, M. F. 2001. Reconstruction of the historical changes in mycorrhizal fungal communities under anthropogenic nitrogen deposition. *Proceedings of the Royal Society of London. Series B: Biological Sciences* 7:2479–2484. doi:10.1098/rspb.2001.1844.

Elphick, C. S. 2000. Functional equivalency between rice fields and seminatural wetland habitats. *Conservation Biology* 14:181–191. doi:10.1046/j.1523-1739.2000.98314.x.

Elphick, C. S. and Oring, L. W. 1998. Winter management of Californian rice fields for waterbirds. *Journal of Applied Ecology* 35:95–108. doi:10.1046/j.1365-2664.1998.00274.x.

Emberson, L. D., Büker, P., Ashmore, M. R., Mills, G., Jackson, L., Agrawal, M., Atikuzzaman, M. D. et al. 2009. A comparison of North American and Asian exposure response data for ozone effects on crop yields. *Atmospheric Environment* 43:1945–1953.

Emerson, R. W. 1849. *Nature*. J. Munroe and Company, Boston, MA.

Emmerich, S. D. 2009. Fostering environmental responsibility among watermen of Chesapeake Bay: a faith and action research approach, in Heie, H. and King, M. A. (Eds): *Mutual Treasure: Seeking Better Ways for Christians and Culture to Converse*. Cascadia Publishing House, Telford, PA.

EPA (US Environmental Protection Agency). 1989. *Health Issue Assessment: Summary Review Of Health Effects Associated With Ammonia (No. PEA/600/8-89-052F)*. Office of Health and Environmental Assessment, Environmental Criteria and Assessment Office, Washington, DC.

EPA. 1990. *Final Drinking Water Criteria Document on Nitrate/Nitrite (No. PB91-142836)*. EPA, Washington, DC.

EPA. 1997. *Retrospective Study: 1970 to 1990. Benefits and Costs of the Clean Air Act*. Accessed June 7, 2015. http://www.epa.gov/air/sect812/retro.html.

EPA. 1999a. *1999 Update of Ambient Water Quality Criteria for Ammonia*. EPA, Washington, DC.

EPA. 1999b. *Technical Bulletin: Nitrogen Oxides (NO$_x$). Why and How They Are Controlled*. EPA, Washington, DC.

EPA. 2002. *Onsite Wastewater Treatment Systems Manual*, EPA/625/R-00/008. EPA, Washington, DC.

EPA. 2004a. *National Emission Inventory: Ammonia Emissions from Animal Husbandry Operations (Draft Report)*. EPA, Washington, DC.

EPA. 2004b. *National Emission Inventory: Ammonia Emissions from Animal Husbandry Operations (Draft Report) – Population*, English edition. EPA, Washiington, DC.

EPA. 2006a. *Conceptual Model for Nutrients in the Central Valley and Sacramento–San Joaquin Delta: Final Report*. EPA, Washington, DC.

EPA. 2006b. *Air Quality Criteria for Ozone and Related Photochemical Oxidants*. National Center for Environmental Assessment-RTP Division, Research Triangle Park, NC.

EPA. 2007. *National Estuary Program Coastal Condition Report*. EPA, Washington, DC.

EPA. 2008a. *Municipal Nutrient Removal Technologies Reference Document, Volume 1*, Technical Report EPA 832-R-08-006.

EPA. 2008b. 2005 *National Emissions Inventory Data & Documentation*. Available at: http://www.epa.gov/ttn/chief/net/2005inventory.html#inventorydata.

EPA. 2008c. *Integrated Science Assessment for Oxides of Nitrogen: Health Criteria*. Research Triangle Park, Research Triangle Park, NC.

EPA. 2008d. *Risk and Exposure Assessment to Support the Review of the NO$_2$ Primary National Ambient Air Quality Standard*. Office of Air Quality Planning and Standards, Research Triangle Park, NC.

EPA. 2008e. *Smart Energy Resources Guide*. EPA, Washington, DC.

EPA. 2009a. *National Drinking Water Regulations MCL Booklet*. EPA, Washington, DC.

EPA. 2009b. *Integrated Science Assessment for Particulate Matter (Final Report)*. EPA, Washington, DC.

EPA. 2009c. *An Urgent Call to Action. Report of the State-EPA Nutrient Innovations Task Group*. EPA, Washington, DC.

EPA. 2010. *Factsheet: Final Revisions to the National Ambient Air Quality Standards for Nitrogen Dioxide*. EPA, Washington, DC.

EPA. 2012. *National Ambient Air Quality Standards*. Research Triangle Park, NC.

EPA. 2013. *Clean Air Act*. Available at: http://www.epa.gov/air/caa/.

EPA. 2014. *Drinking Water Contaminants*. Accessed June 1, 2013. http://water.epa.gov/drink/contaminants/index.cfm.

EPA. 2015. *Air Quality System (AQS) Data Mart [internet database]. Technology Transfer Network (TTN)*. Accessed June 14, 2015. http://www.epa.gov/ttn/airs/aqsdatamart.

EPA SAB (US Environmental Protection Agency Science Advisory Board). 2011. *Reactive Nitrogen in the United States: An Analysis of Inputs, Flows, Consequences, and Management Options: A Report of the EPA Science Advisory Board*. EPA SAB, Washington, DC.

Erisman, J. W., Galloway, J. N., Seitzinger, S., Bleeker, A., and Butterbach-Bahl, K. 2011. Reactive nitrogen in the environment and its effect on climate change. *Current Opinion in Environmental Sustainability, Carbon and Nitrogen Cycles* 3:281–290. doi:10.1016/j.cosust.2011.08.012.

Erisman, J. W., Sutton, M. A., Galloway, J. N., Klimont, Z., and Winiwarter, W. 2008. How a century of ammonia synthesis changed the world. *Nature Geoscience* 1:636–639. doi:10.1038/ngeo325.

Escamilla-Nuñez, M. C., Barraza-Villarreal, A., Hernandez-Cadena, L., Moreno-Macias, H., Ramirez-Aguilar, M.,

Sienra-Monge, J.J., Cortez-Lugo, M., Texcalac, J.L., del Rio-Navarro, B., and Romieu, I. 2008. Traffic-related air pollution and respiratory symptoms among asthmatic children, resident in Mexico City: the EVA cohort study. *Respiratory Research* 9. doi:10.1186/1465-9921-9-74.

European Food Safety Authority. 2008. Nitrate in vegetables: Scientific Opinion of the Panel on Contaminants in the Food chain. *The EFSA Journal* 689: 1–79. doi:10.2903/j.efsa.2008.689.

European Food Safety Authority. 2010. Statement on possible public health risks for infants and young children from the presence of nitrates in leafy vegetables. *The EFSA Journal* 8:1935–1977. doi:10.2903/j.efsa.2010.1935.

Evans, R., Dodge, L., and Newman, J. 2007. *Nutrient Management in Nursery and Floriculture: Publication No. 8221.* University of California Division of Agriculture and Natural Resources, Davis.

Faith-350. 2013. *Faith Action Round-up.* Available at: http://350.org/en/faith.

Fan, A.M., Willhite, C.C., and Book, S.A. 1987. Evaluation of the nitrate drinking water standard with reference to infant methemoglobinemia and potential reproductive toxicity. *Regulatory Toxicology and Pharmacology* 7:135–148.

Feigin, A., Letey, J., and Jarrell, W.M. 1982a. Nitrogen utilization efficiency by drip irrigated celery receiving preplant or water applied N fertilizer. *Agronomy Journal* 74:978–983. doi:10.2134/agronj1982.00021962007400060012x.

Feigin, A., Letey, J., and Jarrell, W.M. 1982b. Celery response to type, amount, and method of N-fertilizer application under drip irrigation. *Agronomy Journal* 74:971–977.

Felzer, B.S., Cronin, T., Reilly, J.M., Melillo, J.M., and Wang, X. 2007. Impacts of ozone on trees and crops. *Comptes Rendus Geoscience* 339:784–798.

Felzer, B.S., Kicklighter, D.W., Melillo, J.M., Wang, C., Zhuang, Q., and Prinn, R.G. 2004. Ozone effects on net primary production and carbon sequestration in the conterminous United States using a biogeochemistry model. *Tellus* 56:230–248.

Felzer, B.S., Reilly, J.M., Melillo, J., Kicklighter, D.W., Sarofim, M., Wang, C., Prinn, R.G., and Zhuang, Q. 2005. Future effects of ozone on carbon sequestration and climate change policy using a biogeochemistry model. *Climatic Change* 73:345–373.

Feng, Z.Z. and Kobayashi, K. 2009. Assessing the impacts of current and future concentrations of surface ozone on crop yield with meta-analysis. *Atmospheric Environment* 43:1510–1519.

Feng, Z.Z., Kobayashi, K., and Ainsworth, E.A. 2008. Impact of elevated ozone concentration on growth, physiology, and yield of wheat (Triticum aestivum L.): a meta-analysis. *Global Change Biology* 14:2696–2708.

Fenn, M.E., Allen, E.B., and Geiser, L.H. 2011.Mediterranean California. Chapter 13 in *Assessment of Nitrogen Deposition Effects and Empirical Critical Loads of Nitrogen for Ecoregions of the United States.* USDA Forest Service, Northern Research Station, Newtown Square, PA, pp.143–169.

Fenn, M.E., Allen, E.B., Weiss, S.B., Jovan, S., Geiser, L.H., Tonnesen, G.S., Johnson, R.F. et al. 2010. Nitrogen critical loads and management alternatives for N-impacted ecosystems in California. *Journal of Environmental Management* 91:2404–2423. doi:10.1016/j.jenvman.2010.07.034.

Fenn, M.E., Baron, J.S., Allen, E.B., Rueth, H.M., Nydick, K.R., Geiser, L., Bowman, W.D. e t al. P. 2003. Ecological effects of nitrogen deposition in the Western United States. *BioScience* 53:404–420. doi:10.1641/0006-3568(2003)053[0404:EEONDI]2.0.CO;2.

Fenn, M.E., Jovan, S., Yuan, F., Geiser, L., Meixner, T., and Gimeno, B.S. 2008. Empirical and simulated critical loads for nitrogen deposition in California mixed conifer forests. *Environmental Pollution* 155:492–511. doi:10.1016/j.envpol.2008.03.019.

Fenn, M.E., Poth, M.A., and Johnson, D.W. 1996. Evidence for nitrogen saturation in the San Bernardino Mountains in Southern California. *Forest Ecology and Management* 82:211–230.

Ferrucci, L.M., Sinha, R., Ward, M.H., Graubard, B.I., Hollenbeck, A.R., Kilfoy, B.A., Schatzkin, A., Michaud, D.S., and Cross, A.J. 2010. Meat and components of meat and the risk of bladder cancer in the NIH-AARP Diet and Health Study. *Cancer* 116:4345–4353. doi:10.1002/cncr.25463.

Fewtrell, L. 2004. Drinking-water nitrate, methemoglobinemia, and global burden of disease: a discussion. *Environmental Health Perspectives* 112:1371–1374. doi:10.1289/ehp.7216.

Fick, G.W. 2008. *Food, Farming, and Faith.* SUNY Press, Albany, NY.

Firestone, L. 2009. *Guide to Community Drinking Water Advocacy.* Community Water Center, Visalia, CA.

Firestone, L., Kaswan, A., and Meraz, S. 2006. Environmental justice: access to clean drinking water. *Hastings Law Journal* 57:1367–1386.

Fischer, A.C. 2006. Determinants of California farmland values and potential impacts of climate change. *Agricultural and Resource Economics Update* 9:5–8.

Florez, D. 203. *California Senate Bill 700: Native Americans: Agriculture & Air Quality Summary and Implementation.* Available at: http://www.arb.ca.gov/ag/sb700/sb700.pdf.

Foe, C., Ballard, A., and Fong, S. 2010. *Nutrient Concentrations and Biological Effects in the Sacramento-San Joaquin Delta.* Central Valley RWQCB, Rancho Cordova, CA.

Fogg, G.E., Rolston, D.E., Decker, D. I., Louie, D.T., and Grismer, M.E. 1998. Spatial Variation in nitrogen isotope values beneath nitrate contamination sources. *Ground Water* 36:418–426. doi:10.1111/j.1745-6584.1998.tb02812.x.

Forster, P., Ramaswamy, V., Artaxo, P., Berntsen, T., Betts, R., Fahey, D.W., Haywood, J. et al. 2007. Changes in atmospheric constituents and in radiative forcing, in Solomon, S., Qin, D., Manning, M., Chen, Z., Marquis, M., Averyt, K.B., Tignor, M., and Miller, H.L. (Eds): *Climate Change 2007: The Physical Science Basis. Contribution of Working Group I to the Fourth Assessment Report of the Intergovernmental Panel on Climate Change.* Cambridge University Press, Cambridge, UK.

Franco, J., Schad, S., and Cady, C.W. 1994. California's experience with a voluntary approach to reducing nitrate contamination in groundwater: the fertilizer research and education program (FREP). *Journal of Soil and Water Conservation* 49:76–81.

FRAP (Fire Resource and Assessment Program). 2010. Wildfire threat to ecosystem health and community safety. Chapter 2.1 in *The 2010 Forest and Range Assessment: Final Document.* CAL FIRE, Sacramento, CA.

Fritschi, F.B., Roberts, B.A., Rains, D.W., Travis, R.L., and Hutmacher, R.B. 2005. Recovery of residual fertilizer-n and cotton residue-N by Acala and Pima cotton. *Soil Science Society of America Journal* 69:718–728. doi:10.2136 /sssaj2003.0340.

Fritschi, F.B., Roberts, B.A., Travis, R.L., Rains, D.W., and Hutmacher, R.B. 2004. Seasonal nitrogen concentration, uptake, and partitioning pattern of irrigated Acala and Pima cotton as influenced by nitrogen fertility level. *Crop Science* 44:516–527.

Fuhrer, J. 2003. Agroecosystern responses to combinations of elevated CO_2, ozone, and global climate change. *Agriculture Ecosystems and Environment* 97:1–20.

Fuller, R.A., Irvine, K.N., Devine-Wright, P., Warren, P.H., and Gaston, K.J. 2007. Psychological Benefits of Greenspace Increase with Biodiversity. *Biology Letters* 3:390–394. doi:10.1098/rsbl.2007.0149.

Galloway, J.N. 1998. The global nitrogen cycle: changes and consequences. *Environmental Pollution, Nitrogen, the Confer-N-s First International Nitrogen Conference* 102:15–24. doi:10.1016/S0269-7491(98)80010-9.

Galloway, J.N., Aber, J.D., Erisman, J.W., Seitzinger, S.P., Howarth, R.W., Cowling, E.B., and Cosby, B.J. 2003. The nitrogen cascade. *Bioscience* 53:341–356.

Galloway, J.N. and Cowling, E.B. 2002. Reactive nitrogen and the world: 200 years of change. *AMBIO: A Journal of the Human Environment* 31:64–71. doi:10.1579/0044-7447-31.2.64.

Galloway, J.N., Dentener, F.J., Capone, D.G., Boyer, E.W., Howarth, R.W., Seitzinger, S.P., Asner, G.P. et al. 2004. Nitrogen cycles: past, present, and future. *Biogeochemistry* 70:153–226. doi:10.1007/s10533-004-0370-0.

Galloway, J.N., Townsend, A.R., Bekunda, M., Cai, Z., Freney, J.R., Martinelli, L.A., Seitzinger, S.P., and Sutton, M.A. 2008. Transformation of the nitrogen cycle: recent trends, questions, and potential solutions. *Science* 320:889–892. doi:10.1126/science.1136674.

Gardner, J.B. and Drinkwater, L.E. 2009. The fate of nitrogen in grain cropping systems: a meta-analysis of 15N field experiments. *Ecological Applications* 19:2167–2184.

Garr, J.D. 2002. *Rice Enhancement Project Annual Report.* Ducks Unlimited, Inc., Western Regional Office, Rancho Cordova, CA.

Garten, C.T., Wullschleger, S.D., and Classen, A.T. 2011. Review and model-based analysis of factors influencing soil carbon sequestration under hybrid poplar. *Biomass and Bioenergy* 35:214–226. doi:10.1016/j.biombioe.2010.08.013.

Gaskell, M., Fouche, B., Koike, S., Lanini, T., Mitchell, J., and Smith, R. 2000. Organic vegetable production in California: science and practice. *HortTechnology* 10:699–713.

Gatto, M. and Alejo, L. 2014. *California Assembly Bill 52: Native Americans: California Environmental Quality Act.*

Gauderman, W.J., Avol, E., Gilliland, F., Vora, H., Thomas, D., Berhane, K., McConnell, R. et al. 2004. The effect of air pollution on lung development from 10 to 18 years of age. *New England Journal of Medicine* 351:1057–1067. doi:10.1056/ NEJMoa040610.

Gauderman, W.J., Avol, E., Lurmann, F., Kuenzli, N., Gilliland, F., Peters, J., and McConnell, R. 2005. Childhood asthma and exposure to traffic and nitrogen dioxide. *Epidemiology* 16:737–743. doi:10.1097/01.ede.0000181308.51440.75.

Gauderman, W.J., Gilliland, G.F., Vora, H., Avol, E., Stram, D., McConnell, R., Thomas, D. et al. 2002. Association between air pollution and lung function growth in Southern California children: results from a second cohort. *American Journal of Respiratory and Critical Care Medicine* 166:76–84. doi:10.1164/rccm.2111021.

Gelperin, A., Moses, V.K., and Bridger, C. 1975. Relationship of high nitrate community water supply to infant and fetal mortality. *IMJ Illinois Medical Journal* 147:155–157, 186.

Gerland, P., Raftery, A.E., Ševčíková, H., Li, N., Gu, D., Spoorenberg, T., Alkema, L. et al. 2014. World population stabilization unlikely this century. *Science* 346:234–237. doi:10.1126/science.1257469.

Gilchrist, M., Shore, A.C., and Benjamin, N. 2011. Inorganic nitrate and nitrite and control of blood pressure. *Cardiovascular Research* 89:492–498. doi:10.1093/cvr/cvq309.

Glibert, P.M. 2010. Long-term changes in nutrient loading and stoichiometry and their relationships with changes in the food web and dominant pelagic fish species in the San Francisco estuary, California. *Reviews in Fisheries Science* 18:211–232. doi:10.1080/10641262.2010.492059.

Goolsby, D.A., Battaglin, W.A., Lawrence, G.B., Artz, R.S., Aulenbach, B.T., Hooper, R.P., Keeney, D.R., and Stensland, G.J. 1999. *Flux and Sources of Nutrients in the Mississippi-Atchafalaya River Basin: Topic 3 – Report for the Integrated Assessment on Hypoxia in the Gulf of Mexico, Decision Analysis Series No. 17.* NOAA Coastal Ocean Program, NOAA, Washington, DC.

Gould, B.W., Sharpe, W.E., and Klemme, R.M. 1989. Conservation tillage: the role of farm and operator characteristics and the perception of soil erosion. *Land Economics* 85:167–82.

Governor's Delta Vision Blue Ribbon Task Force. 2008. *Blue Ribbon Task Force Final Vision Report.* State of California Resources Agency, Sacramento, CA.

Granett, A.L. and Musselman, R.C. 1984. Simulated acidic fog injures lettuce. *Atmospheric Environment* 18:887–890.

Grantz, D.A. 2003. Ozone impacts on cotton: towards an integrated mechanism. *Environmental Pollution* 126:331–344. doi:10.1016/S0269-7491(03)00246-X.

Grantz, D.A. and Shrestha, A. 2005. Ozone reduces crop yields and alters competition with weeds such as yellow nutsedge. *California Agriculture* 59:137–143.

Grantz, D.A. and Shrestha, A. 2006. Tropospheric ozone and interspecific competition between yellow nutsedge and Pima cotton. *Crop Science* 46:1879–1889.

Green, C.T., Fisher, L.H., and Bekins, B.A. 2008a. Nitrogen fluxes through unsaturated zones in five agricultural settings across the United States. *Journal of Environment Quality* 37:1073–1085. doi:10.2134/jeq2007.0010.

Green, C.T., Puckett, L.J., Böhlke, J.K., Bekins, B.A., Phillips, S.P., Kauffman, L.J., Denver, J.M., and Johnson, H.M. 2008b. Limited occurrence of denitrification in four shallow aquifers in agricultural areas of the United States. *Journal of Environment Quality* 37:994–1009. doi:10.2134 /jeq2006.0419.

Greenblatt, M. and Mirvish, S.S. 1973. Dose-response studies with concurrent administration of piperazine and sodium nitrite to strain A mice. *Journal of the National Cancer Institute* 50:119–124.

Greenblatt, M., Mirvish, S., and So, B.T. 1971. Nitrosamine studies: induction of lung adenomas by concurrent administration of sodium nitrite and secondary amines in Swiss mice. *Journal of National Cancer Institute* 46:1029–1034.

Griesenbeck, J. S., Brender, J. D., Sharkey, J. R., Steck, M. D., Huber, J. C., Rene, A. A., and McDonald, T. J. et al. 2010. Maternal characteristics associated with the dietary intake of nitrates, nitrites, and nitrosamines in women of child-bearing age: a cross-sectional study. *Environmental Health* 9:10. doi:10.1186/1476-069X-9-10.

Griffin, D. and Anchukaitis, K. J. 2014. How unusual is the 2012–2014 California drought? *Geophysical Research Letters* 41:9017–9023. doi:10.1002/2014GL062433.

Grigg, D. 1995. The pattern of world protein consumption. *Geoforum* 26:1–17. doi:10.1016/0016-7185(94)00020-8.

Griliches, Z. 1958. The demand for fertilizer: an economic interpretation of a technical change. *American Journal of Agricultural Economics* 40:591–606. doi:10.2307/1235370.

Grimm, K. A., Blanck, H. M., Scanlon, K. S., Moore, L. V., and Grummer-Strawn, L. M. 2010. State-specific trends in fruit and vegetable consumption among adults: united States, 2000–2009. *Center for Disease Control* 59:1125–1130.

Grimm, N. B., Grove, J. G., Pickett, S. T. A., and Redman, C. L. 2000. Integrated approaches to long-term studies of urban ecological systems urban ecological systems present multiple challenges to ecologists: pervasive human impact and extreme heterogeneity of cities, and the need to integrate social and ecological approaches, concepts, and theory. *BioScience* 50:571–584. doi:10.1641/0006-3568(2000)050[0571:IATLTO]2.0.CO;2.

Grineski, S. E., Staniswalis, J. G., Peng, Y., and Atkinson-Palombo, C. 2010. Children's asthma hospitalizations and relative risk due to nitrogen dioxide (NO_2): effect modification by race, ethnicity, and insurance status. *Environmental Research* 110:178–188. doi:10.1016/j.envres.2009.10.012.

Groffman, P. M., Law, N. L., Belt, K. T., Band, L. E., and Fisher, G. T. 2004. Nitrogen fluxes and retention in urban watershed. *Ecosystems* 7:393–403. doi:10.1007/s10021-003-0039-x.

Gruber, N. and Galloway, J. N. 2008. An earth-system perspective of the global nitrogen cycle. *Nature* 451: 293–296.

Grulke, N. E., Andersen, C. P., Fenn, M. E., and Miller, P. R. 1998. Ozone exposure and nitrogen deposition lowers root biomass of ponderosa pine in the San Bernardino Mountains, California. *Environmental Pollution* 103:63–73.

Gu, B., Chang, J., Ge, Y., Ge, H., Yuan, C., Peng, C., and Jiang, H. 2009. Anthropogenic modification of the nitrogen cycling within the Greater Hangzhou Area system, China. *Ecological Applications* 19:974–988. doi:10.1890/08-0027.1.

Gupta, P., Christopher, S. A., Wang, J., Gehrig, R., Lee, Y. C., and Kumar, N. 2006. Satellite remote sensing of particulate matter and air quality over global cities. *Atmospheric Environment* 40:5880–5892.

Gupta, S. K., Fitzgerald, J. F., Chong, S. K. F., Croffie, J. M., and Garcia, J. G. N. 1998. Expression of inducible nitric oxide synthase (iNOS) mRNA in inflamed esophageal and colonic mucosa in a pediatric population. *American Journal of Gastroenterology* 93:795–798. doi:10.1111/j.1572-0241.1998.227_a.x.

Hackes, B. L., Shanklin, C. W., Kim, T., and Su, A. Y. 1997. Tray service generates more food waste in dining areas of a continuing-care retirement community. *Journal of the American Dietetic Association* 97:879–882. doi:10.1016/S0002-8223(97)00213-7.

Haden, V. R., Dempsey, M., Wheeler, S., Salas, W., and Jackson, L. E. 2013. Use of local greenhouse gas inventories to prioritise opportunities for climate action planning and voluntary mitigation by agricultural stakeholders in California. *Journal of Environmental Planning and Management* 56:553–571. doi:10.1080/09640568.2012.689616.

Hajrasuliha, S., Rolston, D. E., and Louie, D. T. 1998. Fate of 15N fertilizer applied to trickle-irrigated grapevines. *American Journal of Enology and Viticulture* 49:191–198.

Hall, J. V., Brajer, V., and Lurmann, F. W. 2003. Economic valuation of ozone-related school absences in the South Coast air basin of California. *Contemporary Economic Policy* 21:407–417.

Hall, J. V., Brajer, V., and Lurmann, F. W. 2006. The health and related economic benefits of attaining healthful air in the San Joaquin Valley. California State University-Fullerton, Institute for Economic and Environmental Studies, Fullerton.

Hall, J. V., Brajer, V., and Lurmann, F. W. 2008a. *The Benefits of Meeting Federal Clean Air Standards in the South Coast and San Joaquin Valley Air Basins.* California State University-Fullerton, Institute for Economic and Environmental Studies, Fullerton.

Hall, J. V., Brajer, V., and Lurmann, F. W. 2008b. Measuring the gains from improved air quality in the San Joaquin Valley. *Journal of Environmental Management* 88:1003–1015. doi:10.1016/j.jenvman.2007.05.002.

Hall, J. V., Brajer, V., and Lurmann, F. W. 2010. Air pollution, health and economic benefits—lessons from 20 years of analysis. *Ecological Economics* 69:2590–2597. doi:10.1016/j.ecolecon.2010.08.003.

Hall, J. V., Winer, A. M., Kleinman, M. T., Lurmann, F. W., Brajer, V., and Colome, S. D. 1992. Valuing the health benefits of clean air. *Science* 255:812–817.

Hall, K. D., Guo, J., Dore, M., and Chow, C. C. 2009. The progressive increase of food waste in America and its environmental impact. *PLOS ONE* 4:e7940. doi:10.1371/journal.pone.0007940.

Hallock, R. J., Elwell, R. F., and Fry, D. H. J. 1970. *Fish Bulletin 151. Migrations of Adult King Salmon Oncorhynchus tshawytscha in the San Joaquin Delta as Demonstrated by the Use of Sonic Tags.* Scripps Institution of Oceanography Library, La Jolla, CA.

Halonen, J. I., Lanki, T., Tiittanen, P., Niemi, J. V., Loh, M., and Pekkanen, J. 2010. Ozone and cause-specific cardiorespiratory morbidity and mortality. *Journal of Epidemiology and Community Health* 64(9):814–820. doi:10.1136/jech.2009.087106.

Han, H. and Allen, J. D. 2008. Estimation of nitrogen inputs to catchments: comparison of methods and consequences for riverine export prediction. *Biogeochemistry* 91:177–199.

Han, Y., Li, X., and Nan, Z. 2011. Net anthropogenic nitrogen accumulation in the Beijing metropolitan region. *Environmental Science and Pollution Research* 18:485–496.

Hanley, N. 1990. The economics of nitrate pollution. *European Review of Agricultural Economics* 17:129–151. doi:10.1093/erae/17.2.129.

Hanley, N., Shogren, J. F., and White, B. 1997. *Environmental Economics in Theory and Practice.* Oxford University Press, Oxford/New York.

Hanson, B. R. 1995. Water resources engineering, Volume 2: practical potential irrigation efficiencies. Paper presented at

The 1st International Conference, American Society of Civil Engineers, San Antonio, TX, pp.1580–1584.

Hanson, B.R., Hopmans, J.W., and Šimůnek, J. 2008. Leaching with subsurface drip irrigation under saline, shallow groundwater conditions. *Vadose Zone Journal* 7:810–818. doi:10.2136/vzj2007.0053.

Hanson, B.R., May, D.E., Simünek, J., Hopmans, J.W. and Hutmacher, R.B. 2009. Drip irrigation provides the salinity control needed for profitable irrigation of tomatoes in the San Joaquin Valley. *California Agriculture* 63:131–136. doi:10.3733/ca.v063n03p131.

Hanson, B.R., Schwankl, L.J., Schulbach, K.F., and Pettygrove, G.S. 1997. A comparison of furrow, surface drip, and subsurface drip irrigation on lettuce yield and applied water. *Agricultural Water Management* 33:139–157. doi:10.1016/S0378-3774(96)01289-9.

Hanukoglu, A. and Danon, P.N. 1996. Endogenous methemoglobinemia associated with diarrheal disease in infancy. *Journal of Pediatric Gastroenterology and Nutrition* 23:1–7.

Harrison, J.A., Maranger, R.J., Alexander, R.B., Giblin, A.E., Jacinthe, P.A., Mayorga, E., Seitzinger, S.P., Sobota, D.J., and Wollheim, W.M. 2008. The regional and global significance of nitrogen removal in lakes and reservoirs. *Biogeochemistry* 93:143–157. doi:10.1007/s10533-008-9272-x.

Harrod, H.L. 2000. *The Animals Came Dancing: Native American Sacred Ecology and Animal Kinship*. University of Arizona Press, Tucson.

Harter, T. 2009. Nitrates in groundwater: Agricultural impacts on groundwater nitrate. *Southwest Hydrology* 8(4):22–24.

Harter, T., Davis, H., Mathews, M.C., and Meyer, R.D. 2002. Shallow groundwater quality on dairy farms with irrigated forage crops. *Journal of Contaminant Hydrology* 55:287–315. doi:10.1016/S0169-7722(01)00189-9.

Harter, T. and Lund, J.R. 2012. Project and technical report outline: Technical Report 1, in *Addressing Nitrate in California's Drinking Water with A Focus on Tulare Lake Basin and Salinas Valley Groundwater*. Report for the State Water Resources Control Board Report to the Legislature. Center for Watershed Sciences, University of California, Davis.

Harter, T., Lund, J.R., Darby, J., Fogg, G.E., Howitt, R.E., Jessoe, K., Pettygrove, G.S. et al. 2012. *Addressing Nitrate in California's Drinking Water: With a Focus on Tulare Lake Basin and Salinas Valley Groundwater*. Report for the State Water Resources Control Board Report to the Legislature. Center for Watershed Sciences, University of California, Davis.

Hart, J.F. 2001. Half a century of cropland change. *Geographical Review* 91:525–543. doi:10.1111/j.1931-0846.2001.tb00239.x.

Hart, J.F. 2003. *The Changing Scale of American Agriculture*. University of Virginia Press, Charlottesville.

Hartz, T.K., Bendixen, W.E., and Wierdsma, L. 2000. The value of presidedress soil nitrate testing as a nitrogen management tool in irrigated vegetable production. *HortScience* 35:651–656.

Hartz, T.K. and Bottoms, T.G. 2010. Humic substances generally ineffective in improving vegetable crop nutrient uptake or productivity. *HortScience* 45:906–910.

Hartz, T.K. and Johnstone, P.R. 2006. Nitrogen availability from high-nitrogen-containing organic fertilizers. *HortTechnology* 16:39–42.

Hartz, T.K., Johnstone, P.R., and Nunez, J.J. 2005. Production environment and nitrogen fertility affect carrot cracking. *HortScience* 40:611–615.

Hartz, T.K., Le Strange, M., May, D.M. 1993. Nitrogen requirements of drip-irrigated peppers. *HortScience* 28:1097–1099.

Hartz, T.K., Mitchell, J.P., and Giannini, C. 2000. Nitrogen and carbon mineralization dynamics of manures and composts. *HortScience* 35:209–212.

Hartz, T.K., Le Strange, M., and May, D.M. 1994. Tomatoes respond to simple drip irrigation schedule and moderate nitrogen inputs. *California Agriculture* 48(2):28–31.

Hartz, T.K., Reade, C., Wierdsma, L., Costa, F., Benedixen, W., and Kitinoja, L. 1994a. *Nitrogen Management through Intensive On-Farm Monitoring FREP Contract # 94-036*. Fertilizer Research and Education Program (FREP), Sacramento, CA.

Hartz, T.K., Smith, R., Schulbach, K.F., and Le Strange, M. 1994b. On-farm nitrogen tests improve fertilizer efficiency, protect groundwater. *California Agriculture* 48:29–32. doi:10.3733/ca.v048n04p29.

Hassan, I.A., Bender, J., and Weigel, H.J. 1999. Effects of ozone and drought stress on growth, yield and physiology of tomatoes (*Lycopersicon esculentum* Mill. cv. Baladey). *Gartenbauwissenschaft* 64:152–157.

Hauser, H., Gordon, M., Engel, K., Langenback, A., Harryman, W., Hernandez, C., and Verma, P. 2010. *California Ocean Wastewater Discharge Report and Inventory*. Heal the Ocean, Santa Barbara, CA.

Hayhoe, K., Cayan, D., Field, C.B., Frumhoff, P.C., Maurer, E.P., Miller, N.L., Moser, S.C. et al. 2004. Emissions pathways, climate change, and impacts on California. *Proceedings of the National Academy of Sciences of the United States of America* 101:12422–12427. doi:10.1073/pnas.0404500101.

He, F.J., Nowson, C.A., Lucas, M., and MacGregor, G.A. 2007. Increased consumption of fruit and vegetables is related to a reduced risk of coronary heart disease: meta-analysis of cohort studies. *Journal of Human Hypertension* 21:717–728. doi:10.1038/sj.jhh.1002212.

Health Canada. 2013. *Guidelines for Canadian Drinking Water Quality: Guideline Technical Document – Nitrate and Nitrite*. Health Canada, Ottawa, ON.

Heck, W.W., Taylor, O.C., Adams, R., Bingham, G., Miller, J., Preston, E., and Weinstein, L. 1982. Assessment of crop loss from ozone. *Journal of the Air Pollution Control Association* 32:353–362.

Hedin, L.O., Armesto, J.J., and Johnson, A.H. 1995. patterns of nutrient loss from unpolluted, old-growth temperate forests: evaluation of biogeochemical theory. *Ecology* 76:493–509. doi:10.2307/1941208.

Hegesh, E. and Shiloah, J. 1982. Blood nitrates and infantile methemoglobinemia. *Clinica Chimica Acta* 125:107–115.

Helfand, G.E. and House, B.W. 1995. Regulating non-point source pollution under heterogeneous conditions. *American Journal of Agricultural Economics* 77:1024–1032.

Henrichs, T., Zurek, M., Eickhout, B., Kok, K., Raudsepp-Hearne, C., Ribeiro, T., van Vuuren, D., and Volkery, A. 2010. Scenario development and analysis for forward-looking ecosystem assessments, in Ash, N., Blanco, H., Brown, C., Garcia, K., Henrichs, T., Lucas, N., Raudsepp-Hearne, C., Scholes, R., and Simpson, R.D. (Eds): *Ecosystems and Human Well-Being: A Manual for Assessment Practitioners*. Island Press, Washington, DC.

Henze, M. 1991. Capabilities of biological nitrogen removal processes from wastewater. *Water Science & Technology* 23:669–679.

Henze, M. 2008. *Biological Wastewater Treatment: Principles, Modelling and Design*. IWA Publishing, London, UK.

Heres-Del-Valle, D. and Niemeier, D.A. 2011. CO_2 emissions: are land-use changes enough for California to reduce VMT? Specification of a two-part model with instrumental variables. *Transportation Research Part B: Methodological* 45:150–161. doi:10.1016/j.trb.2010.04.001.

Hesterberg, T.W., Bunn, W.B., McClellan, R.O., Hamade, A.K., Long, C.M., and Valberg, P.A. 2009. Critical review of the human data on short-term nitrogen dioxide (NO_2) exposures: evidence for NO_2 no-effect levels. *Critical Reviews in Toxicology* 39:743–781. doi:10.3109/10408440903294945.

Hidy, G.M., Brook, J.R., Chow, J.C., Green, M., Husar, R.B., Lee, C., Scheffe, R.D., Swanson, A., and Watson, J.G. 2009. Remote sensing of particulate pollution from space: have we reached the promised land? *Journal of the Air and Waste Management Association* 59:1130–1139.

Hill, J.E., Williams, J.F., Mutters, R.G., and Greer, C.A. 2006. The California rice cropping system: agronomic and natural resource issues for long-term sustainability. *Paddy and Water Environment* 4:13–19. doi:10.1007/s10333-005-0026-2.

Hills, F.J., Broadbent, F.E., and Fried, M. 1978. Timing and rate of fertilizer nitrogen related to nitrogen uptake and pollution potential. *Journal of Environment Quality* 7:368–372.

Hills, F.J., Broadbent, F.E., and Lorenz, O.A. 1983. Fertilizer nitrogen utilization by corn, tomato, and sugarbeet. *Agronomy Journal* 75:423–426. doi:10.2134/agronj1983.0002 1962007500030002x.

Hinkle, N.C. and Hickle, L.A. 1999. California caged layer pest management evaluation. *Journal of Applied Poultry Research* 8:327–338. doi:10.1093/japr/8.3.327.

Hitzhusen, G.E., Fick, G.W., and Moore, R.H. 2013. Theological and religious approaches to soil stewardship. Chapter 12 in Lal, R. and Stewart, B.A. (Eds): *Advances in Soil Sciences: Principles of Sustainable Soil Management in Agroecosystems*. Taylor & Francis Group, Boca Raton, FL.

Hoefer, M., Rytina, N., and Baker, B. 2012. *Estimates of the Unauthorized Immigrant Population Residing in the United States*. Department of Homeland Security, Office of Immigration Statistics, Washington, DC.

Hoek, G. and Brunekreef, B. 1994. Effects of low-level winter air pollution concentrations on respiratory health of Dutch children. *Environmental Research* 64:136–150. doi:10.1006/enrs.1994.1012.

Hoek, G., Brunekreef, B., Goldbohm, S., Fischer, P., and Van Den Brandt, P.A. 2002. Association between mortality and indicators of traffic-related air pollution in the Netherlands: a cohort study. *Lancet* 360:1203–1209. doi:10.1016/s0140-6736(02)11280-3.

Hollander, A. 2010. *The California Augmented Multisource Landcover Map*. Accessed January 27, 2015. http://climate.calcommons.org/dataset/california-augmented-multisource-landcover-map-caml-2010.

Holmes, C.D., Prather, M.J., Søvde, O.A., and Myhre, G. 2013. Future methane, hydroxyl, and their uncertainties: key climate and emission parameters for future predictions. *Atmospheric Chemistry and Physics* 13:285–302.

Holmes, G.J. and Schultheis, J.R. 2003. Sensitivity of watermelon cultigens to ambient ozone in North Carolina. *Plant Disease* 87:428–434.

Holtby, C.E., Guernsey, J.R., Allen, A.C., Vanleeuwen, J.A., Allen, V.M., and Gordon, R.J. 2014. A population-based case-control study of drinking-water nitrate and congenital anomalies using geographic information systems (GIS) to develop individual-level exposure estimates. *International Journal of Environmental Research and Public Health* 11:1803–1823. doi:10.3390/ijerph110201803.

Holz, O., Mücke, M., Paasch, K., Böhme, S., Timm, P., Richter, K., Magnussen, H., and Jörres, R.A. 2002. Repeated ozone exposures enhance bronchial allergen responses in subjects with rhinitis or asthma. *Clinical and Experimental Allergy* 32:681–689. doi:10.1046/j.1365-2222.2002.01358.x.

Home, R., Bauer, N., and Hunziker, M. 2010. Cultural and biological determinants in the evaluation of urban green spaces. *Environment and Behavior* 42:494–523. doi:10.1177/0013916509338147.

Honeycutt, K., Canada, H.E., Jenkins, M.W., and Lund, J.R. 2012. Alternative water supply options for nitrate contamination. Technical Report 7, in *Addressing Nitrate in California's Drinking Water with A Focus on Tulare Lake Basin and Salinas Valley Groundwater*. Report for the State Water Resources Control Board Report to the Legislature.. Center for Watershed Sciences, University of California, Davis.

Hord, N.G., Tang, Y., and Bryan, N.S. 2009. Food sources of nitrates and nitrites: the physiologic context for potential health benefits. *American Journal of Clinical Nutrition* 90:1–10. doi:10.3945/ajcn.2008.27131.

Horwath, W. and Burger, M. 2012. *Assessment of Baseline Nitrous Oxide Emissions in California Cropping Systems*, Contract No. 08-324. California Air Resources Board, Sacramento.

Houlton, B.Z., Boyer, E.W., Finzi, A., Galloway, J.N., Leach, A., Liptzin, D., Melillo, J., Rosenstock, T.S., Sobota, D., and Townsend, A.R. 2012 The US nitrogen inventory: N-use efficiency among economic sectors and N_x climate risks nationwide. . Chapter 2 in Suddick, E.C. and Davidson, E.A. (Eds): *The Role of Nitrogen in Climate Change and the Impacts of Nitrogen-Climate Interactions on Terrestrial and Aquatic Ecosystems, Agriculture, and Human Health in the United States*, A Technical Report Submitted to the US National Climate Assessment. North American Nitrogen Center of the International Nitrogen Initiative (NANC-INI), Falmouth, MA.

Houlton, B.Z., Boyer, E.W., Finzi, A., Galloway, J.N., Leach, A., Liptzin, D., Melillo, J., Rosenstock, T.S., Sobota, D., and Townsend, A.R. 2013. Intentional versus unintentional nitrogen use in the United States: trends, efficiency and implications. *Biogeochemistry* 114:11–23.

Howarth, R., Swaney, D., Billen, G., Garnier, J., Hong, B., Humborg, C., Johnes, P., Mörth, C.M., and Marino, R. 2012. Nitrogen fluxes from the landscape are controlled by net anthropogenic nitrogen inputs and by climate. *Frontiers in Ecology and the Environment* 10:37–43. doi:10.1890/100178.

Howarth, R.W. 2004. Human acceleration of the nitrogen cycle: drivers, consequences, and steps toward solutions. *Water Science and Technology* 49:7–13.

Howarth, R. W., Billen, G., Swaney, D., Townsend, A., Jaworski, N., Lajtha, K., Downing, J. A. et al. 1996. Regional nitrogen budgets and riverine N & P fluxes for the drainages to the North Atlantic Ocean: natural and human influences, in Howarth, R. W. (Ed.): *Nitrogen Cycling in the North Atlantic Ocean and Its Watersheds*. Springer Netherlands, Heidelberg, pp.75–139.

Howe, G. R., Jain, M., and Miller, A. B. 1990. Dietary factors and risk of pancreatic cancer: results of a Canadian population-based case-control study. *International Journal of Cancer* 45:604–608.

Howitt, R. E. 2014. Water, climate change, and California agriculture. *Agricultural And Resource Economics Update* 18:13–15.

Howitt, R. E., Adams, R. M., and Gossard, T. W. 1984. Effects of alternative ozone concentrations and response data on economic assessments: the case of California crops. *Journal of the Air Pollution Control Association* 34:1122–1127.

Howitt, R. E. and Goodman, C. 1989. *The Economic Assessment of California Field Crop Losses Due to Air Pollution: Final Report*, Contract #A5-105-32. California Air Resources Board, Sacramento.

Howley, P. 2011. Landscape aesthetics: assessing the general publics' preferences towards rural landscapes. *Ecological Economics* 72:161–169. doi:10.1016/j.ecolecon.2011.09.026.

Hristov, A. N., Hanigan, M., Cole, A., Todd, R., McAllister, T. A., Ndegwa, P. M., and Rotz, A. 2011. Review: ammonia emissions from dairy farms and beef feedlots. *Canadian Journal of Animal Science* 91:1–35. doi:10.4141/CJAS10034.

Hu, S., Fruin, S., Kozawa, K., Mara, S., Winer, A. M., and Paulson, S. E. 2009. Aircraft emission impacts in a neighborhood adjacent to a general aviation airport in southern California. *Environmental Science & Technology* 43:8039–8045. doi:10.1021/es900975f.

Huang, W. 2009. *Factors Contributing to the Recent Increase in U.S. Fertilizer Prices, 2002-08 (Outlook No. AR-33)*. USDA ERS, Washington, DC.

Huang, W., Magleby, R., and Christensen, L. 2005. Economic impacts of EPA's manure application regulations on dairy farms with lagoon liquid systems in the southwest region. *Journal of Agricultural and Applied Economics* 37:209–227.

Huenneke, L. F., Hamburg, S. P., Koide, R., Mooney, H. A., and Vitousek, P. M. 1990. Effects of soil resources on plant invasion and community structure in Californian serpentine grassland. *Ecology* 71:478–491.

Hughes, J. E., Knittel, C. R., and Sperling, D. 2008. Evidence of a shift in the short-run price elasticity of gasoline demand. *The Energy Journal* 29:113–134.

Hughes, L. S., Allen, J. O., Salmon, L. G., Mayo, P. R., Johnson, R. J., and Cass, G. R. 2002. Evolution of nitrogen species air pollutants along trajectories crossing the Los Angeles area. *Environmental Science & Technology* 36:3928–3935. doi:10.1021/es0110630.

Hummels, D. 1999. *Have International Transportation Costs Declined? (No. GTAP Resource #1158)*. Purdue University Global Trade Analysis Project.

Hummels, D. 2007. Transportation costs and international trade in the second era of globalization. *The Journal of Economic Perspectives* 21:131–154.

Hundley, N. 1992. *The Great Thirst: Californians and Water, 1770s-1990s*. University of California Press, Berkeley.

Hutmacher, R. B., Travis, R. L., Rains, D. W., Vargas, R. N., Roberts, B. A., Weir, B. L., Wright, S. D., Munk, D. S., Marsh, B. H., and Keeley, M. P. 2004. Response of recent Acala cotton varieties to variable nitrogen rates in the San Joaquin Valley of California. *Agronomy Journal* 96: 48–62.

Hutson, S., Barber, N., Kenny, J. F., Linsey, K. S., Lumia, D., and Maupin, M. 2004. *Estimated Use of Water in the United States in 2000*. USGS Circular 1268. USGS, Reston, VA.

IARC (International Agency for Research on Cancer). 2014. *IARC Monographs on the Evaluation of Carcinogenic Risks to Humans*. World Health Organization, Geneva.

Idaho DWR (Department of Water Resources). 2013. *Mapping Evapotranspiration*. Idaho DWR, Boise.

IFA (International Fertilizer Industry Association). 2010. Fertilizer outlook 2010–2014. Paper presented at *The 78th IFA Annual Conference*, 31 May–June 2, Paris, France.

Innes, R. and Cory, D. 2001. The economics of safe drinking water. *Land Economics* 77:94–117. doi:10.2307/3146983.

Inoue-Choi, M., Ward, M. H., Cerhan, J. R., Weyer, P. J., Anderson, K. E., and Robien, K. 2012. Interaction of nitrate and folate on the risk of breast cancer among postmenopausal women. *Nutrition and Cancer* 64:685–694. doi:10.1080/01 635581.2012.687427.

Integrated Waste Management Consulting. 2010. *Third Assessment of California's Compost- and Mulch-Producing Infrastructure: Management Practices and Market Conditions, Waste Management*. CalRecycle, Sacramento, CA.

Iowa State University. 2004. *Practices to Reduce Ammonia Emissions from Livestock Operations*. Iowa State University, Ames.

IPCC (Intergovernmental Panel on Climate Change). 1999. *Appendix A to the Principles Governing IPCC Work*. IPCC, Geneva, Switzerland.

IPCC. 2000. *Special Report on Emissions Scenarios*. Cambridge University Press, Cambridge.

IPCC. 2001. *Climate Change 2001: The Scientific Basis. Contribution of Working Group I to the Third Assessment Report of the Intergovernmental Panel on Climate Change*. Cambridge University Press, Cambridge, UK/New York.

IPCC. 2006. *2006 IPCC Guidelines for National Greenhouse Gas Inventories, Prepared by the National Greenhouse Gas Inventories Program*. IGES, Kanagawa, Japan.

IPCC. 2007a. *Climate Change 2007: Synthesis Report. Contribution of Working Groups I, II and III to the Fourth Assessment Report of the Intergovernmental Panel on Climate Change*. IPCC, Geneva, Switzerland.

IPCC. 2007b. *Climate Change 2007: Mitigation of Climate Change: Working Group III Contribution to the Fourth Assessment Report of the IPCC*. Cambridge University Press, Cambridge, UK.

IPCC. 2007c. Policies, instruments, and co-operative arrangements. Chapter 13 in Metz, B., Davidson, O. R., Bosch, P. R., Dave, R., and Meyer, L. A. (Eds): *Contribution of Working Group III to the Fourth Assessment Report of the Intergovernmental Panel on Climate Change, 2007*. Cambridge University Press, UK/New York.

IPCC. 2014. *Climate Change 2014: Synthesis Report. Contribution of Working Groups I, II, and III to the Fifth Assessment Report of the Intergovernmental Panel on Climate Change*. IPCC, Geneva, Switzerland.

Irving, P.M. 1983. Acidic precipitation effects on crops: a review and analysis of research. *Journal of Environmental Quality* 12:442–453.

Islam, T., Berhane, K., McConell, R., Gauderman, W.J., Avol, E., Peters, J.M., and Gilliand, F.D. 2009. Glutathione-S-transferase (GST) P1, GSTM1, exercise, ozone and asthma incidence in school children. *Thorax* 64:197–202. doi:10.1136/thx.2008.099366.

Islam, T., Gauderman, W.J., Berhane, K., McConell, R., Avol, E., Peters, J.M., and Gilliand, F.D. 2007. Relationship between air pollution, lung function and asthma in adolescents. *Thorax* 62:957–963. doi:10.1136/thx.2007.078964.

Islam, T., McConell, R., Gauderman, W.J., Avol, E., Peters, J.M., and Gilliand, F.D. 2008. Ozone, oxidant defense genes, and risk of asthma during adolescence. *American Journal of Respiratory and Critical Care Medicine* 177:388–395. doi:10.1164/rccm.200706-863OC.

IUCN (International Union for Conservation of Nature). 2013. *IUCN Red List of Threatened Species.* Accessed June 16, 2013. www.iucnredlist.org.

Jackson, L.E. 2000. Fates and losses of nitrogen from a nitrogen-15-labeled cover crop in an intensively managed vegetable system. *Soil Science Society of America Journal* 64:1404. doi:10.2136/sssaj2000.6441404x.

Jackson, L.E., Haden, V.R., Hollander, A., Lee, H., Lubell, M., Mehta, V.K., O'Geen, T., and Niles, M.T. 2012a. *Adaptation Strategies for Agricultural Sustainability in Yolo County: A White Paper from the California Energy Commission's California Climate Change Center (No. 500-2012-032).* University of California, Davis.

Jackson, L.E., Haden, V.R., Wheeler, S., Hollander, A.D., Perlman, J., O'Geen, T., Mehta, V., Clark, V., and Williams, J. 2012b. *Vulnerability and Adaptation to Climate Change in California Agriculture: A White Paper from the California Energy Commission's California Climate Change Center (No. CEC-500-2012-031).* University of California, Davis.

Jackson, L.E., Ramirez, I., Yokota, R., Fennimore, S.A., Koike, S.T., Henderson, D.M., Chaney, W.E., and Klonsky, K.M. 2003. Scientists, growers assess trade-offs in use of tillage, cover crops and compost. *California Agriculture* 57:48–54.

Jackson, L.E., Santos-Martin, F., Hollander, A.D., Horwath, W.R., Howitt, R.E., Kramer, J.B., O'Geen, A.T. et al. 2009. *Potential for Adaptation to Climate Change in an Agricultural Landscape in the Central Valley of California (No. CEC-500-2009-044-F).* Prepared for the California Energy Commission, Sacramento.

Jackson, L.E., Stivers, L.J., Warden, B.T., and Tanji, K.K. 1994. Crop nitrogen utilization and soil nitrate loss in a lettuce field. *Fertilizer Research* 37:93–105. doi:10.1007/BF00748550.

Jacobson, J.S. 1984. Effects of acidic aerosol, fog, mist and rain on crops and trees. *Philosophical Transactions of the Royal Society B* 305:327–338.

Jaffé, E.R. 1993. Impact of the environment on blood and blood-forming tissues. *Otolaryngoly: Head and Neck Surgery* 109:806–807.

Jakszyn, P. and Gonzalez, C.A. 2006. Nitrosamine and related food intake and gastric and oesophageal cancer risk: a systematic review of the epidemiological evidence. *World Journal of Gastroenterology* 12:4296–4303.

Jalaludin, B.B., O'Toole, B.I., and Leeder, S.R. 2004. Acute effects of urban ambient air pollution on respiratory symptoms, asthma medication use, and doctor visits for asthma in a cohort of Australian children. *Environmental Research* 95:32–42. doi:10.1016/s0013-9351(03)00038-0.

Jalaludin, B., Morgan, G., Lincoln, D., Sheppeard, V., Simpson, R., and Corbett, S. 2006. Associations between ambient air pollution and daily emergency department attendances for cardiovascular disease in the elderly (65 + years), Sydney, Australia. *Journal of Exposure Science and Environmental Epidemiology* 16:225–237. doi:10.1038/sj.jea.7500451.

Jansen, K.L., Larson, T.V., Koenig, J.Q., Mar, T.F., Fields, C., Stewart, J., and Lippmann, M. 2005. Associations between health effects and particulate matter and black carbon in subjects with respiratory disease. *Environmental Health Perspectives* 113:1741–1746.

Janssens, I.A., Dieleman, W., Luyssaert, S., Subke, J.-A., Reichstein, M., Ceulemans, R., Ciais, P. et al. 2010. Reduction of forest soil respiration in response to nitrogen deposition. *Nature Geoscience* 3:315–322. doi:10.1038/ngeo844.

Janzen, H.H., Beauchemin, K.A., Bruinsma, Y., Campbell, C.A., Desjardins, R.L., Ellert, B.H., and Smith, E.G. 2003. The fate of nitrogen in agroecosystems: an illustration using Canadian estimates. *Nutrient Cycling in Agroecosystems* 67:85–102. doi:10.1023/A:1025195826663.

Jarrell, W.M. and Beverly, R.B. 1981. The dilution effect in plant nutrition studies. *Advances in Agronomy* 34:197–224.

Jassby, A. 2008. Phytoplankton in the Upper San Francisco Estuary: recent biomass trends, their causes and their trophic significance. *San Francisco Estuary and Watershed Science* 6:1–24.

Jassby, A. and Van Nieuwenhuyse, E.E. 2005. Low dissolved oxygen in an estuarine channel (San Joaquin River, California): mechanisms and models based on long-term time series. *San Francisco Estuary and Watershed Science* 3(2).

Jaworska, G. 2005. Nitrates, nitrites, and oxalates in products of spinach and New Zealand spinach: Effect of technological measures and storage time on the level of nitrates, nitrites, and oxalates in frozen and canned products of spinach and New Zealand spinach. *Food Chemistry* 93:395–401. doi:10.1016/j.foodchem.2004.09.035.

Jensen, L.S., Schjoerring, J.K., van der Hoek, K.W., Poulsen, H.D., Zevenbergen, J.F., Pallière, C., Lammel, J. et al. 2011. Benefits of nitrogen for food, fibre and industrial production, in Sutton, M.A., Howard, C.M., Erisman, J.W., Billen, G., Bleeker, A., Grennfelt, P., van Grinsven, H., and Grizzetti, B. (Eds): *The European Nitrogen Assessment: Sources, Effects and Policy Perspectives.* Cambridge University Press, Cambridge.

Jensen, V.B., Darby, J., Seidel, C., and Gorman, C. 2012. *Drinking Water Treatment for Nitrate.* Technical Report 6, in *Addressing Nitrate in California's Drinking Water with a Focus on Tulare Lake Basin and Salinas Valley Groundwater. Report for the State Water Resources Control Board Report to the Legislature.* Center for Watershed Sciences, University of California, Davis.

Jensen, V.B., Darby, J.L., Seidel, C., and Gorman, C. 2014. Nitrate in potable water supplies: alternative management strategies. *Critical Reviews in Environmental Science and Technology* 44:2203–2286. doi:10.1080/10643389.2013.828272.

Johnson, T. V. 2009. Review of diesel emissions and control. *International Journal of Engine Research* 10:275–285. doi:10.1243/14680874JER04009.

Johnston, W. E. and McCalla, A. F. 2004. *Whither California Agriculture: Up, Down, or Out? Some Thoughts about the Future.* Giannini Foundation of Agricultural Economics, Oakland, CA.

Jordan, T. E., Correll, D. L., and Weller, D. E. 1997a. Effects of agriculture on discharges of nutrients from coastal plain watersheds of Chesapeake Bay. *Journal of Environment Quality* 26:836–848. doi:10.2134/jeq1997.00472425002600030034x.

Jordan, T. E., Correll, D. L., and Weller, D. E. 1997b. Nonpoint source discharges of nutrients from Piedmont Watersheds of Chesapeake Bay. *JAWRA Journal of the American Water Resources Association* 33:631–645. doi:10.1111/j.1752-1688.1997.tb03538.x.

Jordan, T. E. and Weller, D. E. 1996. Human contributions to terrestrial nitrogen flux. *BioScience* 46:655–664. doi:10.2307/1312895.

Jörres, R., Nowak, D., Magnussen, H., Speckin, P., and Koschyk, S. 1996. The effect of ozone exposure on allergen responsiveness in subjects with asthma or rhinitis. *American Journal of Respiratory and Critical Care Medicine* 153:56–64.

Just, J., Ségala, C., Sahraoui, F., Priol, G., Grimfeld, A., and Neukirch, F. 2002. Short-term health effects of particulate and photochemical air pollution in asthmatic children. *European Respiratory Journal* 20:899–906. doi:10.1183/09031936.02.00236902.

Kaduwela, A. 2007. *Central California Ozone Study*, Report No. CEC-500-2006-087-APA. California Energy Commission, Sacramento.

Kaffka, S. R. 2009. Can feedstock production for biofuels be sustainable in California? *California Agriculture* 63:202–207.

Kallenbach, C. M., Rolston, D. E., and Horwath, W. R. 2010. Cover cropping affects soil N_2O and CO_2 emissions differently depending on type of irrigation. *Agriculture, Ecosystems & Environment* 137:251–260. doi:10.1016/j.agee.2010.02.010.

Kampschreur, M. J., Temmink, H., Kleerebezem, R., Jetten, M. S. M., and van Loosdrecht, M. C. M. 2009. Nitrous oxide emission during wastewater treatment. *Water Research* 43:4093–4103. doi:10.1016/j.watres.2009.03.001.

Kang, S., Olmstead, K., Takacs, K., and Collins, J. 2008. *Municipal Nutrient Removal Technologies Reference Document, Vol. 1*, Technical Report. US Environmental Protection Agency, Washington, DC.

Kanter, D., Mauzerall, D. L., Ravishankara, A. R., Daniel, J. S., Portmann, R. W., Grabiel, P. M., Moomaw, W. R., and Galloway, J. N. 2013. A post-Kyoto partner: considering the stratospheric ozone regime as a tool to manage nitrous oxide. *Proceedings of the National Academy of Sciences of the United States of America* 110:4451–4457. doi:10.1073/pnas.1222231110.

Kantor, L. S., Lipton, K., Manchester, A., and Oliveira, V. 1997. Estimating and addressing America's food losses. *Food Review* 20:2–12.

Kaplan, R. and Herbert, E. J. 1987. Cultural and sub-cultural comparisons in preferences for natural settings. *Landscape and Urban Planning* 14:281–293. doi:10.1016/0169-2046(87)90040-5.

Kar, J., Fishman, J., Creilson, J. K., Richter, A., Ziemke, J., and Chandra, S. 2010. Are there urban signatures in the tropospheric ozone column products derived from satellite measurements? *Atmospheric Chemistry and Physics* 10:5213–5222.

Karner, A. A., Eisinger, D. S., and Niemeier, D. A. 2010. Near-roadway air quality: synthesizing the findings from real-world data. *Environmental Science & Technology* 44:5334–5344. doi:10.1021/es100008x.

Katan, M. B. 2009. Nitrate in foods: harmful or healthy? *American Journal of Clinical Nutrition* 90:11–12. doi:10.3945/ajcn.2009.28014.

Katsouyanni, K., Samet, J. M., Anderson, H. R., Atkinson, R., Le Tertre, A., Medina, S., Samoli, E. et al. 2009. Air pollution and health: a European and North American approach (APHENA). *Research Report (Health Effects Institute)* 142:5–90.

Kattan, M., Gergen, P. J., Eggleston, P., Visness, C. M., and Mitchell, H. E. 2007. Health effects of indoor nitrogen dioxide and passive smoking on urban asthmatic children. *Journal of Allergy and Clinical Immunology* 120:618–624. doi:10.1016/j.jaci.2007.05.014.

Kaye, J. P., Groffman, P. M., Grimm, N. B., Baker, L. A., and Pouyat, R. V. 2006. A distinct urban biogeochemistry? *Trends in Ecology & Evolution* 21:192–199. doi:10.1016/j.tree.2005.12.006.

Kean, A. J., Harley, R. A., Littlejohn, D., and Kendall, G. R. 2000. On-road measurement of ammonia and other motor vehicle exhaust emissions. *Environmental Science & Technology* 34:3535–3539. doi:10.1021/es991451q.

Kean, A. J., Littlejohn, D., Ban-Weiss, G. A., Harley, R. A., Kirchstetter, T. W., and Lunden, M. M. 2009. Trends in on-road vehicle emissions of ammonia. *Atmospheric Environment* 43:1565–1570. doi:10.1016/j.atmosenv.2008.09.085.

Kebreab, E., France, J., Beever, D. E., and Castillo, A. R. 2001. Nitrogen pollution by dairy cows and its mitigation by dietary manipulation. *Nutrient Cycling in Agroecosystems* 60:275–285. doi:10.1023/A:1012668109662.

Keeney, D. R. 1979. A mass balance of nitrogen in Wisconsin. *Wisconsin Academy of Sciences, Arts and Letters* 67:95–102.

Kehrl, H. R., Peden, D. B., Ball, B., Folinsbee, L. J., and Horstman, D. 1999. Increased specific airway reactivity of persons with mild allergic asthma after 7.6 hours of exposure to 0.16 ppm ozone. *Journal of Allergy and Clinical Immunology* 104:1198–1204.

Kennedy, C., Cuddihy, J., and Engel-Yan, J. 2007. The changing metabolism of cities. *Journal of Industrial Ecology* 11:43–59. doi:10.1162/jie.2007.1107.

Kenny, J. F., Barber, N., Hutson, S., Linsey, K. S., Lovelace, J. K., and Maupin, M. 2009. *Estimated Use of Water in the United States in 2005*. USGS Circular 1344.

Kevil, C. G. and Lefer, D. J. 2011. Review focus on inorganic nitrite and nitrate in cardiovascular health and disease. *Cardiovascular Research* 89:489–491. doi:10.1093/cvr/cvq409.

Kim, H. J., Howitt, R. E., and Helfand, G. E. 1998. An economic analysis of ozone control in California's San Joaquin Valley. *Journal of Agricultural and Resource Economics* 23:55–70.

Kim, S. W., Heckel, A., Frost, G. J., Richter, A., Gleason, J., Burrows, J. P., McKeen, S., Hsie, E. Y., Granier, C., and Trainer, M. 2009. NO_2 columns in the western United States observed from space and simulated by a regional chemistry model and their implications for NO$_x$ emissions. *Journal of Geophysical Research* 114. doi:10.1029/2008JD011343.

Kim, T., Kim, G., Kim, S., and Choi, E. 2008. Estimating riverine discharge of nitrogen from the South Korea by the mass balance approach. *Environmental Monitoring and Assessment* 136:371–378. doi:10.1007/s10661-007-9692-4.

Kiparsky, M. and Gleick, P. H. 2003. Climate Change and California Water Resources: A Survey and Summary of the Literature (No. CEC-500-04-073). Pacific Institute for Studies in Development, Environment, and Security, Oakland, CA.

Kirchmann, H. and Bergström, L. 2001. Do organic farming practices reduce nitrate leaching? *Communications in Soil Science and Plant Analysis* 32:997–1028. doi:10.1081/CSS-100104101.

Kirchstetter, T. W., Harley, R. A., Kreisberg, N. M., Stolzenburg, M. R., and Hering, S. V. 1999. On-road measurement of fine particle and nitrogen oxide emissions from light- and heavy-duty motor vehicles. *Atmospheric Environment* 33:2955–2968. doi:10.1016/S1352-2310(99)00089-8.

Kleijn, D., Bekker, R. M., Bobbink, R., De Graaf, M. C. C., and Roelofs, J. G. M. 2008. In search for key biogeochemical factors affecting plant species persistence in heathland and acidic grasslands: a comparison of common and rare species. *Journal of Applied Ecology* 45:680–687. doi:10.1111/j.1365-2664.2007.01444.x.

Kliesch, J. and Langer, T. 2006. *Plug-in Hybrids: An environmental and Economic Performance Outlook*, Report No. 61. American Council for an Energy-Efficient Economy, Washington. DC.

Klonsky, K. and Richter, K. 2005. *Statistical Review of California's Organic Agriculture*. UC Agricultural Issues Center, Davis, CA.

Knapp, K. C. and Schwabe, K. A. 2008. Spatial dynamics of water and nitrogen management in irrigated agriculture. *American Journal of Agricultural Economics* 90:524–539. doi:10.1111/j.1467-8276.2007.01124.x.

Knekt, P., Järvinen, R., Dich, J., and Hakulinen, T. 1999. Risk of colorectal and other gastro-intestinal cancers after exposure to nitrate, nitrite and N-nitroso compounds: a follow-up study. *International Journal of Cancer* 80:852–856.

Knobeloch, L., Salna, B., Hogan, A., Postle, J., and Anderson, H. 2000. Blue babies and nitrate-contaminated well water. *Environmental Health Perspectives* 108:675–678.

Knorr, M., Frey, S. D., and Curtis, P. S. 2005. Nitrogen additions and litter decomposition: a meta-analysis. *Ecology* 86:3252–3257.

Knox, G., Jackson, D., and Nevers, E. 1995. *Farm*A*Syst: Progress Report 1991–1994*. University of Wisconsin Cooperative Extension, Madison.

Ko, F. W. S. and Hui, D. S. C. 2010. Effects of air pollution on lung health. *Clinical Pulmonary Medicine* 17:300–304. doi:10.1097/CPM.0b013e3181fa1555.

Kong, A. Y. Y., Fonte, S. J., van Kessel, C., and Six, J. 2009. Transitioning from standard to minimum tillage: trade-offs between soil organic matter stabilization, nitrous oxide emissions, and N availability in irrigated cropping systems. *Soil and Tillage Research* 104:256–262. doi:10.1016/j.still.2009.03.004.

Kongshaug, G. 1998. Energy consumption and greenhouse gas emissions in fertilizer production. Paper presented at *The IFA Technical Conference*, 28 September–1 October, Marrakech, Morocco, 18pp.

Kramer, S. B., Reganold, J. P., Glover, J. D., Bohannan, B. J. M., and Mooney, H. A. 2006. Reduced nitrate leaching and enhanced denitrifier activity and efficiency in organically fertilized soils. *Proceedings of the National Academy of Sciences of the United States of America* 103:4522–4527. doi:10.1073/pnas.0600359103.

Kratzer, C. R., Kent, R. H., Saleh, D. K., Knifong, D. L., Dileanis, P. D., and Orlando, J. L. 2011. *Trends in Nutrient Concentrations, Loads, and Yields in Streams in the Sacramento, San Joaquin, and Santa Ana Basins, California, 1975–2004*. U.S. Geological Survey Scientific Investigations Report 2010-5228.

Kreutzer, K., Butterbach-Bahl, K., Rennenberg, H., and Papen, H. 2009. The complete nitrogen cycle of an N-saturated spruce forest ecosystem. *Plant Biology* 11(5):643–649. doi:10.1111/j.1438-8677.2009.00236.x.

Krewski, D. 2009. Evaluating the effects of ambient air pollution on life expectancy. *New England Journal of Medicine* 360:413–415. doi:10.1056/NEJMe0809178.

Kris-Etherton, P. M., Hu, F. B., Ros, E., and Sabaté, J. 2008. The role of tree nuts and peanuts in the prevention of coronary heart disease: multiple potential mechanisms. *Journal of Nutrition* 138:1746S–1751S.

Kroeber, A. L. 1907. Indian myths of South Central California. *American Archaeology and Ethnology* 4:167–250.

Kroodsma, D. A. and Field, C. B. 2006. Carbon sequestration in California agriculture, 1980–2000. *Ecological Applications* 16:1975–1985.

Krupa, S. V. 2003. Effects of atmospheric ammonia (NH$_3$) on terrestrial vegetation: a review. *Environmental Pollution* 124:179–221.

Kuminoff, N. V. and Sumner, D. A. 2001. Modeling farmland conversion with new GIS data. Paper presented at *The American Agricultural Economics Association Annual Meeting*, Chicago, IL.

Kunkel, K. E., Stevens, L. E., Stevens, S. E., Sun, L., Janssen, E., Wuebbles, D., Redmond, K. T., and Dobson, J. G. 2013. *Regional Climate Trends and Scenarios for the U.S. National Climate Assessment. Part 5. Climate of the Southwest U.S.*, NOAA Technical Report No. NESDIS 142-5.

Kurkalova, L. A., Kling, C. L., and Zhao, J. 2004. Value of agricultural non-point source pollution measurement technology: assessment from a policy perspective. *Applied Economics* 36:2287–2298. doi:10.1080/0003684041233131135 12.

Kushi, L. H., Doyle, C., McCullough, M., Rock, C. L., Demark-Wahnefried, W., Bandera, E. V., Gapstur, S., Patel, A. V., Andrews, K., and Gansler, T. 2012. American Cancer Society guidelines on nutrition and physical activity for cancer prevention. *CA: A Cancer Journal for Clinicians* 62:30–67.

Ladha, J. K., Pathak, H., J. Krupnik, T., Six, J., and van Kessel, C. 2005. Efficiency of fertilizer nitrogen in cereal production: retrospects and prospects, in Sparks, D. L. (Ed.): *Advances in Agronomy*. Academic Press, Vermont, VT, pp.85–156.

Ladha, J. K., Reddy, C. K., Padre, A. T., and van Kessel, C. 2011. Role of nitrogen fertilization in sustaining organic matter in cultivated soils. *Journal of Environment Quality* 40:1756–1766. doi:10.2134/jeq2011.0064.

LaFranchi, B.W., Goldstein, A.H., and Cohen.C, R. 2011. Observations of the temperature dependent response of ozone to NO_x reductions in the Sacramento, CA urban plume. *Atmospheric Chemistry and Physics* 11:6945–6960. doi:10.5194/acp-11-6945-2011.

Lairon, D. 2010. Nutritional quality and safety of organic food: a review. *Agronomy for Sustainable Development* 30:33–41.

Landis, J.D. and Reilly, M. 2004. *How We Will Grow: Baseline Projections of the Growth of California's Urban Footprint through the Year 2100*. Institute of Urban and Regional Development, University of California, Berkeley.

Landon, M.K., Green, C.T., Belitz, K., Singleton, M.J., and Esser, B.K. 2011. Relations of hydrogeologic factors, groundwater reduction-oxidation conditions, and temporal and spatial distributions of nitrate, Central-Eastside San Joaquin Valley, California, USA. *Hydrogeology Journal* 19:1203–1224. doi:10.1007/s10040-011-0750-1

Larson, B.A. and Vroomen, H. 1991. Nitrogen, phosphorus and land demands at the us regional level: a primal approach. *Journal of Agricultural Economics* 42:354–364. doi:10.1111/j.1477-9552.1991.tb00360.x.

LeBauer, D.S. and Treseder, K.K. 2008. Nitrogen limitation of net primary productivity in terrestrial ecosystems is globally distributed. *Ecology* 89:371–379.

Lebby, T., Roco, J.J., and Arcinue, E.L. 1993. Infantile methemoglobinemia associated with acute diarrheal illness. *American Journal of Emergency Medicine* 11:471–472.

Lee, H. 1999. *The Role of Crop Insurance in California: ARE Update*. Giannini Foundation of Agricultural Economics, University of California, Berkeley.

Lee, H. and Sumner, D.A. 2014. *The 2014 Farm Bill, Commodity Subsidies, and California Agriculture: ARE Update*. Giannini Foundation of Agricultural Economics, University of California, Berkeley.

Lehman, P.W., Sevier, J., Giulianotti, J., and Johnson, M. 2004. Sources of oxygen demand in the lower San Joaquin River, California. *Estuaries* 27:405–418. doi:10.1007/BF02803533.

Leip, A., Achermann, B., Billen, G., Bleeker, A., Bouwman, A., de Vries, W., Dragosits, U. et al. 2011. Integrating nitrogen fluxes at the European scale, in Sutton, M., Howard, C., Erisman, J.W., Billen, G., Bleeker, A., Greenfelt, P., van Grinsven, H., and Grizzette, B. (Eds): *The European Nitrogen Assessment*. Cambridge University Press, Cambridge, pp.345–376.

Lenihan, J.M., Bachelet, D., Neilson, R.P., and Drapek, R. 2007. Response of vegetation distribution, ecosystem productivity, and fire to climate change scenarios for California. *Climatic Change* 87:215–230. doi:10.1007/s10584-007-9362-0.

Lenihan, J.M., Drapek, R., Bachelet, D., and Neilson, R.P. 2003. Climate change effects on vegetation distribution, carbon, and fire in California. *Ecological Applications* 13:1667–1681. doi:10.1890/025295.

Leopold, A. 1949. *Sand County Almanac: And Sketches Here and There*. Oxford University Press. Oxford, UK.

Letey, J., Pratt, P.F., and Rible, J.M. 1979. Combining water and fertilizer management for high productivity, low water degradation. *California Agriculture* 33:8–9.

Leverenz, H.L., Haunschild, K., Hopes, G., Tchobanoglous, G., and Darby, J.L. 2010. Anoxic treatment wetlands for denitrification. *Ecological Engineering* 36:1544–1551. doi:10.1016/j.ecoleng.2010.03.014.

Leverenz, H.L., Tchobanoglous, G., and Darby, J.L. 2002. *Review of Technologies for the Onsite Treatment of Wastewater in California*. Center for Environmental and Water Resources Engineering, University of California, Davis.

Levine, J.J., Pettei, M.J., Valderrama, E., Gold, D.M., Kessler, B.H., and Trachtman, H. 1998. Nitric oxide and inflammatory bowel disease: evidence for local intestinal production in children with active colonic disease. *Journal of Pediatric Gastroenterology and Nutrition* 26:34–38.

Lewandowski, A.M., Montgomery, B.R., Rosen, C.J., and Moncrief, J.F. 2008. Groundwater nitrate contamination costs: a survey of private well owners. *Journal of Soil and Water Conservation* 63: 153–161.

Li, C., Narayanan, V., and Harriss, R.C. 1996. Model estimates of nitrous oxide emissions from agricultural lands in the United States. *Global Biogeochemical Cycles* 10:297–306. doi:10.1029/96GB00470.

Liebig, M.A., Gross, J.R., Kronberg, S.L., and Phillips, R.L. 2010. Grazing management contributions to net global warming potential: a long-term evaluation in the Northern Great Plains. *Journal of Environment Quality* 39:799–809. doi:10.2134/jeq2009.0272

Light, A.R. 2010. Reducing nutrient pollution in the everglades agricultural area through best management practices. *Natural Resources & Environment* 25(2):26–30.

Lijinsky, W. 1984. Induction of tumours in rats by feeding nitrosatable amines together with sodium nitrite. *Food and Chemical Toxicology* 22:715–720.

Lillywhite, R., Chandler, D., Grant, W., Lewis, K., Firth, C., Schmutz, U., and Halpin, D. 2007. *Environmental Footprint and Sustainability of Horticulture (including Potatoes) - A Comparison with other Agricultural Sectors*. Warwick HRI, University of Warwick, Warwick, UK.

Lin, S., Bell, E.M., Liu, W., Walker, R.J., Kim, N.K., and Hwang, S.A. 2008. Ambient ozone concentration and hospital admissions due to childhood respiratory diseases in New York State, 1991–2001. *Environmental Research* 108:42–47. doi:10.1016/j.envres.2008.06.007.

Linaker, C.H., Coggon, D., Holgate, S.T., Clough, J., Josephs, L., Chauhan, A.J., and Inskip, H.M. 2000. Personal exposure to nitrogen dioxide and risk of airflow obstruction in asthmatic children with upper respiratory infection. *Thorax* 55:930–933. doi:10.1136/thorax.55.11.930.

Lindemann-Matthies, P., Briegel, R., Schüpbach, B., and Junge, X. 2010. Aesthetic preference for a Swiss Alpine landscape: the impact of different agricultural land-use with different biodiversity. *Landscape and Urban Planning* 98:99–109. doi:10.1016/j.landurbplan.2010.07.015.

Linn, W.S., Shamoo, D.A., Anderson, K.R., Peng, R.C., Avol, E.L., Hackney, J.D., and Gong Jr., H. 1996. Short-term air pollution exposures and responses in Los Angeles area schoolchildren. *Journal of Exposure Analysis and Environmental Epidemiology* 6:449–472.

Linn, W.S., Szlachcic, Y., Henry Jr., G., Kinney, P.L., and Berhane, K.T. 2000. Air pollution and daily hospital admissions in Metropolitan Los Angeles. *Environmental Health Perspectives* 108:427–434.

Linquist, B., van Groenigen, K.J., Adviento-Borbe, M.A., Pittelkow, C., and van Kessel, C. 2012. An agronomic assessment of greenhouse gas emissions from major cereal crops. *Global Change Biology* 18:194–209.

Linquist, B. A., Byous, E., Jones, G., Williams, J. F., Six, J., Horwath, W., and van Kessel, C. 2008. Nitrogen and potassium fertility impacts on aggregate sheath spot disease and yields of rice. *Plant Production Science* 11:260–267. doi:10.1626/pps.11.260.

Linquist, B. A., Hill, J. E., Mutters, R. G., Greer, C. A., Hartley, C., Ruark, M. D., and Van Kessel, C. 2009. Assessing the necessity of surface-applied preplant nitrogen fertilizer in rice systems. *Agronomy Journal* 101:905–915. doi:10.2134/agronj2008.0230x.

Lipman, T. E. and Delucchi, M. A. 2010. Expected greenhouse gas emission reductions by battery, fuel cell, and plug-in hybrid electric vehicles, in Pistoia, G. (Ed.): *Electric and Hybrid Vehicles*. Elsevier, Amsterdam, pp.113–158.

Liu, L. L. and Greaver, T. L. 2009. A review of nitrogen enrichment effects on three biogenic GHGs: the CO_2 sink may be largely offset by stimulated N_2O and CH_4 emission. *Ecology Letters* 12:1103–1117.

Liu, L., Poon, R., Chen, L., Frescura, A. M., Montuschi, P., Ciabattoni, G., Wheeler, A., and Dales, R. 2009. Acute effects of air pollution on pulmonary function, airway inflammation, and oxidative stress in asthmatic children. *Environmental Health Perspectives* 117:668–674. doi:10.1289/ehp11813.

Liu, X., Wang, X., Lin, S., Yuan, J., and Yu, I. T.-S. 2014. Dietary patterns and oesophageal squamous cell carcinoma: a systematic review and meta-analysis. *British Journal of Cancer* 110:2785–2795. doi:10.1038/bjc.2014.172.

Loh, Y. H., Jakszyn, P., Luben, R. N., Mulligan, A. A., Mitrou, P. N., and Khaw, K. T. 2011. N-Nitroso compounds and cancer incidence: the European Prospective Investigation into Cancer and Nutrition (EPIC)-Norfolk Study. *American Journal of Clinical Nutrition* 93:1053–1061. doi:10.3945/ajcn.111.012377.

Lopez, B. 1996. A literature of place, in *A Sense of Place: Regional American Literature*. US Information Agency, Washington, DC, pp.10–12.

Lopus, S. E., Santibáñez, M. P., Beede, R. H., Duncan, R. A., Edstrom, J., Niederholzer, F. J. A., Trexler, C. J., and Brown, P. H. 2010. Survey examines the adoption of perceived best management practices for almond nutrition. *California Agriculture* 64:149–154. doi:10.3733/ca.v064n03p149.

Lorenz, O. A., Weir, B. L., and Bishop, J. C. 1974. Effect of sources of nitrogen on yield and nitrogen absorption of potatoes. *American Journal of Potato Research* 51:56–65. doi:10.1007/BF02858514.

Lowe, K. S., Tucholke, M. B., Tomaras, J. M. B., Conn, K., Hoppe, C., Drewes, J. E., McCray, J. E., and Munakata-Marr, J. 2009. *Influent Constituent Characteristics of the Modern Waste Stream from Single Sources*. IWA Publishing, London, UK.

Lubell, M., Schneider, M., Scholz, J. T., and Mete, M. 2002. Watershed partnerships and the emergence of collective action institutions. *American Journal of Political Science* 46:148–163. doi:10.2307/3088419.

Lund, L. J. 1982. Variations in nitrate and chloride concentrations below selected agricultural fields. *Soil Science Society of America Journal* 46:1062–1066. doi:10.2136/sssaj1982.03615995004600050035x.

Lund, L. J., Page, A. L., and Nelson, C. O. 1976. Nitrogen and phosphorus levels in soils beneath sewage disposal ponds. *Journal of Environment Quality* 5:26–30. doi:10.2134/jeq1976.00472425000500010005x.

Lundberg, J. O., Carlström, M., Larsen, F. J., and Weitzberg, E. 2011. Roles of dietary inorganic nitrate in cardiovascular health and disease. *Cardiovascular Research* 89:525–532. doi:10.1093/cvr/cvq325.

Lundberg, J. O., Feelisch, M., Björne, H., Jansson, E. A., and Weitzberg, E. 2006. Cardioprotective effects of vegetables: is nitrate the answer? *Nitric Oxide* 15:359–362. doi:10.1016/j.niox.2006.01.013.

Lundberg, J. O., Weitzberg, E., Cole, J. A., and Benjamin, N. 2004. Nitrate, bacteria and human health. *Nature Reviews Microbiology* 2:593–602. doi:10.1038/nrmicro929.

Lutsey, N. P. 2008. *Assessment of Out-of-State Heavy-Duty Truck Activity Trends in California (No. 04-328)*. Institute of Transportation Studies, UC Davis.

Lynn, K., Daigle, J., Hoffman, J., Lake, F., Michelle, N., Ranco, D., Viles, C., Voggesser, G., and Williams, P. 2013. The impacts of climate change on tribal traditional foods. *Climate Change* 120:545–556.

Lyon, G. S. and Stein, E. D. 2008. How effective has the Clean Water Act been at reducing pollutant mass emissions to the Southern California Bight over the past 35 years? *Environmental Monitoring Assessment* 154:413–426. doi:10.1007/s10661-008-0408-1.

MA (Millennium Ecosystem Assessment). 2005a. *Ecosystems and Human Well-Being: Current State and Trends. Findings of the Condition and Trends Working Group. Millennium Ecosystem Assessment Series*. Island Press, Washington, DC.

MA. 2005b. *Ecosystems and Human Well-Being: Synthesis*. Island Press, Washington, DC.

Magkos, F., Arvaniti, F., and Zampelas, A. 2003. Organic food: nutritious food or food for thought? A review of the evidence. *International Journal of Food Sciences and Nutrition* 54:357–371.

Marschner, H. 1995. *Mineral Nutrition of Higher Plants*. Academic Press, London, UK.

Marshall, J. D. 2008. Environmental inequality: air pollution exposures in California's South Coast Air Basin. *Atmospheric Environment* 42:5499–5503. doi:10.1016/j.atmosenv.2008.02.005.

Marshall, J. D., Granvold, P. W., Hoats, A. S., McKone, T. E., Deakin, E., and Nazaroff, W. W. 2006. Inhalation intake of ambient air pollution in California's South Coast Air Basin. *Atmospheric Environment* 40:4381–4392. doi:10.1016/j.atmosenv.2006.03.034.

Mar, T. H., Jansen, K., Shepherd, K., Lumley, T., Larson, T. V., and Koenig, J. Q. 2005. Exhaled nitric oxide in children with asthma and short-term $PM_{2.5}$ exposure in Seattle. *Environmental Health Perspectives* 113:1791–1794.

Martin, P. 2001. Farm labor in California: then and now. Paper presented at the *The State of Migrant Labor in the Western United States: Then and Now*. Center for Comparative Immigration Studies, University of California, San Diego.

Martin, P. and Calvin, L. 2011. *Labor Trajectories in California's Produce Industry: ARE Update*. Giannini Foundation of Agricultural Economics, University of California, Davis.

Masumoto, D. M. 1995. *Epitaph for a Peach: Four Seasons on My Family Farm*. HarperCollins, San Francisco, CA.

Matallana González, M. C., Martínez-Tomé, M. J., and Torija Isasa, M. E. 2010. Nitrate and nitrite content in organically cultivated vegetables. *Food Additives & Contaminants: Part B. Surveillance* 3:19–29. doi:10.1080/19440040903586299.

Matson, P. A., Firestone, M., Herman, D., Billow, T., Kiang, N., Benning, T., and Burns, J. 1997. *Agricultural Systems in the San Joaquin Valley: Development of Emissions Estimates for Nitrogen Oxides*, Technical Report No. 94-732). California Air Resources Board, Sacramento.

Matson, P. A., Lohse, K. A., and Hall, S. J. 2002. The globalization of nitrogen deposition: consequences for terrestrial ecosystems. *AMBIO: A Journal of the Human Environment* 31:113–119.

Matthews, W. A., Gabrielyan, G., DaSilva, A., and Sumner, D. A. 2011. *California International Agriculture Exports in 2009*. Agricultural Issues Center, University of California, Davis.

Mayzelle, M. M. and Harter, T. 2011. *Dutch Groundwater Nitrate Contamination Policies*. University of California, Davis.

Mayzelle, M. M., Viers, J. H., Medellin-Azuara, J., and Harter, T. 2015. Economic feasibility of irrigated agricultural land use buffers to reduce groundwater nitrate in rural drinking water sources. *Water* 7:12–27. doi:10.3390/w7010012.

McCalley, C. K. and Sparks, J. P. 2009. Abiotic gas formation drives nitrogen loss from a desert ecosystem. *Science* 326:837–840. doi:10.1126/science.1178984.

McCarty, M. 1981. *The Health Effects of Nitrate, Nitrite, and N-Nitroso Compounds*. National Academy Press, Washington, DC.

McClellan, R. O., Frampton, M. W., Koutrakis, P., McDonnell, W. F., Moolgavkar, S. H., North, D. W., Smith, A. E., Smith, R. L., and Utell, M. J. 2009. Critical considerations in evaluating scientific evidence of health effects of ambient ozone: A conference report. *Inhalation Toxicology* 21:1–36. doi:10.1080/08958370903176735.

McConnell, R., Berhane, K., Gilliland, F., London, S. J., Islam, T., Gauderman, W. J., Avol, E., Margolis, H. G., and Peters, J. M. 2002. Asthma in exercising children exposed to ozone: a cohort study. *Lancet* 359:386–391. doi:10.1016/s0140-6736(02)07597-9.

McConnell, R., Berhane, K., Molitor, J., Gilliland, F., Künzli, N., Thorne, P. S., Thomas, D. et al. 2006. Dog ownership enhances symptomatic responses to air pollution in children with asthma. *Environmental Health Perspectives* 114:1910–1915. doi:10.1289/ehp.8548.

McConnell, R., Islam, T., Shankardass, K., Jerrett, M., Lurmann, F., Gilliland, F., Gauderman, J., Avol, E., Künzli, N., Yao, L., Peters, J., and Berhane, K. 2010. Childhood incident asthma and traffic-related air pollution at home and school. *Environmental Health Perspectives* 118:1021–1026. doi:10.1289/ehp.0901232.

McCune, B. and Geiser, L. 1997. *Macrolichens of the Pacific Northwest*. Oregon State University Press, Corvalis.

McDonald, B. C., Dallmann, T. R., Martin, E. W., and Harley, R. A. 2012. Long-term trends in nitrogen oxide emissions from motor vehicles at national, state, and air basin scales. *Journal of Geophysical Research* 117. doi:10.1029/2012JD018304.

McElroy, J. A., Trentham-Dietz, A., Gangnon, R. E., Hampton, J. M., Bersch, A. J., Kanarek, M. S., and Newcomb, P. A. 2008. Nitrogen-nitrate exposure from drinking water and colorectal cancer risk for rural women in Wisconsin, USA. *Journal of Water and Health* 6:399–409.

McKnight, G. M., Duncan, C. W., Leifert, C., and Golden, M. H. 1999. Dietary nitrate in man: friend or foe? *British Journal of Nutrition* 81:349–358.

McSwiney, C. P. and Robertson, G. P. 2005. Nonlinear response of N_2O flux to incremental fertilizer addition in a continuous maize (*Zea mays* L.) cropping system. *Global Change Biology* 11:1712–1719. doi:10.1111/j.1365-2486.2005.01040.x.

Medellín-Azuara, J., Howitt, R. E., MacEwan, D. J., and Lund, J. R. 2011. Economic impacts of climate-related changes to California agriculture. *Climatic Change* 109:387–405. doi:10.1007/s10584-011-0314-3.

Medellín-Azuara, J., Rosenstock, T. S., Howitt, R. E., Harter, T., Jessoe, K., Dzurella, K., Pettygrove, G. S., and Lund, J. R. 2013. Agroeconomic analysis of nitrate crop source reductions. *Journal of Water Resources Planning and Management* 139:501–511. doi:10.1061/(ASCE)WR.1943-5452.0000268.

Medina-Ramón, M., Zanobetti, A., and Schwartz, J. 2006. The effect of ozone and PM_{10} on hospital admissions for pneumonia and chronic obstructive pulmonary disease: a national multicity study. *American Journal of Epidemiology* 163:579–588. doi:10.1093/aje/kwj078.

Melia, S., Parkhurst, G., and Barton, H. 2011. The paradox of intensification. *Transport Policy* 18:46–52. doi:10.1016/j.tranpol.2010.05.007.

Mellano, V. and Meyer, D. 1996. *Technologies and Management Practices for More Efficient Manure Handling: A Committee Report*. UC Agricultural Issues Center, Davis, CA.

Meng, Y. Y., Rull, R. P., Wilhelm, M., Lombardi, C., Balmes, J., and Ritz, B. 2010. Outdoor air pollution and uncontrolled asthma in the San Joaquin Valley, California. *Journal of Epidemiology and Community Health* 64(2):142–147. doi:10.1136/jech.2009.083576.

Merenlender, A. M. 2000. Mapping vineyard expansion provides information on agriculture and the environment. *California Agriculture* 54:7–12. doi:10.3733/ca.v054n03p7.

Messer, J. and Brezonik, P. L. 1983. Agricultural nitrogen model: a tool for regional. *Environmental Management* 7:177–187. doi:10.1007/BF01867279.

Metcalf & Eddy, Inc. 2003. *Wastewater Engineering: Treatment, Disposal, Reuse*. McGraw-Hill, New York.

Metzger, K. B., Tolbert, P. E., Klein, M., Peel, J. L., Flanders, W. D., Todd, K., Mulholland, J. A., Ryan, P. B., and Frumkin, H. 2004. Ambient air pollution and cardiovascular emergency department visits. *Epidemiology* 15:46–56. doi:10.1097/01.ede.0000101748.28283.97.

Meyer, D. 2000. Dairying and the environment. *Journal of Dairy Science* 83:1419–1427.

Meyer, D., Garnett, I., and Guthrie, J. C. 1997. A survey of dairy manure management practices in California. *Journal of Dairy Science* 80:1841–1845. doi:10.3168/jds.S0022-0302(97)76119-8.

Meyer, D., Price, P. L., Rossow, H. A., Silva-del-Rio, N., Karle, B. M., Robinson, P. H., DePeters, E. J., and Fadel, J. G. 2011. Survey of dairy housing and manure management practices in California. *Journal of Dairy Science* 94:4744–4750. doi:10.3168/jds.2010-3761.

Meyer, R. D. and Marcum, D. B. 1998. Potato yield, petiole nitrogen, and soil nitrogen response to water and nitrogen. *Agronomy Journal* 90:420–429.

Mikkelsen, D. 1987. Nitrogen budgets in flooded soils used for rice production. *Plant and Soil* 100:71–97. doi:10.1007/BF02370933.

Mikkelsen, D. and Rauschkolb, R. 1978. *Survey of Fertilizer Use in California: 1973.* University of California Division of Agricultural Sciences, Oakland.

Mikkelsen, R. and Hartz, T. 2008. Nitrogen sources for organic crop production. *Better Crops* 92:16–19.

Miller, C. and Urban, D.L. 1999. A model of surface fire, climate and forest pattern in the Sierra Nevada, California. *Ecological Modelling* 114:113–135. doi:10.1016/S0304-3800(98)00119-7.

Miller, J.D., Safford, H.D., Crimmins, M., and Thode, A.E. 2008. Quantitative evidence for increasing forest fire severity in the Sierra Nevada and Southern Cascade Mountains, California and Nevada, USA. *Ecosystems* 12:16–32. doi:10.1007/s10021-008-9201-9.

Miller, J.D., Skinner, C.N., Safford, H.D., Knapp, E.E., and Ramirez, C.M. 2012. Trends and causes of severity, size, and number of fires in northwestern California, USA. *Ecological Applications* 22:184–203. doi:10.1890/10-2108.1.

Miller, K.A., Siscovick, D.S., Sheppard, L., Shepherd, K., Sullivan, J.H., Anderson, G.L., and Kaufman, J.D. 2007. Long-term exposure to air pollution and incidence of cardiovascular events in women. *New England Journal of Medicine* 356:447–458. doi:10.1056/NEJMoa054409.

Miller, R.J., Rolston, D.E., Rauschkolb, R.S., and Wolfe, D.W. 1981. Labeled nitrogen uptake by drip-irrigated tomatoes. *Agronomy Journal* 73:265. doi:10.2134/agronj1981.00021962007300020006x.

Miller, R.J. and Smith, R.B. 1976. Nitrogen balance in the Southern San Joaquin Valley. *Journal of Environmental Quality* 5:274–278. doi:10.2134/jeq1976.00472425000500030011x.

Millock, K., Sunding, D., and Zilberman, D. 2002. Regulating pollution with endogenous monitoring. *Journal of Environmental Economics and Management* 44:221–241.

Mills, G., Buse, A., Gimeno, B., Bermejo, V., Holland, M., Emberson, L., and Pleijel, H. 2007. A synthesis of AOT40-based response functions and critical levels of ozone for agricultural and horticultural crops. *Atmospheric Environment* 41:2630–2643.

Millstein, D.E. and Harley, R.A. 2010. Effects of retrofitting emission control systems on in-use heavy diesel vehicles. *Environmental Science and Technology* 44:5042–5048. doi:10.1021/es1006669.

Millstein, J., Gilliland, F., Berhane, K., Gauderman, W.J., McConnell, R., Avol, E., Rappaport, E.B., and Peters, J.M. 2004. Effects of ambient air pollutants on asthma medication use and wheezing among fourth-grade school children from 12 Southern California communities enrolled in the children's health study. *Archives of Environmental Health* 59:505–514.

Milly, P.C.D., Betancourt, J., Falkenmark, M., Hirsch, R.M., Kundzewicz, Z.W., Lettenmaier, D.P., and Stouffer, R.J. 2008. Stationarity is dead: whither water management? *Science* 319:573–574. doi:10.1126/science.1151915.

Mirvish, S.S., Grandjean, A.C., Reimers, K.J., Connelly, B.J., Chen, S.C., Morris, C.R., Wang, X., Haorah, J., and Lyden, E.R. 1998. Effect of ascorbic acid dose taken with a meal on nitrosoproline excretion in subjects ingesting nitrate and proline. *Nutrition and Cancer* 31:106–110. doi:10.1080/01635589809514688.

Mitchell, J.P., Klonsky, K., Shrestha, A., Fry, R., DuSault, A., Beyer, J., and Harben, R. 2007. Adoption of conservation tillage in California: current status and future perspectives. *Australian Journal of Experimental Agriculture* 47:1383. doi:10.1071/EA07044.

Modell, B. and Darlison, M. 2008. Global epidemiology of haemoglobin disorders and derived service indicators. *Bulletin of the World Health Organization.* Available at: http://www.who.int/bulletin/volumes/86/6/06-036673/en/.

Møller, H. 1997. Work in agriculture, childhood residence, nitrate exposure, and testicular cancer risk: a case-control study in Denmark. *Cancer Epidemiology, Biomarkers & Prevention* 6:141–144.

Moolgavkar, S.H. 2003. Air pollution and daily mortality in two U.S. counties: season-specific analyses and exposure-response relationships. *Inhalation Toxicology* 15:877–907.

Moore, E. and Matalon, E. 2011. *The Human Costs of Nitrate-contaminated Drinking Water in the San Joaquin Valley.* The Pacific Institute, Oakland, CA.

Moran, J.E., Esser, B.K., Hillegonds, D., Holtz, M., Roberts, S.K., Singleton, M.J., and Visser, A. 2011. *California GAMA Special Study: Nitrate Fate and Transport in the Salinas Valley, Final Report for the California State Water Resources Control Board (No. LLNL - TR - 484186).* Lawrence Livermore National Laboratory, Livermore, CA.

Morello-Frosch, R. and Lopez, R. 2006. The riskscape and the color line: Examining the role of segregation in environmental health disparities. *Environmental Research* 102:181–196. doi:10.1016/j.envres.2006.05.007.

Morrison, F.B. and Henry, W.A. 1940. *Feeds and Feeding: A Handbook for the Student and Stockman.* The Morrison Pub. Co., Ithaca, NY.

Morse, D. 1995. Environmental considerations of livestock producers. *Journal of Animal Science* 73:2733–2740.

Mortimer, K.M., Neas, L.M., Dockery, D.W., Redline, S., and Tager, I.B. 2002. The effect of air pollution on inner-city children with asthma. *European Respiratory Journal* 9:699–705. doi:10.1183/09031936.02.00247102.

Moshammer, H., Hutter, H.P., Hauck, H., and Neuberger, M. 2006. Low levels of air pollution induce changes of lung function in a panel of schoolchildren. *European Respiratory Journal* 27:1138–1143.

Mosier, A., Bleken, M.A., Chaiwanakupt, P., Ellis, E.C., Freney, J.R., Howarth, R.B., Matson, P.A. et al. 2002. Policy implications of human-accelerated nitrogen cycling, in Boyer, E.W. and Howarth, R.W. (Eds): *The Nitrogen Cycle at Regional to Global Scales.* Springer, New York, pp.477–516.

Mosier, A., Kroeze, C., Nevison, C., Oenema, O., Seitzinger, S., and van Cleemput, O. 1998. Closing the global N_2O budget: nitrous oxide emissions through the agricultural nitrogen cycle. *Nutrient Cycling in Agroecosystems* 52:225–248. doi:10.1023/A:1009740530221.

Mosier, A., Syers, J.K., and Freney, J.R. 2001. Nitrogen fertilizer: an essential component of increased food, feed, and fiber production, in Mosier, A., Syers, J.K., and Freney, J.R. (Eds): *Agriculture and the Nitrogen Cycle: Assessing the Impacts of Fertilizer Use on Food Production and the Environment.* Island Press, Washington, DC, pp.3–18.

Moxey, A. and White, B. 1994. Efficient compliance with agricultural nitrate pollution standards. *Journal of Agricultural Economics* 25:27–37.

Mozafar, A. 1996. Decreasing the NO_3 and increasing the vitamin C contents in spinach by a nitrogen deprivation method. *Plant Foods for Human Nutrition* 49:55–162.

Mudway, I.S. and Kelly, F.J. 2004. An investigation of inhaled ozone dose and the magnitude of airway inflammation in healthy adults. *American Journal of Respiratory and Critical Care Medicine* 169:1089–1095.

Mueller, B.A., Searles Nielsen, S., Preston-Martin, S., Holly, E.A., Cordier, S., Filippini, G., Peris-Bonet, R., and Choi, N.W. 2004. Household water source and the risk of childhood brain tumours: results of the SEARCH International Brain Tumor Study. *International Journal of Epidemiology* 33:1209–1216. doi:10.1093/ije/dyh215.

Muga, H.E. and Mihelcic, J.R. 2008. Sustainability of wastewater treatment technologies. *Journal of Environmental Management* 88:437–447. doi:10.1016/j.jenvman.2007.03.008.

Muir, J. 1894. *The Mountains of California*. Century, New York.

Muir, J. 1901. *Our National Parks*. Houghton Mifflin, Boston, MA.

Muir, J. 1997. The wild parks and forest reservations of the West, in Cronon, W. (ED.): *Nature Writings*. Library of America, New York.

Mullens, B.A., Hinkle, N.C., Szijj, C.E., and Kuney, D.R. 2001. Managing manure and conserving predators helps control flies in caged-layer poultry systems. *California Agriculture* 55:26–30.

Muller, N.Z. and Mendelsohn, R. 2007. Measuring the damages of air pollution in the United States. *Journal of Environmental Economics and Management* 54:1–14.

Muller, N.Z. and Mendelsohn, R. 2009. Efficient pollution regulation: getting the prices right. *The American Economic Review* 99:1714–1739.

Muller, N.Z., Mendelsohn, R., and Nordhaus, W. 2011. Environmental accounting for pollution in the United States economy. *The American Economic Review* 101:1649–1675.

Munger, J.W., Jacob, D.J., Waldman, J.M., and Hoffmann, M.R. 1983. Fogwater chemistry in an urban environment. *Journal of Geophysical Research* 88:5109–5121.

Murphy, J.G., Day, D.A., Cleary, P.A., Wooldridge, P.J., Millet, D.B., Goldstein, A.H., and Cohen, R.C. 2007. The weekend effect within and downwind of Sacramento – Part 1: observations of ozone, nitrogen oxides, and VOC reactivity. *Atmospheric Chemistry and Physics Discussions* 6:11427–11464.

Murphy, J.G., Day, D.A., Cleary, P.A., Wooldridge, P.J., Millet, D.B., Goldstein, A.H., and Cohen, R.C. 2006. The weekend effect within and downwind of Sacramento – Part 2: observational evidence for chemical and dynamical contributions. *Atmospheric Chemistry and Physics Discussions* 6:11971–12019.

Murphy, J.J., Delucchi, M.A., McCubbin, D.R., and Kim, H.J. 1999. The cost of crop damage caused by ozone air pollution from motor vehicles. *Journal of Environmental Management* 55:273–289.

Murray, A.B. and Morrison, B.J. 1986. The effect of cigarette smoke from the mother on bronchial responsiveness and severity of symptoms in children with asthma. *Journal of Allergy and Clinical Immunology* 77:575–581.

Muth, M.K., Karns, S.A., Nielsen, S.J., Buzby, J.C., and Wells, H.F. 2011. *Consumer-Level Food Loss Estimates and Their Use in the ERS Loss-Adjusted Food Availability*. USDA, Washington, DC.

Mutters, R. and Soret, S. 1998. *Statewide Potential Crop Yield Losses from Ozone Exposure (No. 94-345)*. Division of Agricultural and Natural Resources, North Region, University of California, Davis.

Nafstad, P., Haheim, L.L., Oftedal, B., Gram, F., Holme, I., Hjermann, I., and Leren, P. 2003. Lung cancer and air pollution: a 27 year follow up of 16 209 Norwegian men. *Thorax* 58:1071–1076. doi:10.1136/thorax.58.12.1071.

Nagy, K.A., Henen, B.T., and Vyas, D.B. 1998. Nutritional quality of native and introduced food plants of wild desert tortoises. *Journal of Herpetology* 32:260–267. doi:10.2307/1565306.

Nahm, K.H. 2002. Efficient feed nutrient utilization to reduce pollutants in poultry and swine manure. *Critical Reviews in Environmental Science and Technology* 32:1–16. doi:10.1080/10643380290813435.

Naidoo, R. and Adamowicz, W.L. 2005. Biodiversity and nature-based tourism at forest reserves in Uganda. *Environment and Development Economics* 10:159–178.

Nakamura, G. 1996. Harvesting forest biomass reduces wildfire fuel. *California Agriculture* 50:13–16. doi:10.3733/ca.v050n02p13.

Navarro Silvera, S.A., Mayne, S.T., Risch, H.A., Gammon, M.D., Vaughan, T., Chow, W.H., Dubin, J.A. et al. 2011. Principal component analysis of dietary and lifestyle patterns in relation to risk of subtypes of esophageal and gastric cancer. *Annals of Epidemiology* 21:543–550. doi:10.1016/j.annepidem.2010.11.019.

NCHS (National Center for Health Statistics) 2011. Asthma prevalence, health care use, and mortality: United States, 2005–2009, in *National Health Statistics Reports*. NCHS, Hyattsville, MD.

Ndegwa, P.M., Hristov, A.N., Arogo, J., and Sheffield, R.E. 2008. A review of ammonia emission mitigation techniques for concentrated animal feeding operations. *Biosystems Engineering* 100:453–469. doi:10.1016/j.biosystemseng.2008.05.010.

Neary, D.G., Klopatek, C.C., DeBano, L.F., and Ffolliott, P.F. 1999. Fire effects on belowground sustainability: a review and synthesis. *Forest Ecology and Management* 122:51–71. doi:10.1016/S0378-1127(99)00032-8.

Neijat, M., House, J.D., Guenter, W., and Kebreab, E. 2011. Calcium and phosphorus dynamics in commercial laying hens housed in conventional or enriched cage systems. *Poultry Science* 90:2383–2396.

NEJAC (National Environmental Justice Advisory Council). 2002. *Integration of Environmental Justice in Federal Agency Programs*. United States Environmental Protection Agency, Washington, DC.

NHLBI (National Heart, Lung, and Blood Institute). 2007. *Expert Panel Report 3: Guidelines for the Diagnosis and Management of Asthma*. US Department of Health and Human Services, National Institutes of Health, National Asthma Education and Prevention Program, Bethesda, MD.

NHLBI. 2010. *What Is COPD?* Available at: http://www.nhlbi .nih.gov/health/public/lung/copd/what-is-copd/index.htm.

Nichols, F. H., Cloern, J. E., Luoma, S. N., and Peterson, D. H. 1986. The modification of an estuary. *Science* 231:567–573.

Niederholzer, F. J. A., DeJong, T. M., Saenz, J.-L., Muraoka, T. T., and Weinbaum, S. A. 2001. Effectiveness of fall versus spring soil fertilization of field-grown peach trees. *Journal of the American Society for Horticultural Science* 126:644–648.

Niemeier, D. A. and Rowan, D. 2009. From kiosks to megastores: the evolving carbon market. *California Agriculture* 63:96–103. doi:10.3733/ca.v063n02p96.

NOAA (National Oceanic and Atmospheric Administration). 2006. *National Atmospheric Deposition Program Annual Summary.* NOAA, Silver Springs, MD.

Noble, R., Weisberg, S., Leecaster, M., McGee, C., Dorsey, J., Vainik, P., and Orozco-Borbon, V. 2003. Storm effects on regional beach water quality along the Southern California Shoreline. *Journal of Water Health* 1:23–31.

Nohl, W. 2001. Sustainable landscape use and aesthetic perception–preliminary reflections on future landscape aesthetics. *Landscape and Urban Planning* 54:223–237. doi:10.1016/S0169-2046(01)00138-4.

Nowak, J. B., Neuman, J. A., Bahreini, R., Middlebrook, A. M., Holloway, J. S., McKeen, S. A., Parrish, D. D., Ryerson, T. B., and Trainer, M. 2012. Ammonia sources in the California South Coast Air Basin and their impact on ammonium nitrate formation. *Geophysical Research Letters* 39(7). doi:10.1029/2012GL051197.

Nowak, P. J., O'Keefe, G., Bennett, C., Anderson, S., and Trumbo, C. 1997. *Communication and Adoption of USDA Water Quality Demonstration Projects (Evaluation Report).* USDA CSREES, Washington, DC.

NRC (National Research Council). 1972. *Accumulation of Nitrate.* Committee on Nitrate Accumulation. Agriculture Board, National Academy of Sciences, Washington, DC.

NRC. 1994. *Nutrient Requirements of Poultry,* 9th rev. ed. National Academies Press, Washington, DC.

NRC. 2001. *Nutrient Requirements of Dairy Cattle,* 7th ed. National Academies Press, Washington, DC.

NRC. 2008. *Estimating Mortality Risk Reduction and Economic Benefits from Controlling Ozone Air Pollution.* National Academies Press, Washington, DC.

NRC. 2011. *Renewable Fuel Standard: Potential Economic and Environmental Effects of U.S. Biofuel Policy.* National Academies Press, Washington, DC.

NRCan (Natural Resources Canada). 2007. *Canadian Ammonia Producers: Benchmarking Energy Efficiency and Carbon Dioxide Emissions.* Canadian Industry Program for Energy Efficiency, Ottowa, ON.

Nyberg, F., Gustavsson, P., Järup, L., Bellander, T., Berglind, N., Jakobsson, R., and Pershagen, G. 2000. Urban air pollution and lung cancer in Stockholm. *Epidemiology* 11:487–495. doi:10.1097/00001648-200009000-00002.

O'Brien, D., Shalloo, L., Patton, J., Buckley, F., Grainger, C., Wallace, M. 2012. A life cycle assessment of seasonal grass-based and confinement dairy farms. *Agricultural Systems* 107:33–46. doi.org/10.1016/j.agsy.2011.11.004.

OECD (Organisation for Economic Co-operation and Development). 2001. *OECD National Soil Surface Nitrogen Balances: Explanatory Notes.* OECD, Paris.

OECD. 2007. Instrument mixes addressing non-point sources of water pollution, in: *Instrument Mixes for Environmental Policy.* OECD, Paris.

Oenema, O., Bannink, A., Sommer, S. G., Van Groenigen, J. W., and Velthof, G. L. 2008. Gaseous nitrogen emissions from livestock farming systems, in Hatfield, J. L. and Follett, R. F. (Eds): *Nitrogen in the Environment: Sources, Problems, & Management.* Academic Press, New York, pp.395–442.

Oenema, O., Bleeker, A., Braathen, N. A., Budnakova, M., Bull, K., Cermak, P., Geupel, M. et al. 2011a. Nitrogen in current European policies. Chapter 4 in *The European Nitrogen Assessment.* Cambridge University Press, Cambridge, UK.

Oenema, O., Salomez, J., Branquinho, C., Budnáková, M., Cermák, P., Geupel, M., Johnes, P. et al. 2011b. Developing integrated approaches to nitrogen management. Chapter 23 in *The European Nitrogen Assessment.* Cambridge University Press, Cambridge, UK.

Oenema, O. and Tamminga, S. 2005. Nitrogen in global animal production and management options for improving nitrogen use efficiency. *Science China Life Sciences* 48:871–887. doi:10.1007/BF03187126.

Oftedal, B., Brunekreef, B., Nystad, W., Madsen, C., Walker, S. E., and Nafstad, P. 2008. Residential outdoor air pollution and lung function in schoolchildren. *Epidemiology* 19:129–137. doi:10.1097/EDE.0b013e31815c0827.

Ogden, J. and Anderson, L. 2011. *Sustainable Transportation Energy Pathways: A Research Summary for Decision Makers.* Accessed April 21, 2015. http://steps.ucdavis.edu/files /09-06-2013-STEPS-Book-A-Research-Summary-for-Decision-Makers-Sept-2011.pdf.

Okey, T. A. 2003. Macrobenthic colonist guilds and renegades in Monterey canyon (USA) drift algae: partitioning multidimensions. *Ecological Monographs* 73:415–440.

O'Neil, C. E., Keast, D. R., Nicklas, T. A., and Fulgoni III, V. L. 2011. Nut consumption is associated with decreased health risk factors for cardiovascular disease and metabolic syndrome in U.S. adults: NHANES 1999–2004. *Journal of the American College of Nutrition* 30:502–510. doi:10.1080/07315 724.2011.10719996.

Orians, G. H. and Heerwagen, J. H. 1992. Evolved responses to landscapes, in: Barkow, J. H., Cosmides, L., and Tooby, J. (Eds): *The Adapted Mind: Evolutionary Psychology and the Generation of Culture.* Oxford University Press, New York, pp.555–579.

Ostro, B., Roth, L., Malig, B., and Marty, M. 2009. The effects of fine particle components on respiratory hospital admissions in children. *Environmental Health Perspectives* 117:475–480. doi:10.1289/ehp.11848.

Ostro, B. D., Tran, H., and Levy, J. I. 2006. The health benefits of reduced tropospheric ozone in California. *Journal of the Air and Waste Management Association* 56:1007–1021.

Ostrom, E. 1990. *Governing the Commons: The Evolution of Institutions for Collective Action.* Cambridge University Press, Cambridge, UK.

P. E. International. 2009. *GaBi 4 Professional.* P. E. International, Stuttgart.

Padgett, P. E. and Allen, E. B. 1999. Differential responses to nitrogen fertilization in native shrubs and exotic annuals common to Mediterranean Coastal Sage Scrub of California. *Plant Ecology* 144:93–101. doi:10.1023/A: 1009895720067.

Padgett, P. E., Allen, E. B., Bytnerowicz, A., and Minich, R. A. 1999. Changes in soil inorganic nitrogen as related to atmospheric nitrogenous pollutants in southern California. *Atmospheric Environment* 33:769–781. doi:10.1016/S1352-2310(98)00214-3.

Pan, Y., Birdsey, R., Hom, J., and McCullough, K. 2009. Separating effects of changes in atmospheric composition, climate and land-use on carbon sequestration of U.S. Mid-Atlantic temperate forests. *Forest Ecology and Management* 259:151–164. doi:10.1016/j.foreco.2009.09.049.

Pang, X. P., Letey, J., and Wu, L. 1997. Irrigation quantity and uniformity and nitrogen application effects on crop yield and nitrogen leaching. *Soil Science Society of America Journal* 61:257–261. doi:10.2136/sssaj1997.03615995006100010036x.

Parfitt, R. L., Schipper, L. A., Baisden, W. T., and Elliott, A. H. 2006. Nitrogen inputs and outputs for New Zealand in 2001 at national and regional scales. *Biogeochemistry* 80:71–88. doi:10.1007/s10533-006-0002-y.

Parker, A. E., Dugdale, R. C., and Wilkerson, F. P. 2012. Elevated ammonium concentrations from wastewater discharge depress primary productivity in the Sacramento River and the Northern San Francisco Estuary. *Marine Pollution Bulletin* 64:574–586. doi:10.1016/j.marpolbul.2011.12.016.

Parker, J. D., Woodruff, T. J., Basu, R., and Schoendorf, K. C. 2005. Air pollution and birth weight among term infants in California. *Pediatrics* 115:121–128. doi:10.1542/peds.2004-0889.

Park, S., Croteau, P., Boering, K. A., Etheridge, D. M., Ferretti, D., Fraser, P. J., Kim, K. R. et al. 2012. Trends and seasonal cycles in the isotopic composition of nitrous oxide since 1940. *Nature Geoscience* 5:261–265.

Parrish, D. D., Singh, H. B., Molina, L., and Madronich, S. 2011. Air quality progress in North American megacities: a review. *Atmospheric Environment* 45:7015–7025.

Parry, M. L., Canziani, O. F., Palutikof, J. P., van der Linden, P. J., and Hanson, C. E. (Eds). 2007. *IPCC: Climate Change 2007: Impacts, Adaptation and Vulnerability. Contribution of Working Group II to the Fourth Assessment Report of the Intergovernmental Panel on Climate Change.* Cambridge University Press, Cambridge.

Parsons, R. and Daniel, T. C. 2002. Good looking: in defense of scenic landscape aesthetics. *Landscape and Urban Planning* 60:43–56. doi:10.1016/S0169-2046(02)00051-8.

Paulson, L. and Baker, J. 1980. *Nutrient Interactions among Reservoirs on the Colorado River.* Publications (WR), Littleton, CO.

Peacock, W. L., Christensen, L. P., and Hirschfelt, D. J. 1991. Influence of timing of nitrogen fertilizer application on grapevines in the San Joaquin Valley. *American Journal of Enology and Viticulture* 42:322–326.

Pearson, J. and Stewart, G. R. 1993. The deposition of atmospheric ammonia and its effects on plants. *New Phytologist* 125:283–305. doi:10.1111/j.1469-8137.1993.tb03882.x.

Peel, J. L., Tolbert, P. E., Klein, M., Metzger, K. B., Flanders, W. D., Todd, K., Mulholland, J. A., Ryan, P. B., and Frumkin, H. 2005. Ambient air pollution and respiratory emergency department visits. *Epidemiology* 16:164–174. doi:10.1097/01.ede.0000152905.42113.db.

Pellerin, B. A., Downing, B. D., Kendall, C., Dahlgren, R. A., Kraus, T. E. C., Saraceno, J., Spencer, R. G. M., and Bergamas-

chi, B. A. 2009. Assessing the sources and magnitude of diurnal nitrate variability in the San Joaquin River (California) with an in situ optical nitrate sensor and dual nitrate isotopes. *Freshwater Biology* 54:376–387. doi:10.1111/j.1365-2427.2008.02111.x.

Pelletier, N. 2008. Environmental performance in the US broiler poultry sector: life cycle energy use and greenhouse gas, ozone depleting, acidifying and eutrophying emissions. *Agricultural Systems* 98:67–73. doi:10.1016/j.agsy.2008.03.007.

Pelletier, N., Audsley, E., Brodt, S., Garnett, T., Henriksson, P., Kendall, A., Kramer, K. J., Murphy, D., Nemecek, T., and Troell, M. 2011. Energy intensity of agriculture and food systems. *Annual Review of Environment and Resources* 36:223–246.

Pelletier, N., Pirog, R., and Rasmussen, R. 2010. Comparative life cycle environmental impacts of three beef production strategies in the Upper Midwestern United States. *Agricultural Systems* 103:380–389. doi:10.1016/j.agsy.2010.03.009.

Peters, G. P., Minx, J. C., Weber, C. L., and Edenhofer, O. 2011. Growth in emission transfers via international trade from 1990 to 2008. *Proceedings of the National Academy of Sciences of the United States of America* 108:8903–8908. doi:10.1073/pnas.1006388108.

Pettygrove, G. S., Putnam, D. H., and Meyer, D. 2003. *Integrating Forage Production with Dairy Manure Management in the San Joaquin Valley Final Report.* SAREP (Sustainable Agriculture Research and Education Program). University of California, Davis.

Pickett, S. T. A., Cadenasso, M. L., Grove, J. M., Groffman, P. M., Band, L. E., Boone, C. G., Burch, W. R. et al. 2008. Beyond urban legends: an emerging framework of urban ecology, as illustrated by the Baltimore ecosystem study. *BioScience* 58:139–150. doi:10.1641/B580208.

Pidgeon, N. and Fischhoff, B. 2011. The role of social and decision sciences in communicating uncertain climate risks. *Nature Climate Change* 1:35–41.

Pilotto, L., Nitschke, M., Smith, B. J., Pisaniello, D., Ruffin, R. E., McElroy, H. J., Martin, J., and Hiller, J. E. 2004. Randomized controlled trial of unflued gas heater replacement on respiratory health of asthmatic schoolchildren. *International Journal of Epidemiology* 33:208–214. doi:10.1093/ije/dyh018.

Pinder, R. W., Bettez, N. D., Bonan, G. B., Greaver, T. L., Wieder, W. R., Schlesinger, W. H., and Davidson, E. A. 2012a. Impacts of human alteration of the nitrogen cycle in the US on radiative forcing. *Biogeochemistry* 114:25–40. doi:10.1007/s10533-012-9787-z.

Pinder, R. W., Davidson, E. A., Goodale, C. L., Greaver, T. L., Herrick, J. D., and Liu, L. 2012b. Climate change impacts of US reactive nitrogen. *Proceedings of the National Academy of Sciences of the United States of America* 109:7671–7675. doi:10.1073/pnas.1114243109.

Pliss, G. B. and Frolov, A. G. 1991. Sodium nitrate as a possible promoter of bladder carcinogenesis in rats. *Voprosy Onkologii* 37:203–206.

Pollack, E. S. and Pollack, C. V. 1994. Incidence of subclinical methemoglobinemia in infants with diarrhea. *Annals of Emergency Medicine* 24:652–656.

Pollan, M. 2008. *In Defense of Food: An Eater's Manifesto.* Penguin, New York.

Popp, D. 2010. Exploring links between innovation and diffusion: adoption of NO_x control technologies at US coal-fired power plants. *Environmental and Resource Economics* 45:319–352. doi:10.1007/s10640-009-9317-1.

Posas, P.J. 2007. Roles of religion and ethics in addressing climate change. *Ethics in Science and Environmental Politics* 6:31–49.

Posey, D.A. (Ed.). 1999. *Cultural and Spiritual Values of Biodiversity*. Intermediate Technology Publications, London.

Post, W.M. and Mann, L.K. 1990. Changes in soil organic carbon and nitrogen as a result of cultivation, in Bouwman, A.F. (Ed.): *Soils and the Greenhouse Effect*. John Wiley & Sons, New York, pp.401–406.

Postma, D.S. and Boezen, H.M. 2004. Rationale for the Dutch hypothesis: allergy and airway hyperresponsiveness as genetic factors and their interaction with environment in the development of asthma and COPD. *Chest* 126. doi:10.1378/chest.126.2_suppl_1.96S.

Potter, C. 2010. The carbon budget of California. *Environmental Science & Policy* 13:373–383. doi:10.1016/j.envsci.2010.04.008.

Potter, C.S., Matson, P.A., Vitousek, P.M., and Davidson, E.A. 1996. Process modeling of controls on nitrogen trace gas emissions from soils worldwide. *Journal of Geophysical Research* 101:1361–1377. doi:10.1029/95JD02028.

Poudel, D.D., Horwath, W.R., Lanini, W.T., Temple, S.R., and van Bruggen, A.H.C. 2002. Comparison of soil N availability and leaching potential, crop yields and weeds in organic, low-input and conventional farming systems in northern California. *Agriculture, Ecosystems & Environment* 90:125–137. doi:10.1016/S0167-8809(01)00196-7.

Powell, J.M., Gourley, C.J.P., Rotz, C.A., and Weaver, D.M. 2010. Nitrogen use efficiency: a potential performance indicator and policy tool for dairy farms. *Environmental Science & Policy* 13:217–228. doi:10.1016/j.envsci.2010.03.007.

Powers, R.F. and Reynolds, P.E. 1999. Ten-year responses of ponderosa pine plantations to repeated vegetation and nutrient control along an environmental gradient. *Canadian Journal of Forest Research* 29:1027–1038.

Powlson, D.S., Addiscott, T.M., Benjamin, N., Cassman, K.G., de Kok, T.M., van Grinsven, H., L'Hirondel, J.L., Avery, A.A., and van Kessel, C. 2008. When does nitrate become a risk for humans? *Journal of Environmental Quality* 37:291–295. doi:10.2134/jeq2007.0177.

PPIC. 2008. *Just the Facts: California's Future Population*. Available at: http://www.ppic.org/content/pubs/jtf/JTF_FuturePopulationJTF.pdf.

Prasad, S. and Chetty, A.A. 2008. Nitrate-N determination in leafy vegetables: study of the effects of cooking and freezing. *Food Chemistry* 106:772–780. doi:10.1016/j.foodchem.2007.06.005.

Prasad, V.K., Badarinath, K.V.S., Yonemura, S., and Tsuruta, H. 2004. Regional inventory of soil surface nitrogen balances in Indian agriculture (2000–2001). *Journal of Environmental Management* 73:209–218. doi:10.1016/j.jenvman.2004.06.013.

Pratt, P.F. 1984. Salinity, sodium, and potassium in an irrigated soil treated with bovine manure. *Soil Science Society of America Journal* 48:823–828.

Pratt, P.F.. and Castellanos, J.Z. 1981. Available nitrogen from animal manures. *California Agriculture* 35:24–24.

Preston-Martin, S., Pogoda, J., Mueller, B.A., Holly, E., Lijinsky, W., and Davis, R.L. 1996. Maternal consumption of cured meats and vitamins in relation to pediatric brain tumors. *Cancer Epidemiology Biomarkers & Prevention* 8:599–605.

Pulkrabek, W. 2003. *Engineer Fundamentals of the Internal Combustion Engine*, 2nd. ed. Prentice Hall, Upper Saddle River, NJ.

Purkey, D.R., Huber-Lee, A., Yates, D.N., Hanemann, M., and Herrod-Julius, S. 2007. Integrating a climate change assessment tool into stakeholder-driven water management decision-making processes in California. *Water Resources Management* 21:315–329. doi:10.1007/s11269-006-9055-x.

Pusede, S.E. and Cohen, R.C. 2012. On the observed response of ozone to NO_x and VOC reactivity reductions in San Joaquin Valley California 1995–present. *Atmospheric Chemistry and Physics* 12:8323–8339.

Quynh, L.T.P., Billen, G., Garnier, J., Théry, S., Fézard, C., and Minh, C.V. 2005. Nutrient (N, P) budgets for the Red River basin (Vietnam and China). *Global Biogeochemical Cycles* 19:GB2022. doi:10.1029/2004GB002405.

Rabalais, N.N. 1998. *Oxygen Depletion in Coastal Waters. National Oceanic and Atmospheric Administration's State of the Coast Report*. NOAA, Silver Spring, MD.

Rabotyagov, S., Valcu, A., Dampbell, T., Jha, M.K., Gassman, P., and Kling, C.L. 2012. Using a coupled simulation-optimization approach to design cost-effective reverse auctions for watershed nutrient reductions. Paper presented at *The Agricultural and Applied Economics Association Annual Meeting*, Seattle, WA, pp.173–184.

Raciti, S.M., Groffman, P.M., Jenkins, J.C., Pouyat, R.V., Fahey, T.J., Pickett, S.T.A., and Cadenasso, M.L. 2011. Accumulation of carbon and nitrogen in residential soils with different land-use histories. *Ecosystems* 14:287–297. doi:10.1007/s10021-010-9409-3.

Rademacher, J.J., Young, T.B., and Kanarek, M.S. 1992. Gastric cancer mortality and nitrate levels in Wisconsin drinking water. *Archives of Environmental Health* 47:292–294. doi:10.1080/00039896.1992.9938364.

Rae, A.N. and Strutt, A. 2004. Agricultural trade reform and nitrogen pollution in OECD countries. Paper presented at *The New Zealand Association of Economists Conference*, Wellington.

Rajsic, P., Weersink, A., and Gandorfer, M. 2009. Risk and nitrogen application levels. *Canadian Journal of Agricultural Economics/Revue canadienne d'agroeconomie* 57:223–239. doi:10.1111/j.1744-7976.2009.01149.x.

Ramos, D.E. 1997. *Walnut Production Manual*. University of California Division of Agriculture and Natural Resources, Davis.

Randall, D.J. and Tsui, T.K.N. 2002. Ammonia toxicity in fish. *Marine Pollution Bulletin* 45:17–23.

Randall, G.W., Delgado, J.A., and Schepers, J.S. 2008. Nitrogen management to protect water resources, in Schepers, J.S. and Raun, W.R. (Eds): *Nitrogen in Agricultural Systems*. Agronomy monographs 49. ASA, CSSA, and SSA, Madison, WI.

Ravishankara, A.R., Daniel, J.S., and Portmann, R.W. 2009. Nitrous oxide (N_2O): the dominant ozone-depleting substance emitted in the 21st century. *Science* 326:123–125. doi:10.1126/science.1176985.

Reed, K., Horwath, W.R., Kaffka, S.R., Denison, R.F., Bryant, D., and Kabir, Z. 2006. Long-term comparison of yield and nitrogen use in organic, winter legume cover crop and conventional farming systems. *SAFS Newsletters* 6(2):1–2.

Reganold, J.P., Andrews, P.K., Reeve, J.R., Carpenter-Boggs, L., Schadt, C.W., Alldredge, J.R., Ross, C.F., Davies, N.M., and Zhou, J. 2010. Fruit and soil quality of organic and conventional strawberry agroecosystems. *PLOS ONE* 5:e12346. doi:10.1371/journal.pone.0012346.

Reganold, J.P., Glover, J.D., Andrews, P.K., Hinman, H.R. 2001. Sustainability of three apple production systems. *Nature* 410:926–930. doi:10.1038/35073574.

Reid, S., Sullivan, D., Penfold, B.M., Raffuse, S.M., and Funk, T.H. 2007. Activity trends for key emission sources in California's San Joaquin Valley, 1970–2030. Paper presented at *The Environmental Protection Agency's 16th Annual International Emission Inventory Conference*, May 14–17, Raleigh.

Reish, D.J. 1955. The relation of polychaetous annelids to harbor pollution. *Public Health Reports* 70:1168–1174.

Reish, D.J. 2000. The seasonal settlement of polychaete larvae before and after pollution abatement in Los Angeles-Long Beach Harbors, California. *Bulletin of Marine Science* 67:672.

Rembialkowska, E. 2007. Quality of plant products from organic agriculture. *Journal of the Science of Food and Agriculture* 87:2757–2762.

Rettie, D.F. 1996. *Our National Park System: Caring for America's Greatest Natural and Historic Treasures*. University of Illinois Press, Champaign.

Retzlaff, W.A., DeJong, T.M., and Williams, L.E. 1990. Screening almond cultivars for ozone susceptibility. *Oral Presentation at 87th Annual Meeting of the American Society for Horticultural Science, November 4–8, 1990*. Tucson, AZ.

Retzlaff, W.A., Dejong, T.M., and Williams, L.E. 1992. Photosynthesis and growth-response of almond to increased atmospheric ozone partial pressures. *Journal of Environmental Quality* 21:208–216.

Ribaudo, M. and Agapoff, J. 2005. Importance of cost offsets for dairy farms meeting a nutrient application standard. *Agricultural and Resource Economics Review* 34:173–184.

Ribaudo, M., Delgado, J., Hansen, L., Livingston, M., Mosheim, R., and Williamson, J. 2011. *Nitrogen in Agricultural Systems: Implications for Conservation Policy (No. 127)*. USDA ERS (Economic Research Service), Washington, DC.

Ribaudo, M., Gollehon, N., Aillery, M., Kaplan, J., Johansson, R., Agapoff, J., Christensen, L., Breneman, V., and Peters, M. 2003. *Manure Management for Water Quality: Costs to Animal Feeding Operations of Applying Manure Nutrients to Land (Agricultural Economic Report No. AER824)*. USDA ERS, Washington, DC.

Ribaudo, M., Horan, R.D., and Smith, M.E. 1999. *Economics of Water Quality Protection from Nonpoint Sources (Agricultural Economic Report Number No. 782)*. USDA ERS, Washington, DC.

Richardson, W.F. and Meyer, R.D. 1990. Spring and summer nitrogen applications to Vina walnuts. *California Agriculture* 44:30–32.

Riddell, J., Nash, T.H., III, and Padgett, P. 2008. The effect of HNO₃ gas on the lichen *Ramalina menziesii*. *Flora: Morphology, Distribution, Functional Ecology of Plants* 203:47–54. doi:10.1016/j.flora.2007.10.001.

Rilla, E., Hardesty, S.D., Getz, C., and George, H. 2011. California agritourism operations and their economic potential are growing. *California Agriculture* 65:57–65. doi:10.3733/ca.v065n02p57.

Ritz, B., Wilhelm, M., Hoggatt, K.J., and Ghosh, J.K.C. 2007. Ambient air pollution and preterm birth in the environment and pregnancy outcomes study at the University of California, Los Angeles. *American Journal of Epidemiology* 166:1045–1052. doi:10.1093/aje/kwm181.

Ritz, B., Zhao, Y., and Wilhelm, M. 2006. Air pollution and infant death in Southern California, 1989–2000. *Pediatrics* 118:493–502. doi:10.1542/peds.2006-0027.

Robertson, G.P. 1982. Regional nitrogen budgets: approaches and problems, in Robertson, G.P., Herrera, R., and Rosswall, T. (Eds): *Nitrogen Cycling in Ecosystems of Latin America and the Caribbean, Developments in Plant and Soil Sciences*. Springer Netherlands, Heidelberg, pp.73–79.

Robertson, G.P. and Rosswall, T. 1986. Nitrogen in West Africa: the regional cycle. *Ecological Monographs* 56:43–72. doi:10.2307/2937270.

Robertson, G.P. and Vitousek, P.M. 2009. Nitrogen in agriculture: balancing the cost of an essential resource. *Annual Review of Environment and Resources* 34:97–125.

Rockström, J., Steffen, W., Noone, K., Persson, A., Chapin, F.S., Lambin, E.F., Lenton, T.M. et al. 2009. A safe operating space for humanity. *Nature* 461:472–475. doi:10.1038/461472a.

Rodriguez, C., Tonkin, R., Heyworth, J., Kusel, M., Klerk, N.D., Sly, P.D., Franklin, P., Runnion, T., Blockley, A., Landau, L., and Hinwood, A.L. 2007. The relationship between outdoor air quality and respiratory symptoms in young children. *International Journal of Environmental Health Research* 17:351–360. doi:10.1080/09603120701628669.

Rogers, M.A., Vaughan, T.L., Davis, S., and Thomas, D.B. 1995. Consumption of nitrate, nitrite, and nitrosodimethylamine and the risk of upper aerodigestive tract cancer. *Cancer Epidemiology, Biomarkers & Prevention* 4:29–36.

Rojas-Martinez, R., Perez-Padilla, R., Olaiz-Fernandez, G., Mendoza-Alvarado, L., Moreno-Macias, H., Fortoul, T., McDonnell, W., Loomis, D., and Romieu, I. 2007. Lung function growth in children with long-term exposure to air pollutants in Mexico City. *American Journal of Respiratory and Critical Care Medicine* 176:377–384. doi:10.1164/rccm.200510-1678OC.

Roland-Holst, D. and Zilberman, D. 2006. *How Vulnerable is California Agriculture to Higher Energy Prices?: ARE Update*. Giannini Foundation of Agricultural Economics, University of California, Davis.

Romieu, I., Ramirez-Aguilar, M., Sienra-Monge, J.J., Moreno-Macias, H., del Rio-Navarro, B., David, G., Marzec, J., Hernandez-Avila, M., and London, S. 2006. GSTM1 and GSTP1 and respiratory health in asthmatic children exposed to ozone. *European Respiratory Journal* 28:953–959. doi:10.1183/09031936.06.00114905.

Romley, J.A., Hackbarth, A., and Goldman, D.P. 2010. *The Impact of Air Quality on Hospital Spending Monica, Calif. (Technical Report No. TR-777-WFHF)*. RAND Corporation, Santa Monica, CA.

Roosevelt, T. 1985. *An Autobiography*. Da Capo Press, New York.

Rosecrance, R., Faber, B., and Lovatt, C. 2012. Patterns of nutrient accumulation in "Hass" avocado fruit. *Better Crops* 96:12–13.

Rosenstock, T.S., Liptzin, D., Dzurella, K., Fryjoff-Hung, A., Hollander, A., Jensen, V., King, A. et al. 2014. Agriculture's contribution to nitrate contamination of Californian groundwater (1945–2005). *Journal of Environment Quality* 43:895. doi:10.2134/jeq2013.10.0411.

Rosenstock, T.S., Liptzin, D., Six, J., and Tomich, T.P. 2013. Nitrogen fertilizer use in California: assessing the data, trends and a way forward. *California Agriculture* 67:68–79. doi:10.3733/ca.E.v067n01p68.

Rotz, C.A. 2004. Management to reduce nitrogen losses in animal production. *Journal of Animal Science* 82: E119–E137.

Rotz, C.A., Soder, K.J., Skinner, R.H., Dell, C.J., Kleinman, P.J., Schmidt, J.P., and Bryant, R.B. 2009. Grazing can reduce the environmental impact of dairy production systems. *Forage and Grazinglands*. doi:10.1094/FG-2009-0916-01-RS.

Rowe, R.D. and Chestnut, L.G. 1985. Economic assessment of the effects of air pollution on agricultural crops in the San Joaquin Valley. *Journal of the Air Pollution Control Association* 35:728–734.

Ruehl, C.R., Fisher, A.T., Los Huertos, M., Wankel, S.D., Wheat, G.C., Kendall, C., Hatch, C.E., and Shennan, C. 2007. Nitrate dynamics within the Pajaro River, a nutrient-rich, losing stream. *Journal of the North American Benthological Society* 26:191–206. doi:http://dx.doi.org/10.1899/0887-3593(2007)26[191:NDWTPR]2.0.CO;2.

Russell, A.G., Winner, D.A., Harley, R.A., McCue, K.F., and Cass, G.R. 1993. Mathematical modeling and control of the dry deposition flux of nitrogen-containing air pollutants. *Environmental Science & Technology* 27:2772–2782. doi:10.1021/es00049a016.

Russell, A.R., Valin, L.C., Bucsela, E.J., Wenig, M.., and Cohen, R.C. 2010. Space-based constraints on spatial and temporal patterns of NO_x emissions in California, 2005,àí2008. *Environmental Science & Technology* 44:3608–3615. doi:10.1021/es903451j.

Russelle, M.P., Entz, M.H., and Franzluebbers, A.J. 2007. Reconsidering integrated crop–livestock systems in North America. *Agronomy Journal* 99:325. doi:10.2134/agronj2006.0139.

Ryden, J.C., Lund, L.J., and Whaley, S.A. 1981. Direct measurement of gaseous nitrogen losses from an effluent irrigation area. *Journal (Water Pollution Control Federation)* 53:1677–1682.

Ryerson, T.B., Andrews, A.E., Angevine, W.M., Bates, T.S., Brock, C.A., Cairns, B., Cohen, R.C. et al. 2013. The 2010 California Research at the Nexus of Air Quality and Climate Change (CalNex) field study. *Journal of Geophysical Research: Atmospheres* 118:5830–5866. doi:10.1002/jgrd.50331.

Rynk, R., van de Kamp, M., Willson, G.B., Singley, M.E., Richard, T.L., Kolega, J.J., Gouin, F.R. et al. 1992. *On-Farm Composting Handbook*. Northeast Regional Agricultural Engineering Service, Ithaca, NY.

Sadeq, M., Moe, C.L., Attarassi, B., Cherkaoui, I., Elaouad, R., and Idrissi, L. 2008. Drinking water nitrate and prevalence of methemoglobinemia among infants and children aged 1–7 years in Moroccan areas. *International Journal of Hygiene and Environmental Health* 211:546–554. doi:10.1016/j.ijheh.2007.09.009.

Saenz, J.L., DeJong, T.M., and Weinbaum, S.A. 1997. Nitrogen stimulated increases in peach yields are associated with extended fruit development period and increased fruit sink capacity. *Journal of the American Society for Horticultural Science* 122(6):772–777.

Salam, M.T., Millstein, J., Li, Y.F., Lurmann, F.W., Margolis, H.G., and Gilliland, F.D. 2005. Birth outcomes and prenatal exposure to ozone, carbon monoxide, and particulate matter: results from the Children's Health Study. *Environmental Health Perspectives* 113:1638–1644. doi:10.1289/ehp.8111.

Salo, T., Lemola, R., and Esala, M. 2008. National and regional net nitrogen balances in Finland in 1990–2005. *Agricultural and Food Science* 16:366–375.

Samani Majd, A.M. and Mukhtar, S. 2013. Ammonia recovery enhancement using a tubular gas-permeable membrane system in laboratory and field-scale studies. *Transactions of the ASABE* 56:1951–1958.

Samaras, C. and Meisterling, K. 2008. Life cycle assessment of greenhouse gas emissions from plug-in hybrid vehicles: implications for policy. *Environmental Science and Technology* 42:3170–3176. doi:10.1021/es702178s.

Samoli, E., Aga, E., Touloumi, G., Nisiotis, K., Forsberg, B., Lefranc, A., Pekkanen, J. et al. 2006. Short-term effects of nitrogen dioxide on mortality: an analysis within the APHEA project. *European Respiratory Journal* 27:1129–1138. doi:10.1183/09031936.06.00143905.

San Joaquin Valley Dairy Manure Technology Feasibility Assessment Panel. 2005a. *An Assessment of Technologies for Management and Treatment of Dairy Manure in California's San Joaquin Valley.* CARB (California Air Resources Board), Sacramento.

San Joaquin Valley Dairy Manure Technology Feasibility Assessment Panel. 2005b. *An Assessment of Technologies for Management and Treatment of Dairy Manure in California's San Joaquin Valley.* San Joaquin Valley Dairy Manure Technology Feasibility Assessment Panel, San Francisco, CA.

Sandermann, H. 1996. Ozone and plant health. *Annual Review of Phytopathology* 34:347–366.

Sanger, D.M., Arendt, M.D., Chen, Y., Wenner, E.L., Holland, A.F., Edwards, D., and Caffrey, J. 2002. *A Synthesis of Water Quality Data: National Estuarine Research Reserve System-Wide Monitoring Program (1995–2000) (No. 500)*, National Estuarine Research Reserve Technical Report Series. South Carolina Department of Natural Resources, Marine Resources Division, Charleston.

SAREP (Sustainable Agriculture Research and Education Program). 2004. *Dairy Manure Nutrient Management Survey Results*. Unpublished. University of California, Davis.

Sawada, H. and Kohno, Y. 2009. Differential ozone sensitivity of rice cultivars as indicated by visible injury and grain yield. *Plant Biology* 11:70–75.

Sawyer, R.F., Harley, R.A., Cadle, S.H., Norbeck, J.M., Slott, R., and Bravo, H.A. 2000. Mobile sources critical review: 1998 NARSTO assessment. *Atmospheric Environment* 34:2161–2181. doi:10.1016/S1352-2310(99)00463-X.

Schaap, M., Timmermans, R.M.A., Roemer, M., Boersen, G.A.C., Builtjes, P., Sauter, F., Velders, G., and Beck, J. 2008. The LOTOS-EUROS model: description, validation and latest developments. *International Journal of Environment and Pollution* 32:270–290.

Schaefer, S. and Alber, M. 2007. Temperature controls a latitudinal gradient in the proportion of watershed nitrogen exported to coastal ecosystems. *Biogeochemistry* 85:333–346.

Schaefer, S. C., Hollibaugh, J. T., and Alber, M. 2009. Watershed nitrogen input and riverine export on the west coast of the US. *Biogeochemistry* 93:219–233. doi:10.1007/s10533-009-9299-7.

Schildcrout, J. S., Sheppard, L., Lumley, T., Slaughter, J. C., Koenig, J. Q., and Shapiro, G. G. 2006. Ambient air pollution and asthma exacerbations in children: an eight-city analysis. *American Journal of Epidemiology* 164:505–517. doi:10.1093/aje/kwj225.

Schindler, C., Keidel, D., Gerbase, M. W., Zemp, E., Bettschart, R., Brändli, O., Brutsche, M. H. et al. 2009. Improvements in PM_{10} exposure and reduced rates of respiratory symptoms in a cohort of Swiss adults (SAPALDIA). *American Journal of Respiratory and Critical Care Medicine* 179:579–587. doi:10.1164/rccm.200803-388OC.

Schindler, C., Künzli, N., Bongard, J. P., Leuenberger, P., Karrer, W., Rapp, R., Monn, C., and Ackermann-Liebrich, U. 2001. Short-term variation in air pollution and in average lung function among never-smokers: the Swiss Study on Air Pollution and Lung Diseases in Adults (SAPALDIA). *American Journal of Respiratory and Critical Care Medicine* 163:356–361.

Schlesinger, W. H. 2009. On the fate of anthropogenic nitrogen. *Proceedings of the National Academy of Sciences of the United States of America* 106:203–208. doi:10.1073/pnas.0810193105.

Schlosser, E. 2001. *Fast Food Nation: The Dark Side of the All-American Meal*. Harper Perennial, New York.

Schwartz, J., Dockery, D. W., Neas, L. M., Wypij, D., Ware, J. H., Spengler, J. D., Koutrakis, P., Speizer, F. E., and Ferris Jr., B. G. 1994. Acute effects of summer air pollution on respiratory symptom reporting in children. *American Journal of Respiratory Critical Care Medicine* 150:1234–1242.

Schwartz, P. 1996. *The Art of the Long View: Planning for the Future in an Uncertain World*. Doubleday, New York.

Schweizer, S., Davis, S., and Thompson, J. L. 2013. Changing the conversation about climate change: a theoretical framework for place-based climate change engagement. *Environmental Communication: A Journal of Nature and Culture* 7:42–62. doi:10.1080/17524032.2012.753634.

Searchinger, T., Heimlich, R., Houghton, R. A., Dong, F., Elobeid, A., Fabiosa, J., Tokgoz, S., Hayes, D., and Yu, T. H. 2008. Use of U.S. croplands for biofuels increases greenhouse gases through emissions from land-use change. *Science* 319:1238–1240. doi:10.1126/science.1151861.

Seidel, C., Gorman, C., Darby, J., and Jensen, V. B. 2011. *An Assessment of the State of Nitrate Treatment Alternatives*. American Water Works Association, West Sacramento, CA.

Seinfeld, J. H. and Pandis, S. N. 1998. *Atmospheric Chemistry and Physics: From Air Pollution to Climate Change*. Wiley-Interscience Publication, New York, NY. USA.

Seitzinger, S., Harrison, J. A., Böhlke, J. K., Bouwman, A. F., Lowrance, R., Peterson, B., Tobias, C., and Van Drecht, G. 2006. Denitrification across landscapes and waterscapes: a synthesis. *Ecological Applications* 16:2064–2090. doi:10.1890/1051-0761(2006)016[2064:DALAWA]2.0.CO;2.

Selman, M., Greenhalgh, S., Taylor, M., and Guiling, J. 2008. *Paying for Environmental Performance: Potential Cost Savings Using a Reverse Auction in Program Sign-Up*. World Resources Institute, Washington, DC.

Service, R. F. 2014. New recipe produces ammonia from air, water, and sunlight. *Science* 345:610.

Shah, P. S. and Balkhair, T. 2011. Air pollution and birth outcomes: a systematic review. *Environment International* 37:498–516. doi:10.1016/j.envint.2010.10.009.

Shank, R. C. and Newberne, P. M. 1976. Dose-response study of the carcinogenicity of dietary sodium nitrite and morpholine in rats and hamsters. *Food and Cosmetics Toxicology* 14:1–8.

Shaver, T. M., Khosla, R., and Westfall, D. G. 2011. Evaluation of two crop canopy sensors for nitrogen variability determination in irrigated maize. *Precision Agriculture* 12:892–904. doi:10.1007/s11119-011-9229-2.

Shindell, D. T., Faluvegi, G., Koch, D. M., Schmidt, G. A., Unger, N., and Bauer, S. E. 2009. Improved attribution of climate forcing to emissions. *Science* 326:716–718.

Shine, K. P., Berntsen, T. K., Fuglestvedt, J. S., Skeie, R. B., and Stuber, N. 2007. Comparing the climate effect of emissions of short and long lived climate agents. *Philosophical Transactions of the Royal Society A* 365:1903–1914.

Shine, K. P., Fuglestvedt, J. S., Hailemariam, K., and Stuber, N. 2005. Alternatives to the global warming potential for comparing climate impacts of emissions of greenhouse gases. *Climatic Change* 68:281–302.

Shortle, J. S., Ribaudo, M., Horan, R. D., and Blandford, D. 2012. Reforming agricultural nonpoint pollution policy in an increasingly budget-constrained environment. *Environmental Science & Technology* 46:1316–1325.

Shrestha, A. and Grantz, D. A. 2005. Ozone impacts on competition between tomato and yellow nutsedge: above- and below-ground. *Crop Science* 45:1587–1595.

Shuval, H. I. and Gruener, N. 1972. Epidemiological and toxicological aspects of nitrates and nitrites in the environment. *American Journal of Public Health* 62: 1045–1052.

Silkoff, P. E., Zhang, L., Dutton, S., Langmack, E. L., Vedal, S., Murphy, J., and Make, B. 2005. Winter air pollution and disease parameters in advanced chronic obstructive pulmonary disease panels residing in Denver, Colorado. *The Journal of Allergy and Clinical Immunology* 115:337–344. doi:10.1016/j.jaci.2004.11.035.

Silverman, R. A. and Ito, K. 2010. Age-related association of fine particles and ozone with severe acute asthma in New York City. *Journal of Allergy and Clinical Immunology* 125:367–373. doi:10.1016/j.jaci.2009.10.061.

Simpson, R., Williams, G., Petroeschevsky, A., Best, T., Morgan, G., Denison, L., Hinwood, A., and Neville, G. 2005a. The short-term effects of air pollution on hospital admissions in four Australian cities. *Australian and New Zealand Journal of Public Health* 29:213–221.

Simpson, R., Williams, G., Petroeschevsky, A., Best, T., Morgan, G., Denison, L., Hinwood, A., Neville, G., and Neller, A. 2005b. The short-term effects of air pollution on daily mortality on four Australian cities. *Australian and New Zealand Journal of Public Health* 29:205–212.

Sinclair, A., Edgerton, E. S., Wyzga, R., and Tolsma, D. 2010. A two-time-period comparison of the effects of ambient air pollution on outpatient visits for acute respiratory illnesses.

Journal of the Air and Waste Management Association 60:163–175.

Singer, M.J. 2001. Looking back 60 years, California soils maintain overall chemical quality. *California Agriculture* 57:2–5.

Sitch, S., Cox, P.M., Collins, W.J., and Huntingford, C. 2007. Indirect radiative forcing of climate change through ozone effects on the land-carbon sink. *Nature* 448:791–794.

Six, J., Ogle, S.M., Breidt, F.J., Conant, R.T., Mosier, A.R., and Paustian, K. 2004. The potential to mitigate global warming with no-tillage management is only realized when practised in the long term. *Global Change Biology* 10: 155–160. doi:10.1111/j.1529-8817.2003.00730.x.

Skrivan, J. 1971. Methemoglobinemia in pregnancy (clinical and experimental study). *Acta Universitatis Carolinae Medica (Praha)* 17:123–161.

Sleeter, B.M., Wilson, T.S., Soulard, C.E., and Liu, J. 2010. Estimation of late twentieth century land-cover change in California. *Environmental Monitoring and Assessment* 173:251–266. doi:10.1007/s10661-010-1385-8.

Small, K.A. and Van Dender, K. 2007. Fuel efficiency and motor vehicle travel: the declining rebound effect. *The Energy Journal* 28:25–51.

Smart, D.R., Alsina, M.M., Wolff, M.W., Matiasek, M.G., Schellenberg, D.L., Edstrom, J.P., Brown, P.H., and Scow, K.M. 2011a. N$_2$O emissions and water management in California perennial crops. Chapter 12 in Guo, L., Gunasekara, A., and McConnell, L. (Eds): *Understanding Greenhouse Gas Emissions from Agricultural Management.* American Chemical Society, Washington, DC, pp. 227–255.

Smart, J.C.R., Hicks, K., Morrissey, T., Heinemeyer, A., Sutton, M.A., and Ashmore, M. 2011b. Applying the ecosystem service concept to air quality management in the UK: a case study for ammonia. *Environmetrics* 22:649–661. doi:10.1002/env.1094.

Smil, V. 1999. Nitrogen in crop production: an account of global flows. *Global Biogeochemical Cycles* 13:647–662. doi:10.1029/1999GB900015.

Smil, V. 2001. *Feeding the World: A Challenge for the Twenty-First Century,* Reprint ed. The MIT Press, Cambridge, MA.

Smith, K.A., Mosier, A.R., Crutzen, P.J., and Winiwarter, W. 2012. The role of N$_2$O derived from crop-based biofuels, and from agriculture in general, in Earth's climate. *Philosophical Transactions of the Royal Society B: Biological Sciences* 367:1169–1174.

Smith, P.D., Martino, Z., Cai, D., Janzen, H., Kumar, P., McCarl, B., Ogle, S., O'Mara, F., Rice, C., Scholes, B., and Sirotenko, O. 2007. Agriculture, in *Climate Change 2007: The Physical Science Basis. Contribution of Working Group I to the Fourth Assessment Report of the Intergovernmental Panel on Climate Change.* Cambridge University Press, Cambridge, UK/New York.

Smith, W.N., Grant, B.B., Desjardins, R.L., Worth, D., Li, C., Boles, S.H., and Huffman, E.C. 2010. A tool to link agricultural activity data with the DNDC model to estimate GHG emission factors in Canada. *Agriculture, Ecosystems & Environment* 136:301–309. doi:10.1016/j.agee.2009.12.008.

Smukler, S.M., Jackson, L.E., Murphree, L., Yokota, R., Koike, S.T., and Smith, R.F. 2008. Transition to large-scale organic vegetable production in the Salinas Valley, California.

Agriculture, Ecosystems & Environment 126:168–188. doi:10.1016/j.agee.2008.01.028.

Snyder, C.S. and Bruulsema, T.W. 2007. *Nutrient Use Efficiency and Effectiveness in North America: Indices of Agronomic and Environmental Benefits.* International Plant Nutrition Institute, Peachtree Corners, GA.

Snyder, C.S., Bruulsema, T.W., Jensen, T.L., and Fixen, P.E. 2009. Review of greenhouse gas emissions from crop production systems and fertilizer management effects. *Agriculture, Ecosystems & Environment* 133:247–266. doi:10.1016/j.agee.2009.04.021.

Snyder, G. 1993. Coming in to the watershed: biological and cultural diversity in the California habitat. *Chicago Review* 39:75–86. doi:10.2307/25305721.

Sobota, D.J., Harrison, J.A., and Dahlgren, R.A. 2009. Influences of climate, hydrology, and land use on input and export of nitrogen in California watersheds. *Biogeochemistry* 94:43–62. doi:10.1007/s10533-009-9307-y.

Söderlund, R. and Svensson, B.H. 1976. The global nitrogen cycle, in Svensson, B.H. and Soderlund, R. (Eds): *Nitrogen, Phosphorus and Sulphur—Global Cycles. Ecological Bulletin* 22:23–73.

Sokolow, A.D. 2006. A *National View of Agricultural Easement Programs: Measuring Success in Protecting Farmland, Report 4.* American Farmland Trust, Dekalb, IL.

Solomon, S., Qin, D., Manning, M., Marquis, M., Averyt, K., Tignor, M., Miller, H.L., and Chen, Z. (Eds). 2007. *Climate Change 2007: The Physical Science Basis. Contribution of Working Group I to the Fourth Assessment Report of the Intergovernmental Panel on Climate Change.* Intergovernmental Panel on Climate Change, Cambridge, UK.

Sommer, S.G. and Hutchings, N.J. 2001. Ammonia emission from field applied manure and its reduction—invited paper. *European Journal of Agronomy* 15:1–15. doi:10.1016/S1161-0301(01)00112-5.

Southwick, S.M., Rupert, M.E., Yeager, J.T., Lampinen, B.D., Dejong, T.M., and Weis, K.G. 1999. Effects of nitrogen fertigation on fruit yield and quality of young "French" prune trees. *Journal of Horticultural Science and Biotechnology* 74:187–195.

Srivastava, P., Day, R.L., Robillard, P.D., and Hamlett, J.M. 2001. AnnGIS: integration of GIS and a continuous simulation model for non-point source pollution assessment. *Transactions in GIS* 5:221–234.

Staley, B. and Barlaz, M. 2009. Composition of municipal solid waste in the United States and implications for carbon sequestration and methane yield. *Journal of Environmental Engineering* 135:901–909. doi:10.1061/(ASCE)EE.1943-7870 .0000032.

Stallman, H.R. 2011. Ecosystem services in agriculture: determining suitability for provision by collective management. *Ecological Economics* 71:131–139. doi:10.1016/j. ecolecon.2011.08.016.

Stedman, R.C. 2002. Toward a social psychology of place predicting behavior from place-based cognitions, attitude, and identity. *Environment and Behavior* 34:561–581. doi:10.1 177/0013916502034005001.

Steenwerth, K.L. and Belina, K.M. 2010. Vineyard weed management practices influence nitrate leaching and nitrousoxide emissions. *Agriculture, Ecosystems and Environment* 138:127–131.

Steffen, W., Persson, A., Deutsch, L., Zalasiewicz, J., Williams, M., Richardson, K., Crumley, C. et al. 2011. The Anthropocene: from global change to planetary stewardship. *Ambio* 40:739–761.

Stehfest, E. and Bouwman, L. 2006. N$_2$O and NO emission from agricultural fields and soils under natural vegetation: summarizing available measurement data and modeling of global annual emissions. *Nutrient Cycling in Agroecosystems* 74:207–228. doi:10.1007/s10705-006-9000-7.

Stern, P. C. 2011. Design principles for global commons: natural resources and emerging technologies. *International Journal of the Commons* 5:213–232.

Sterner, T. S. 2003. *Policy Instruments for Environmental and Natural Resource Management*. Resources for the Future, Washington, DC.

Stewart, I. T., Cayan, D. R., and Dettinger, M. D. 2005a. Changes toward earlier streamflow timing across Western North America. *Journal of Climate* 18:1136–1155. doi:10.1175/JCLI3321.1.

Stewart, W. M., Dibb, D. W., Johnston, A. E., and Smyth, T. J. 2005b. The contribution of commercial fertilizer nutrients to food production. *Agronomy Journal* 97:1. doi:10.2134/agronj2005.0001.

Stieb, D. M., Judek, S., and Burnett, R. T. 2003. Meta-analysis of time-series studies of air pollution and mortality: update in relation to the use of generalized additive models. *Journal of the Air and Waste Management Association* 53:258–261.

Stieb, D. M., Szyszkowicz, M., Rowe, B. H., and Leech, J. A. 2009. Air pollution and emergency department visits for cardiac and respiratory conditions: a multi-city time-series analysis. *Environmental Health: A Global Access Science Source* 8:25. doi:10.1186/1476-069x-8-25.

Stillerman, K. P., Mattison, D. R., Giudice, L. C., and Woodruff, T. J. 2008. Environmental exposures and adverse pregnancy outcomes: a review of the science. *Reproductive Sciences* 15:631–650. doi:10.1177/1933719108322436.

Su, J. G., Morello-Frosch, R., Jesdale, B. M., Kyle, A. D., Shamasunder, B., and Jerrett, M. 2009. An index for assessing demographic inequalities in cumulative environmental hazards with application to Los Angeles, California. *Environmental Science and Technology* 43:7626–7634. doi:10.1021/es901041p.

Suddick, E. C., Steenwerth, K., Garland, G. M., Smart, D. R., and Six, J. 2011. Discerning agricultural management effects on nitrous oxide emissions from conventional and alternative cropping systems: a California case study, in *Understanding Greenhouse Gas Emissions from Agricultural Management, ACS Symposium Series*. American Chemical Society, Washington, DC, pp.203–226.

Suding, K. N., Collins, S. L., Gough, L., Clark, C., Cleland, E. E., Gross, K. L., Milchunas, D. G., and Pennings, S. 2005. Functional- and abundance-based mechanisms explain diversity loss due to N fertilization. *Proceedings of the National Academy of Sciences of the United States of America* 102:4387–4392. doi:10.1073/pnas.0408648102.

Sugihara, N. G., van Wagtendonk, J. W., Shaffer, K. E., Fites-Kaufman, J., Thode, A. E., and Agee, J. K. (Eds). 2006. *Fire in California's Ecosystems*. University of California Press, Berkeley.

Sulc, R. M. and Tracy, B. F. 2007. Integrated crop–livestock systems in the U.S. corn belt. *Agronomy Journal* 99:335. doi:10.2134/agronj2006.0086.

Summers, C. G. and Putnam, D. H. 2008. *Irrigated Alfalfa Management for Mediterranean and Desert Zones (Publication 3512)*. Division of Agriculture and Natural Resources, University of California, Davis.

Sumner, D. A. and Rosen-Molina, J. T. 2010. Impacts of AB 32 on Agriculture, ARE Update. Giannini Foundation of Agricultural Economics, University of California, Davis.

Sunyer, J., Puig, C., Torrent, M., Garcia-Algar, O., Calicó, I., Muñoz-Ortiz, L., Barnes, M., and Cullinan, P. 2004. Nitrogen dioxide is not associated with respiratory infection during the first year of life. *International Journal of Epidemiology* 33:116–120. doi:10.1093/ije/dyh037

Sutton, M. A., Howard, C. M., Erisman, J. M., Billen, G., Bleeker, A., Grennfelt, P., van Grinsven, H., and Grizzetti, B. 2011a. *The European Nitrogen Assessment: Sources, Effects and Policy Perspectives*. Cambridge University Press, Cambridge.

Sutton, M. A., Oenema, O., Erisman, J. W., Leip, A., van Grinsven, H., and Winiwarter, W. 2011b. Too much of a good thing. *Nature* 472:159–161.

Sutton, M. A., Simpson, D., Levy, P. E., Smith, R. I., Reis, S., van Oijen, M., and de Vries, W. 2008. Uncertainties in the relationship between atmospheric nitrogen deposition and forest carbon sequestration. *Global Change Biology* 14:2057–2063.

SWRCB (State Water Resources Control Board). 2010. *California Polluted Runoff Reduction or Nonpoint Source (NPS) Success Stories*. SWRCB, Sacramento, CA.

SWRCB. 2015. *Water Quality Control Policy for Siting, Design, Operation and Maintenance of Onsite Wastewater Treatment Systems (OWTS Policy)*. SWRCB, Sacramento, CA.

Syakila, A. and Kroeze, C. 2011. The global nitrous oxide budget revisited. *Greenhouse Gas Measurement and Management* 1:17–26.

Tabacova, S., Baird, D. D., and Balabaeva, L. 1998. Exposure to oxidized nitrogen: lipid peroxidation and neonatal health risk. *Archives of Environmental Health* 53:214–221. doi:10.1080/00039899809605698.

Tabacova, S., Balabaeva, L., and Little, R. E. 1997. Maternal exposure to exogenous nitrogen compounds and complications of pregnancy. *Archives of Environmental Health* 52:341–347. doi:10.1080/00039899709602209.

Takemoto, B. K., Bytnerowicz, A., and Fenn, M. E. 2001. Current and future effects of ozone and atmospheric nitrogen deposition on California's mixed conifer forests. *Forest Ecology and Management* 144(1–3):159–173. doi:10.1016/S0378-1127(00)00368-6.

Takemoto, B. K., Bytnerowicz, A., and Olszyk, D. M. 1988. Depression of photosynthesis, growth, and yield of field-grown green pepper (*Capsicum annum* L.) exposed to acidic fog and ambient ozone. *Plant Physiology* 88: 477–481.

Tamme, T., Reinik, M., Roasto, M., Meremäe, K., and Kiis, A. 2010. Nitrate in leafy vegetables, culinary herbs, and cucumber grown under cover in Estonia: content and intake. *Food Additives & Contaminants: Part B. Surveillance* 3:108–113. doi:10.1080/19440041003725944.

Tanaka, Y., Sato, Y. 2005. Farmers managed irrigation districts in Japan: assessing how fairness may contribute to sustainability. *Agricultural Water Management* 77: 196–209.

Tang, Y., Jiang, H., and Bryan, N. S. 2011. Nitrite and nitrate: cardiovascular risk-benefit and metabolic effect. *Current*

Opinion in Lipidology 22:11–15. doi:10.1097/MOL. 0b013e328341942c.

Tangermann, S., Blandford, D., and Meilke, K.D. 2010. Agricultural Trade 1980 vs. 2010: some progress, but still so far to go. Chapter 4 in *Proceedings Issues, 2010: Trade in Agriculture. Much Done, So Much More to Do.* International Agricultural Trade Research Consortium, St. Paul, MN.

Tannenbaum, S.R., Fett, D., Young, V.R., Land, P.D., and Bruce, W.R. 1978. Nitrite and nitrate are formed by endogenous synthesis in the human intestine. *Science* 200:1487–1489.

Tate, K.W., Lancaster, D.L., Morrison, J.A., Lile, D.F., Sado, Y., and Huang, B. 2005. Monitoring helps reduce water-quality impacts in flood-irrigated pasture. *California Agriculture* 59:168–175.

Taylor, B. 2004. A green future for religion? *Futures* 36:991–1008. doi:10.1016/j.futures.2004.02.011.

Taylor, B. 2005. *Encyclopedia of Religion and Nature.* Continuum, New York/London.

Taylor, B. 2009. *Dark Green Religion: Spirituality and the Planetary Future.* University of California Press, Berkeley.

Tchobanoglous, G., Stensel, H.D., Tsuchihashi, R., and Burton, F. 2013. *Wastewater Engineering: Treatment and Resource Recovery*, 5th ed. McGraw-Hill Education, New York.

Temple, P.J., Lennox, R.W., Bytnerowicz, A., and Taylor, O.C. 1987. Interactive effects of simulated acidic fog and ozone on field-grown alfalfa. *Environmental and Experimental Botany* 27:409–417.

Thayer, R.L. 2003. *LifePlace: Bioregional Thought and Practice.* University of California Press, Berkeley.

Thomas, M.D. 1951. Gas damage to plants. *Annual Review of Plant Physiology* 2:293–322.

Thomas, R.Q., Canham, C.D., Weathers, K.C., and Goodale, C.L. 2010. Increased tree carbon storage in response to nitrogen deposition in the US. *Nature Geoscience* 3:13–17. doi:10.1038/ngeo721.

Tietenberg, T.H. and Sterner, T.S. 1998. Disclosure strategies for pollution control, in Sterner, T. (Ed.): *The Market and the Environment.* Edward Elgar, Cheltenham, UK.

Tilman, D., Cassman, K.G., Matson, P.A., Naylor, R., and Polasky, S. 2002. Agricultural sustainability and intensive production practices. *Nature* 418:671–677. doi:10.1038/nature01014.

Timonen, K.L., Pekkanen, J., Tiittanen, P., and Salonen, R.O. 2002. Effects of air pollution on changes in lung function induced by exercise in children with chronic respiratory symptoms. *Occupational and Environmental Medicine* 59:129–134. doi:10.1136/oem.59.2.129.

Tolbert, P.E., Klein, M., Peel, J.L., Sarnat, S.E., and Sarnat, J.A. 2007. Multipollutant modeling issues in a study of ambient air quality and emergency department visits in Atlanta. *Journal of Exposure Science and Environmental* 17(Suppl 2):S29–S35. doi:10.1038/sj.jes.7500625.

Tolley, G., Randall, A., Blomquist, G., Fabian, R., Fishelson, G., Frankel, A., Hoehn, J., Krumm, R., Mensah, E., and Smith, T. 1986. *Establishing and Valuing the Effects of Improved Visibility in the Eastern United States (Report to the Environmental Protection Agency).* EPA, Washington, DC.

Townsend-Small, A. and Czimczik, C.I. 2010. Carbon sequestration and greenhouse gas emissions in urban turf.

Geophysical Research Letters 37:L02707. doi:10.1029/2009GL041675.

Townsend-Small, A., Pataki, D.E., Czimczik, C.I., and Tyler, S.C. 2011a. Nitrous oxide emissions and isotopic composition in urban and agricultural systems in southern California. *Journal of Geophysical Research* 116:G01013. doi:10.1029/2010JG001494.

Townsend-Small, A., Pataki, D.E., Tseng, L.Y., Tsai, C.Y., and Rosso, D. 2011b. Nitrous oxide emissions from wastewater treatment and water reclamation plants in Southern California. *Journal of Environment Quality* 40:1542. doi:10.2134/jeq2011.0059.

Tress, B., Tress, G., Décamps, H., and d'Hauteserre, A.-M. 2001. Bridging human and natural sciences in landscape research. *Landscape and Urban Planning* 57:137–141. doi:10.1016/S0169-2046(01)00199-2.

Tyler, K.B., Broadbent, F.E., and Bishop, J.C. 1983. Efficiency of nitrogen uptake by potatoes. *American Potato Journal* 60:261–269. doi:10.1007/BF02854276.

Tyler, K.B., May, D.M., Guerard, J.P., Ririe, D., and Hatakeda, J.J. 1988. Diagnosing nutrient needs of garlic. *California Agriculture* 42:28–29.

UC Davis Agricultural and Resource Economics. n.d. *Current Cost and Return Studies.* Accessed June 2, 2012. coststudies.ucdavis.edu.

UN FAO (Food and Agriculture Organization of the United Nations). n.d. *FAOSTAT.* Accessed June 5, 2015. http://faostat.fao.org/site/291/default.aspx.

United Nations. 2011. *World Population Prospects: The 2010 Revision.* Department of Economic and Social Affairs, Population Division, New York.

United Nations. 2013. *World Population Prospects: The 2012 Revision.* Department of Economic and Social Affairs, Population Division, New York.

United States Census Bureau. 2012. *Historical Census of Housing Tables: Sewage Disposal.* Available at: http://www.census.gov/housing/census/data/sewage.html.

United States Census Bureau. 2013. *2010 Census Data. United States Census 2010.* Accessed July 10, 2015. http://www.census.gov/2010census/data/.

United States District Court for the Middle District of Pennsylvania. 2013. *American Farm Bureau Federation et al. v. United States Environmental Protection Agency et al.*

Unkovich, M.J., Baldock, J., Peoples, M.B. 2010. Prospects and problems of simple linear models for estimating symbiotic N_2 fixation by crop and pasture legumes. *Plant Soil* 329:75–89. doi:10.1007/s11104-009-0136-5.

US BEA (Bureau of Economic Analysis). 2015. *Table 1.1.9. Implicit Price Deflators for Gross Domestic Product.* US BEA, Washington, DC.

US EIA (Energy Information Administration). 2015. *Short-Term Energy Outlook.* Accessed February 15, 2015. http://www.eia.gov/forecasts/steo/.

US General Accounting Office. 2003. *Natural Gas: Domestic Nitrogen Fertilizer Production Depends on Natural Gas Availability and Prices (No. GAO-03-1148).* US General Accounting Office, Washington, DC.

US Patent and Trademark Office. 2014. *Patenting By Geographic Region (CALIFORNIA), Breakout By Technology Class, CY1963-2014 Utility Patent Grants Distributed By Year of*

Grant. Accessed January 5, 2015. http://www.uspto.gov/web/offices/ac/ido/oeip/taf/stctca/castcl_gd.htm.

USDA (US Department of Agriculture). 2009a. *2007 Census of Agriculture United States (No. AC-07-A-51).* USDA, Washington, DC.

USDA. 2009b. *Income from Farm-Related Sources.* USDA, Washington, DC.

USDA. 2010. *2007 Census of Agriculture: Organic Production Survey (2008) (No. AC-07-SS-2).* USDA, Washington, DC.

USDA. 2011a. *Crop Production 2010 Summary. January 2011. Cr Pr 2-1 (11) (No. Cr Pr 2-1 (11)).* USDA, Washington, DC.

USDA. 2011b. *Dietary Guidelines for Americans, 2010,* 7th ed. US Government Printing Office, Washington, DC.

USDA. 2012. *Crop Nutrient Tool.* Accessed July 1, 2015. http://plants.usda.gov/npk/main.

USDA ERS (US Department of Agriculture Economic Research Service). 2006. *Balancing the Multiple Objectives of Conservation Programs (No. ERR-19).* USDA ERS, Washington, DC.

USDA ERS. 2009. *2008 Farm Bill Side-By-Side Comparison.* Accessed June 7, 2015. http://webarchives.cdlib.org/wayback.public/UERS_ag_1/20111128201038/http://ers.usda.gov/FarmBill/2008/.

USDA ERS. 2012. *ARMS Farm Financial and Crop Production Practices: Tailored Reports.* United States Department of Agriculture Economic Research Service, Washington, DC.

USDA ERS. 2013. *Food Availability (Per Capita) Data System.* Accessed December 18, 2014. http://www.ers.usda.gov/data-products/food-availability-(per-capita)-data-system/.aspx.

USDA ERS. n.d. *Food Expenditures.* Accessed December 18, 2014a. http://www.ers.usda.gov/data-products/food-expenditures.aspx.

USDA ERS. n.d. *Feed Grains: Yearbook Tables.* Accessed February 20, 2015b. http://www.ers.usda.gov/data-products/feed-grains-database/feed-grains-yearbook-tables.aspx.

USDA ERS. n.d. *Agricultural Exchange Rate Data Set.* Accessed January 6, 2015c. http://www.ers.usda.gov/data-products/agricultural-exchange-rate-data-set.aspx.

USDA ERS. n.d. *Farm Income and Wealth Statistics.* Accessed January 5, 2015d. http://www.ers.usda.gov/data-products/farm-income-and-wealth-statistics/government-payments-by-program.aspx.

USDA ERS. n.d. *Fertilizer Use and Price.* Accessed January 5, 2015e. http://www.ers.usda.gov/data-products/fertilizer-use-and-price.aspx.

USDA FAS (US Department of Agriculture Foreign Agricultural Service). n.d. *Production, Supply, and Distribution Online Database.* Available at: http://apps.fas.usda.gov/psdonline/.

USDA FSA (US Department of Agriculture Farm Service Agency). 2007. *Conservation Reserve Program: Summary and Enrollment Statistics.* USDA FSA, Washington, DC.

USDA NASS (US Department of Agriculture National Agricultural Statistics Service). 2006. *California Agricultural Statistics, 2005.* USDA NASS, Washington, DC.

USDA NASS. 2010. *Agricultural statistics 2010.* United States Government Printing Office, Washington, DC.

USDA NASS. 2011a. *California Agricultural Production Statistics: 2010 Crop Year.* USDA NASS, Washington, DC.

USDA NASS. 2011b. *California Walnut Acreage Report.* United States Department of Agriculture, Sacramento, CA.

USDA NASS. 2012. *County Agricultural Commissioner's Data.* Accessed July 1, 2012. http://www.nass.usda.gov/Statistics_by_State/California/Publications/AgComm/Detail/index.asp.

USDA NASS. 2014a. *California Agricultural Production Statistics.* California Department of Food and Agriculture, Sacramento.

USDA NASS. 2014b. *Land Values 2014 Summary.* USDA NASS, Washington, DC.

USDA NASS. 2015a. *Farms and Land in Farms: 2014 Summary.* USDA NASS, Washington, DC.

USDA NASS. 2015b. *Agricultural Prices.* Accessed March 2, 2015. http://usda.mannlib.cornell.edu/MannUsda/viewDocumentInfo.do?documentID=1002.

USDA NASS. n.d. *Historical Data.* Accessed July 1, 2012a. http://www.nass.usda.gov/Statistics_by_State/California/Historical_Data/.

USDA NASS. n.d. *Historical Data.* Accessed January 6, 2015b. http://www.nass.usda.gov/Statistics_by_State/California/Historical_Data/index.asp.

USDA NASS. n.d. *Surveys: Agricultural Chemical Use Program.* Accessed June 2, 2012c. http://www.nass.usda.gov/Surveys/Guide_to_NASS_Surveys/Chemical_Use/.

USDT (US Department of Transportation). 2013. *Public Use Waybill Sample.* USDT, Washington, DC.

USGS (US Geological Survey). 2012. *Mineral Commodity Summaries: Nitrogen (Fixed). Ammonia.* USGS, Reston, VA.

Valiela, I. and Bowen, J.L. 2002. Nitrogen sources to watersheds and estuaries: role of land cover mosaics and losses within watersheds. *Environmental Pollution* 118:239–248. doi:10.1016/S0269-7491(01)00316-5.

Van der Schans, M.L., Harter, T., Leijnse, A., Mathews, M.C., and Meyer, R.D. 2009. Characterizing sources of nitrate leaching from an irrigated dairy farm in Merced County, California. *Journal of Contaminant Hydrology* 110:9–21. doi:10.1016/j.jconhyd.2009.06.002.

Van Drecht, G., Bouwman, A.F., Knoop, J.M., Beusen, A.H.W., Meinardi, C.R. 2003. Global modeling of the fate of nitrogen from point and nonpoint sources in soils, groundwater, and surface water. *Global Biogeochemical Cycles* 17:1115. doi:10.1029/2003GB002060.

Van Grinsven, H.J.M., Rabl, A., and de Kok, T.M. 2010. Estimation of incidence and social cost of colon cancer due to nitrate in drinking water in the EU: a tentative cost-benefit assessment. *Environmental Health* 9:58. doi:10.1186/1476-069X-9-58.

Van Grinsven, H.J.M., ten Berge, H.F.M., Dalgaard., T., Fraters, B., Durand, P., Hart, A., Hofman., G. et al. 2012. Management, regulation and environmental impacts of nitrogen fertilization in northwestern Europe under the Nitrates Directive; a benchmark study. *Biogeosciences* 9:5143.

Van Groenigen, J.W., Velthof, G.L., Oenema, O., Van Groenigen, K.J., and Van Kessel, C. 2010. Towards an agronomic assessment of N_2O emissions: a case study for arable crops. *European Journal of Soil Science* 61:903–913. doi:10.1111/j.1365-2389.2009.01217.x.

Van Houtven, G., Loomis, R., Baker, J., Beach, R., and Casey, S. 2012. *Nutrient Credit Trading for the Chesapeake Bay.* Chesapeake Bay Commission, Annapolis, MD.

Van Kessel, C., Clough, T., and van Groenigen, J.W. 2009. Dissolved organic nitrogen: an overlooked pathway of nitrogen loss from agricultural systems? *Journal of Environmental Quality* 38:393–401. doi:10.2134/jeq2008 .0277.

Van Loon, A.J., Botterweck, A.A., Goldbohm, R.A., Brants, H.A., van Klaveren, J.D., and van den Brandt, P.A. 1998. Intake of nitrate and nitrite and the risk of gastric cancer: a prospective cohort study. *British Journal of Cancer* 78:129–135.

VandeHaar, M.J. and St-Pierre, N. 2006. Major advances in nutrition: relevance to the sustainability of the dairy industry. *Journal of Dairy Science* 89:1280–1291. doi:10.3168/ jds.S0022-0302(06)72196-8.

VanDerslice, J. 2009. *Well water quality and infant health study: Final Report.* EPASTAR Grant #82978101 (unpublished).

VanDerslice, J. 2011. Drinking water infrastructure and environmental disparities: evidence and methodological considerations. *American Journal of Public Health* 101(Suppl 1):S109–S114. doi:10.2105/AJPH.2011.300189.

Velders, G.J.M., Andersen, S.O., Daniel, J.S., Fahey, D.W., and McFarland, M. 2007. The importance of the Montreal Protocol in protecting climate. *Proceedings of the National Academy of Sciences of the United States of America* 104:4814–4819. doi:10.1073/pnas.0610328104.

Velmurugan, A., Dadhwal, V.K., and Abrol, Y.P. 2008. Regional nitrogen cycle: an Indian perspective. *Current Science* 94:1455–1468.

Velthof, G.L., Oudendag, D., Witzke, H.P., Asman, W.A.H., Klimont, Z., and Oenema, O. 2009. Integrated assessment of nitrogen losses from agriculture in EU-27 using MITERRA-EUROPE. *Journal of Environment Quality* 38:402. doi:10.2134/ jeq2008.0108.

Venterea, R.T., Bijesh, M., and Dolan, M.S. 2011. Fertilizer source and tillage effects on yield-scaled nitrous oxide emissions in a corn cropping system. *Journal of Environmental Quality* 40:1521–1531.

Vermeer, I.T., Moonen, E.J., Dallinga, J.W., Kleinjans, J.C., and van Maanen, J.M. 1999. Effect of ascorbic acid and green tea on endogenous formation of N-nitrosodimethyl-amine and N-nitrosopiperidine in humans. *Mutation Research* 428:353–361.

Vermeer, I.T., Pachen, D.M., Dallinga, J.W., Kleinjans, J.C., and van Maanen, J.M. 1998. Volatile N-nitrosamine formation after intake of nitrate at the ADI level in combination with an amine-rich diet. *Environmental Health Perspectives* 106:459–463.

Vermeulen, S.J., Campbell, B.M., and Ingram, J.S.I. 2012. Climate change and food systems. *Annual Review of Environment and Resources* 37:195–222. doi:10.1146/ annurev-environ-020411-130608.

Viers, J.H., Liptzin, D., Rosenstock, T.S., Jensen, V.B., Hollander, A.D., McNally, A., King, A.M. et al. 2012. *Nitrogen Sources and Loading to Groundwater. Technical Report 2, in: Addressing Nitrate in California's Drinking Water with a Focus on Tulare Lake Basin and Salinas Valley Groundwater.* Report for the State Water Resources Control Board Report to the Legislature. Center for Watershed Sciences, University of California, Davis.

Vitousek, P.M., Aber, J.D., Howarth, R.W., Likens, G.E., Matson, P.A., Schindler, D.W., Schlesinger, W.H., and Tilman, D.G. 1997. Human alteration of the global nitrogen cycle: sources and consequences. *Ecological Applications* 7:737–750. doi:10.1890/1051-0761(1997)007[0737:HAOTGN] 2.0.CO;2.

Volkmar, E.C. and Dahlgren, R.A. 2006. Biological oxygen demand dynamics in the lower San Joaquin River, California. *Environmental Science & Technology* 40:5653–5660. doi:10.1021/es0525399.

Von Klot, S., Peters, A., Aalto, P., Bellander, T., Berglind, N., D'Ippoliti, D., Elosua, R. et al. 2005. Ambient air pollution is associated with increased risk of hospital cardiac readmissions of myocardial infarction survivors in five European cities. *Circulation* 112:3073–3079. doi:10.1161 /circulationaha.105.548743.

Vorne, V., Ojanperä, K., De Temmerman, L., Bindi, M., Högy, P., Jones, M.B., Lawson, T., and Persson, K. 2002. Effects of elevated carbon dioxide and ozone on potato tuber quality in the European multiple-site experiment 'CHIP-project. *European Journal of Agronomy* 17:369–381.

Wagner, D.A., Young, V.R., Tannenbaum, S.R., Schultz, D.S., and Deen, W.M. 1984. Mammalian nitrate biochemistry: metabolism and endogenous synthesis. *IARC Scientific Publications* 54:247–253.

Wainger, L.A. and Shortle, J.S. 2013. Local innovations in water protection: experiments with economic incentives. *Choices* 28.

Wakida, F.T. and Lerner, D.N. 2005. Non-agricultural sources of groundwater nitrate: a review and case study. *Water Research* 39:3–16. doi:10.1016/j.watres.2004.07.026.

Waldman, J.M., Munger, J.W., Jacob, D.J., Flagan, C., and Hoffmann, M.R. 1982. Chemical composition of acid fog. *Science* 218:677–680.

Walters, S.P., Thebo, A.L., and Boehm, A.B. 2011. Impact of urbanization and agriculture on the occurrence of bacterial pathogens and stx genes in coastal waterbodies of central California. *Water Research* 45:1752–1762. doi:10.1016/j. watres.2010.11.032.

Walton, G. 1951. Survey of literature relating to infant methemoglobinemia due to nitrate-contaminated water. *American Journal of Public Health and the Nation's Health* 41:986–996.

Wang, J. 2012. *Policies for Controlling Groundwater Pollution from Concentrated Animal Feeding Operations*, Doctoral Dissertation. Department of Environmental Sciences, University of California, Riverside.

Wang, J. and Baerenklau, K.A. 2014. How inefficient are nutrient application limits? A dynamic analysis of ground-water nitrate pollution from concentrated animal feeding operations. *Applied Economic Perspectives and Policy* 37:130–150. doi:10.1093/aepp/ppu023.

Wang, X. and Taub, D.R. 2010. Interactive effects of elevated carbon dioxide and environmental stresses on root mass fraction in plants: a meta-analytical synthesis using pairwise techniques. *Oecologia* 163:1–11.

Ward, D.J., Roberts, K.T., Jones, N., Harrison, R.M., Ayres, J.G., Hussain, S., and Walters, S. 2002. Effects of daily variation in outdoor particulates and ambient acid species in normal and asthmatic children. *Thorax* 57:489–502. doi:10.1136/thorax.57.6.489.

Ward, M.H., Cantor, K.P., Riley, D., Merkle, S., and Lynch, C.F. 2003. Nitrate in public water supplies and risk of bladder cancer. *Epidemiology* 14:183–190. doi:10.1097/01. EDE.0000050664.28048.DF.

Ward, M.H., Cerhan, J.R., Colt, J.S., and Hartge, P. 2006. Risk of non-Hodgkin lymphoma and nitrate and nitrite from drinking water and diet. *Epidemiology* 7:375–382. doi:10.1097/01.ede.0000219675.79395.9f.

Ward, M.H., Heineman, E.F., Markin, R.S., and Weisenburger, D.D. 2008. Adenocarcinoma of the stomach and esophagus and drinking water and dietary sources of nitrate and nitrite. *International Journal of Occupational and Environmental Health* 14:193–197. doi:10.1179/oeh.2008.14.3.193.

Ward, M.H., Heineman, E.F., McComb, R.D., and Weisenburger, D.D. 2005. Drinking water and dietary sources of nitrate and nitrite and risk of glioma. *Journal of Occupational and Environmental Medicine* 47:1260–1267.

Ward, M.H., Kilfoy, B.A., Weyer, P.J., Anderson, K.E., Folsom, A.R., and Cerhan, J.R. 2010. Nitrate intake and the risk of thyroid cancer and thyroid disease. *Epidemiology* 21:389–395. doi:10.1097/EDE.0b013e3181d6201d.

Ward, M.H., Mark, S.D., Cantor, K.P., Weisenburger, D.D., Correa-Villaseñor, A., and Zahm, S.H. 1996. Drinking water nitrate and the risk of non-Hodgkin's lymphoma. *Epidemiology* 7:465–471.

Ward, M.H., Rusiecki, J.A., Lynch, C.F., and Cantor, K.P. 2007. Nitrate in public water supplies and the risk of renal cell carcinoma. *Cancer Causes Control* 18:1141–1151. doi:10.1007/s10552-007-9053-1.

Warneke, C., de Gouw, J.A., Holloway, J.S., Peischl, J., Ryerson, T.B., Atlas, E., Blake, D., Trainer, M., and Parrish, D.D. 2012. Multiyear trends in volatile organic compounds in Los Angeles, California: five decades of decreasing emissions. *Journal of Geophysical Research* 117:D00V17. doi:10.1029/2012JD017899.

Webb, A.J., Patel, N., Loukogeorgakis, S., Okorie, M., Aboud, Z., Misra, S., Rashid, R. et al. 2008. Acute blood pressure lowering, vasoprotective, and antiplatelet properties of dietary nitrate via bioconversion to nitrite. *Hypertension* 51:784–790. doi:10.1161/HYPERTENSIONAHA.107.103523.

Weinbaum, S.A., Klein, I., Broadbent, F.E., Micke, W.C., and Muraoka, T.T. 1984. Use of isotopic nitrogen to demonstrate dependence of mature almond trees on annual uptake of soil nitrogen. *Journal of Plant Nutrition* 7:975–990. doi:10.1080/01904168409363258.

Weinbaum, S.A., Niederholzer, F.J.A., Ponchner, S., Rosecrance, R.C., Carlson, R.M., Whittlesey, A.C., and Muraoka, T.T. 1994a. Nutrient uptake by cropping and defruited fieldgrown "French" prune trees. *Journal of the American Society for Horticultural Science* 119:925–930.

Weinbaum, S.A., Picchioni, G.A., Muraoka, T.T., Ferguson, L., and Brown, P.H. 1994b. Fertilizer nitrogen and boron uptake, storage, and allocation vary during the alternate-bearing cycle in pistachio trees. *Journal of the American Society of Horticultural Science* 119:24–31.

Weinbaum, S.A., Uriu, K., Micke, W.C., and Meith, H.C. 1980. Nitrogen fertilization increases yield without enhancing blossom receptivity in almond. *HortScience* 15:78–79.

Weinbaum, S. and Van Kessel, C. 1998. Quantitative estimates of uptake and internal cycling of 14N-labeled fertilizer in mature walnut trees. *Tree Physiology* 18:795–801. doi:10.1093/treephys/18.12.795.

Weinmayr, G., Romeo, E., De Sario, M., Weiland, S.K., and Forastiere, F. 2009. Short-term effects of PM_{10} and NO_2 on respiratory health among children with asthma or asthma-like symptoms: a systematic review and meta-analysis. *Environmental Health Perspectives* 118:449–457. doi:10.1289/ehp.0900844.

Weiss, S.B. 1999. Carros, Vacas, y Mariposas: Deposición De Nitrógeno y Manejo De Pastisales Pobres En Nitrógeno Para Una Especie Amenazada. *Conservation Biology* 13:1476–1486. doi:10.1046/j.1523-1739.1999.98468.x. (In Spanish)

Welch, N.C., Tyler, K.B., and Ririe, D. 1979. Nitrogen stabilization in the Pajaro Valley in lettuce, celery, and strawberries. *California Agriculture* 33:12–13.

Welch, N.C., Tyler, K.B., Ririe, D., and Broadbent, F. 1983. Lettuce efficiency in using fertilizer nitrogen. *California Agriculture* 37:18–19.

Welch, N.C., Tyler, K.B., Ririe, D., and Broadbent, F. 1985. Nitrogen uptake by cauliflower. *California Agriculture* 39:12–13.

Wellenius, G.A., Burger, M.R., Coull, A.B., Schwartz, J., Suh, H.H., Koutrakis, P., Schlaug, G., Gold, D.R., and Mittleman, M.A. 2012. Ambient air pollution and the risk of acute ischemic stroke. *Archives of Internal Medicine* 172:229–234. doi:10.1001/archinternmed.2011.732.

Weng, H.H., Tsai, S.S., Wu, T.N., Sung, F.C., and Yang, C.Y. 2011. Nitrates in drinking water and the risk of death from childhood brain tumors in Taiwan. *Journal of Toxicology and Environmental Health, Part A* 74:769–778. doi:10.1080/15287394.2011.567951.

Westhoek, H., van den Berg, R., de Hoop, W., and van der Kamp, A. 2004. Economic and environmental effects of the manure policy in The Netherlands: synthesis of integrated ex-post and ex-ante evaluation. *Water Science & Technology* 49:109–16.

Weyer, P.J., Cerhan, J.R., Kross, B.C., Hallberg, G.R., Kantamneni, J., Breuer, G., Jones, M.P., Zheng, W., and Lynch, C.F. 2001. Municipal drinking water nitrate level and cancer risk in older women: the Iowa Women's Health Study. *Epidemiology* 12:327–338.

Wheeler, S.M. 2008. State and municipal climate change plans: the first generation. *Journal of the American Planning Association* 74:481–496.

White, A., Cannell, M.G.R., and Friend, A.D. 1999. Climate change impacts on ecosystems and the terrestrial carbon sink: a new assessment. *Global Environmental Change* 9:21–30.

Whitledge, T.E. 1985. *Nationwide Review of Oxygen Depletion and Eutrophication in Estuarine and Coastal Waters: Northeast Region.* Report to U.S. Department of Commerce, NOAA, National Ocean Service. NOAA, Rockville, MD.

Wickham, J.D., Wade, T.G., and Riitters, K.H. 2008. Detecting temporal change in watershed nutrient yields. *Environmental Management* 42:223–231. doi:10.1007/s00267-008-9120-8.

Wiek, A., Binder, C., and Scholz, R.W. 2006. Functions of scenarios in transition processes. *Futures* 38:740–766.

Wiesen, P. 2010. Abiotic nitrous oxide sources: chemical industry and mobile and stationary combustion systems, in Smith, K. (Ed.): *Nitrous Oxide and Climate Change.* Routledge, London, UK.

Wilhelm, M. and Ritz, B. 2003. Residential proximity to traffic and adverse birth outcomes in Los Angeles County,

California, 1994–1996. *Environmental Health Perspectives* 111:207–216.

Wilkens, L. R., Kadir, M. M., Kolonel, L. N., Nomura, A. M., and Hankin, J. H. 1996. Risk factors for lower urinary tract cancer: the role of total fluid consumption, nitrites and nitrosamines, and selected foods. *Cancer Epidemiology, Biomarkers & Prevention* 5:161–166.

Williams, C. M. 2002. Nutritional quality of organic food: shades of grey or shades of green? *Proceedings of the Nutritional Society* 61:19–24.

Williams, J. W., Seabloom, E. W., Slayback, D., Stoms, D. M., Viers, J. H. 2005. Anthropogenic impacts upon plant species richness and net primary productivity in California. *Ecology Letters* 8:127–137.

Williams, L. E. 1991. Vine nitrogen requirements: utilization of N sources from soils, fertilizers, and reserves, in *Proceedings of the International Symposium on Nitrogen in Grapes and Wine. American Society of Enology and Viticulture.* pp.62–66.

Winter, T. C., Harvey, J. W., Franke, O. L., and Alley, W. M. 1998. *Ground Water and Surface Water: A Single Resource.* US Geological Survey Circular No. 1139.

Witthöft, T., Eckmann, L., Kim, J. M., and Kagnoff, M. F. 1998. Enteroinvasive bacteria directly activate expression of iNOS and NO production in human colon epithelial cells. *American Journal of Physiology* 275:G564–G571.

Weitzman, M. 1974. Prices vs. quantities. *Review of Economic Studies* 41(4):477–491.

Wohlgemuth, P. M., Hubbert, K., and Arbaugh, M. J. 2006. Fire and physical environment interactions, in Sugihara, N. (Ed.): *Fire in California's Ecosystems.* University of California Press, Berkeley, pp.75–93.

Wong, M., Wolf, C., Collins, N., Guo, L., Metzler, D., and English, P. 2015. Development of a web-based tool to collect and display water system customer service areas for public health action. *Journal of Public Health Management & Practice* 21:S44–S49. doi:10.1097/PHH.0000000000000159.

Wood, S. and Cowie, A. 2004. *A Review of Greenhouse Gas Emission Factors for Fertiliser Production.* Research and Development Division, State Forests of New South Wales, Beecroft, NSW.

Woodruff, T. J., Parker, J. D., and Schoendorf, K. C. 2006. Fine particulate matter ($PM_{2.5}$) air pollution and selected causes of postneonatal infant mortality in California. *Environmental Health Perspectives* 114:786–790. doi:10.1289/ehp.8484.

Woods, M. J. 2009. *Cultivating Soil and Soul: Twentieth-Century Catholic Agrarians Embrace the Liturgical Movement.* Liturgical Press, Collegeville, MN.

World Bank. 1998. Nitrogen oxides: pollution prevention and control, in *Pollution Prevention and Abatement Handbook.* World Bank Group, Washington, DC.

World Bank. 2015. *World Development Indicators: Data.* Accessed January 8, 2015. http://data.worldbank.org/data-catalog/world-development-indicators.

World Health Organization. 1996. *Guidelines for Drinking-Water Quality: Health Criteria and Other Supporting Information,* 2nd ed. WHO, Geneva, Switzerland.

Wossink, A. 2003. The Dutch nutrient quota system: past experience and lessons for the future. Paper presented at *The OECD Workshop: The Ex-Post Evaluation of Tradeable Permit Regimes,* January 21–22, Paris.

Wright, R. O., Lewander, W. J., and Woolf, A. D. 1999. Methemoglobinemia: etiology, pharmacology, and clinical management. *Annals of Emergency Medicine* 34:646–656.

Wu, J. and Babcock, B. 1998. The choice of tillage, rotation, and soil testing: economics and environmental implications. *American Journal of Agricultural Economics* 80:494–511.

Wu, L., Letey, J., French, C., Wood, Y., and Birkle, D. 2005. Nitrate leaching hazard index developed for irrigated agriculture. *Journal of Soil and Water Conservation* 60:90A–95A.

Wuebbles, D. J. 2009. Nitrous oxide: no laughing matter. *Science* 326:56–57.

Wuest, S. B. and Cassman, K. G. 1992. Fertilizer-nitrogen use efficiency of irrigated wheat: I. Uptake efficiency of preplant versus late-season application. *Agronomy Journal* 84:682. doi:10.2134/agronj1992.00021962008400040028x.

Xin, H., Gates, R. S., Green, A. R., Mitloehner, F. M., Moore, P. A., and Wathes, C. M. 2011. Environmental impacts and sustainability of egg production systems. *Poultry Science* 90:263–277. doi:10.3382/ps.2010-00877.

Xuan, W., Peat, J. K., Toelle, B. G., Marks, G. B., Berry, G., and Woolcock, A. J. 2000. Lung function growth and its relation to airway hyperresponsiveness and recent wheeze: results from a longitudinal population study. *American Journal of Respiratory and Critical Care Medicine* 161:1820–1824.

Yang, C., McCollum, D., McCarthy, R., and Leighty, W. 2009. Meeting an 80% reduction in greenhouse gas emissions from transportation by 2050: a case study in California. *Transportation Research Part D: Transport and Environment* 14:147–156. doi:10.1016/j.trd.2008.11.010.

Yang, C. Y., Cheng, M. F., Tsai, S. S., and Hsieh, Y. L. 1998. Calcium, magnesium, and nitrate in drinking water and gastric cancer mortality. *Japanese Journal of Cancer Research* 89:124–130.

Yang, Y. J., Hwang, S. H., Kim, H. J., Nam, S. J., Kong, G., and Kim, M. K. 2010. Dietary intake of nitrate relative to antioxidant vitamin in relation to breast cancer risk: a case-control study. *Nutrition and Cancer* 62:555–566. doi:10.1080/01635581003605557.

Yara. 2009. *Yara Fertilizer Industry Handbook 2009.* Yara, Oslo.

Yara. 2012. *Yara Fertilizer Indusry Handbook February 2012.* Yara, Oslo.

Ye, Y., Liang, X., Chen, Y., Ji, Y., and Zhu, C. 2014. Carbon, nitrogen and phosphorus accumulation and partitioning, and C:N:P stoichiometry in late-season rice under different water and nitrogen managements. *PLOS ONE* 9(7):e101776. doi:10.1371/journal.pone.0101776.

Yeh, S., Rubin, E. S., Taylor, M. R., and Hounshell, D. A. 2005. Technology innovations and experience curves for nitrogen oxides control technologies. *Journal of the Air & Waste Management Association* 55:1827–1838. doi:10.1080/10473289.2005.10464782.

Yeo, B.-L. and Lin, C.-Y. C. 2014. *Optimal Regulation and Management of Nitrogen Emissions to the Atmospheric and Aquatic Ecosystems.* University of California, Davis.

Ying, Q. and Kleeman, M. 2009. Regional contributions to airborne particulate matter in central California during a severe pollution episode. *Atmospheric Environment* 43:1218–1228. doi:10.1016/j.atmosenv.2008.11.019.

Zanobetti, A. and Schwartz, J. 2005. The effect of particulate air pollution on emergency admissions for myocardial infarction: a multicity case-crossover analysis. *Environmental Health Perspectives* 113:978–982. doi:10.1289/ehp.7550.

Zanobetti, A. and Schwartz, J. 2008. Temperature and mortality in nine US cities. *Epidemiology* 19:563–570. doi:10.1097/EDE.0b013e31816d652d.

Zavaleta, E. S., Shaw, M. R., Chiariello, N. R., Mooney, H. A., Field, et al. 2003. Additive effects of simulated climate changes, elevated CO_2, and nitrogen deposition on grassland diversity. *Proceedings of the National Academy of Sciences of the United States of America* 100:7650–7654. doi:10.1073/pnas.0932734100.

Zeegers, M. P., Selen, R. F. M., Kleinjans, J. C. S., Goldbohm, R. A., and van den Brandt, P. A. 2006. Nitrate intake does not influence bladder cancer risk: The Netherlands cohort study. *Environmental Health Perspectives* 114:1527–1531.

Zeman, C. L., Kross, B., and Vlad, M. 2002. A nested case-control study of methemoglobinemia risk factors in children of Transylvania, Romania. *Environmental Health Perspectives* 110:817–822.

Zhang, Q. and Anastasio, C. 2003. Free and combined amino compounds in atmospheric fine particles ($PM_{2.5}$) and fog waters from Northern California. *Atmospheric Environment* 37:2247–2258.

Zhang, R., El-Mashad, H. M., Hartman, K., Wang, F., Liu, G., Choate, C., and Gamble, P. 2007. Characterization of food waste as feedstock for anaerobic digestion. *Bioresource Technology* 98:929–935. doi:10.1016/j.biortech.2006.02.039.

Zhu, Y., Hinds, W. C., Kim, S., and Sioutas, C. 2002. Concentration and size distribution of ultrafine particles near a major highway. *Journal of the Air & Waste Management Association* 52:1032–1042. doi:10.1080/10473289.2002.1047 0842.

Zia, H., Harris, N. R., Merrett, G. V., Rivers, M., and Coles, N. 2013. The impact of agricultural activities on water quality: A case for collaborative catchment-scale management using integrated wireless sensor networks. *Computers and Electronics in Agriculture* 96:126–138.

INDEX

Note: Locators followed by (b), (f), (m), (t) refer to boxes, figures, maps, and tables, respectively.